THE ROUTLEDGE HANDBOOK
OF TOURISM
AND INDIGENOUS PEOPLES

The Routledge Handbook of Tourism and Indigenous Peoples presents an up-to-date, critical and comprehensive overview of established and emerging themes around Indigeneity and connections between Indigenous peoples and tourism development.

Offering socio-cultural perspectives and multidisciplinary insights from leading Indigenous and non-Indigenous scholars and tourism practitioners, the book explores contemporary issues, challenges and trends. Organised into six sections, the handbook explores Indigenous community involvement in tourism, Indigenous entrepreneurship and innovation, Indigenous tourism policies and politics, and the complexities of colonialism and decolonisation issues. This text focuses on the active role that Indigenous peoples have in the industry and uses international case studies and experiences to explore the global context of Indigenous tourism.

This handbook fills a notable gap by offering a critical and detailed understanding of the role of Indigenous practitioners and societies in tourism and how they interact within the tourism nexus. It will be of interest to scholars, students, tourism practitioners and policymakers working in tourism, development studies, anthropology, human geography and sociology.

Richard Butler is Emeritus Professor of Tourism at Strathclyde University, Glasgow, Scotland. He has taught at the University of Western Ontario in London, Canada, at the universities of Surrey and Strathclyde in the UK and held visiting professorships in Australia, Austria, Italy, Hong Kong and the Netherlands.

Anna Carr (Ngāpuhi, Ngati Ruanui, Ngāruahine) is an Associate Professor, co-director of the Centre for Recreation Research and Head of Department at the Department of Tourism, University of Otago, New Zealand.

THE ROUTLEDGE HANDBOOK OF TOURISM AND INDIGENOUS PEOPLES

Edited by Richard Butler and Anna Carr

Routledge
Taylor & Francis Group

LONDON AND NEW YORK

Designed cover image: Jad Davenport

First published 2025
by Routledge
4 Park Square, Milton Park, Abingdon, Oxon, OX14 4RN

and by Routledge
605 Third Avenue, New York, NY 10158

Routledge is an imprint of the Taylor & Francis Group, an informa business

British Library Cataloguing-in-Publication Data
A catalogue record for this book is available from the British Library

Library of Congress Cataloging-in-Publication Data
Names: Butler, Richard, 1943- editor. | Thompson-Carr, Anna, editor.
Title: The Routledge handbook of tourism and Indigenous peoples / edited by
Richard Butler and Anna Carr.
Other titles: Handbook of tourism and Indigenous peoples
Description: Abingdon, Oxon ; New York, NY : Routledge, 2024. | Includes
bibliographical references and index.
Identifiers: LCCN 2024015666 (print) | LCCN 2024015667 (ebook) |
ISBN 9781032136547 (hardback) | ISBN 9781032136578 (paperback) |
ISBN 9781003230335 (ebook)
Subjects: LCSH: Culture and tourism. | Tourism—Social aspects. |
Indigenous peoples—Social conditions. | Economic anthropology.
Classification: LCC G156.5.H47 R68 2024 (print) | LCC G156.5.H47 (ebook) |
DDC 306.4/819—dc23/eng20240416
LC record available at https://lccn.loc.gov/2024015666
LC ebook record available at https://lccn.loc.gov/2024015667

ISBN: 9781032136547 (hbk)
ISBN: 9781032136578 (pbk)
ISBN: 9781003230335 (ebk)

DOI: 10.4324/9781003230335

Typeset in Times New Roman
by codeMantra

In Memoriam

Johnny Edmonds, (Ngāpuhi), Co-founder and former Director of the World Indigenous Tourism Alliance (WINTA).
Professor David Harrison, University of the South Pacific and universities of Middlesex and North London.
Emeritus Professor Keith Hollinshead, University of Bedfordshire, United Kingdom.
Ben Sherman (Oglala Lakota (Sioux) Nation), former Chairperson of the WINTA board.
Emerita Professor Valene Smith, the pioneer of tourism and Indigenous people's tourism, University of Chico, California.
All five were highly respected colleagues, leaders and mentors to many Indigenous tourism practitioners, professionals and academics. Their absence will be felt deeply.

CONTENTS

Contents

FIGURES

TABLES

CONTRIBUTORS

Editors

Richard Butler is Emeritus Professor of Tourism at Strathclyde University, Glasgow, Scotland. He has taught at the University of Western Ontario in London, Canada, at the universities of Surrey and Strathclyde in the UK and held visiting professorships in Australia, Austria, Italy, Hong Kong and the Netherlands. He was a founding member and former President of the International Academy for the Study of Tourism, has given presentations in some 30 countries and has acted as consultant for numerous agencies, governments and for the United Nations World Tourism Organisation. He has supervised over 40 doctoral students, published 28 books and over 200 papers and chapters in books. His main research areas are destination development, sustainability, tourism war and political change, Indigenous tourism and tourism in peripheral areas. In 2016, he was awarded the United Nations World Tourism Organisation (UNWTO) Ulysses medal for 'excellence in the creation and dissemination of knowledge'. He currently lives in the traditional and unceded territories of the Ts'msyen and Sm'algyax speaking peoples in northern British Columbia, Canada.

Anna Carr (Ngāpuhi, Ngati Ruanui, Ngāruahine) is an Associate Professor, co-director of the Centre for Recreation Research and Head of Department at the Department of Tourism, University of Otago, New Zealand. Her research interests focus on interdisciplinary research approaches to ecotourism, protected areas, cultural landscapes and Indigenous community development. She contributed to the 2020-2023 advisory group developing the World Indigenous Tourism Alliance (WINTA) Noongar Indigenous Research Protocol, presented at the World Indigenous Tourism Summit, Perth, Australia, in March 2023. She has undertaken research and community service with Indigenous peoples in the Pacific and New Zealand, including four years as a board member for Ngāi Tahu developing Te Ana Māori Rock Art Centre (New Zealand).

Contributors

Clive Allanso is an Economic and Social Research Council/South Coast Doctoral Training Partnership candidate and visiting lecturer at the University of Brighton. He is developing a critical understanding of the 'value of arts' and the role of art champions in the development of the creative sector SMEs and value chain connection with the tourism sector in West Africa. He is a business professional with an extensive background in governance and risk management and has worked in both local and central governments at various levels and with SMEs and NGOs in Nigeria and Ghana. He has mixed Nigerian/British heritage and is passionate about people and cultures.

Christine Ampumuza is a Lecturer in the Department of Tourism and Hospitality, Kabale University, Uganda. She holds a PhD in Tourism from Wageningen University, The Netherlands.

Rhonda Appo is a Mamu woman from the coastal rainforest region south of Cairns, in northern Queensland. She has been the Indigenous tourism programme manager at the Queensland Tourism Industry Council for over a decade, leading the Indigenous Champions Network.

Easnin Ara is an Associate Professor at the Bangladesh University of Professionals and has a PhD in non-human agency in tourism businesses from the University of Otago, New Zealand. Her research interests are multidisciplinary, particularly in developing contexts and focus on tourism product development, ethnic tourism, tourism micro-businesses, social sustainability and community-based tourism (CBT).

Sophie Auckram is a research assistant at Massey University (New Zealand) and is currently studying her master's (Political Science: Environmental Governance and Behavior) education at the University of Gothenburg (Sweden).

Tracy Berno is Professor at the Auckland University of Technology in Auckland. She lived and worked in the Pacific Islands for many years and has researched and taught on aspects of tourism in the South Pacific for over 30 years. She contributed to the 1996 and 2007 volumes on Indigenous Tourism edited by Butler and Hinch. Tracy has Sami ancestry (Finland) and has a particular interest in how indigenous foods and cuisines are represented in the context of tourism.

Brenda Boonabaana is a Lecturer in the Department of Forestry, Biodiversity and Tourism, Makerere University, Uganda, and a Provost Early Career Fellow, University of Texas, Austin, USA. She holds a PhD in Tourism from the University of Otago, New Zealand.

Kelly Bricker is a Professor and the Director of Hainan University–Arizona State University Joint International Tourism College. She completed her PhD with Penn State University and was Co-founder of Rivers Fiji. She has authored books on sustainability, highlighting case studies that address environmental and societal issues. She serves on the boards of the Global Sustainable Tourism Council, Travel Unity, and Tourism and Protected Area Specialists Group of the International Union for Conservation of Nature.

Caroline Butler is a cultural anthropologist whose academic research has focused on Indigenous fisheries, local ecological knowledge, fisheries privatisation and research processes and methods. She has a PhD in Anthropology from the University of British Columbia. Caroline began collaborating with the Gitxaała Nation in 2001 and worked directly for the Gitxaała government from 2009 to 2023, coordinating marine planning initiatives, environmental assessment research, language documentation and repatriation.

Abraham Cáceres Cabana has a bachelor of history, with a master's degree in sustainable tourism in process at the Universidad Nacional Mayor de San Agustín Arequipa, Peru. He is a specialist in rural community tourism, organisation and management models, with experience in public and private entities in the field of tourism and culture.

Cameron Costello is a Quandamooka man from Moreton Bay on the coast of Brisbane in south-east Queensland. He is Chair of the Queensland First Nations Tourism Council and board member of the Queensland Industry Tourism Council. Cameron is a law graduate from the University of Queensland and holds a bachelor of arts in Leisure Management from Griffith University. He was formerly the CEO for the Quandamooka Yoolooburrabee Aboriginal Corporation.

Nicole Curtin holds a bachelor of psychological science (Honours) and a PhD in community psychology from Northern Institute, Charles Darwin University, Larrakia Country, Northern Territory. Nicole is collaborating with Aboriginal tourism operators across Western Australia in support of ecological and social justice. Her research explores reconciliation pathways through Indigenous tourism. She has experience across the non-for-profit community services sector in mental health programme delivery, co-design and evaluation.

Gebeyaw Ambelu Degarege has a PhD from the University of Otago and is a Senior Lecturer, Department of Tourism Sport & Society, Lincoln University (New Zealand). His research interests explore the wider field of tourism and sustainable development using a cross-disciplinary lens with particular focus on sustainable tourism planning and management, food tourism, poverty, food security, sustainable livelihoods and sustainable development. He has undertaken research in Africa and New Zealand, examining sustainable tourism development and management concerns employing both quantitative and qualitative methods.

Atefeh Ahmadi Dehrashid is currently an Assistant Professor in the faculty of Natural Resources and the Department of Climatology at the University of Kurdistan in Iran. Prior to her recent appointment in the position, she was a Lecturer at Kharazmi University in Tehran. Dr. Ahmadi Dehrashid received her PhD in Geography and Rural Planning from Kharazmi University. She has published articles in referred journals and chapters in books. Her research interests include tourism spatial planning, community tourism and operational strategic planning.

Johan Edelheim is a Professor of Tourism and Media at the *Research Faculty of Media and Communication* at *Hokkaido University*, Japan. He has interests in tourism, education, philosophy and cultural studies. Behind most of his research lies a deeply rooted aim for mutuality.

Dionne Fonoti is a Senior Lecturer at the Centre for Samoan Studies at the National University of Samoa. She is highly regarded as a Samoan anthropologist and filmmaker and studied as a PhD candidate in Cultural Anthropology at Victoria University of Wellington (NZ).

Anne Ford is an Associate Professor at the Department of Anthropology and Archaeology, University of Otago. She specialises in the archaeology of Aboriginal Australia and Papua New Guinea and has an interest in science communication and community involvement in archaeology.

Zahed Ghaderi is attached to the Department of Tourism, College of Arts and Social science, Sultan Qaboos University, Muscat, Oman. Zahed has over two decades of experience in the field and has published extensively in top-tier tourism and hospitality journals including Tourism Management, Journal of Travel Research, International Journal of Hospitality Management and Journal of Sustainable Tourism. His research interests include cultural aspects of tourism, HR in hospitality and tourism, destination management, organisational learning, host–guest relationship, sustainable tourism and tourism crisis management.

Dawn Gibson is a Senior Lecturer in the School of Tourism and Hospitality Management, University of the South Pacific (USP), Fiji. She has a PhD in Tourism Studies awarded by the USP. Her PhD examined challenges that Indigenous owned businesses face delivering consistent quality services in the backpacker/budget accommodation sector in the Yasawas Islands in Fiji. Her research interests span entrepreneurship, CBT, cross-cultural management and employee empowerment, especially concerning South Pacific communities.

Sonya Graci is an Associate Professor at the Ted Rogers School of Hospitality and Tourism Management at Toronto Metropolitan University, University in Toronto, Ontario. She is Director of the Hospitality and Tourism Research Institute and has worked on numerous projects around the world related to sustainable tourism development and community capacity building including Honduras, Indonesia, Canada, Fiji and China.

Bryan Grimwood is Settler Canadian and an Associate Professor in the Department of Recreation and Leisure Studies at the University of Waterloo, Canada. His research analyses human–nature relationships and advocates social justice and sustainability in contexts of tourism, leisure and livelihoods.

Atsuko Hashimoto is Professor and Graduate Studies Director (MA Geography programme) in the Department of Geography and Tourism Studies at Brock University, Canada.

Suzanne Hepi is a professional with a background in research and mentoring. Her whakapapa roots are deeply embedded in Aotearoa, with iwi and hapū affiliations spanning to Ngāpuhi, Ngāti Rāhiri, Te Kapotai, Waikato-Tainui, Tūwharetoa, Ngāti Maniapoto, Ngāti Whitikaupeka, Te Āti Haunui-a-Pāpārangi, Ngai Tukairangi and Ngāti Pāhauwera. Her rich cultural and academic background give her a unique perspective and valuable skills, contributing to her role as Strategic Lead Māori Economy at the economic development agency in Whanganui, New Zealand.

Freya Higgins-Desbiolles is Adjunct Associate Professor with the Department of Recreation and Leisure, University of Waterloo, Canada, on the traditional territory of the Neutral, Anishinaabeg and Haudenosaunee peoples. She is also Visiting Professor with the Centre for Research and Innovation in Tourism, Taylor's University of Malaysia; and Adjunct with UniSA Business (unceded Kaurna Country), University of South Australia. She has worked with communities, non-governmental organisations and businesses on research into social justice, human rights and sustainability issues in tourism. She is also a Co-founder of the Tourism Alert and Action Forum, which advocates for community rights in tourism.

Emily Höckert works as a postdoctoral researcher in the project 'Intra-living in the Anthropocene' (Academy of Finland) at the University of Lapland, Finland. Her research and publications, such as *Negotiating Hospitality* (Routledge, 2018), explore relational ethics in tourism settings.

Afiya Holder is a Lecturer/Assistant Professor within The Business School and Tourism Research Centre at Edinburgh Napier University. She has a PhD and master's from the University of Queensland and held academic roles there and at Griffith University. Her research focuses on tourist behaviour and social sustainability. She held senior professional roles at Tobago Division of Tourism, UNWTO, Brisbane Economic Development Agency and Brisbane City Council (formerly Brisbane Marketing).

Keith Hollinshead was Emeritus Professor at the University of Bedfordshire, alongside appointments at Texas A and M University and the University of Technology, Sydney. He worked on matters of public culture, public heritage and public nature from postdisciplinary outlooks, often from post-qualitative standpoints. He was Vice President of the International Sociological Association (for International Tourism) for 16 years and contributed to the 1996 and 2007 editions of Butler and Hinch.

Md Ariful Hoque is an Associate Professor in the Department of Marketing, Jahangirnagar University, Bangladesh. He holds a PhD with a focus on NGO–tourism–Indigenous poverty nexus from the University of Otago, New Zealand. His research interests are multidisciplinary and primarily centred on Indigenous/marginal CBT businesses, NGO facilitation in community-centred tourism businesses and community empowerment through tourism businesses' involvement.

Gregory Jackmond is a researcher and Archaeologist at the National University of Samoa's Center of Samoan Studies. He specialises in LIDAR technology.

Helen Jennings currently has a research fellowship at the University of the Arctic, Norway, where she completed her PhD. She was part of the international research group 'Indigenous Religion(s): Local Grounds, Global Networks', and during her PhD studies, she was a guest researcher at the University of Victoria, British Columbia. Her main research interests include Indigenous tourism, the critical study of religion, gender and methodology, and decolonisation.

Jason W. Johnston (Anishinaubae from Chippewas of Nawash Unceded First Nation (Neyaashiinigmiing) has a master of science in Environmental Science from Thompson Rivers University. He is the Interpretative Programme Coordinator for Cape Croker Park in Neyaashiinigmiing and serves on the Executive Board of Directors – Indigenous Tourism Association of Canada (2020–2024). He has worked for Parks Canada and has been a professional interpreter at Spirit Bear Lodge and Knight Inlet Lodge and a wilderness guide leading nature tours across Canada.

Outi Kugapi works as a Coordinator in Educational and Development Services and wrote her PhD in tourism research at University of Lapland, Finland. Her research interests are tourism experiences in craft and creative tourism.

Taiwo Temitope Lasisi is a postdoctoral researcher at the Department of Recreology and Tourism, Faculty of Informatics and Management in University of Hradec Králové, Czech Republic. Her research interest includes tourism management, tourist behaviour, human resource management and environmental sustainability.

Jean-Philippe Le Moigne is a professional in Tourism and Environmental Sciences, with qualifications as a Forestry Engineer and a Master in Development & Management of Tourism (Research) from the Pantheon-Sorbonne University in France. He has managed numerous ecotourism and Indigenous tourism development and training programmes in North, Central and Latin America, Europe, Asia and Oceania. Jean-Philippe is a Latin American agent for WINTA.

Monika Lüthje works as a Senior Lecturer in tourism research at the University of Lapland, Finland. Her research has focused on host and guest experiences of tourism from a cultural studies perspective.

Heather Mair is a Professor in the Department of Recreation and Leisure Studies at the University of Waterloo in Waterloo, Canada. Heather has developed a programme of research that has been focused on the ways sport, leisure and tourism and contributed to rural community development and wellbeing in Canada.

Courtney W. Mason is an Associate Professor and Canada Research Chair in Rural Livelihoods and Sustainable Communities at Thompson Rivers University. He has held SSHRC and Health

Canada-funded research awards, and his collaborative research supports community-driven initiatives that enhance local food security and tourism development, while supporting cultural continuities. He completed his PhD at the University of Alberta where he investigated the displacement of Indigenous peoples in the formation of Banff National Park. He also held a postdoctoral research fellowship with the Indigenous Health Research Group at the University of Ottawa.

Gabrielle McGinnis has research interests in Indigenous methodologies, biocultural heritage conservation and sustainable tourism development using digital technologies. She has a PhD from the University of Newcastle, Australia, and is CEO and Founder of BrodiMapi LLC providing consultations, digital mapping and marketing services to those who wish to preserve, conserve and share biocultural heritage for the long term. She has recently worked as the Tourism Development Coordinator for the New Mexico Tourism Department. She resides in Copenhagen, Denmark, with her husband and son.

Alison McIntosh is a Professor at Auckland University of Technology. As a critical tourism scholar, she has published extensively in leading international tourism and hospitality journals. A central theme of her research is that experiential, qualitative and social justice analyses reveal subjective, emotional, spiritual and neglected aspects of tourism experiences. In 2011, Alison co-founded the interdisciplinary journal, Hospitality and Society, looking at critical elements of hospitality, and in 2017, she co-founded the online open-access journal, Hospitality Insights, to provide access to academic research by the hospitality industry and community.

Jason Mika (Tūhoe, Ngāti Awa, Whakatōhea, Ngāti Kahungunu) is an Associate Professor at Te Raupapa Waikato Management School and Te Kotahi Research Institute, University of Waikato, New Zealand. His research centres on Indigenous business philosophy in multiple sites, sectors and scales, including Indigenous trade, tourism, agribusiness and the marine economy. In 2019, Jason was a Fulbright-Ngā Pae o Te Māramatanga senior scholar at Stanford University's Woods Institute for the Environment and the University of Arizona's Native Nations Institute.

Nyssa Mildwaters is Head of Collections (Access) at the Victoria and Albert Museum (London). She was previously a conservation manager at Otago Museum and has an Ma/MSc from University College London. She has worked in museums, on archaeological sites and delivered conservation training in Europe, the Pacific and New Zealand, particularly around community engagement with cultural heritage and the conservation of indigenous materials.

Apisalome Movono is an indigenous Fijian from Buca, Natewa district, hailing from a long line of navigators, fishermen and warriors of the sea. He is an academic, a conservationist, an activist and an oceans advocate. His PhD sought to understand the livelihoods, resilience and complex adaptive systems of Indigenous Fijian communities engaging in tourism. He is a Senior Lecturer in Development Studies at Massey University and holds a Royal Society Te Apārangi Marsden grant for research which seeks to find resilient pathways for Pacific tourism futures. He is also the Co-founder of the Laucala Beach Sustainability Society.

Marina Novelli is Professor of Marketing and Tourism at the University of Nottingham Business School (UK). She is a globally renowned tourism and sustainable development expert who has worked in the field of international tourism policy, planning and development in Africa, Europe and Asia for institutions such as the World Bank, the EU, the UN, the Commonwealth Secretariat, the Millennium Challenge Corporation, the British Council, National

Ministries and Tourism Boards, Regional Development Agencies, private sectors and NGOs. Marina is a *Fellow of the Academy of Social Sciences* and *Fellow of the International Academy for the Study of Tourism*, Alternate Member of the UNWTO World Committee on Tourism Ethics and *Member of the World Economic Forum Global Future Council for Sustainable Tourism.*

Amos Ochieng is a Lecturer in the Department of Forestry, Biodiversity and Tourism, Makerere University, Uganda. He holds a PhD in Tourism, Conservation and Development from Wageningen University, The Netherlands.

Mayumi Okada is an Associated Professor in the Office of Ainu Relations and Initiatives and is affiliated with the Centre for Advanced Tourism Studies at Hokkaido University. Her research concerns the involvement of cultural heritage in the expression of identity, social empowerment and economic development in contemporary contexts.

Martina Pásková is Associate Professor at the Department of Recreology and Tourism, Faculty of Informatics and Management in the University of Hradec Králové, Czech Republic. Her research domains are human geography, ethno-ecology and geo-anthropology, including topics of geoheritage interpretation, tourism sustainability, geoparks management and Indigenous and local knowledge regarding nature and Earth.

Ashley Puriri (Ngāti Kahungunu, Ngāti Rongomai-wahine and Ngāti Porou) has a PhD from the University of Waikato where he utilised a Kaupapa Māori research methodology to develop distinctive cultural frameworks and paradigms. He is a fluent speaker of 4 different languages and has over 35 years of international business experience, with government and industry leadership globally. He is Director and CEO of Te Mana Consultancy Ltd (New Zealand).

Carina Ren is Associate Professor at the Department of Culture and Learning at Aalborg University. She is interested in how tourism interferes with other fields of society through cultural innovation and explores new ways in which tourism is developed, organised and valued. Her research often takes place in research collaborations with tourism organisations and industry, citizens and students.

Lisa Ruhanen is a Professor in Tourism and the Director of Education with the University of Queensland Business School. She has undertaken more than 30 academic and consultancy research projects in Australia and overseas in the areas of Indigenous tourism, sustainable tourism and policy, planning and governance. Lisa has more than 100 academic publications, and in 2017, she and colleagues co-edited a book on *Indigenous Tourism: Cases from Australia and New Zealand.* She has worked with the United Nations World Tourism Organization, including as an Advisory Board member and Auditor for the UNWTO's TedQual accreditation programme.

Regina Scheyvens is a Professor of International Development at Massey University, where her research focuses on tourism, community wellbeing and sustainable development, focusing on the South Pacific. She has Slovenian and Dutch heritage. Her books include 'Tourism for Development: Empowering Communities', 'Tourism and Poverty' and 'Inclusive Tourism Development'. Her recent collaborations with Indigenous scholars have led to work on the Sustainable Development Goals (see www.reimaginingsouthpacifictourism.com). She holds a James Cook Fellowship, Te Apārangi/Royal Society of New Zealand.

Ben Sherman was a founding member of the Native Tourism Alliance, Business Enterprises for Sustainable Travel, First Peoples Fund, American Indian/Alaska Native Tourism Association and a foundation member of the WINTA board.

Bill Snow is Acting Director of Consultation at Stoney Tribal Administration and member of the Wesley First Nation, Stoney Nakoda / Yuma Quecha. He has published and undertaken research projects around First Nations and natural resource management, notably incorporating Traditional Ecological Knowledge and Cultural Monitoring. Bill is a graduate of the University of Lethbridge, Business Administration programme. He is a member of the Canadian Mountain Network, and his work includes Bison Reintroduction at Banff National Park & Elk Island National Park.

Glenn Summerhayes (OL FSA FLS FRAI) is a Professor in the Archaeology department at the University of Otago. He is the University's Foundation Chair of anthropology, an honorary professor at both the Australian National University and the University of Queensland. He has worked in Papua New Guinea in the field of archaeology and history for more than 35 years. He has a research interest in cultural heritage management alongside international expertise and recognition or his specialty in Pacific archaeology.

Kasimiro Taukeinikoro is Fijian and an owner and managing director of Rivers Fiji (Upper Navua Conservation Area) and Co-chair of the Duavata Sustainable Tourism Collective, Fiji. He holds a bachelor of arts in business management (1999) from the School of Business Management, USP, Fiji.

Kylik Kisoun Taylor is of Inuvialuit, Gwich'in and Scandinavian descent. He is founding owner and CEO of Tundra North Tours (est. 2006). He graduated from the University of Victoria and serves as a member of the Board of Directors on both the Indigenous Tourism Association of Canada and Northwest Territories Tourism.

Robert Taylor, a Nanda Yamaji man from the mid-west of WA and experienced CEO of WAITOC, has transformed the organisation into a leader in Aboriginal tourism in Western Australia, adding business development to its core functions. With over 38 years in the hospitality and tourism industries, he has a diverse entrepreneurial background, crossing into various sectors. Since becoming CEO in 2015, Robert has significantly grown Aboriginal tourism, creating 89 new businesses and 244 full-time jobs and securing over $40 million in federal government funding. He was pivotal in developing the JINA Aboriginal tourism action plan in partnership with Tourism WA, a $20 million investment in Aboriginal tourism.

Rerekura Teaurere (Tongareva, Cook Islands/Pakeha) is a lecturer in hospitality and tourism at Auckland University of Technology (New Zealand), where she teaches under the Faculty of Culture and Society. She completed her PhD with the USP which focused on corporate environmental responsibility in the accommodation industry of Rarotonga, Cook Islands. Her research interests sit within sustainable development, ecotourism, corporate environmental responsibility, traditional ecological knowledge and Indigenous forms of environmental management.

Arvid Viken is Professor in Tourism, UiT Arctic University of Norway, Norway. His research interests include destination development, Indigenous tourism and tourism–community interaction.

Clinton Walker is a Ngarluma and Yindjibarndi man, the Traditional Owners of the coastal and inland areas of the West Pilbara region. Through his award-winning tourism business Ngurrangga Tours (Ngarluma and Yindjibarndi Country, Western Australia), Clinton aims to educate and immerse people in the ways of his culture and history. Clinton has sat on the boards of the Ngarluma Yindjibarndi Foundation, West Australian Indigenous Tourism Operators Council, Ngarluma Aboriginal Corporation and the Karratha Visitor Centre.

Bruce Watkinson is a member of the Gitxaala Nation. He belongs to the Gitnagun'aks House which is part of the Gisputwada (Killer Whale) Clan and bears the name Wudimes. Bruce graduated from the University of Victoria with a bachelor of science (biology) degree in 1996 and has spent his entire career working for north coast First Nation government organisations. His career has spanned across multiple aspects of the natural resource sector, including scientific research, fisheries management, marine and terrestrial planning, collaborative governance, and strategic planning. His research is focused on the application of traditional governance in the contemporary context.

Michelle Whitford is an Associate Professor and Director (Curriculum innovation) at Griffith University. She is a member of the Griffith Institute of Tourism. Her research expertise is in the field of policy, planning, Indigenous tourism and events with a focus on supply and demand, capacity development, entrepreneurship, authenticity and commodification and management. She has co-coordinated research projects for various organisations including the Sustainable Tourism Cooperative Research Centre, the Australian Institute of Aboriginal and Torres Strait Island Studies, Indigenous Business Australia, National Environmental Science Program and numerous Australian Federal State and Local Government agencies.

Kelly Whitney-Gould is an independent scholar of tourism and cross-cultural studies, focusing on local Indigenous languages in tourism settings. She has a PhD from the University of Otago and is an associate faculty member at Royal Roads University (Canada). She is semi-retired and lives and works on the islands of Haida Gwaii. Her research interests include the revitalisation of local languages, community-centric development and sustainable tourism practices.

Tracy Woodroffe is a research active Senior Lecturer in the Faculty of Arts and Society specialising in Teacher Education, Teaching Indigenous Learners and Indigenous Knowledge in Education, Northern Institute, Charles Darwin University, Larrakia Country, Northern Territory. She is a Warumungu Luritja woman with extensive experience in Early Childhood, Primary and Secondary classrooms. Tracy is interested in Educational Pedagogy and the use of Indigenous Knowledge to improve opportunity in Indigenous academic achievement. Her work includes Indigenous methodology in examining the Australian education system through an Indigenous Women's Standpoint.

Xiaotao Yang is a lecturer in the School of Tourism and Urban–Rural Planning, Zhejiang Gongshang University, China. Primarily interested in social justice, her interests lie in indigenous knowledge, poverty alleviation and knowledge production.

Jianhong Zhou is a PhD candidate at the Graduate School of International Media, Communication, and Tourism Studies, Hokkaido University, Sapporo, Japan.

ACKNOWLEDGEMENTS

First and foremost, our thanks go to the many contributors to this volume. We appreciate very much their efforts and their patience over the time taken to complete the book. Many suffered through COVID-19 and other problems during that time, particularly from the impact of lockdowns on the tourism and education sectors worldwide. We are grateful for their continued support through such difficult times and their positive responses to our requests for expansions, reductions, corrections and clarifications. We hope they approve of our ordering of chapters and our editing and that we have produced a volume worthy of their contributions. **Thank you for sharing your research knowledge and practitioner insights in this volume**.

We owe a considerable debt of gratitude to Harriet Cunningham and Emma Travis for their continued support. We are most grateful for their help in making the completion of the book much easier than it might have been and for the assistance of their colleagues at Taylor and Francis in the final stages of publication.

Finally, we wish to acknowledge the Indigenous families, communities and tourism businesses around the world who either participate in or have been affected by the tourism industry. Many have patiently worked with or alongside researchers and are increasingly collaborators in the research journey. We recognise their resilience, perseverance and commitment to keeping their cultures alive and hope this book can improve their experiences of tourism in some way. We acknowledge and thank the members and board of WINTA and other Indigenous tourism organisations, some of whom are contributors to this book, in particular the Indigenous Tourism Association of Canada for permission to include excerpts from their report *Building Back Better*. We also thank the board of WINTA, in particular the late Ben Sherman (Chair of WINTA), John Barrett (Kapiti Island Nature Tours, WINTA), Robert Taylor (Chair of WAITOC) and Dr Damien Jacobsen. They have mentored so many, including contributors, and permitted us to include WINTA's Noongar Indigenous Tourism Research Protocol Declaration and Recommendations as an appendix.

1

INTRODUCTION

Revisiting Tourism and Indigenous People

Anna Carr and Richard Butler

Introduction

Indigenous tourism has continued to grow and integrate within the global tourism industry as Indigenous peoples and communities become increasingly involved as providers and participants in tourism. The introduction outlines the content of the Handbook and provides an overview of some of the key changes affecting Indigenous tourism over the past few decades.

This Handbook presents research findings, focusing on the relationships between tourism and Indigenous peoples, and seeks to continue the discussions around Indigenous tourism that have arisen since the 1996 and 2007 volumes edited by Butler and Hinch. Since 2007, there have been several edited volumes about Indigenous communities and tourism, including a 2016 special issue of the *Journal of Sustainable Tourism* (24: 8 & 9), Viken and Müller (2017), Whitford, Ruhanen and Carr (2017) and Graburn and Bunten (2018). This edited Handbook's aim is to further explore and critically examine current debates, controversies and questions through contributions from leading researchers and tourism practitioners from a range of disciplinary backgrounds and geographical regions. The book encourages dialogue across disciplinary boundaries and areas of study through its coverage and contributors. It is international in its focus, emphasising that the issues of Indigenous tourism not only are localised but transcend national boundaries that sometimes require both international and global responses.

Both editors, one of whom identifies as Indigenous and the other as a non-Indigenous person, are conscious of the problems around researching and writing about Indigenous tourism and Indigenous communities, particularly if researchers lack the personal knowledge that being Indigenous brings to any such work. At the 2023 World Indigenous Tourism Summit, delegates were able to contribute to final revisions of the World Indigenous Tourism Alliance's (WINTA's) Noongar Indigenous Tourism Research Protocol Declaration and Recommendations. We were privileged to gain WINTA's permission to publish the Declaration in the book (see Addendum). This Declaration complements the WINTA 2012 Larrakia Declaration and sets clear expectations and research guidelines for future researchers from academia, consultancies, government organisations or the tourism sector itself.

We have endeavoured to locate and encourage scholars and practitioners who identify as Indigenous, or have Indigenous ancestry, to contribute to the Handbook. This volume has 25

DOI: 10.4324/9781003230335-1

contributors with Indigenous ancestry. Other contributors identify as non-western or from ethnic minority groups – far more so than was the case in the earlier volumes with the same primary title. We, particularly Butler, are cognisant of the points made by McCartan, Brimblecombe and Adams (2022, 10) about what they call "methodological tensions" faced by non-Indigenous writers and the need for the consideration of three areas, "critical reflexivity, colonial power analysis and demonstrable anti-racism action". We hope that readers of this volume will be satisfied with the steps we have taken to ensure that Indigenous themes, including voices, beliefs, actions and policies, have been treated with the respect and emphasis which they deserve. In summarising contributions and formulating general conclusions, we have borne these points in mind and elsewhere included a specific item related to critical reflexivity in the context of one of the chapters. It is almost inevitable that some readers will disagree with some of the content of the volume, for unanimity of opinion on such topics included in this volume is never likely to be reached, but hopefully, the volume will serve to highlight problems and issues related to Indigenous peoples' involvement with tourism and may even provide material and examples useful for those who teach and research in this area.

The Handbook is divided into six sections, each preceded by a reflective introduction, with the chapters focusing on diverse issues and conceptual approaches around Indigenous tourism. Whilst there are many synergies between the papers, we have divided the Handbook into six sections that explore dominant themes. Section 1 presents chapters that address a critical aspect of Indigenous involvement in tourism, the essential question of Indigeneity. Section 2, Indigenous Viewpoints, addresses issues such as problems arising from a heritage of colonialism, inappropriate tourism development, frustration with imagery and the longstanding problem of lack of control over the development and resulting impacts of tourism in their communities. Chapters in Section 3, Indigenous Alternatives, explore, through uniquely Indigenous approaches to tourism development, how Indigenous communities can address the frustrations and challenges discussed in Sections 1 and 2. Indigenous Knowledge and Rights is the theme of chapters in Section 4, which explores how the involvement of communities and incorporation of Indigenous knowledge in tourism development are essential for achieving the desired relationships between Indigenous communities, tourists and the tourism industry. Section 5 presents case studies from Indigenous communities and entrepreneurs where the critical importance of culturally sensitive policies and practices during the innovation process is essential to ensure the resulting experiences are not appropriated, distorted or divorced from the original cultural traditions or meanings.

Chapters in the final section of the Handbook explore the scale, role and problems arising during the planning and development of Indigenous tourism. These include reflections from the Western Australia Indigenous Tourism Operators' Council (WAITOC) and WINTA board member, Robert Taylor. As has been noted in several earlier chapters in the Handbook, this section focusses on how Indigenous communities have achieved a much greater role in decision-making and policy formulation related to the introduction and expansion of tourism in Indigenous communities. Communication, networks and Indigenous-led governance are key.

The Indigenous Tourism Setting

Since the publication of Butler and Hinch (1996 and 2007), Indigenous tourism experiences have continued to increase in diversity, quantity, quality and geographical distribution. Businesses have modernised with advances in technology. There is increasing activity in urban settings, and Indigenous tourism providers and workers themselves are travelling extensively and collaborating with people from other Indigenous cultures. The Internet-enabled websites and social media profiles

Table 1.1 Examples of Indigenous tourism organisations

World Indigenous Tourism Alliance (WINTA)	https://www.facebook.com/WINTA.community
NZ Māori Tourism, New Zealand	https://maoritourism.co.nz/
Indigenous Tourism Alberta	https://indigenoustourismalberta.ca/
Indigenous Tourism BC	https://www.indigenousbc.com/corporate/
Indigenous Tourism Association of Canada (ITAC)	https://indigenoustourism.ca/
Indigenous Tourism Collaborative of the Americas	https://indigenoustourism.net/
Novia Scotia Indigenous Tourism Enterprise Network	https://www.nsiten.com/
Travel Nunavat	https://travelnunavut.ca/
Aboriginal Tourism Association (ATA) Australia	https://www.isx.org.au/aboriginal-tourism-association
Queensland First Nations Tourism Council (QFNTC)	https://www.qfntc.com.au/
Western Australia Indigenous Tourism Operators' Council (WAITOC) Australia	https://www.waitoc.com/
Native Hawaiian Hospitality Association	https://www.nahha.com/

Source: Authors.

have brought Indigenous tourism into potential visitors' homes, raising awareness of the abundant opportunities to be experienced. Indigenous tourism operations have utilised social media to gain worldwide followers and provide opportunities to engage virtually as they grow awareness of their experiences online, for instance the food truck startup of Mr Bannock in Vancouver (https://www.mrbannock.com/). Controlling social media profiles on TikTok, Instagram, Facebook and the like was not a concern for Indigenous tourism providers when the previous volumes were published.

In 2020, the global COVID-19 pandemic posed a challenge to Indigenous communities and tourism businesses as travel mobilities paused. For some, this pause provided an opportunity to re-envision their businesses or community aspirations; for others, there was a cessation of tourism activities. The Internet enhanced connectivity and enabled marketing to continue during the pandemic so those businesses that had remained could resume their offerings as the world re-opened, depending on local circumstances such as communities' willingness to host guests. The post-pandemic desire to connect with nature and venture into natural environments complements increased visitor demand for Indigenous tourism experiences.

Prior to the pandemic, the 1990s and 2000s saw the emergence of Indigenous tourism and business networks (see Table 1.1). WINTA was established by Indigenous tourism leaders in 2012, under the then-director Johnny Edmonds. It was then led by Chair Ben Sherman (Oglala Lakota (Sioux) Nation) and a board of Indigenous tourism operators. In 2024 Chief Frank Antoine (Bonaparte First Nations, Canada) became Board Chair. WINTA is notable for enabling and committing to enhancing the prospects of Indigenous communities in the tourism space, regularly partnering with the World Tourism Organization to advance Indigenous tourism activities. WINTA has focused on the advancement of Indigenous peoples in the tourism industry with reference to the *United Nations Declaration on the Rights of Indigenous Peoples*. According to its profile,

> The purpose of WINTA is to provide a forum for Indigenous peoples seeking to draw upon and share their traditional experiences and universal Indigenous values, and through tourism, seek to address the need for balance and harmony both between different peoples and between people and environment.
>
> (WINTA, 2023a)

WINTA has also been active in developing international codes such as the Larrakia Declaration and the 2023 Noongar Indigenous Research Protocol Declaration and Recommendations (WINTA, 2023b). Such activities are vital in showing the way forward to decolonising tourism and tourism research/education spaces.

Elsewhere, some Indigenous tourism associations may predominantly focus on the marketing of tourism activities, whilst others have a broader purpose; for instance, The Indigenous Tourism Collaborative of the Americas is "a network of representatives from Indigenous organizations and tourism industry organizations including travel companies, ministries of tourism, state tourism offices, tourism nonprofits, Tribal colleges and academia". Another example of a country-wide Indigenous tourism association is the Indigenous Tourism Association of Canada (ITAC), which represents around 200 member organisations (see Butler, Chapter 31). The United Nations World Tourism Organization (UNWTO, 2023) has recognised that Indigenous tourism and its associated networks are pivotal to the post-pandemic recovery of the tourism industry.

Another trend has been the development of urban Indigenous experiences. These include Indigenous-owned galleries and cultural heritage centres drawing on Indigenous arts and cultural heritage or opportunities to experience Indigenous cuisine. Indigenous values inform the Minneapolis restaurant and social-justice-inspired Indigenous Food Lab (NATIF) developed by acclaimed chef Sean Sherman (Oglala Lakota Sioux) – see https://seansherman.com/. New Zealand chef Monique Fiso (Samoan Māori) is renowned for her dishes of foraged, natural ingredients, associated with Māori and Pacific cuisine, experienced in her restaurant Hiakai in Wellington – see https://www.hiakai.co.nz/.

These experiences illustrate, as Butler (2021) notes in reference to the 1996 matrix (Figure 1.1, Butler & Hinch, 1996), an expansive diversity of Indigenous "control" of tourism worldwide. There has been a corresponding rise between Indigenous tourism experiences and visitor demand as visitors have a greater choice of locations and experience types. An observable strengthening of Indigenous identities – the way that Indigenous peoples see themselves, their ancestral connections, and relationships to their wider communities and the "other" – is the underpinning cultural values that guide the ways they live their lives when participating in tourism. The "crude" stereotyping of Indigenous peoples as noted by Hollinshead (1996) has been very much laid to rest by a diversity of Indigenous involvement in the tourism sector and Indigenous tourism experiences reflecting both heritage and modern aspirations of the people.

Indigenous Tourism as Regenerative Tourism

Another major change in recent years is the heightened public awareness of climate change and the reality of the negative impacts of increased drought, heatwave, flooding and storm events on the environment. The natural world is in crisis. As Pollock (2019) noted, many wicked problems affect our travel behaviours. The privilege of travel comes at a cost to those living in the most affected regions of the world – travel is still a luxury of the wealthy, and many Indigenous peoples lack wealth. The global and local sectors of the tourism industry have been continually engaging in discussions and activities about the Sustainable Development Goals (SDGs) and sustainability, alongside rejuvenating, revisioning, reconnecting and revitalising communities, ecosystems and travel itself through regenerative tourism. The travel sector's reliance on fossil fuels for transport means many Indigenous communities are more exposed to the tragic consequences of climate change. Indigenous tourism providers need to navigate these challenging times of "climate boiling", pandemic recovery and increasing political global instability.

Nature-centric Indigenous tourism has been observed to strengthen Indigenous peoples' relationships to their cultural landscapes and natural world. Researchers such as Becken and Kaur (2022) and Pollock (2019) acknowledged that the movement towards "regenerative tourism" planning and management processes are inspired by and draw heavily from the practices and wisdom of Indigenous peoples' communities around caring for nature. Tourism outcomes that reflect local community values and have positive environmental outcomes are increasingly sought by Indigenous entrepreneurs and collective entrepreneurs. Tourism destination and conservation managers, practitioners and planners are increasingly inspired by Indigenous traditional knowledge to inform solutions that address the challenges arising from environmental and climate change (Latip, Rasoolimanesh, Jaafar, Marzuki & Umar, 2018; Fusté-Forné & Hussain, 2022; Mason, Carr, Snow, Vandermale & Philipp, 2022). The involvement of, and leadership by, local Indigenous peoples in conservation and ecotourism settings has been pivotal for advancing successful nature restoration and economic and socio-cultural outcomes (Carr, 2007; Amoamo, Ruckstuhl & Ruwhiu, 2018). By incorporating and prioritising Indigenous knowledge systems, values and practices in western management approaches, it has been argued that regenerative tourism can be transformative and less exploitative, can enhance the environment and can be more acceptable to all host communities (Matunga, Matunga & Urlich, 2020; Dredge, 2022).

Indigenous values have informed tourist pledges to guide visitors' behaviours, for instance the Haida Gwaii Pledge (https://haidagwaiipledge.ca/) and Aotearoa New Zealand's Tiaki Promise (https://www.tiakinewzealand.com/en_NZ/). Pledges are a timely move towards including cultural values and environmentally appropriate behaviour guidelines for both operators and visitors. Other initiatives in Indigenous arts, ecocultural/nature-centric tourism and events have developed that connect locals and visitors to ancestral landscapes, skyscapes and waterscapes (Higgins-Desbiolles, 2016; Walters & Ruwhiu, 2021). Community festivals aimed at locals (not tourists) have also evolved; for example, in 2022, Matariki was declared the first Indigenous public holiday to be celebrated nationwide by the New Zealand government. The Matariki holiday recognises the Māori calendar's New Year when the Matariki star constellation becomes visible in the eastern night skies at dawn (Matariki is known elsewhere as *Subaru*, the Seven Sisters or Pleiades).

Engaging Indigenous Scholarship in Tourism

Finally, there is a necessity to comment on the need to decolonise tourism education and research in academic and training spaces. In recent decades, universities in the Americas, Aotearoa New Zealand, the Pacific, Australia, Canada, Asia, the African continent and Scandinavia have strategised to increase Indigenous student and academic cohorts and to support them to be active researchers (preferably in consultation or partnership with local Indigenous peoples). Australia has taken one such initiative through the formation of a First Nations Tourism Mentoring Programme, delivered by the National Indigenous Australians Agency (NIAA).

It is vital that Indigenous tourism scholars and researchers can inform Indigenous decision-making and also take on educational, training, governance and management responsibilities that affect tourism (and other) policy thus enabling diverse Indigenous worldviews (Whitford & Ruhanen, 2010; Carr, Ruhanen, & Whitford 2016). Indigenous researchers, teachers and student representatives on the executives of editorial boards, committees and associations in the academic tourism and hospitality space would be desirable as well as advisable. There is no global research organisation dedicated to monitoring and collecting data and research relevant to Indigenous tourism that could provide comparisons and insights for Indigenous businesses. An internationally dedicated organisation created to collect such statistical data would inform future Indigenous tourism management. The UNWTO, working alongside WINTA, has the potential to include datasets monitoring visitation to Indigenous experience providers and communities which would inform the management of tourism in a beneficial manner.

Conclusion

Indigenous peoples' relationships to tourism go beyond the dimensions of being a business or a tourism attraction. Their place within the global tourism sector is increasingly one of leadership and connecting communities. For many, Indigenous peoples' involvement in tourism can be a means of strengthening or connecting with their personal identity or ancestry. There is a growing Indigenous diaspora that travels to visit family and rediscover their cultural heritage (Hall & Duval, 2004). The relationship between Indigenous peoples and tourism goes beyond business and entrepreneurial activity. Indigenous tourism enables relational interactions that can enrich society by improving livelihoods and raising awareness around social justice, human rights and ways to decolonise the worlds.

Enhancing Indigenous presence in the tourism sector and connecting visitors to Indigenous tourism activities within Indigenous communities are dependent on planning approaches that centre on Indigenous voices and local values that are crucial for the prospects of Indigenous tourism. There is a moral obligation on governments and NGOs worldwide to support Indigenous employees and to ensure Indigenous peoples lead tourism planning approaches that incorporate local Indigenous values and aspirations for future generations. Indigenous peoples have inherent endurance and resilience strengthened from enduring centuries of colonisation. For Indigenous travellers themselves, global experiences of other Indigenous cultures will enrich their own cross-cultural understandings and creativity (crucial for innovations that will be responsive to the challenges ahead). Acts of decolonisation enable people to be strongly situated within Indigenous lands; strengthening cultural place identity, drawing upon cultural and ancestral values and embracing Indigenous knowledge are all ways of building resilience and empowerment. Underpinning Indigenous and cultural tourism activities with political and economic systems that have Indigenous representation and leadership, informed by research led by, with or for Indigenous peoples will ensure future management that magnifies the promises of tourism.

References

Amoamo, M., Ruckstuhl, K. & Ruwhiu, D. (2018). Balancing Indigenous values through diverse economies: A case study of Māori ecotourism. *Tourism Planning & Development*, 15(5): 478–495.

Becken, S. & Kaur, J. (2022). Anchoring "tourism value" within a regenerative tourism paradigm – A government perspective. *Journal of Sustainable Tourism*, 30(1): 52–68. https://doi.org/10.1080/09669582.2021.1990305

Butler, R. (2021). Research on tourism, Indigenous peoples and economic development: A missing component. *Land*, 10(12): 1329. https://doi.org/10.3390/land10121329

Butler, R.W. & Hinch, T.D. (1996). *Tourism and Indigenous Peoples*. London: International Thomson Business Press.

Butler, R.W. & Hinch, T.D. (Eds.) (2007). *Tourism and Indigenous Peoples*. London: Routledge.

Carr, A. (2007). Strengthening of identity through participation in tourism guiding, pp. 113–127 in R. Butler & T. Hinch (Eds.) *Tourism and Indigenous Peoples*, 2nd Edition. Oxford: Butterworth-Heinemann.

Carr, A., Ruhanen, L. & Whitford, M. (2016). Indigenous Peoples and tourism: The challenges and opportunities for sustainable tourism. *Journal of Sustainable Tourism*, 24(8&9): 1047–1069. https://doi.org/10.1080/09669582.2016.1206112

Dredge, D. (2022). Regenerative tourism: Transforming mindsets, systems and practices. *Journal of Tourism Futures*, 8(3): 269–281.

Fusté-Forné, F. & Hussain, A. (2022). Regenerative tourism futures: A case study of Aotearoa New Zealand. *Journal of Tourism Futures*, 8(3): 346–351.

Graburn, N.H.H. & Bunten, A.C. (Eds.) (2018). *Indigenous Tourism Movements*. Toronto: University of Toronto Press.

Hall, C.M. & Duval, D.T. (2004). Linking diasporas and tourism: Transnational mobilities of Pacific Islanders resident in New Zealand, pp. 78–94 in T. Coles & D.J. Timothy (Eds.) *Tourism, Diasporas and Space*. London: Routledge.

Higgins-Desbiolles, F. (2016). Sustaining spirit: A review and analysis of an urban Indigenous Australian cultural festival. *Journal of Sustainable Tourism*. https://doi.org/10.1080/09669582.2016.1149184

Hollinshead, K. (1996). Marketing and metaphysical realism: The disidentification of aboriginal life and traditions through tourism, pp. 308–348 in R. Butler & T. Hinch (Eds.) *Tourism and Indigenous Peoples*. London: International Thomson Business Press.

Indigenous Tourism Collaborative of the Americas (2023). Who we are. Available online at https://indigenoustourismamericas.org/about-us/ (accessed 13 October 2023).

Latip, N.A., Mostafa Rasoolimanesh, S., Jaafar, M., Marzuki, A. & Umar, M. (2018). Indigenous participation in conservation and tourism development: A case of native people of Sabah, Malaysia. *International Journal of Tourism Research*, 20(3): 400–409. https://doi.org/10.1002/jtr.2191

Mason, C., Carr, A., Snow, W., Vandermale, E. & Philipp, L. (2022). Rethinking the role of Indigenous knowledge in sustainable mountain park development in Canada and Aotearoa New Zealand. *Mountain Research and Development* (Special Issue: Weaving together knowledges - Collaborations in support of the wellbeing of mountain peoples and regions), 42(4): A1–A9. https://doi.org/10.1659/mrd.2022.00016

Matunga, H., Matunga, H. & Urlich, S. (2020). From exploitative to regenerative tourism. *MAI Journal: A New Zealand Journal of Indigenous Scholarship*, 9(3): 295–308. https://doi.org/10.20507/MAIJournal.2020.9.3.10

McCartan, J., Brimblecombe, J.K. & Adams, K. (2022). Methodological tensions for non-Indigenous people in Indigenous research: A critique of critical discourse analysis in the Australian context. *Social Sciences & Humanities Open*, 6(1): 100282. https://doi.org/10.1016/j.ssaho.2022.100282

Pollock, A. (2019). *Regenerative Tourism: The Natural Maturation of Sustainability*. Available online at https://medium.com/activate-the-future/regenerative-tourism-the-natural-maturation-of-sustainability-26e6507d0fcb (accessed 15 July 2023).

UNWTO (2023). Available online at https://www.unwto.org/news/empowering-indigenous-communities-to-drive-tourism-s-recovery (accessed 2 September 2023).

Viken, A. & Müller, D. (Eds.) (2017). *Tourism and Indigeneity in the Arctic*. Bristol: Channel View Publications.

Walters, T. & Ruwhiu, D. (2021). Navigating by the stars: A critical analysis of Indigenous events as constellations of decolonization. *Annals of Leisure Research*, 24(1): 132–149. https://doi.org/10.1080/11745398.2019.1643749

Whitford, M. & Ruhanen, L. (2010). Australian Indigenous tourism policy: Practical and sustainable policies? *Journal of Sustainable Tourism*, 18(4): 475–496.

Whitford, M., Ruhanen, L. & Carr, A. (Eds.) (2017). *Indigenous Tourism: Cases from Australia and New Zealand*. Oxford: Goodfellow Publishers.

WINTA (2023a). World Indigenous Tourism Alliance Facebook 'About'. Available online at https://www.facebook.com/WINTA.community (accessed 18 June 2023).

WINTA (2023b). World Indigenous Tourism Alliance LinkedIn. Available online at https://www.linkedin.com/company/world-indigenous-tourism-alliance/ (accessed 18 June 2023).

SECTION 1

Indigeneity

The first section of this volume addresses a particularly sensitive and critical aspect of Indigenous involvement in tourism, the essential question of Indigeneity. The chapters that follow in this book reveal considerable variation in what is meant by the term Indigenous and by implication, Indigeneity, which is why the chapter by Ren was selected to be the first of the contributions. Its very title, "Are we not all Indigenous?" puts that question front and centre, and while there is no clear answer to that question, the context in which the question was put to the author by the person making it, showed clearly that the issue is not simple. The United Nations' definition of Indigenous is not necessarily appropriate now in the current period, and as becomes clear from the following chapters, the term is not accepted by some national governments, in some cases because of concerns over domestic security and the integrity of the nation state, in others because the term "ethnic" is preferred because it carries less "baggage" in political dialogue. Ren's discussion of how Indigeneity is being used by Greenland in developing its case for independence and home rule reveals some of the politics that involve tourism when tourism and related development are seen as empowering and strengthening the case for statehood. Hollinshead, in what is one of the last of a long stream of insightful and thoughtful commentaries on Aboriginality expands on the need for dialogue, understanding, and the use of new terminology and ways of thinking about this difficult subject. His use of examples from Indigenous writers links with similar critical commentaries in the section on Indigenous Voices, such as those of Higgins-Desbiolles and Jennings and provides conceptual arguments on understanding and accepting the term.

Edelheim and Zhou explore the meaning of Indigenity through a critical realist appraisal of the term in the context of an Indigenous community in the Peoples; Republic of China and discuss the term through the concrete universal perspectives involved. They argue that it is a relational representation tied closely to the nature of the region involved and the common components such as resilience and adaptation shared by many ethnic groups, and conclude that Indigeneity is fluid in terms of the context rather than simply expressed by identity, blood and place ties. Difficulties in accepting the concept of Indigeneity and defining peoples in those terms are found in a number of countries and societies, (including China, and Iran as illustrated by Dehrashi and Ghader) and Japan is relatively recent in recognising the Ainu as being an Indigenous people. This has resulted in it proving necessary to enact new legislation to ensure the appropriate treatment of the Ainu

DOI: 10.4324/9781003230335-2

and protect their involvement in developing their rights and portrayal through major tourist infra-structure as Okada and Edelheim record. Some, although not all of the Ainu traditional activities and rights in Hokkaido, their traditional homeland, have been protected, and the Ainu language, a key element in the survival of the culture is utilised increasingly through the new legislation in those developments. The issues revealed by these new arrangements are examined further in a later chapter by Hashimoto. The many questions arising from the varying concepts of Indigeneity and its implications run throughout the volume and will doubtless remain issues for some time to come.

2

"ARE WE NOT ALL INDIGENOUS?" NEGOTIATING INDIGENEITY IN GREENLANDIC TOURISM

Carina Ren

Introduction

This contribution explores the shifting and negotiated meanings of the concept of Indigeneity in the development of Greenlandic tourism, an area that is gaining growing attention in Greenlandic society and politics. Tourism is used as a prism to explore and trace Indigeneity in public identity discourses as Greenland moves towards statehood and full independence from the Danish Kingdom. This chapter describes and discusses the coexistence of and friction between Indigenous, national and hybrid identities within the areas of product development and marketing of Greenlandic tourism. It ends by arguing how the situated experience with Indigeneity from Greenland tourism enriches as well as destabilizes Indigeneity as a more general concept.

The only way to access Greenland (*Kalaallit Nunaat*) with trans-Atlantic or similar aeroplanes is through Kangerlussuaq, an airport established as an airbase by the Americans during WW2. Kangerlussuaq serves as a hub for connecting flights to other parts of the Greenlandic West Coast and is home to most of the nation's 56,000 inhabitants (stat.gl, 2021), a vast majority of whom are Greenlandic Inuit. As part of the *Culturally Sensitive Tourism in the Arctic* (ARCTISEN) project, in this chapter I am reporting on a study trip with a group of Sami tourism entrepreneurs and Nordic tourism scholars who met with a range of small Greenlandic tourism businesses. The goal of the trip was to learn more about and benchmark their tourism products and experiences.

The project is in its second year. Already, an initial desk study has been conducted to map the meanings, concerns and controversies around cultural sensitivity in Greenlandic, Canadian Inuit and Nordic Sami tourism. It revealed that vast differences exist in how Indigenous culture and identity are understood and play out across Canadian, Greenlandic Inuit and Sami territories (Olsen et al., 2019). According to the report, "Something that sets Greenland apart from the Indigenous populations in the Nordic countries is the idea of balancing an Indigenous identity and a national identity" (Olsen et al., 2019, p. 17), thus suggesting that Indigeneity plays out differently in Greenland than in Sapmí.

As argued by Thisted (2013), a reason why an ethnos-based identity – and hence Indigeneity – plays out differently in Greenland than that in other Indigenous territories is the wish for future independence, which ranks high on the political agenda in Greenland. Since the

DOI: 10.4324/9781003230335-3

instigation of Home-Rule (in 1979) and later Self-Rule (2009), the Greenlandic government has been pursuing a trajectory supported by most Greenlanders towards full independence from Denmark. For that reason, the term 'Indigenous' has strategically been avoided in the Act on Greenland Self-Government, as according to Thisted (2013), Indigeneity counters, rather than promotes, the Greenlandic desire to establish their own state sometime in future. For instance, the United Nations Declaration on the Rights of Indigenous Peoples (UNDRIP) declaration (UN, 2007) does not give Indigenous peoples rights to establish their own independent states.

Other scholars have also pointed to the coexistence of a strong national identity in Greenland next to an Indigenous one, explaining this situation by pointing out Greenlanders have always constituted a majority in their own country. While no ethnic registration exists in Greenland, 89% of the population today has been born in Greenland (stat.gl, 2021) and is for that reason considered Inuit or European-Inuit. Also, and in contrast to Sapmí (see Chapter 19, editors' note) and other Indigenous territories, the land of Greenlanders was never nationalized, confiscated or expropriated by settlers. Land in Greenland cannot be owned and is today in the hands of the Greenlandic government. In the context of tourism, this means that a common way to stimulate tourism development is through the use of concessions, whereby investors may gain exclusive rights to develop services, facilities and infrastructure, for instance, heli-skiing, sports fishing or musk ox hunting in a particular area.

The fact that land cannot be owned by outsiders does not deny that Greenland was indeed colonized. In the late 10th century, Norse settlements were first established in South Greenland but vanished again in the early 15th century. In 1720, Danish-Norwegian missionary Hans Egede arrived in Greenland, marking the beginning of Greenlandic colonization, which officially ended in 1953, when Greenland was instead granted the status of a county (Danish: *amt*). Unlike development in other Indigenous lands, Danish colonists never managed – or aimed – to become a settler majority. Throughout the centuries, most Danes working in Greenland, in both white- or blue-collar jobs, returned to Denmark after a few years, and many of those who stayed established themselves with Greenlandic partners to continue a long tradition of inter-ethnic marriages and families. Regardless, Danes arriving in Greenland received, until recently, and – some would argue – still receive, privileges in Greenlandic society such as higher salaries, quick access to accommodation and yearly trips to Denmark. According to Rud (2017), the recent progressed status of Greenlandic Self-Rule has been instrumental in nourishing a naïve Danish manifestation of itself as a 'benevolent' colonizer "that the rest of the world could learn from" (Rud, 2017, p. 132). However, statistics display in all clarity the many social problems such as alcohol abuse and one of the highest suicide rates in the world that are rooted in the colonial legacy. Also, debates and testimonies continually unravel the pain, shame and taboos connected to a colonial past. It is therefore unsurprising that many Greenlanders are looking forward to gaining full sovereignty.

One major obstacle towards this goal is the yearly block grant of 3.9 billion Danish Crowns (approximately 524 million Euros) from Denmark. The block grant currently covers approximately half of Greenland's budget. Denmark also manages costly public services (Rosing et al., 2014) within the island nation of approximately 56,000 inhabitants. Consequently, and according to Agneman (2021, p. 3), "near-future political independence resulting in an immediate reduction or stop of Danish economic support would imply either dramatic cuts in welfare provision, sharp increases in tax rates, or both".

In order to develop the needed economic robustness for gaining economic self-sustainability and to counter a possible cut in welfare support and care in a future sovereign state, tourism has been pinpointed as a viable opportunity for development since the inauguration of the Home-Rule and more intensively during the last decade (Ren & Hall, 2021). As a "third economic pillar" (Bjørst & Ren, 2015), tourism is envisioned as an income source able to supplement revenues from the currently dominant fisheries industry and the oil, gas and mining industries currently either shelved or kept on hold as part of Greenland's political aim to initiate a transition to a greener economy. In this context, tourism is seen as not just an economic activity, but also a political one, because of "tourism's economic role embedded (in discourse) in the ongoing nation building process in Greenland" (Bjørst & Ren, 2015, p. 92).

To stimulate tourism and regional development, the Greenlandic government has initiated massive airport construction in the capital Nuuk and in the tourism capital Ilulissat, the home of the UNESCO World Heritage Site of the iconic Ilulissat Ice fjord. One hope is that direct international flights will boost tourism numbers that – until COVID-19 – were numbering around 100,000 yearly arrivals (Visit Greenland, 2019). In this context, of Greenland preparing for increased tourism and moving towards full independence from the Danish Kingdom, this chapter discusses how the Indigenous reality plays out (or not) in Greenlandic tourism through the sensitive and powerful entanglement of (geo)politics, history and culture. This contribution describes and discusses the coexistence of, and friction between, different identities in tourism and argues how the situated experience with Indigeneity from Greenland tourism enriches as well as destabilizes Indigeneity as a universal category and a stable analytical concept.

A Different Kind of Cultural Sensitivity in Greenland

While some argue that increasing flights counteract environmentally sustainable development in Greenland and accelerate climate change, which are felt at an accelerated rate in the Arctic, the general view is that the air-transport-related constructions will bring development and opportunities to airport towns. As pointed out by Bjørst (2018), this aligns with the argument of favouring economic sustainability above climate change in the national political context. Ren and Bjørst (2017) have shown how global discourses on the Arctic as a victim of climate change are reconfigured in a local, Greenlandic context, "seeing receding ice as an opportunity for industrial and tourism development" (p. 501). Climate change is seen by many in Greenland as an opportunity rather than an obstacle, as for instance, the changes extend the fishing, sailing and cruise tourism seasons.

Discussions in Kangerlussuaq

For now, awaiting the 'game changers' that new airports will supposedly be for Greenlandic tourism according to many, we are here, in Kangerlussuaq, waiting for our connecting flight to Nuuk. Besides being an airport and flight hub, Kangerlussuaq is also the home of a few hundred residents, mostly working in the airport and the tourism industry and mostly catering for tourists that stop over here for a few hours or days. To make the most of our time before our departure to Nuuk, we cross the gravel courtyard behind the airport to a range of small shops selling souvenirs to tourists. In one of the shops, we meet the owner, Niví, a Danish-speaking Greenlandic woman in her thirties of half-Greenlandic, half-Danish descent. In her gift shop, Niví sells a variety of products made of local products and traditional material: reindeer tusk and bone, glass beads and musk ox wool. She also produces jewellery, artwork and musk wool products herself. Niví lives in Kangerlussuaq with her Greenlandic husband, an adventure tourism entrepreneur, and their children. I have met and talked with her before, in her shop in Kangerlussuaq and at other tourism events in Nuuk and online. I introduce her to some of the Sami visitors, and as they proceed around the shop to enjoy the beautiful products on sale, we continue to talk.

I first explain the idea behind bringing the Sami entrepreneurs to Greenland and continue to talk about the ARCTISEN project, elaborating on the concept of cultural sensitivity, hoping that she would be interested in joining different project activities later on. As I had already experienced on other occasions and having now worked a year on the ARCTISEN project, I know that talking about cultural sensitivity and Indigeneity in Greenland is not straightforward. Just as in other Nordic languages, the Danish term for sensitive – *følsom* – carries… well, sensitive connotations. As explained in a footnote in Olsen (2020, p. 17), the term was also received with some hesitation in Norway. A suggested explanation for the reluctance against the term was the Norwegian connotations (similarly to Danish) alluding to how Sámi, like many minorities around the world, "are often accused of being touchy (Norw. *hårsår*) and easily upset" (p. 17).

In our initial conversations about the project with tourism stakeholders in Greenland and about the concept of cultural sensitivity, many participants had objected to the relevance of the term for discussion in a cultural context. Many pointed to how Greenlandic culture and language were vigorous and robust, thus discursively contrasting it to 'sensitive' in the sense of 'in need of protection'. Some, as mentioned in the Greenlandic ARCTISEN project report, seemed to interpret cultural sensitivity as equivalent to cultural frailty (Ren et al., 2019, pp. 26–27). Contrary to the concerns voiced in the material from Sapmí, Greenlandic participants were generally not very concerned when asked about the impact of tourism on Greenlandic culture, as illustrated by this quote by a local cultural entrepreneur:

> Here in Sisimiut and Qeqqata municipality, (…) people are trying to make the tourism industry local. So it is coming from within the culture. I don't see very many things like that [insensitive ways of developing tourism, ed.] because people that are selling and creating the products are from within the culture.

In the capital of Nuuk, a destination manager also saw the pressures on culture as minimal and hence not in need of protection:

> We have various offers here in Nuuk in what you would call traditional Inuit culture. However, this is not something that we especially focus on to preserve or to make sure that it is there. I feel confident enough that the operators are offering it. I do not need to make sure that it is happening.

What these quotes illustrate is a relaxed, 'anti-alarmist' and confident view of their culture, which also might 'spill over' into their views on Indigeneity. At least, this was what struck me during my conversation with Niví as it continued. As I described how the project sought to bring together Sami and Inuit stakeholders and work with Indigenous culture (in Danish: *oprindelige* folk), she exclaimed: "Indigenous! But are we not all indigenous?". While most of the Sami visitors had exited the shop by now, I was quite aware at this moment that they would probably not have agreed with that statement. Perhaps, they would even have been provoked or outraged by her proposal, with justification, knowing some of the controversies that have raged over ownership of land and the forceful integration into Norwegian, Finnish or Swedish national identity. Claiming or granting someone an Indigenous identity is not something done lightly, but requires taxing, sensitive and often painful processes of translation, negotiation and, in some cases, certification (García-Rosell, 2016; De Bernardi, 2019).

Continuing her argument, Niví pointed to the fact that we all had an 'origin' (*oprindelse* drawing on the Danish word for Indigenous peoples, *oprindelige folk*). She pointed out how she rather preferred seeing the practices of herself and her family as *naturfolk*, hence referring to a now outdated and politically incorrect Danish term stemming from the German *Naturvolk* (nature people).

While I interpreted the now abandoned term as insensitive and based on an evolutionary and racist understanding of Indigenous people as more 'primitive', she saw it as quite descriptive of the life as carried out by herself and her family *in* and *with* nature.

In her later contributions to the ARCTISEN project, Niví displayed a highly developed sense of cultural sensitivity and reflexivity through her creative practices, in which she connected culture, local materials and traditional knowledge with innovative techniques. I became convinced that the (*de facto* politically incorrect) term of 'nature people' was suited very well to how she perceived her daily life. But the way in which she had questioned or disregarded the term and simultaneously (in my view) challenged Indigenous identity as a meaningful category (in the sense that she saw it as an 'all-inclusive' human classification) seemed puzzling to me. How, I asked myself, could Niví be so carefree in inviting everyone into Indigeneity, thus deconstructing it as a useful category? Or what was I missing here? To pursue these questions, let us try to get a better understanding of how Indigeneity works in Greenland and step closer to how discussions around it unfold in Greenland and Greenlandic tourism.

Strategic Indigeneity – Greenlandic Identities in and around Tourism

Many political scientists have argued that Greenland draws on (at least) two sets of identities in navigating international politics: a national one and an Indigenous one (Sowa, 2013). According to Strandsbjerg (2014), "two different identity complexes face each other. On the one hand, we have a territorially defined political society, and on the other hand, there is an ethnically defined Inuit identity" (p. 264). Greenland has no pre-existing model to draw on in international politics, and with no majority Indigenous nation state currently in existence, the principles of statehood and Indigenous rights are currently most often perceived as mutually exclusive.

As multiple foreign policy examples illustrate, playing the role of an Indigenous rather than a sovereign nation has diminished Greenland's influence and opportunities for representation. Consequently, Indigeneity is often downplayed in Greenland's current politics when they relate to matters of independence. According to Gad (2009),

> the essentialist ethno-nationalist revivalism may be said to have been a wise strategic 'choice' for the Greenlandic elite to further the de-colonization process: it has served as an argument for transferring sovereignty and resources from Denmark to Greenland much more rapidly than anyone imagined possible in the 1960s and 1970s.
>
> (p. 138)

However, foregrounding Indigenous Inuit identity, culture and way of life have also been used in cases of negotiating whaling and sealing rights to secure Indigenous rights. Thus, "The logic of ethnicity also encompasses a goal of achieving increased independence and representation in matters relating to Greenland's interests, not necessarily with an emphasis on formal state formation, but rather in a process of decolonisation" (Ren & Abildgaard, 2021, p. 189). In other words, and depending on the situation, different identity positions serve different strategic purposes.

According to Ren and Abildgaard (2021), the national and Indigenous identities described in the section above follow two contrasting logics that also transcend and translate into tourism. In the first *modern* tourism logic, tourism is inscribed in the building of a strong national identity enabled through infrastructure development on and mobility within Greenlandic territory. In line with Franklin (2003), who discussed the strong and integral connections between tourism and nation-building, a *modern* tourism initiative is for instance the *Nunarput Nuan* 'staycation' campaign launched by Visit Greenland that revolved around citizens encountering and exploring their own country.

In the second identified *ethnic* logic leading to strengthened Inuit identity, tourism is about immersion in Indigenous tradition and ethnic expressions and translates into the growing offer of tourism products and experiences that focus on Inuit cultural artefacts and performances, heritage and food culture. Such a logic is mirrored in reoccurring and heated political and public debates on what defines a 'real' Greenlander (Brøns et al., 2022) and in the reinvigoration of, for instance, Inuit drum dancing and tattoos (Thisted & Ren, 2021). These contrasting, yet coexisting, tourism logics are instructive as analytical tools to show how identity informs or entangles with current and future tourism development in Greenland.

Tourism of Greenlanders – or Tourism Colonialism?

As in other territories with colonial pasts, ethnic tensions, mutual distrust and latent racism still exist in Greenlandic politics, business and everyday life – and tourism is no exception. Issues uncovered as part of the ARCTISEN project included discontent with foreign, often Danish, guides coming to Greenland for the summer with no prior experience about the place and culture and foreign tourism operators 'moving in' with their own gear and guests before the summer, only to depart again after the short season. During a tourism stakeholder seminar and workshop, this led a Greenlandic entrepreneur to label such activities as tourism colonialism (see Ren & Abildgaard, 2021). Local concerns around how foreign businesses, investors and tourists comport themselves when operating in or visiting Greenland not only point to the potentially serious challenges awaiting Greenlandic communities and businesses as tourism develops, but also display how past and present (post-)colonial ties and breaches linger on within tourism.

Post-colonial tensions and traumas in Greenland reflect issues common to Indigenous tourism. But we also find features that are more distinctive to Greenland and related to the nation's advanced Self-Rule. A particularity, which sets Greenlandic tourism aside from Sami tourism in Arctic Europe, is that the Greenlandic government owns most of the tourism infrastructure and logistics. This includes *Air Greenland*, Greenlandic airports and harbours, the travel agency *Greenland Travel*, a coast/cruise ship, many large hotels such as the iconic *Hotel Arctic* in Ilulissat, and the *World of Greenland* tourism agency, which operates luxurious flagship lodges at Eqi and Ilimanaq. In addition, the tourism strategies of the government, Visit Greenland and regional Destination Management Organisations (DMOs) are developed by Greenlandic political and/or business organizations.

The *Pioneering Nation* brand developed by Visit Greenland, for instance, has played an important part in creating and constructing the global imaginary of Greenland. Thisted (2015) argued how the development of the *Pioneering Nation* brand is illustrative of a positive transition from Southern representations of the Arctic, as "the people of the Arctic are now to a much larger degree representing themselves" (p. 23). In contrast to the situation found in many other places, Greenland could therefore be argued to already possess Indigenous involvement and leadership in developing, delivering and storying tourism experiences and material objects, which Higgins-Desbiolles (2003) identifies as central in empowering reconciliation in and through tourism.

However, a strong Greenlandic representation and empowerment may also lead, as identified in other spheres of Greenlandic society, to an even more nationalist position, where the right to represent and 'sell' Greenland and Greenlandic culture in tourism is determined not based on Indigeneity, but rather on 'Greenlandicness', and even what Gad (2009) identifies as an essentialist conceptualization of Greenlandic identity. An example of such 'reversed traumas' is language, where Greenlanders who do not speak Greenlandic experience shaming or other kinds of public or concealed exclusion. Such feelings were expressed during a later online ARCTISEN workshop with Sami and Greenlandic entrepreneurs, Arctic DMOs and tourism researchers, where Niví, after similar thoughts from Sami entrepreneurs, shared her thoughts on coming to terms with her inability to speak Greenlandic and related thoughts on gaining legitimacy as a cultural producer.

Hybrid and Local Sensitivity

Not long after our passing through Kangerlussuaq, Niví decided to close her shop. Shortly afterwards, she relocated with her family to Norway. From there, she continues to share knowledge about and sell her handcrafted wool products and jewellery online using Greenlandic and increasingly Norwegian natural materials. Thus, she aligns with other creative Greenlandic entrepreneurs who experiment with and reinterpret cultural practices, narratives and materials. Within the literature, arts and broader cultural economy, we find many examples that draw on Inuit heritage while generously blending it with innovative materials and hybrid symbolism. These include the highly original *Bestiarium Groenlandica*, reinterpreting Greenlandic myths and creatures, the fashion brand *Inuit*, and the hand tattoo-like jewellery by *PanikPanik*.

As already displayed in countless research studies and also represented in this volume (see for example, Chapters 20, 25, 19), tourism is entangled in multiple ways with Arctic and other Indigenous identities and cultures. In their overview article on Arctic and Indigenous cultures in tourism, Ren et al. (2021) argue that "In some places and under certain conditions, the marketing and development of tourism is part of creating and exacerbating essentialization while it in other instances supports and strengthens local and indigenous heritage and traditions" (p. 115). The hybrid cultural products offer mundane and un-exotifying representation and interpretation of modern Arctic everyday life. An example is Visit Greenland, whose marketing and branding work has served, according to Ren et al. (2021), as "a central player in (re)defining the representation (…) of Greenlanders and Greenlandic everyday life (…) in which nature, culture and modern everyday life in the Arctic is blended into a hybrid, playful and often humoristic imagery" (p. 115).

Together with the excerpts from our interviews on cultural sensitivity and the role of the Indigenous in Greenland, such examples point towards a coexistence or blending of modern and ethnic logics in tourism development. These hybrid cultural products and practices also express how tourism has, to a great extent, been able to 'rewire' the usual post-colonial discussions, for instance around language (Gad, 2009), by more often using English – as opposed to Greenlandic or Danish – as a *lingua franca*. Experiences from the project, among them the sharing of stories from Niví and others, illustrate that tourism can serve as a cultural engine to empower and boost cultural heritage but that it can also stir and ignite controversy around identity, belonging and ownership.

Conversations in the ARCTISEN project with stakeholders showed that many dismiss or challenge an ethnic identity as relevant or recognizable, rather than pointing to Greenlandic culture as strong and in no need of 'sensitive treatment'. However, amidst emerging concerns in tourism around access to land, the right to guide and who has the right to tell – and sell – stories about Greenland, many tourism entrepreneurs expressed a will to strengthen 'Greenlandic' and local ways of doing things. As argued by a manager of a DMO:

> I just hope that it is always the Greenlanders that define the roles, what is right and what is wrong and what to do. (…) If Greenland wants tourism, it should be under our rules. I am not a fan of copying a system from outside. We say that we want to be independent, but we

keep on copying from Denmark. Denmark is different. Many things are different and so we should do it from the inside.

As argued in Ren et al. (2021) "the conceptualizations and practices of cohabitation and reconciliation are ongoing within most Arctic nations, [but] the forms that these struggles take vary" (p. 117). A question is how may Greenland develop in its move towards independence in ways that acknowledge and better cultivate sensitivity towards cultures (in plural), whether Indigenous, traditional or simply local? Much may be learned from tourism, which, as seen, increasingly accompanies the usual national and Indigenous Greenlandic identities through lively and vibrant imagery, products and experiences that playfully curate objects, stories and materials towards new tourism experiences.

Concluding Remarks

The Greenlandic case of Indigeneity in tourism outlines the contours of new identity positions and different and contrasting ways in which the Indigenous is reconfigured as Greenland moves towards independence. It suggests that Indigeneity – and Indigenous tourism – is not a *universal category* but always a *situated outcome* of powerful political, social and historical (and other) forces (as discussed in Chapter 3, this volume). As argued by Strandsbjerg (2014) "geopolitics has always been part of indigenous people's colonial history" (p. 261), and tourism is no exception. While the definition of Indigenous peoples remains controversial, it plays out differently in different countries or destinations (Ren et al., 2021).

According to Sowa (2013), Indigenous peoples and the identity politics they engage in may only find international recognition by reproducing (outdated) preconceived and projected images of representation accepted in Euro-American societies. This, he argues, results in a 'museumification' of Indigeneity that excludes social transformation and change. By turning to Greenland and to the tourism practices taking place there, we see the multiple ways in which Inuit culture ongoingly builds on, reinvents, juxtaposes, amplifies and innovates its cultural practices through colonial, global and tourism encounters, thus challenging and adding to the ways in which Indigenous nations may be represented. I am not sure that I agree with Niví in that we are, or should, all be Indigenous, but in *Kalaallit Nunaat*, Indigeneity and its trajectories are lively, on the move and far from museumified.

Acknowledgements

The author would like to acknowledge the EU-Northern Periphery Arctic programme for financing the project *ARCTISEN: Culturally Sensitive Tourism in the Arctic* and thus for enabling the research behind this contribution. Also a warm thank you to Niví (anonymized) and other Indigenous, Greenlandic and Arctic tourism entrepreneurs for sharing and enlightening me as well as other project members on our journey towards more sensitive tourism in the Arctic.

References

Agneman, G. (2021). How economic expectations shape preferences for national independence: Evidence from Greenland. *European Journal of Political Economy*, 72, 102112.
Bjørst, L. R., & Ren, C. (2015). Steaming up or staying cool? Tourism development and Greenlandic futures in the light of climate change. *Arctic Anthropology*, 52(1), 91–101.

Bjørst, L. R. (2018). The right to 'sustainable development' and Greenland's lack of a climate policy. In Gad, U. and Strandsbjerg, J. (Eds.), *The Politics of Sustainability in the Arctic* (pp. 121–135). London: Routledge.

Brøns, M., Kristiansen, I. & Holm, M. (2022). Før debatten om hvem, der er grønlænder: Vi skal gøre plads til alle i vores samfund (Before the debate on who is Greenlandic: We have to make room for everyone in our society). *Sermitsiaq*. Accessed 7 June 2022: https://knr.gl/da/nyheder/f%C3%B8r-debatten-om-hvem-der-er-gr%C3%B8nl%C3%A6nder-vi-skal-g%C3%B8re-plads-til-alle-i-vores-samfund

De Bernardi, C. (2019). A critical realist appraisal of authenticity in tourism: The case of the Sámi. *Journal of Critical Realism, 18*(4), 437–452.

Franklin, A. (2003). *Tourism: An Introduction*. London: Sage.

Gad, U. P. (2009). Post-colonial identity in Greenland?: When the empire dichotomizes back—bring politics back in. *Journal of Language and Politics, 8*(1), 136–158.

García-Rosell, J. C. (2016). Certification of indigenous cultures in tourism-The case of the Sápmi Experience Quality Mark. *Matkailututkimus, 12*(2), 60–62.

Higgins-Desbiolles, F. (2003). Reconciliation tourism: Tourism healing divided societies! *Tourism Recreation Research, 28*(3), 35–44.

Olsen, N. (2010). Kalaalimernit – an intangible cultural heritage. *The Journal Greenland, 3*, 218.

Olsen, K. O., Abildgaard, M. S., Brattland, C., Chimirri, D., De Bernardi, C., Edmonds, J. & Viken, A. (2019). Looking at Arctic tourism through the lens of cultural sensitivity: ARCTISEN–a transnational baseline report. urn:nbn:se:umu:diva-166664.

Ren, C. & Abildgaard, M. S. (2021). Greenlandic independence and tourism futures: Exploring modern and ethnic logics. In Hall, D. R. (Ed.), *Tourism, Climate Change and a Geopolitics of Arctic Development: The Critical Case of Greenland* (pp. 178–186). Wallingford: CABI Publishing.

Ren, C., Jóhannesson, G. T., Kramvig, B., Pashkevich, A. & Höckert, E. (2021). 20 years of research on Arctic and Indigenous cultures in Nordic tourism: A review and future research agenda. *Scandinavian Journal of Hospitality and Tourism, 21*(1), 111–121.

Ren, C. B. & Bjørst, L. R. (2017). Cold facts, hot topics and uncertain futures: Political and industry responses to climate changes in Greenland. In DellaSalla, D. and Goldstein, M. (Eds.), *Encyclopedia of the Anthropocene* (pp. 501–503). Munchen: Elsevier.

Ren, C., Gad, U. P., & Bjørst, L. R. (2019). Branding on the Nordic margins: Greenland brand configurations. In Cassinger, C., Lucarelli, A., & Gyimóthy, S. (Eds.), *The Nordic Wave in Place Branding: Poetics, Practices, Politics* (pp. 160–174). Cheltenham: Edward Elgar Publishing.

Ren, C., & Hall, D. (2021). Fem årtiers turismeudvikling i Grønland–selvstændighed, geopolitik og nye paradokser (Five years of tourism development in Greenland - independence, geopolitics and new paradoxes). Samfundsøkonomen, 2021(4), 84–91.

Rosing, M., Mosbech, A. & Mortensen, B. O. G. (2014). *To the Benefit of Greenland. The Committee for Greenlandic Mineral Resources to the Benefit of Society*. Nuussuaq and Copenhagen: Ilisimatusarfik/University of Greenland and University of Copenhagen.

Rud, S. (2017). *Colonialism in Greenland: Tradition, Governance and Legacy*. New York: Springer.

Sowa, F. (2013). Relations of power and domination in a world polity: The politics of indigeneity and national identity in Greenland. In *Arctic Yearbook* (pp. 184–198). Akureyri: Northern Research Forum.

Stat.gl. (2021). Grønland i tal 2021 (Greenland in numbers 2021). Accessed 17 April 2021: https://bank.stat.gl

Strandsbjerg, J. (2014). Making sense of contemporary Greenland: Indigeneity, resources and sovereignty. In Powell, R. and Dodds, K. (Eds.), *Polar Geopolitics? Knowledges, Resources and Legal Regimes* (pp. 259–276). Cheltenham: Edward Elgar Publishing.

Thisted, K. (2013). Discourses of indigeneity: Branding Greenland in the age of self-government and climate change. In Sörlin, S. (Ed.), *Science, Geopolitics and Culture in the Polar Region, Norden. Beyond Borders* (pp. 227–258). London: Routledge.

Thisted, K. (2015). Pioneering nation: New narratives about Greenland and Greenlanders launched through arts and branding. In Evengård, B., Larsen, J-N. and Paasche, O. (Eds.), *The New Arctic* (pp. 23–38). Cham: Springer.

United Nations. (2007). *Universal Declaration on the Rights of Indigenous Peoples*. https://www.un.org/esa/socdev/unpfii/documents/DRIPS_en.pdf

Visit Greenland. (2019). *Tourism Statistics Report 2019 Greenland*. Accessed March 18 https://tourismstat.gl/wp-content/uploads/2023/03/Tourism-Statistics-Report-North-Greenland-2019.pdf

3

THE CONSTANT CREATION OF ABORIGINALITY

A Commentary on Indigenous Being and Becoming

Keith Hollinshead

Introduction

In many senses, this chapter on ethnocentric learning (or, rather, on the necessity for cross-cultural *unlearning*) is a call for many non-Indigenous practitioners and researchers to think otherwise and cultivate something of a dissident imagination which recognises that Indigenous realms are not static and inert domains but are rather intricate webs of interlaced strands and bonds which are always in motion and which serve as vital relationalities (i.e. as generative forces of flux and becoming) that sustain those people, not only in their longtime traditionalities but in their emerging transitionalities. Thus, this chapter warns about the danger of Western-centric practitioners in tourism (and in related fields of inscription) in complicitly servicing – often in unsuspected/under-suspected ways – ontologically weak non-Indigenous categorisations that are often issued via a dictatorship of mainstreamed/outsider vision. In this regard, this critical chapter calls for important forms of *new sense thinking* about the diachronic of Indigenous relationalities today – that is, of how Aboriginal/First Nation/First People populations inevitably change through time (as do all populations) and flourish via both tradition-informed and transition-responsive forms of 'becoming'.

Indigenous Voice and Identity Today

This chapter generally concerns the sorts of imaginations that regulate what is seen and known through tourism (and tourism studies) and through related declarative industries. Specifically, the critique contained in it distils the ways in which the Indigenous 'Other' has been conceived, communicated, and metabolised through the cultural representations of tourism (and those related inscriptive industries/fields). The particular critical assessment within this chapter inspects the sorts of representations of 'the other'/'the Other' which operate in and across Indigenous spheres today as they function to translate, to exclude, to silence, or to overdraw Indigenous populations. In working to the aforesaid overall research agenda's guiding premise that "the Other is never simply given, simply found, or simply encountered but is [always] made" (Hallam and Street 2000:1), this chapter examines the commonplace ways in which dominant non-Indigenous/'Western-centric' representations actually work to signify 'the Other', a phenomenon that has been given a very

DOI: 10.4324/9781003230335-4

20

useful foundational treatment by Hall (1997). Thus, this chapter explores the ideological preconceptions that seemingly underline the act of Othering, or what one could call the Western-centric 'selfing' of the Indigenous 'Other', as non-Indigenous practitioners and researchers in tourism project their own ontological assumptions about matters of being and variously produce/reproduce their own so-regularly-dominant fixed sense of and about Indigenous 'identity'. Such are the outsider/non-Indigenous representations and misrepresentations that Indigenous groups, communities, and businesses are substantively railing against today.

In thereby seeking to uncover what conceivably constitutes some of the representational forms, frameworks, codes, and conventions which characterise this common essentialisation of Indigenous peoples across the world, this chapter is predicated on the view of Jordanova (2000:246) "that [the absolute] understanding other cultures in their own terms is impossible", a view supported by Escobar (2020). Clearly, there will be many Indigenous observers out there – and many old-school ethnographers also – who will be thoroughly dubious that anything positive could ever immediately stem from this effort to cleanse representations of 'the other'. As Jordanova (2000:46) has reasoned, such an interlocutory quest to closely get to know anOther/anyOther culture is both an intellectual and a political necessity: but it does require considerable courage and perspicacity to inquire into these tensions of recognition and signification. As Law (2000, cited in Escobar 2020:14–15) put it, when examining even a singular Indigenous or 'other' population, it is not a question of how to capture the distinct 'real' there but the many live 'reals' there. Such is the fluidity today through which the members of Indigenous groups and communities relate differentially to the human and nonhuman world within and across settler-cum-Western-centric societies and to the changing dependencies and interdependencies which they beget as 'othered peoples' (or, rather, as 'sub-othered peoples', after Mignolo and Walsh 2018:92).

The Ethnoaesthetics of Cross-Cultural 'Listening'

"Indigenous people have spent a long time looking at and resisting the powerful. We have become extremely knowledgeable about white Australia [for instance] in ways that are unknown to most white 'settlers'" (Aileen Moreton-Robinson 2015:127, an activist and academic on matters of native title, whiteness, race, and feminism, based at University of Queensland).

In considering the ways in which other cultures are researched and represented today, Atkinson, Delamont, and Coffey (2003) consider that too many individual investigators and commentators and too many projective and promotional agencies all-too-readily collapse the various social terms they encounter in those other cultures into one aesthetic mode or otherwise convey them via one undifferentiated symbolic genre universalised across the world. In calling for much more awareness about the hurt and harm that such research and such reductionist communication can generate at each specific place and for each specific population, Atkinson et al. follow Blumer (1954) and Hammersley (1989) in demanding that painstaking steps should be taken to ensure that such research and such representational action should be faithful to the culture/subculture under investigation. This chapter thereby seeks to translate the demand of Atkinson et al. (2003) for much more aware and informed plural differentiation and multilayering of 'other'/'Other' cultures by examining the projective role and reach of tourism in communicating who is who and what is what for supposedly removed and distant cultures. Thus, this commentary serves as an examination of the quality of ethnography that is purposely or accidentally conveyed through tourism – that is, through tourism studies research and through everyday projective action in tourism management/tourism development. It constitutes a critique of the found common overdeterminations of tourism (Horne 1992; Meethan 2001) as it follows Atkinson et al. (2003) in demanding much more ethnoaesthetic sensitivity to other cultures and thereby much more diligent attention to local/ Indigenous systems of identity, self-representation, and action. But in these regards, such matters of fluidity and uncertainty, perhaps the focus of the chapter should even be shifted from its outlook upon a or the sole other (and the sub-other) towards an inlook upon the hybrid (Boehmer 2016:310–312), viz., upon the many hybrids.

In particularly examining the contemporary representations of 'Aboriginal' Australia in the light of Atkinson et al. (2003), this chapter comprises an inspection of the authorial privilege of tourism (within research and management action) in communicating about the Indigenous imaginal in its throbbing multiplicities. It questions the agency and the reach of tourism in communicating what is tacit to Aboriginal populations there, and it registers the necessity for those who research tourism – conceivably the world's largest industry of communication about culture (Rojek and Urry 1997) – to not just pay respect to the need to ground their data in an Indigenous frame of reference (singular), but to help in faithfully capturing and faithfully reconstructing the many different imagined worlds, the many different changing reals (after Law 2000), which are constructed day by day and communicated to them (i.e. to these researchers). This chapter is therefore predicated upon the view that there is not only a multiplicity of different imagined Indigenous realms in a given country like Australia – as there are in and across each and all Indigenous societies across the world – but there is a pulsing simultaneity to those Indigenous cultural frames of order and action across Australia, and indeed across anywhere else. In so doing, this chapter does not call for any comprehensiveness or completeness of account of and about the Indigenous worlds but demands much more informed perspicacity in the ways in which those who research tourism and those who project and communicate through tourism indeed open themselves up to receiving Indigenous constructions of self. In a nutshell, this chapter, therefore, calls for much more sagacious discrimination in taking on board what Aboriginal people are saying is so about themselves: it thereby fundamentally calls for much more serious and fastidious listening on the part of non-Indigenous institutions and interest groups (see also Whitney-Gold chapter, editors' comment).

In New Zealand, the Māori educator/academic Linda Tuhiwai Smith (1999) has also clamoured for non-Indigenous researchers across the globe to adopt not so much a set of new technical methods but a whole new and serious sensate orientation in the effort to meaningfully capture – or rather, to receive – the contemporary canons of rhetoric and self-construction of the Indigenous communities they deal with. While Smith's primary aim is to encourage individuals and groups in Indigenous societies to disrupt forms of external/etic research that are exploitative and 'deeply colonial' (Smith 2005:87, 91), she clearly also advocates the presence of non-Indigenous researchers in Indigenous communities provided that they are genuinely affianced with the said Indigenous populations. To Smith, this crucially means that such interested non-Indigenous researchers on the cusp of Indigenous societies can and ought to be encouraged there if they are found to be demonstrably able to build up and engage in dialogue which acceptably (to them) helps those communities reclaim their knowledge (i.e. their own self-recognised endangered authenticities) <u>AND</u> consolidate and confirm their other/new/emergent ways of being and knowing. But the efforts to successfully disrupt representations of being are difficult matters to get right, and to whose satisfaction will they ever succeed? (See Hollinshead 1992; Munoz 1999; O'Leary 1998; Schöttler 2013; and Braidotti 2019 on the subject of the 'disidentification' of representations of being and becoming.)

Thus, to Smith (2005), while the main impulse – in terms of research into Indigenous communities – must be the transformed facilitation of the native's capacity to be self-defining and self-naming as they claim the right to tell stories in their own ways (Smith 2005:89), the transformed research genres are now necessarily blurred (Smith 2005:90), and new translators or interlocutors are required on both sides of the Indigenous–mainstream dialogue (Smith 2005:96), viz., that of the Indigenous and the non-Indigenous. There is nothing axiomatically wrong with Indigenous populations using non-Indigenous translators or interpreters: they can serve as well-pointed 'allies' in the efforts to benefit Indigenous communities, whether the sought amelioratory-cum-corrective acts be counter-identified or more pungent/attritional disidentified ones (Hollinshead 1992).

To Smith (2005), the communicative situation is contextually complex. While Indigenous populations wish to escape external definition, contemporary identities are becoming multiple and shifting as they (ceaselessly?) develop and change. While much Indigenous aspiration is still based almost exclusively on ancient memories, other self-defined identities are based on new vistas, on new attitudes, and on new dilemmas. As 'the other'/'the Other' is reworked, extending from reifications of ancient narratives to enfolding visions based on emergent voices, emergent leaders, and emergent organisations, there is scope for a myriad of new translators and interpreters of being whether they actually be Indigenous or non-Indigenous. The critical thing is whether these various interpreters of being and meaning can indeed privilege the particular Indigenous voice and support the involved Indigenous community aspiration (Smith 2005:90). The challenge over these difficult efforts to map and monitor diversity is thereby to generate 'cross-border' or 'cross-cultural' respect in the form of dialogic conversations which permit Indigenous groups and individuals to meaningfully articulate their own narratives of representation or resistance and their own poetically integrated storylines of vibrant being. To Smith, it is necessarily incumbent upon Indigenous communities in this ongoing communicative effort – drawing from LeCompte et al. (1992) – to encourage researchers from mainstream locales or non-Indigenous origins in order that they dutifully learn how to listen to the voices of the formerly silenced host Indigenous populations and thereby recognise or downplay what might ordinarily be their own conceivable researcher-as-expert assumptive standpoints (Smith 2005:101).

Russell Bishop (a Māori educationalist) is another commentator on Indigenous Affairs in New Zealand who thinks there is a specific place for the right sort of non-Indigenous ally to help Indigenous groups and communities retrieve their own projective space in the communication of culture and becoming. Bishop (2005) is articulate about the plethora of often small but perniciously aggregate ways in which the colonial paradigm has not only belittled Māori knowledge over the centuries, but also continues to undersuspectingly threaten Māori cultural epistemologies and cosmologies today in ways that non-Indigenous bodies and organisations are often unaware of or only half-alert to. He is keen to wrest matters of representation (and thereby of the legitimisation) of culture and nation back under Indigenous control, notably in and through the domain of research (Bishop 1996), and he notes (as cited in Merriam, et al. 2001) how

> it has commonly been assumed that being an [Indigenous] insider means easy access [to host groups/communities], the ability to ask more meaningful questions and real non-verbal cues, and most importantly be able to project a more truthful, authentic understanding of the culture under study.
>
> (Merriam, et al. 2001:411)

But he recognises that the total reliance upon the Indigenous-only communicators may be unhealthy: he is concerned that insiders may be "inherently biased, or ... too close to the culture to ask critical questions" (Bishop 2005:111). To Bishop, it is clear that there is a place in communication of and about culture and nature for non-Indigenous researchers as significant invitees provided that, drawing from Tillman (2002), he/she can be seen to faithfully interpret and contextually sustain the Indigenous phenomenon or outlook in question by that inspected or engaged population (Bishop 2005:113) and that Māori people (in the particular arena) remain in control of the overall research agenda involved.

To Bishop, as to Smith (1999) in New Zealand, and to Jacobs-Huey (2002) on a broader front, what counts is the capacity of the non-Indigenous or Indigenous researcher to harness the power of critical reflexivity, and it is unwise to rely only upon Indigenous voices if those researchers/writers/communicators operate non-critically as they privilege their own insider status. But we do have to watch our words here, for if a cultural scenario is to be accurately interpreted, it implies that there is indeed a single interpretation out there ready to be captured. And if that scenario is to be contextually validated, it implies that there is a single cultural or axiological standard out there that can be not only readily harnessed but simply understood by etic (outsider) observers. Such visions of 'accuracy' and 'validated' legitimacy are highly interpretive matters of perspective and presence and are often of an alternative ontology and an imaginative geography (Huggan 2016). But, in the case of Māori communities and collectives in New Zealand, non-Indigenous institutions and individuals must keep foremost the reminder that such Indigenous groups do not need *Pakeha* (i.e. external 'white' observers) to tell them who they, the Indigenous people themselves, are (Huggan 2016:178).

In this sense, Bishop (2005), Smith (1999), and Jacobs-Huey (2002) each mirror Narayan (1993:671) who argued that it is no longer indefinitely useful to label researchers via hard-line classifications as 'insiders' or 'outsiders' for all Indigenous communities are these days the sites of not only fast-shifting identifications but fresh presentations/interpretations of culture and power. Many Indigenous individuals thereby consist of a constellation of mixed-up and infeeding emic and extrinsic etic outlooks. Culture today, arguably, culture anytime, is not a fixed edifice but a realm of mobile influences that have to be investigated in specific but fluid situational circumstances, and often along uncertain-cum-inelegant trajectories. It thus considerably helps to have in the research fold individuals who can bring witness to those infusions and imbrications from several perspectives: all Indigenous works are now almost unavoidably permeative in this fashion, to some degree (Bhabha 1994; Bridges 2001).

Today, it is the quality of the cross-cultural or multiple relations that frequently counts, and the quality of both the inlooks and outlooks involved: essentialist (or highly traditional and distinctive) understandings have somehow to be bedded in with emergent hybrid viewpoints, and intrinsic knowledge has thereby somehow to be conjoined with superadded 'new sense'. The problem with the seemingly faithful effort to capture a or the single Indigenous viewpoint is how one can ever know when one has effectively or roundly captured 'it', if it has ever existed anyway. In so many countries, places, and spaces, one may push the view that Indigenous groups are so moulded within outside/mainstream/'Western-centric' forms of history, national policy, and corporate action that culturally isolationist arguments no longer make pure and incorruptible sense. Hence, for Māori research, a discursive repositioning is often called for these days, that is, of one not totalised by the obligation to secure a single (essentialised) ethnicity (Bishop 2005:115). The research and communication must remain collectivistic, but not built perennially upon hard-and-fast dichotomies that deny the wider revitalisations and permeations of Indigenous life (Hokowhitu and Devadas 2013): it is morbific (i.e. weak/sickly/diseased-in-thought) to expect all Māori to act in absolutely prescribed ways, for all populations are indeed diverse, everywhere. For most/all Indigenous populations, it is important that the 'compassionate consciousness' (after Heshusius 1994) of and about their culture, i.e. consciousness that is embedded in ancient and received traditions, is also ventilated by fresher awarenesses which are seeded in the new fusions of permeative cultural life: the one ought to not overwhelm the other. Propositions for self-determination that seek to maintain Indigenous populations as wholly distant from non-Indigenous society are conceivably unhelpful to the contemporary sustenance or development of Indigenous populations. It is thus important that the effort to secure singular and untainted identities for Indigenous peoples be inter-rupted: such abstractions can too readily be dislocative, disengaged, and disembodied (Mignolo and Walsh 2018:3). Non-Indigenous observers and communicators ought to thereby guard against what the closed circuit of epistemic authority (that produces and reproduces such Western-centric delusions) normalises.

The Experience of Aboriginality in Australia

Having explained the general view that many of the world's populations live in difficult in-between cultural positions, attention will now be turned to the specific consideration of matters of affilia-tion and identity in Aboriginal Australia. A quick review will therefore be provided on Aboriginal identity in general.

Aboriginal Identity Today – An Introduction

Indigenous peoples are so often locked within external visions of an assumed pastness (Vigil 2015:3) that they must match. The general etic quest has been to identify and monitor an almost singular form of traditional Aboriginal identity and to impose that outside characterisation as the yardstick by which to position and judge each and every encountered Aboriginal person/group/community (Stratton and Ang 1998). Such representations and significations tend to only ever deal in boiled-down, over-hardened, or numb mischaracterisations of particular local being. They over-privilege certain external versions of being and condition and overindulge certain external visions of and about the selfhood of that othered population, or of those hybridising individuals liv-ing in difficult Third Space places (after Bhabha 1994). See Taska (2000) on the kinds of onerous psychics involved here and Moreton-Robinson (2015:9) on 'the manufactured character' of such forms of hybrid situatedness within the so-called postcolonial society across Australia. There are also Lee's (2017) coverage of epistemic colonisation in which she evaluates the power of Indig-enous storytelling to generate positive touristic encounters in Tasmania, today, and Jacobsen's (2018) assessment of the worldmaking force of Aboriginal enterprises in the domestic tourism of Australia, as based on customary cultural law today.

Diehard and static Western-centric visions tend to leave many Aboriginal individuals and groups in an unfortunate cultural vacuum (Taylor, et al. 2005: xii; Yunkaporta 2022:132–135). They have been particularly damaging to those Indigenous Australians who lacked the support structures of the so-called traditional Indigenous society and who otherwise were unable to endure successfully within white/Euro-Australian society (Creamer 1994:59). Lately, then, cultural iden-tity has much more certainly become a province of political struggle within and across Aboriginal Australia (Cowlishaw 1994:90; Hornung 2016), and the non-Aboriginal world is presented with the reality that just as Aboriginal people today do not exist as one monolithic group, so the Abo-riginal experience of the past is not singular but is highly variform (Jordan 1994:111; Dodson 2005:31). The traditional has been too readily externally glorified at the expense of the transitional. The alien other has been too hastily externally venerated, albeit so often as the sub-other at the expense of the self-defining 'hybrid', that is of Aboriginal people who can exist in forms of dou-bleness where they might "perform whiteness while [still] being Indigenous" (Moreton-Robinson 2015:11). In these respects, one can note the *trawlwulwuy* woman from tebrakunna country, north-east Tasmania, Emma Lee (2017), with her critique of 'the new knowledge productions' of

Indigenous women, particularly in the projection of 'Black female bodies', especially against the agency of 'Establishment men' who continue to privilege Western-centric epistemologies within the tourism industry.

Over recent decades, social scientists in Australia have grown used to the presence of a medley of voices speaking to and for Aboriginality and have thereby come to acknowledge that Aboriginality is a thing, or rather a mix of entities, in a constant process of creation (Beckett 1994/A:7; Cowlishaw, et al. 2006:2); and, in particular, Fyfe (2021) on the aforesaid work of the academic and activist Emma Lee (2017).

While the recent advancements in confidence of the Indigenous land rights movement have tended to increase the volume and the velocity of Aboriginal calls to define a particular unique identity and, in some circumstances, to define a new overall 'Black' continent-wide selfhood (Jacobs 1994:31; Yunkaporta 2020:11), the new international visibility and significatory value of 'Black Australians' has witnessed the repeated use of Black traditional scenes and images in mainstream and Australia-at-large settings for mere iconographic vividness. It has seen them collectively and graphically used to help compose the image of a new 'timeless' nationalism for whole Australia, that is, for the whole of non-Aboriginal and Aboriginal Australia combined as Beckett (1994/B:225) registered three decades ago. In this longstanding but highly problematic regard, the Aboriginality of people remains a generalised myth in the marketplace of cultural depiction and national projection, where Indigenous Australians remain an undifferentiated population often included for pictorial value amongst the flora and fauna of Australia, thereby being harnessed to speak interestingly and vividly for the new-nation-upon-the-old-continent (Beckett 1994/B:205). In such reactionary light, yesteryear, the Indigenous population (singular) of Australia was repeatedly captured on postage stamps, within art catalogues, and in travel brochures and was blazoned across tourist merchandise merely for its striking difference, i.e. for its bestowed 'culturally-authentic' and also its evidential 'wider-national' representative power (Perkins 2005:97–98). No doubt some of that hackneyed misrepresentation still lingers here and there today.

Understandings of Culture Today

This commentary has sought to capture the ways in which Aboriginality is now projected today. It endeavours to give witness to many of our unfolding notions of and about 'culture' and 'cosmology' today. In relaying the ways in which representations of Aboriginal culture variously operate to suppress, to chastise, to translate, and to caricature contemporary Indigenous reality, it testifies to both the sustained/pervasive and the ephemeral/fast-changing character of particular visions of being and becoming. The nature of social knowledge is a complex thing, and this chapter seeks to codify the realisation that there will always be a multiplicity of subject positions within any single discourse about a found culture. Cross-cultural interactivity is inclined to take place unceasingly over time, and it is impossible to conjure up either a single image for Aboriginal Australia or a single discourse to represent it acceptably on a broad and generalisable front (Dodson 2005:29).

Representations and projections of culture are always evocative, and where a particular population deems itself to have been subjugated or misrepresented by powerful or colonising outsiders, that involved local or Indigenous population will naturally want to let loose distinct sorts of reverse cultural influence (Reynaud 2000:87) to speak back against what is seen to have been such suppressive external significations: and follow (Kovach 2021:242) on capacity-building in the Indigenous framing/reframing of Indigenous storylines. The relentless privileging of certain constructions of culture will hopefully be reduced in effect and then inclined over time eventually to give rise to reverse but positive projections of being and becoming, even though many Indigenous groups and communities will be fast-transitioning or steadily hybridising themselves. Just as otherness is always actively made rather than given (Jordanova 2000:249), so culture itself is not a fixed object to be universally understood and described but is a living and an emergent thing (Hallam 2000:267; Bignall and Rigney 2019:161). It is not only the abstract matter of cultural authority that is changed or consolidated through the everyday politics of 'cultural representation' and of 'reverse cultural representation', but it also is vital/throbbing culture itself that can so change or be spirally reconstituted. Hence, whose longstanding and whose unfolding systems of knowledge are reflected in the industrial representations and the academic predilections of the given day (Kuokkanen 2017:313)? How complicit is tourism (and tourism studies), therefore, in colonialist-cum-neocolonialist acts of ontological foreclosure for the Indigenous peoples it signifies in the given place (Kuokkanen 2017:317)?

A Dialogic Understanding of Being

Initially, this critique was inspired by the writings of Sampson (1993) who has long been concerned with what he calls "Western culture's centuries-long preoccupation with a contained, individualistic, monologic self and its fearful suppression of all that is Other" (Sampson 1993: back cover). What Samson demands in and from the West is sustained engagement with and sincere understanding of others as "viable people in their own right" (Sampson 1993: back cover). Thus, conceptually animated by Sampson, this chapter has sought to create a dialogic base for tourism studies to replace the Other-suppressing purview of Western social science which itself clearly (in Sampson's (1993:80) view) provides the current foundation for so much understanding in the humanities. But Strong (2017:32) warns against over-hasty attempts to cultivate such dialogic communication and insists that the evolving interpretations of Indigeneity really ought to be diachronic and not just dialogic, *ipso facto*, where the cultures and cosmologies being represented are not so much developed through the ages in any linear sense but accounted for within the mobilities and fluidities of the current time. Strong (2017) insists that Indigenous ontologies today ought not just reflect the longtime traditional but should also diachronically capture the changing interconnections of the current moment with its broader and interruptive kinetics. And in thinking beyond the humanities into the conceptual contours of deep posthuman thought, global ecopolitical commentators such as Colebrook (2006) and Braidotti (2013) would no doubt applaud these rich thought lines about diachronic understanding (re: McCullough (2019) regarding unfolding human/nonhuman political spaces).

The improved and enhanced cross-cultural aim is to engage with the other population, with the hybridising population, and both acutely hear and sincerely learn, and not further process epistemologically what one already 'knows' or has 'institutionally assumed' (Kuokkanen 2017:322). Accordingly, for non-Indigenous practitioners and researchers who work in Indigenous settings, the scenarios will (these days) always be fluid and constantly changing (Sillitoe, et al. 2005). The consultative and participative processes engaged in must therefore be pliant and malleable, and the involved non-Indigenous practitioner or researcher must constantly guard against approaches that are unduly inflexible and unduly prescriptive and consonantly against interpretations that are won in over-fast fashion. Non-Indigenous practitioners and researchers should heed Moreton-Robinson (2015:xv) that in these transient globalising and glocalising times, what matters diachronically is not so much how different the culture of one population is from another in any fixed sense, but how densely held is that cultural trait or that adopted behaviour by that population today. To Moreton-Robinson (2015), it is not 'difference' that predominantly counts, but the 'density' in and by which particular different or similar 'things' are owned/practised/indulged in.

The Dialogic or Hopefully Diachronic Imagination of Tourism: The Inscriptions of Industry and Research

Following Bakhtin (1981, 1986), identity has been positioned as a discursive (dialogic) process through which individuals recognise themselves and through which they are re-recognised by others. Hence, identity has been declared to be those projections-cum-idealisations of desire which, grounded in contesting and transforming (historicised) fantasies, "obey the logic of more than one" (Hall 2003:3) at least twice, as they are, firstly, locally issued and as they are, secondly, externally heard. Thus, particular identifications are not a one-way matter of propelled 'speaking', but a double-blessed fictive entity of both encoding and decoding beings.

Existing identities belong not only to the imaginary of the host, but to the imaginary of the recipient. In this light, 'identities' and 'beings' and 'meanings' are far from being unified things and are subject to many sorts of projecting and admitting agencies. Through communication and mediating industries like tourism (and also film production, television broadcasting, and festival management), identities are variously generated, marketed, and sold, within the local area and across the global economy. In this fashion identities are produced, circulated, and exchanged as held or hailed expressions of being and meaning in Bakhtinian regard, they are not merely pre-constructed outside of or prior to these processes of representation in tourism, they are also constructed/deconstructed/reconstructed within the processes of representation there in the vast and almost uncontrollable quotidian (day-by-day/organisation-by-organisation) projectivity of tourism. Accordingly, the communication of being, meaning, and becoming can be a complex act of both accords where, perhaps more commonly, local/Indigenous populations are pleased with the identifications made of and about them and where local populations react in deep and hurtful angst against the felt misrepresentations produced of and about them which in demeaning ways can render them invisible (Langton 2005). Through communication and experience-providing industries like tourism, then, communal identities are not only determined, but (variously) regularly re-determined, regularly under-determined, and regularly over-determined in an unevolved sense (Hokowhitu 2021). Frequently, particular populations, particular places, particular histories are serially overdetermined in ardent-mainstreaming or heavily normalising ways (Hollinshead, et al. 2021).

At times, Indigenous populations will be inscribed through the industry and by the industry in terms of what Bhabha (2011) would describe as a 'non-visible-population', or as a 'too-visible-population', or somewhere between these representations. On so many occasions and in so many contexts, for Indigenous populations, there will be 'a white world'/'a Western world'/'a whatever world' (after Bhabha 2011) between themselves and the world of manifest or potential travellers (as Moreton-Robinson (2015:49–52) on white possessiveness and white writing). Consequently, the identities and the aspirations of Indigenous populations are not only projected on multiple fronts and heard in multiple fashions, they are languaged via the multiple scripted 'accents' of the tourism industry as they also are through the film industry and through the media industry (Smith 2021). It is no wonder, then, that Indigenous populations live in an anxious state of identity (Bhabha 2011) and nowadays often wish to utilise the reach and authority of mediating industries like tourism (the movie industry and the media industry,) to re-articulate themselves and thereby help reclaim their lost territories and help memorise what they regard as lost time

(Wilson-Hokowhitu (2021) and her specific call for grounded and well-founded 'colours of creation' expositions of Indigeneity).

As Hollinshead (1993:273) has shown, echoing the work of Horne (1992), tourism is thus an immense vehicle for such sorts of representative and evocative symbolism. It can indeed be harnessed as a power:

→to project forms of preferred narrative;
→to issue visions of normal culture;
→to suppress unwanted storylines;
→to frame people/ideas/things from particular points of view;
→to celebrate/sanctify/triumphalise particular events;
→to legitimise/authenticate/authorise particular places.

Such is the enunciative or the declarative value of tourism (Hollinshead 2007), a major site and vehicle for dialogic representation and misrepresentation, and potentially a lead vehicle for explanatory and productive matters of diachronic understandings.

Prospect: The Limits of the Western-Centric Imaginary

To sum up, this chapter has addressed the view that colonialism-cum-neocolonialism is not just a proverbial matter of military action, financial control, and state governance, it can also flower in soft conceptual and subtle epistemic ways. It has addressed the notion that much of that seemingly wispy and seemingly incorporeal suppression can come hand-in-glove through the declarative power of tourism (and of tourism studies), where, as Kovach (2016:216) succinctly puts it, the Western gaze only sees what it wants to see. Ergo, seemingly inconsequential outsider projections can indeed cause considerably psychic mortification or cultural agonistics for Indigenous peoples. Hence, what non-Indigenous bodies and organisations tend to register in and through tourism often acutely depends upon what those outsider-institutions and those etic interest groups have been pre-programmed to recognise, name, and 'know' (Mignolo and Walsh 2018:148). The tourism and travel industries, in the main, therefore stand as an immense cipher of colonialist-cum-neocolonialist interpretations which commit(s) everyday mundane but intensive cumulative violence upon the vibrancies of other 'cosmologically-distant' cultures (Clarke 2018:11). In these contexts, the key issue for those who know and manage/administrate/plan in tourism over the coming decades is whether those at the cultural pitface can indeed break with the often-essentialising mind constructs of Western-centric tourism and comfortably work via new sense epistemological imaginaries and fresh political meanings. Regarding Indigenous peoples/places and societies/ spaces, the naturalising narratives of tourism need not just be about the losses and disconnections suffered by those other 'hybrid' populations, they can highlight emergent matters of creativity and diachronic acts of choice, in forward-looking – if often, albeit, in spiralling (i.e. difficult-to-etically-predict but important) – ways (Bignall and Rigney 2019).

For Indigenous populations themselves, the common hope is that the tourism industry will not just bring economic growth but also cross-cultural awakening where the storylines of tourism and travel are not just grounded in sorry tales of deprivation, assimilation, and fracture (duly fragmented via the perennial packaging of the so-called curious objects and exotic entities), but via narratives that speak to vistas of flourishing subjectivity which observantly trace the broader human to non-human relationships enjoyed in Indigenous realms (Kearney 2013:254). Such declarative reloading will often, of course, be painstaking for many non-Indigenous companies and organisations and must always recognise the rights of Indigenous peoples to decide which of their inheritances and their received storylines and how much they are prepared to give, as Higgins-Desbiolles and Akbar (2018:24) have noted vis-à-vis the Yolngu peoples of northern Australia.

Consonantly, this chapter warns those non-Indigenous practitioners and researchers who work in tourism to beware of conceivably remaining embedded, ontologically and epistemologically, within the Western imaginary where they might only deal in what Kovach (2016) styles as stereotypically flattened/stereotypically deficit outsider interpretations. Both tourism and tourism studies need new constancies to these positive ends regarding the flourishing of host Indigenous groups/communities and globe-crossing travellers alike.

As Braidotti (2013:88–89, 167–168) has argued, the acts of disidentification required to project Indigenous populations out-from-under dominant non-Indigenous/Western-centric models of subject-hood will inevitably constitute the quest for lost cherished-habits-of-thought-and-being and may generate further feelings of nostalgic insecurity but they can be diachronically productive in terms of many kinds of fitting new sense becomings.

References

Atkinson, P., Delamont, S., and Coffey, A. (2003) *Key Themes in Qualitative Research.* Walnut Creek, CA: AltaMira.

Bakhtin, M.M. (1981) *The Dialogic Imagination of Tourism.* Austin: Texas UP.

Bakhtin, M.M. (1986) *Speech Genres and Other Late Essays.* Austin: Texas UP.

Beckett, J. (1994/A) "Introduction," in J.R. Beckett (ed.) *Past and Present.* Canberra: Aboriginal Studies Press, 1–10.

Beckett, J. (1994/B) "The Past in the Present," in J.R. Beckett (ed.) *Past and Present.* Canberra: Aboriginal Studies Press, 191–217.

Bhabha, H. (1994) *The Location of Culture.* London: Routledge.

Bhabha, H. (2011) "Cultures In-between," in S. Hall and P. duGay (eds.) *Questions of Cultural Identity.* London: Sage, 53–60.

Bignall, S. and Rigney, D. (2019) "Indigeneity, Posthumanism, and Nomadic Thought," in R. Braidotti and S. Bignall (eds.) *Posthuman Ecologies.* New York: Rowan and Littlefield, 159–182.

Bishop, R. (1996) *Collaborative Research Stories: Whakawanaungatanga.* Palmerston North: Dunmore Press.

Bishop, R. (2005) "Freeing Ourselves from Neo-Colonial Domination in Research," in N.K. Denzin and Y.S. Lincoln (eds.) *Handbook of Qualitative Research.* Thousand Oaks, CA: Sage, 109–138.

Blumer, H. (1954) "What Is Wrong with Social Theory?," *American Sociological Review* 19:3–10.

Boehmer, E. (2016) "Revisiting Resistance: Postcolonial Resistance and the Antecedence of Theory," in G. Huggan (ed.) *The Handbook of Postcolonial Studies.* Oxford: Oxford University Press, 307–323.

Braidotti, R. (2013) *The Posthuman.* Cambridge: Polity.

Braidotti, R. (2019) *Posthuman Knowledge.* Cambridge: Polity.

Bridges, D. (2001) "The Ethics of Outsider Research," *Journal of Philosophy of Education* 35(3):371–386.

Clarke, J. ed. (2018) *Postcolonial Travel Writing.* Cambridge: Cambridge UP.

Colebrook, C. (2006) *Deleuze: A Guide for the Perplexed.* London: Continuum.

Cowlishaw, G.K. (1994) "The Materials for Identity Construction," in J.R. Beckett (ed.) *Past and Present.* Canberra: Aboriginal Studies Press, 87–108.

Cowlishaw, G., Kawal, E., and Lea, T. (2006) "Introduction," in T. Lea, E. Kawal, and G. Cowlishaw (eds.) *Moving Anthropology.* Darwin: C.Darwin UP, 1–16.

Creamer, H. (1994) "Aboriginality in New South Wales," in J.R. Beckett (ed.) *Past and Present.* Canberra: Aboriginal Studies Press, 45–62.

Dodson, M. (2005) "The End in the Beginning," in M. Grossman (ed.) *Blacklines.* Melbourne: Melbourne UP, 25–42.

Escobar, A. (2020) *Pluriversal Politics.* Durham, NC: Duke UP.

Fyfe, M. (2021) "Thankyou for the Genocide," *Melbourne Age* [Good Weekend]. https://www.smh.com.au/national/thank-you-for-the-genocide-the-aboriginal-activist-love-bombing-white-leaders-20201208-p56ljf.html Hall, S. ed. (1997) *Representation.* London: Sage.

Hall, S. (2003) "Introduction," in S. Hall and P. duGay (eds.) *Questions of Cultural Identity.* London: Sage, 1–17.

Hallam, E. (2000) "Texts, Objects and 'Otherness'," in E. Hallam and B.V. Street (eds.) *Cultural Encounters.* London: Routledge, 260–283.

Hallam, E. and Street, B.V. (2000) "Introduction," in E. Hallam and B.V. Street (eds.) *Cultural Encounters.* London: Routledge, 1–10.

Hammersley, M. (1989) *The Dilemma of Qualitative Method.* London: Routledge.

Heshusius, L. (1994) "Freeing Ourselves from Objectivity," *Educational Researcher* 23(3):15–22.

Higgins-Desbiolles, F. and Akbar, S. (2018) "We Will Present Ourselves in Our Ways," in B.S.R. Grimwood, K. Caton, and L. Cooke (eds.) *New Moral Natures in Tourism.* London: Routledge, 13–28.

Hokowhitu, B. (2021) "The Emperor's New Materialisms and 'Disciplinary Colonialism'," in B. Hokowhitu, A. Moreton-Robinson, L.T. Smith, C. Anderson, and S. Larkin (eds.) *Handbook of Critical Indigenous Studies.* London: Routledge, 131–146.

Hokowhitu, B. and Devadas, V. (2013) *The Fourth Eye.* Minneapolis: Minnesota UP.

Hollinshead, K. (1992) "White Gaze, 'Red People' — Shadow Visions: The Disidentification of 'Indians' in Cultural Tourism," *Leisure Studies* 11(1):43–64.

Hollinshead, K. (2007) "'Worldmaking' and the Transformation of Place and Culture," in I. Ateljevic, A. Pritchard, and N. Morgan (eds.) *The Critical Turn in Tourism Studies.* Amsterdam: Elsevier, 165–196.

Hollinshead, K., Suleman, R., and Wang, S. (2021) "Overcoding in Disciplines and Fields (After Deleuze)," *International Journal of Multidisciplinary Research and Analysis* 4(11):1757–1768.

Horne, D. (1992) *The Intelligent Tourist.* McMahon's Point, Australia: M.Gee Publishing.

Hornung, F. (2016) "Indigenous Research within a Cultural Context," in D.M. Mertens, F. Cram, and B. Chilisa (eds.) *Indigenous Pathways into Social Research.* London: Routledge, 133–152.

Huggan, G. (2016) *The Handbook of Postcolonial Studies.* Oxford: Oxford UP.

Jacobsen, D. (2018) "Aboriginal Domestic Tourism Leadership towards Reconciliation in Australia," *Journal of Geographical Research* 1(1):19–31.

Jacobs-Huey, L. (2002) "The Natives Are Gazing and Talking Back," *American Anthropologist* 104(3):791–804.

Jordan, D.F. (1994) "Aboriginal Identity," in J.R. Beckett (ed.) *Past and Present.* Canberra: Aboriginal Studies Press, 109–130.

Jordanova, L. (2000) "History, 'Otherness' and Display," in E. Hallam and B.V. Street (eds.) *Cultural Encounters.* London: Routledge, 245–259.

Kearney, A. (2013) "Emerging Ethnicities and Instrumental Identities in Australia and Brazil," in M. Harris, M. Nakarta, and B. Carlson (eds.) *The Politics of Identity.* Sydney: UTS Press, 240–257.

Kovach, M. (2016) "Doing Indigenous Methodologies," in N.K. Denzin and Y.S. Lincoln (eds.) *Handbook of Qualitative Research.* Los Angeles, CA: Sage, 214–234.

Kovach, M. (2021) *Indigenous Methodologies.* Toronto: Toronto UP.

Kuokkanen, R. (2017) "Indigenous Epistemes," in I. Szeman, S. Blacker, and J. Sully (eds.) *A Companion to Critical and Cultural Theory.* Hoboken, NJ: Wiley, 313–326.

Langton, M. (2005) "Aboriginal Art and Film," in M. Grossman (ed.) *Blacklines.* Melbourne: Melbourne UP, 104–124.

Law, J. (2000) *"What Is Wrong With a One-World World?,"* [Presentation for The Humanities] Wesleyan University.

LeCompte, M, Milroy, W., and Preissle, T. eds. (1992) *Handbook of Qualitative Research in Education.* San Diego, CA: Academic Press.

Lee, E. (2017) "Performing Colonisation: The Manufacture of Black Female Bodies in Tourism Research," *Annals of Tourism Research* 66:95–104.

McCullough, S. (2019) "Heterogenous Collectivity and the Capacity to Act," in R. Braidotti and S. Bignall (eds.) *Posthuman Ecologies.* New York: Rowan and Littlefield, 141–157.

Meethan, K. (2001) *Tourism in Global Society.* Basingstoke: Palgrave.

Merriam, S.B., Johnson-Bailey, J., Lee, M-Y., Kee, Y., Ntseare, G., and Muhamad, M. (2001) "Power and Positionality," *International Journal of Lifelong Education* 20(5):405–416.

Mignolo, W.D. and Walsh, C.E. eds. (2018) *On Decoloniality*. Durham, NC: Duke UP.

Moreton-Robinson, A. (2015) *The White Possessive*. Minneapolis: Minnesota UP.

Munoz, J.E. (1999) *Disidentifications*. Oxford: Blackwell.

Narayan, K. (1993) "How Native Is a 'Native' Anthropologist?," *American Anthropologist* 95:671–686.

O'Leary, J.E. (1998) "Counteridentification or Counterhegemony?," *Women and Politics* 18(3):45–72. https://doi.org/10.1300/J014v18n03_04 Ol

Perkins, H. (2005) "Seeing and Seeming," in M. Grossman (ed.) *Blacklines*. Melbourne: Melbourne UP, 97–103.

Reynaud, A. (2000) "Sandra Kogut's What Do You Think People Think Brazil Is?," in E. Hallam and B.V. Street (eds.) *Cultural Encounters*. London: Routledge, 72–88.

Rojek, C. and Urry, J. (1997) "Transformations of Travel and Theory," in C. Rojek and J. Urry (eds.) *Touring Cultures*. London: Routledge, 1–20.

Sampson, E.E. (1993) *Celebrating the Other* New York: Harvester-Wheatsheaf.

Schöttler, P. (2013) "Introduction to Michael Pêcheux's 'Dare to Think and Rebel'," *Décolages* 1:4 (article 11).

Sillitoe, P., Dixon, P., and Parr, J. (2005) *Indigenous Knowledge Inquiries* Rugby: ITDG Publishing.

Smith, J. (2021) "Indigenous Insistence on Film," in B. Hokowhitu, A. Moreton-Robinson, L.T. Smith, C. Anderson, and S. Larkin (eds.) *Handbook of Critical Indigenous Studies*. London: Routledge, 488–500.

Smith, L.T. (1999) *Decolonizing Methodologies*. London: Zed.

Smith, L.T. (2005) "On Tricky Ground," in N.K. Denzin and Y.S. Lincoln (eds.) *Handbook of Qualitative Research*. Thousand Oaks, CA: Sage, 85–107.

Stratton, J. and Ang, I. (1998) "Multicultural Imagined Communities," in D. Bennett (ed.) *Multicultural States*. London: Routledge, 135–162.

Strong, P.T. (2017) "History, Anthropology, Indigenous Studies," in C. Andersen and J.M. O'Brien (eds.) *Sources and Methods in Indigenous Studies*. London: Routledge, 31–40.

Taska, L. (2000) "Like a Bicycle, Forever Teetering between Individualism and Collectivism," *Labour History* 78:7–32.

Taylor, L., Ward, G.K., Henderson, G., Davis, R., and Wallin, L.A. (2005) "Introduction," in L. Taylor, G.K. Ward, G. Henderson, R. Davis, and L.A. Wallis (eds.) *The Power of Knowledge*. Canberra: Aboriginal Studies Press, xi-xiii.

Vigil, K.M. (2015) *Indigenous Intellectuals*. New York: Cambridge UP.

Wilson-Hokowhitu, N. (2021) "Colours of Creation," in B. Hokowhitu, A. Moreton-Robinson, L.T. Smith, C. Anderson, and S. Larkin (eds.) *Handbook of Critical Indigenous Studies*. London: Routledge, 114–128.

Yunkaporta, T. (2020). *Sand Talk: How Indigenous Thinking Can Save the World*, Text Publishing Company.

4

A CRITICAL REALIST APPRAISAL OF INDIGENEITY

The Case of Miao Peoples in a Tourism Village in China

Jianhong Zhou and Johan Edelheim

Introduction

As the concern for human-induced climate change is gaining attention with Indigenous knowledge often at the forefront of the debate, the concept of indigeneity gains new relevancy. Historically, indigeneity was used to describe Indigenous peoples/First Peoples who were oppressed by colonisers in history or now. Land rights and moral claims were the foci of such discussion. There have been acceptances, rejections and strategic uses of the concept of indigeneity; for example, it has been rejected by states, including China, who claim there was no substantial colonisation in China (e.g. Merlan 2009). However, when Indigenous knowledge was acknowledged as an invaluable strategy in response to environmental and other forms of change in the Fourth Assessment Report of the Intergovernmental Panel on Climate Change (IPCC 2007), the value of indigeneity has been more widely recognised (Chandler & Reid 2018). As is exemplified in China, the concept of indigeneity is accepted to describe ethnic minorities' ecological knowledge by government officials (Hathaway 2016). Accordingly, the concept of indigeneity deserves to be reconsidered. Relational definitions that emphasise close interactions between native peoples and environments expand the scope of indigeneity. Issues such as land rights, moral claims, the capacity of resilience and adaption are outcomes of diverse interactions and practices, and these issues affect not only Indigenous peoples, but also many ethnic groups. Thus, indigeneity is a powerful term for ethnic peoples, including, on the one hand, the historical Indigenous power with maintained political rights and moral claims and, on the other hand, Indigenous knowledge as a new form of governing in the modern and global world. However, existing research focuses mainly on studying the indigeneity of Indigenous peoples and connects seldom with ethnic minorities. It is necessary to appraise the representation of ethnic minorities in tourism based on the perspective of indigeneity as a means of empowerment.

Upper Langde Miao village is part of Langde Town, which lies in the north of Leishan County in Qiandongnan Miao and Dong Autonomous Prefecture in Guizhou Province, China. Miao peoples in this village are recognised as the holders of decision-making power in tourism. Their ecological worldview, sense of geographical belonging, as well as political and cultural systems, form

DOI: 10.4324/9781003230335-5

Table 4.1 Appraising the indigeneity of Miao peoples in tourism in China

Macro-layer – Universal dimension of indigeneity		
Indigenous knowledge	Belonging to a place	Political and historical movement
Resilience	Adaption	Rebellion
Meso-layer – Concrete dimension of indigeneity of Miao people in China		
Miao knowledge	Cultural adaption and identification Informal system	
Micro-layer – Concrete dimension of indigeneity of Miao people in the tourism village in China		
Valuing nature and culture	Keeping diasporic longing	Insisting on self-governance

the basis for tourism development there. Tourism in this village is an ideal example of ethnic village tourism presenting the indigeneity of local Miao villagers (Chen et al. 2017).

Focusing on this case, this chapter aims to use concrete universals as a theoretical framework by which to appraise the indigeneity of Miao peoples in tourism in China from different dimensions at different layers (see Table 4.1). We first discuss the theory of critical realism (CR) and how to use concrete universals as a theoretical framework to guide research. We then review the abstract conceptualisation of indigeneity, which is the same for all ethnic groups at a macro-level. Following this, we analyse the universal and concrete representations of indigeneity for Miao groups in China based on a meso-geo-historical background. This is followed by a short methods section in which we describe how data were collected and how we as authors are positioned relative to the results we present. Finally, we appraise the indigeneity of the Miao group and individual Miao people based on their tourism practices in the micro-layer of a tourism village.

Critical Realism Theory

CR has been offered as a good alternative to constructivist and positivist approaches in tourism studies (Platenkamp & Botterill 2013; de Bernardi 2019; Gao 2021). To be more specific, both constructivist and positivist research approaches reduce reality to empirically accessible human knowledge. The former diminishes reality to the meaning entailed in common-sensical discourses, while the latter reduces reality to experiment results (Gao 2021). However, knowledge is always potentially fallible, and epistemological truth claims do not always reside within a single ontological reality (de Bernardi 2019). CR was originally constructed to solve this epistemic fallacy. According to Bhaskar (1978), the social world is stratified into three layers: they are (1) surface or experiential knowledge perceived at the *empirical layer*, (2) events that happen whether we experience them or not at the *actual layer* and (3) mechanisms hidden in a *real layer* that produce the events in the world. The stratified and differentiated understanding of reality in CR provides an ontology that accommodates plural epistemologies for researching the targeted social subject (Gao 2021).

The concrete universal, as one of the main concepts pertaining to CR, will provide us with access to multidimensional contents of the domain of indigeneity and understanding thereof. According to Shillcock (2014), two sorts of universals exist: an abstract universal and a concrete universal, both available in the philosophical literature. The former implies that individual elements of a group can be connected to the whole group by some universal characteristic. In contrast, the latter is a material entity that can appear within the set of entities it describes, of which it represents the essential, paradigmatic case (Shillcock 2014; Bhaskar 2016). This means that concrete universals have a universal component, which has "transfactually applicable properties and laws", as well as particular mediations (Bhaskar 2016: 130). Irreducibility is also emphasised here as a very important concept of CR. It means that even though the universal category is abstracted from different instances, it is not possible to reduce the single instance to the general category (Bhaskar 2016). On the contrary, the universal category should be understood to be located in a particular spatial and temporal context, mediated by social positionality, and concretised in the life experiences of individuals (Martinez Dy et al. 2014). In using the plural form of people in this chapter, we aim to show that there is not one version of reality, conversely that every people or tribe has a subjective way of realising their indigeneity, which makes reality rich in its particularity (Watson 2014). Therefore, the concrete universal provides a multidimensional position on universals which can give us a fuller philosophical picture.

Many scholars use this theory to research problems related to ethnic peoples. For example, Smallwood (2015) advocates for viewing human rights from a universal perspective, but also advocates for an understanding of different structures that permeate the different relationships between people, such as racism and discrimination. de Bernardi (2019) proposes that Sámi authenticity is the concrete representation of universal authenticity in society which therefore supports Sámi entrepreneurs and lightens interferences by hegemonic discourses. The theory of the concrete universal can also be applied to the concept of indigeneity. As Müller and Viken (2017) pointed out, there is a diverse indigeneity to celebrate, since places and people are different. In order to analyse indigeneity, it is then necessary to understand its universal characteristics, particular mediations and geo-historical context.

The Universal Representation of Indigeneity

Under traditional colonial and modernist discourses, the view of indigeneity as primitive, timeless or a circular situation of stasis would have been starkly counterposed to Western or Eurocentric notions of progress or development (Sabaratnam 2013). As such, indigeneity would facilitate an invitation to intervene, colonise, develop or civilise (Chandler & Reid 2018), but, according to Bryan (2009), indigeneity becomes respected when it helps to solve a problem of governance, and The ecological value of Indigenous knowledge makes the term more widely accepted. In addition, according to Matute (2020: 5), for local peoples, the term 'indigeneity' could be a powerful spiritual bond to "strengthen an existing ethnic group that confronts racial, ethnic, cultural and social hierarchies and struggle to keep control over 'development', and thus, over their lives and territories".

However, the definition of indigeneity was for a long time based on 'criterial' qualities to mainly describe the identity of Indigenous peoples (Merlan 2009). These refer to (1) having a historical continuity with pre-invasion and pre-colonial societies; (2) self-identification and recognition as a distinct collectivity by others, including the state; (3) perpetuation of cultural distinctiveness; and (4) an experience of marginalisation and dispossession either now or in the past (OHCHR 2013: 6). Bryan (2009) argued that indigeneity in this criterial definition works as a residual category because this definition mainly emphasises everything that existed prior to what is Western or modern. It asserts the essential connection between place and identity and represents the outcomes of specific historical and geographical processes as timeless facts. Such an assumed concept obscures social processes and mobility and therefore makes the term 'indigeneity' fixed and exclusive (Bryan 2009; Ludlow et al. 2016). For example, Miao peoples, as the original ethnic groups in the mountains of southwest China, cannot be described by indigeneity according to this definition (Schein 2007). The reason is that the Miao identities encompass long histories of movement which is said to remain "contrapuntal to those collectivities whose identity claims are based on rootedness" (Schein 2007: 242). Thus, Bryan (2009: 25) argued that indigeneity describes a relationship rather than an objective fact, emphasising the importance of understanding what work those concepts do in terms of the relationships they make possible and what forms of knowledge they produce.

Resilience and adaptation are actually outcomes of close interactions between local peoples and nature and society (Rival 2009; Viken & Muller 2017). According to Rival (2009), the epistemology contained in indigeneity focuses on understanding human physical presence on earth in terms of relative intensity. Instead of treating nature as a passive object, it sees nature as having more power and agency than humans and better ways of knowing and calculating. Such epistemology cultivates the capacity of resilience to cope with environmental disasters by listening to the perceived needs of nature. For example, Miao peoples effectively cope with rocky desertification by observing the behaviour of bird species feeding on certain seeds or the function of various plants, such as moss, pteridophyte and stringy stonecrop herb, in accelerating the dissolution rate of limestone (Luo & Shao 2011). There is little doubt that the Miao peoples use a complex system of ecological indicators to deal with fluctuations in natural factors and to defuse risks.

The capacity of adaption is embedded in their social processes. Different from the essential relation of 'belonging **in** a place' claimed in the 'criterial' definition, the capacity of adaption highlights a fluid relation of 'belonging **to** a place' (Watson 2014) (emphasis added). As Kingsbury(1998) pointed out, Indigenous groups have often been displaced from traditional lands and have formed their lives in new circumstances. However, their great passion about their ties to land, nature and culture does not disappear. On the contrary, these characteristics will adapt to the new place to foster continuity. Accordingly, 'belonging to a place' on the one hand refers to transcending the bounds of the original place and space; on the other hand, it also underlines affective relations with the original place.

Further, it is self-evident that indigeneity represents the defence of land rights, equal and inclusionary treatment, if interpreted through political and historical movements. The appropriation of Indigenous lands and the dispossession of local peoples have not led simply to arguments for the return of those lands to their original owners (Chandler & Reid 2018), but made indigeneity a sphere of commonality among those holding moral claims, and indigeneity then becomes a representation of rebellion for those making such claims.

Accordingly, indigeneity as a relational representation attends to natural, social, historical and political relations (de La Cadena & Starn 2009; Watson 2014; Matute 2020). The universal representation of indigeneity is constructed as the capacity of resilience and adaptability, simultaneously as the consciousness of rebellion. Indigeneity is not reduced to a concrete identity but can include universal ethnic peoples who have these relational representations and who have knowledge produced through these relations (Trigger & Dalley 2010; Müller & Viken 2017). However, these universal representations are always mediated by their concrete geo-historical context in order to retain its active, mobile, relational nature (Radcliffe 2015). As this formulation makes clear, appraising the indigeneity of Miao peoples should consider the universal representations under the concrete contexts.

The Indigeneity of Miao Peoples in China

As Schein (2007: 242) pointed out, the past two thousand years of Miao history is one long skein of rebellion, defeat, migration and flight. According to a mythical legend, the original conflict between Miao peoples and Han peoples was a fight for the control of lowlands between the Yellow River basin and the Yangtze River basin. When the Miao peoples' ancestors, the Chi You tribe (Indigenous peoples of the middle reaches of the Yellow River), were defeated, they were forced to migrate south (Shi 1995; Yang 2010). During this long migration process, some chose to be absorbed as subjects of the Han state and hence disappeared as a distinguishable Miao identity; some rebelled and were ready to fight, while others chose to move to a new administrative area or to another country (Scott 2009). This is why Miao peoples have evolved from one ethnic group in mainland China to a cross-border global ethnic group (the name of Miao peoples abroad is Hmong). Similarly, Miao peoples in mainland China have also formed different branches, such as 'white Miao', 'flower Miao' and 'black Miao' during migration. These tribal peoples adapted their culture according to their specific living environment and finally developed different religious beliefs, customs, dress habits and even languages within Miao groups (Chen 2007).

Although Miao peoples scattered in different places became accustomed to adapting to new environments, a strong cultural identification also emerged in these different Miao groups. According to Scott (2009), Han officials used the term 'Miao' to designate any groups that were rebellious and primitive and refused to submit to the Han administration. However, Miao peoples believe that the term 'Miao' is a sacred symbol, representing the 'heart' of the maple that gave birth to their ancestor and can also protect future Miao generations (Chen 2007; Xing 2007). Singing and embroidery are shared ways to record their religious beliefs, moral ethics and social processes, showing their common cultural identification (Zou 2019). For example, a large number of 'migration songs' have been handed down from generation to generation by Miao peoples in different places. Dragons, butterflies, maple leaves and ox horns act as totems of Miao culture and are generally present in the patterns embroidered by Miao peoples. Thus, through these sources, Miao peoples can represent indigeneity in terms of their capacity to adapt well, transcending spatial bounds and adapting to new environments, while also maintaining strong affective relations with their original place.

The characteristics of resilience and rebellion of Miao peoples are also discussed in relation to Miao ecological worldviews and knowledge, and informal political systems (Ma 2018; Luo & Shao 2011; Xing 2019). As remote and hidden places, mountains have been important refuges for these state-fleeing Miao peoples. Due to their lengthy interaction with natural environments, a common ecological worldview has also appeared among Miao groups. The relationship between peoples and nature in Miao peoples' minds is that humans are an integral part of nature, which is the primary source of all life that nourishes, supports and teaches (Ma 2018). All activities of Miao peoples are centred on building a harmonious ecological chain. They believe that if they violate this principle, they will be restricted or punished by a god. Their ecological worldview and knowledge in turn contribute to protecting natural environments. For example, China's oldest existing maple trees appear in Miao villages (Xing 2007). The problems of rocky desertification were also solved by the ecological knowledge of Miao peoples, rather than by the government technicians (Luo & Shao 2011). In addition, strictly informal political systems are consistently retained in different Miao groups today. For example, a '寨老' (village elder), a '巫师' (Shaman) and a '议榔' (colloquium) as a traditional self-governance system coexists with formal top-down systems, such as village committee organisations introduced by the state (Xing 2019).

In addition to these universal representations of adaption, resilience and rebellion, Miao peoples have also formed concrete representations based on geo-political contexts in China. One important characteristic discussed in the literature is the identity of Miao peoples. It is argued that Miao peoples could be seen to be linked with a global Indigenous identity through a history of expropriation of once native lands (Siu-woo 2003). However, Miao peoples do not identify with the idea of rootedness in place, instead sharing "narratives of displacement" (Schein 2007: 225). They understand that it is easier to obtain China's official recognition as an ethnic minority, compared to being sanctioned on the grounds of threatening national unity as an Indigenous community (Yeh 2007;

Siu-woo 2003). Additionally, traditionally the Miao have been regularly on the move, eschewing the identity of First Peoples (Lu & Lv 2012). This implies another concrete characteristic of Miao peoples that is a diasporic longing, which is generated by their long migration history (Schein 2007). It does not necessarily refer to a desire for movement to different places, but an identification with the scattered and free lifestyle in the hills, whereby they can keep a distance from the authorities (Scott 2009). It was recorded that during the migrations, Miao peoples also assimilated other ethnic groups into Miao groups (Wen 2018), showing that the continuity of Miao meaning in Miao peoples' minds may lie more powerfully in a shared history, culture and practices than in any presumed essential claim to ancestral blood, identity and land ties.

Discussions documented in these sources revolve around the characteristics of Miao peoples in their relationships between history, land, nature and society (Zou 2019; Ma 2018; Luo & Shao 2011). Miao peoples' common cultural identification, ecological knowledge and self-governing consciousness act as outcomes that can satisfy universal representations of indigeneity reviewed above. Their ethnic minority identity and diasporic longing with diversified and inclusive cultures formulated under the concrete geo-historical context can also reinforce the universal concrete representation of indigeneity (Chen 2007; Scott 2009; Wen 2018). These features from the literature were taken as starting points for the fieldwork described below.

Method

This study is based on qualitative data obtained from multidimensional methods, which include off-site critical literature review, as well as on-site participant observations, and semi-structured interviews in the field. According to Jesson and Lacey (2006), to transform existing knowledge into new theories and foster a deeper understanding, literature reviews should take a critical approach. We use a critical literature review to explore deeply how indigeneity is conceptualised in current studies and how Miao peoples' characteristics are conceptualised in relevant studies that satisfy dimensions of indigeneity. In terms of appraising the representation of Miao peoples' indigeneity in tourism, data were collected during three extended periods in September 2019, August 2021 and November 2021 at Upper Langde Miao village. During these periods, the first author lived in the village, conducting participant observation and unstructured interviews. Miao peoples are often regarded as a reserved group seldom voicing their concerns directly with members beyond their community (Wen 2018). Thus, formal interviews and structured questionnaires were not considered. Interviews were conducted in the Upper Langde Dalu Eco-Tourism Company (ULDEC), the village committee, at performance venues and in the homes of villagers. Informants include four leaders of the village as well as 16 members of the village comprising both male and female adults (totalling 20 persons, all being Miao peoples, except for one villager of Mongolian ethnicity). The questions examined views on tourism development, their social system, lifestyle and livelihoods. Data collected during the fieldwork were analysed using a qualitative content analysis, according to themes that were previously reviewed in studies. The first author is partially of Miao descent, whereas the second author was brought up in the culture of a minority language group. Such cooperation can facilitate emic and etic perspectives to objectively understand the indigeneity of ethnic minorities better.

The Indigeneity of Miao Peoples in the Tourism Village

At the micro-level of the tourism village, indigeneity is represented in universal practices that are in relation to valuing nature and culture, keeping diasporic longing and insisting on self-governance.

Figure 4.1 A view of the Upper Langde Miao Village (10 November 2021).

For Miao individuals, indigeneity is concretely represented in their personal position, relationships and daily practices.

Located on the Leigong mountain (雷公山麓) slope by the Wangfeng River (望丰河), Upper Langde Miao village has a dense mountain forest as a green barrier for the village (Figure 4.1). The roads in the village are paved with pebbles or bluestones. Some main paths in the village are accompanied by running water from ancient wells on the hillside. The pleasant sound of water hitting stones and birds chirping can also be heard clearly in this village. Many eco-trails leading to the mountain have been built, and a number of rest spots, viewing platforms and interpretive signage for landscape and history set up on the way. Fishing, loach catching, and fruit, vegetable and paddy planting experiences along the river are open for tourists without charge. Tourism performances are performed by local peoples at two set periods each day. The repertoire of performances has not changed for about three years (personal interview, 2 August 2021). Almost all the peoples in the village participate in these performances. However, it is rare to see local peoples wandering in the village after performances. Most villagers will either go to the mountains to do farm work or remain at home to do some housework or handicraft.

The tourism company built by local villagers is named the Upper Langde Dalu Eco-Tourism Company (ULDEC). Dalu is the name of a famous hero of Miao peoples who was born in this village and who was important in resisting the oppression of the Qing Dynasty. The name of this

39

company reflects that 'eco-tourism' is the core theme by which local villagers want to attract tourists. In Miao peoples' minds, the ecology of the environment should be preserved, while their culture should be adapted by themselves according to the specific situation. Therefore, in order to protect the environment and mitigate the interference of tourists in Miao peoples' living, they try to shift tourists' attention from 'exotic' cultural encounters to an appreciation of the natural habitat.

According to the village Party Secretary, the first and most important task for the village is to expand a new village nearby. This reflects the desire of the local Miao peoples to avoid too much interference from tourism development by relocating to another place. Built at the end of the Yuan and beginning of the Ming Dynasty (approximately 1400 CE), the Upper Langde Miao Village has a history of over 600 years (Leishan County Annals Compiling Committee 1992: 688–689). In 1987, in the first batch of ethnic villages to develop tourism in China, the development attracted the attention of the central government (Oakes 1998). In 2001, the State Council listed the village as a protected national heritage site. Once a place is so identified, it is often constrained by the discourse of 'protection' while developing its tourism in the name of heritage (An et al. 2021). While obtaining the economic benefits from heritage, local residents can become dissatisfied with the constraints of heritage protection. For example, local villagers are not allowed to expand houses or build new houses in the village (personal interview, 15 November 2021). However, unlike most Han families with only one child, most Miao peoples have at least two or three children because they lived in remote mountains and avoided the one-child policy in the late 1970s (personal interview, 15 November 2021). By 2019, this village had 213 households with a population of 871 (Feng & Li 2020), so with providing accommodation for tourists, the living space for villagers is very limited. According to a villager in his thirties, there are six persons with only three bedrooms in his household. He got married last year and now plans to be adopted into his wife's family in another village (personal interview, 15 November 2021). "Although the plan to expand a new village nearby has encountered many difficulties, we will insist on applying until it is approved", the village Party Secretary said (personal interview, 18 November 2021). This ambition and villagers' attitudes towards it embody a diasporic longing in the tourism context, reflecting past migrations and longing to remove constraints created by tourism and to live freely.

In contradistinction to diasporic longing, local Miao peoples generally represent a strong sense of cultural identity and belonging when interacting with tourists. The souvenir shops in the village resemble family museums. In addition to small souvenirs bought wholesale, many Miao costumes and embroidered pieces displayed in the shops are handmade. Some of these are handed down from their great-grandmothers, some being a century old. As well as trying to sell goods, they are also very happy to teach tourists to recognise different embroidery stitches and the meaning of different patterns on clothes (Figure 4.2). One seller said:

Embroidery is our important wealth handed down from generation to generation. We are proud to let more peoples know about it. At the same time, embroidering a cloth is very time consuming and expensive. I am not willing to bargain with tourists. It is good if someone can buy one, but I am also happy to leave this wealth to our future generations.

(personal interview, 1 August 2021)

The tourism performances are selected by village elders from Miao folk customs including the drinking culture, drum culture and the migration history of Miao suffering. All performances are consistent with the original without any adaptation. According to one village elder, ancient songs cannot be adapted, no matter whether tourists can understand them or not since, once adapted, these ancient songs would lose the charm inherited from thousands of years in history (personal

Figure 4.2 A vendor of a souvenir shop and a tourist (20 August 2021).

interview, 15 August 2021). The significant beneficial distribution system in this village is a work-points system created by the village elders and contains the Miao peoples' traditional cultural values. Even though this system was described as backward egalitarianism, local Miao peoples still maintain it (Chen et al. 2017).

The Upper Langde Miao Village has already had about 25 years of involvement with tourism, and the core principle is that "everyone has contributed to the construction and protection of villages, and everyone should benefit" (personal interview, 15 August 2021). This principle has been at the forefront of tourism policy, despite government officials' criticism that while this principle emphasises fairness, it hurts efficiency (Chen et al. 2017). Miao peoples' proud attitude towards their traditional culture when in contact with tourism and their insistence on self-governance are formed in and by their shared identification.

According to a village elder, the county government was once hosting some provincial officials visiting this village. However, the county government at the same time arranged for the County Music and Dance Troupe to replace local villagers to perform for provincial officials. Instead of obedience, local villagers refused to participate in other welcome activities because this arrangement both reduced the benefits and hurt the feelings of local villagers. In a similar vein in 2007, in order to prevent the county government from reconstructing commercial activities in the tourism village, the village elders directly contacted the National Cultural Heritage Administration in

Beijing for help. As a result, the county government turned their investment to Xijiang (another Miao village nearby) in 2008 (field note, 12 August 2021).

However, in 2015, the national tourism administration put forward the concept of 'regional tourism'. This concept requires including a single scenic spot in the regional development plan to promote the all-round development of tourism harmoniously. The county government took this opportunity and became a presence in the village by creating a tourism company named Langde Culture Tourism Company (LCTC) in 2015. In order to ensure a fair negotiation relationship with the LCTC, the ULDEC was founded in 2018 by local villagers. Although the co-governing model limits the power of local peoples in tourism, the ULDEC has sought to ensure that villagers are the main stakeholders in tourism development management. Up to now, the businesses on the land of the village have basically been operated by local peoples. Only a few silver jewellery shops are operated by Miao peoples from other villages, all of which were introduced after getting the consent of local Miao peoples (field note, 15 August 2021).

"Compared with the top-down tourism development of Xijiang, the land is still our homeland not tourists' playground, although tourism growth is slow. We will insist on self-governance in tourism, since we need to make sure that tourism mainly services our benefits not others", a village elder said (personal interview, 16 August 2021). The political representation of indigeneity is not simply to ask for land rights, but to strategically maintain self-governance in the concrete tourism context. These strategies include proper compromise and resistance, therefore making this tourism village a good learning example of empowering locals in tourism.

Different Miao individuals have different concrete representations based on their specific situation. However, even if some concrete representations seem to deviate from the universal themes, they are strategies chosen by individuals in order to better maintain their indigeneity. For example, the village committee, as the official rural organisation, is supervised by the township-level party committee, both being made up of local Miao villagers. These young village cadres are obviously supposed to adopt the county government's ideology of tourism development. In order to assist future regional tourism planning, the LCTC started the requisition of as much farmland as possible from individual villagers. These village cadres are actually the main force that the LCTC relied on to persuade the villagers. However, the plan of expanding the new village is also an important project that the village cadres are striving for. Accordingly, the good relationship between these young village cadres and the county government is interpreted as a strategy to make a space for peaceful negotiations.

Another case is an ethnically Mongolian woman who has been married to a local man for 25 years. She successfully applied to be the leader of the Miao dance team because she performs Miao songs and dances very well. She also dared to publicly express villagers' smouldering dissatisfaction and moral claims to officials. "Local villagers are all Miao relatives. Some words are inconvenient to speak out. However, I am not afraid.", she said. In terms of blood relationship and ethnicity, she is an outsider in the village, but she is accepted as a member of the Miao village because her tourism practices closely relate to Miao cultural identification. In contrast, a ten-year-old 'left-behind' Miao girl (an expression used to describe children who are cared for by their grandparents since their parents work as migrant workers in urban areas of China) said in Mandarin that she could not speak the Miao language. She also said that "our school is not a circus" when she was asked whether the school offered courses related to Miao song and dance or handicrafts (personal interview, 20 November 2021). She seems to be more interested in knowing the 'tourists' world' than discussing her own daily reality. However, when she talked about her family members, she revealed that her grandmother is suffering from a degenerative eye disease, but still embroidered a lot of clothes for her. She also happily showed her father singing Miao songs stored on her mobile phone. It seemed that in her eyes Miao culture is sacred and can be shared through close daily interaction with family members instead of being an acrobatic skill to entertain tourists. Both the visible tourism practices of the Mongolian wife and the secret cultural representation of the left-behind Miao girl are examples of concrete representations of universal indigeneity mediated by personal position, relationships and daily tourism practices.

Conclusion

This chapter commenced with a consideration of the theory of CR and how concrete universals as a main concept of CR are applied in this chapter. The 'criterial' and 'relational' concepts were discussed as two main definitions of indigeneity. The former limits indigeneity to Indigenous peoples because of advocating essential criteria, while the latter emphasises the fluid relation between native peoples with land, nature, culture and social processes. Issues such as land rights, moral claims, and capacity of resilience and adaption as outcomes do not only appear among Indigenous peoples, but also in many ethnic groups. This expands the scope of indigeneity and therefore illustrates these outcomes as universal representations for ethnic groups around the world. Miao peoples' common characteristics, including cultural identification, ecological knowledge and self-governing consciousness, can satisfy the universal representations of indigeneity. Likewise, these universal representations of Miao peoples are also mediated by a 'geo-historical trajectory' that makes Miao peoples unique compared to other ethnic groups sharing the universal components. For example, the diasporic longing and acceptance of officially recognised ethnic minority identity are concrete representations of Miao peoples' indigeneity in China. Equally, in the more concrete context of Upper Langde, preferentially valuing nature and culture rather than business, keeping diasporic longing and insisting on self-governance are common representations of Miao peoples' indigeneity in tourism. Personal position, relationships and daily tourism practices also make representations of indigeneity diverse. Both the common and concrete representations of indigeneity provide authority for Miao peoples to represent their dignification in tourism. Therefore, indigeneity as a relational concept can be applied to ethnic minority identities and as a way of empowerment in ethnic tourism.

Indigeneity, as a relational representation which is proposed here, is therefore an example of the critical realist idea of the concrete universal. This formulation of indigeneity as a relational representation is the first attempt at studying ethnic minorities based on the perspective of indigeneity. Based on this perspective, political rights and Indigenous knowledge possessed by ethnic minorities in China and beyond can be activated in tourism research. In addition, this relational representation provides Miao peoples with a flexible structural framework from which to operate, which is especially important considering the strategic self-governance of ethnic minorities in tourism in other locations.

References

An, C. Y., Li, T. S., & Zhai, Z. Y. (2021) Discourse, Capital and Production of the Heritage Space: A Case Study of Yinxu in Anyang City, *Tourism Tribune* 36(7), 13–26. [安传艳, 李同昇,翟洲燕. 话语、资本与遗址区空间的生产——以安阳市殷墟为例[J]. 旅游学刊, 2021, 36(7): 13–26.].

Bhaskar, R. (1978) *A Realist Theory of Science*, Brighton: Harvester.

Bhaskar, R. (2016) *Enlightened Common Sense: The Philosophy of Critical Realism*, London: Routledge.

Bryan, J. (2009) Where Would We Be without Them? Knowledge, Space and Power in Indigenous Politics, *Futures* 41, 24–32.

Chandler, D., & Reid, J. (2018) 'Being in Being': Contesting the Ontopolitics of Indigeneity, *The European Legacy* 23(3), 251–268.

Chen, Y. C. (2007) *An Ethnological Study of Chinese Miao Culture* (PhD dissertation), Central University for Nationalities Press. [陈昱成.中国苗族文化的民族学研究[D].中央民族大学, 2007].

Chen, Z. Y., Li, L. J., & Li, T. Y. (2017) The Organizational Evolution, Systematic Construction and Empowerment of Langde Miao's Community Tourism, *Tourism Management* 58, 276–285.

de Bernardi, C. (2019) A Critical Realist Appraisal of Authenticity in Tourism: The Case of the Sámi, *Journal of Critical Realism* 18(4), 437–452.

de la Cadena, M., & Starn, O. (Eds.). (2009) *Indigeneidades Contemporáneas: Cultura, Política y Globalización*, Lima: Institut français d'études andines, Instituto de Estudios Peruanos.

Feng, X. H., & Li, Q. Y. (2020) Poverty Alleviation, Community Participation, and the Issue of Scale in Ethnic Tourism in China, *Asian Anthropology* 19(4), 1–24.

Gao, X. X. (2021) *Thinking of Space Relationally-Critical Realism Beyond Relativism –A Manifold Study of the Artworld in Beijing*, Urban Studies. Wetzlar: Majuskel Medienproduktion GmbH.

Hathaway, M. (2016) China's Indigenous Peoples? How Global Environmentalism Unintentionally Smuggled the Notion of Indigeneity into China, *Humanities* 5(3), 54.

IPCC.(2007) *Contribution of Working Groups I, II and III to the Fourth Assessment Report of the Intergovernmental Panel on Climate Change*, Geneva: IPCC.

Jesson, J., & Lacey, F. M. (2006) How to Do (or not to do) a Critical Literature Review, *Pharmacy Education* 6(2), 139–148.

Kingsbury, B. (1998) "Indigenous Peoples" in International Law: A Constructivist Approach to the Asian Controversy, *American Journal of International Law* 92(3), 414–457.

Leishan County Annals Compiling Committee. (1992) *Leishan County Annals,* Guiyang: Guizhou People's Press. [雷山县编年史编纂委员会. 雷山县志 [M]. 贵阳: 贵州人民出版社, 1992.].

Lu, H. F., & Lv, F. H. (2012) On the Motives and Realities of Miao's Spontaneous Migration in Western China: A Case Study in K County of Yunnan Province, *Journal of Original Ecological National Culture* 4, 141–146. [陆海发, 吕付华. 西部地区苗族群众自发迁移的动因及现实条件——基于对云南 K 县的调查[J]. 原生态民族文化学刊, 2012, 4: 141–146.].

Ludlow, F., Baker, L., Broch, S., Hebdon, C., & Dove, P. M. (2016) The Double Binds of Indigeneity and Indigenous Resistance, *Humanities* 53(5), 1–19.

Luo, K. l., & Shao, K. (2011) The Value of the Miao People's Traditional Ecological Knowledge for a Solution to Rocky Desertification in Mashan, *Journal of Resources and Ecology* 2(1), 34–45.

Ma, Q. (2018) The Form and Characteristics of Environmental Protection Customary Law: Based on the Case of Miao Nationality, *Guizhou Ethnic Studies* 39(7), 36–40. [马琼 少数民族环境保护习惯法的表现形式及特征--以苗族为例[J]. 贵州民族研究, 2018, 39(7): 36–40.].

Martinez Dy, A., Martin, L., & Marlow, S. (2014) Developing a Critical Realist Positional Approach to Intersectionality, *Journal of Critical Realism* 13(5), 447–466.

Matute, I. D. (2020) Indigeneity as a Transnational Battlefield: Disputes over Meanings, Spaces and Peoples, *Globalizations*, 18(1), 1–17.

Merlan, F. (2009) Indigeneity: Global and Local, *Current Anthropology* 50(3), 303–333.

Oakes, T. (1998) *Tourism and Modernity in China*, London: Routledge.

Office of the United Nations High Commissioner for Human Rights. (2013) *The United Nations Declaration on the Rights of Indigenous Peoples: A Manual for National Human Rights Institutions*, Geneva.

Platenkamp, V., & Botterill, D. (2013) Critical Realism, Rationality and Tourism Knowledge, *Annals of Tourism Research* 41, 110–129.

Radcliffe, A. S. (2015) Geography and Indigeneity I: Indigeneity, Coloniality and Knowledge, *Progress in Human Geography Progress Reports*, 41(2), 1–11.

Rival, L. (2009) The Resilience of Indigenous Intelligence, in K. Hastrup (ed.) *Question of Resilience: Social Responses to Climate Change*, Copenhagen: The Royal Danish Academy of Sciences and Letters, 293–313.

Sabaratnam, M. (2013) Avatars of Eurocentrism in the Critique of the Liberal Peace, *Security Dialogue* 44(3), 259–278.

Schein, L. (2007) Diasporic Media and Hmong=Miao Formulations of Nativeness and Displacement, in M. de la Cadena & O. Starn (eds.) *Indigenous Experience Today*, London: Routledge, 225–245.

Scott, C. J. (2009) *The Art of Not Being Governed an Anarchist History of Upland Southeast Asia*, New Haven: Yale University Press.

Shi, C. J. (1995) Five Migrations in the History of Miao Peoples, *Journal Guizhou Ethnic Studies* 61(1), 120–128. [石朝江.苗族历史上的五次迁移波[J].贵州民族研究,1995,61(1): 120–128.].

Shillcock, R. (2014) The Concrete Universal and Cognitive Science, *Axiomathes* 24, 63–80.

Siu-woo. C. (2003) Miao Identities, Indigenism and the Politics of Appropriation in Southwest China during the Republican Period, *Asian Ethnicity* 4(1), 85–114.

Smallwood, G. (2015) *Indigenist Critical Realism: Human Rights and First Australians' Wellbeing*, London: Routledge.

Trigger, D. S., & Dalley, C. (2010) Negotiating Indigeneity: Culture, Identity, and Politics, *Reviews in Anthropology* 39(1), 46–65.

Viken, A., & Müller, D. K. (2017) Toward a De-Essentializing of Indigenous Tourism?, in A. Viken & D. K. Müller (eds.) *Tourism and Indigeneity in the Arctic*, Bristol: Channel View Publications.

Watson, M. K. (2014) *Japan's Ainu minority in Tokyo: Diasporic Indigeneity and Urban Politics*, London: Routledge.

Wen, S. X. (2018) Family and Social in Liquidity: The Study of a Miao Family's Migration History, *Guizhou Ethnic Studies* 210(39), 43–49. [温士贤. 流动中的家与社会:一个苗族家庭的迁徙史研究[J].贵州民族研究, 2018, 210(39): 43–49.].

Xing, Q. S. (2007) On the Ecological Culture of Miao Minority in Guizhou Province, *Journal of Qiannan Normal University for Nationalities* 24(2), 24–28. [邢启顺.贵州苗族生态文化简析[J]. 黔南民族师范学院学报, 2007,4(2): 24–28.].

Xing, Y. X. (2019) Practice of Ecological Management of Miao Village from Langyue to Village Regulation, *Yunnan Social Sciences* 2, 120–126. [邢一新. 从榔约到村规苗族村寨生态治理的实践[J]. 云南社会科学, 2019, 2: 120–126.].

Yang, Z. Q. (2010) Chiyou Worship and Ethnic-Identity, *Journal of Qinghai Ethnic Studies* 21(2), 38–45. [杨志强.蚩尤崇拜与民族认同——论当今中国苗族树立精神共祖的过程及背景[J].青海民族研究, 2010，21(2): 38–45.].

Yeh, E. T. (2007) Exile Meets Homeland: Politics, Performance, And Authenticity In The Tibetan Diaspora, *Environment and Planning D: Society and Space* 25 (4): 648–667.

Zou, W. B. (2019) Renewal of Intangible Cultural Heritage in the New Era: Characteristics, Present Situation and Path of Miao Embroidery, *Hundred Schools in Arts* 166(1), 178–196. [邹文兵.新时代非遗苗绣的"活化": 特质、现状与路径[J].艺术百家, 2019, 166(1): 178–196.].

5

POWER, POLICIES, AND CULTURAL SENSITIVITY IN AINU TOURISM

Indigenous Involvement in and Control over Tourism Developments in Hokkaido, Japan

Mayumi Okada and Johan Edelheim

Introduction

One of the recent trends of tourism in Japan, especially in Hokkaido, is the progress of tourism development related to the Ainu people, the Indigenous people of Japan, and to the Ainu culture. There is a range of concurrent factors for this development: first, a worldwide movement of Indigenous peoples reclaiming their rights and sovereignty; and second, two Japanese national initiatives, the enactment of the Ainu Policy Promotion Law in 2019 (see also Chapter 20, editors' note) and the establishment of the first national Ainu facility *"Upopoy-A Symbolic Space for Ethnic Harmony"*, opened in 2020 and perceived as initiating a new era of Indigenous tourism in Japan.

The national and the Hokkaido governments are implementing measures to promote tourism related to the Ainu people/culture (Ainu tourism) across Hokkaido since tourism is expected as a key to balancing the cultural revitalisation and promotion by the Ainu people with economic development. However, viewed from the historical context of the Ainu and tourism, the current discourse does not consider whether the recent Ainu tourism developments have improved the formerly dominant and overwhelmingly unequal relations of the past and instead become a form of tourism with, by, and for the Ainu, as advocated in the Larrakia Declaration (2012).

This chapter aims to situate the contemporary relationship between the Ainu community and tourism in Hokkaido within a broader historical context. We utilise the approach of culturally sensitive tourism as our theoretical framework and as a motivational example through which we suggest future Ainu tourism should be created. Our chapter's empirics come from grasping the current Ainu tourism situation, driven in parts by national Ainu policies. The case studies are from *Upopoy* located in Shiraoi, from Lake Akan Ainu *Kotan* (village), and from other areas of Hokkaido (see Figure 5.1, the map of Hokkaido with case areas indicated). We present challenges that the Ainu people have been facing with related socio-political issues. Following this, we identify the degree of Indigenous control over (or involvement with) recent Ainu tourism initiatives in line with the culturally sensitive tourism concept. Our goal is to learn from good practices in Ainu tourism and to investigate how such practices could be multiplied in an ever more polarised society.

DOI: 10.4324/9781003230335-6

46

Historical Overview of the Ainu People and Tourism

This section gives the historical context of developing Ainu tourism in Hokkaido by discussing three phases based on the interaction between the Ainu people and tourism practices. Historically, the Ainu people lived in a region communally referred to as *ainumosir* – "the land of the Ainu", in the Ainu language (Figure 5.1). It consisted of Japan's northern main island Honshu, Hokkaido

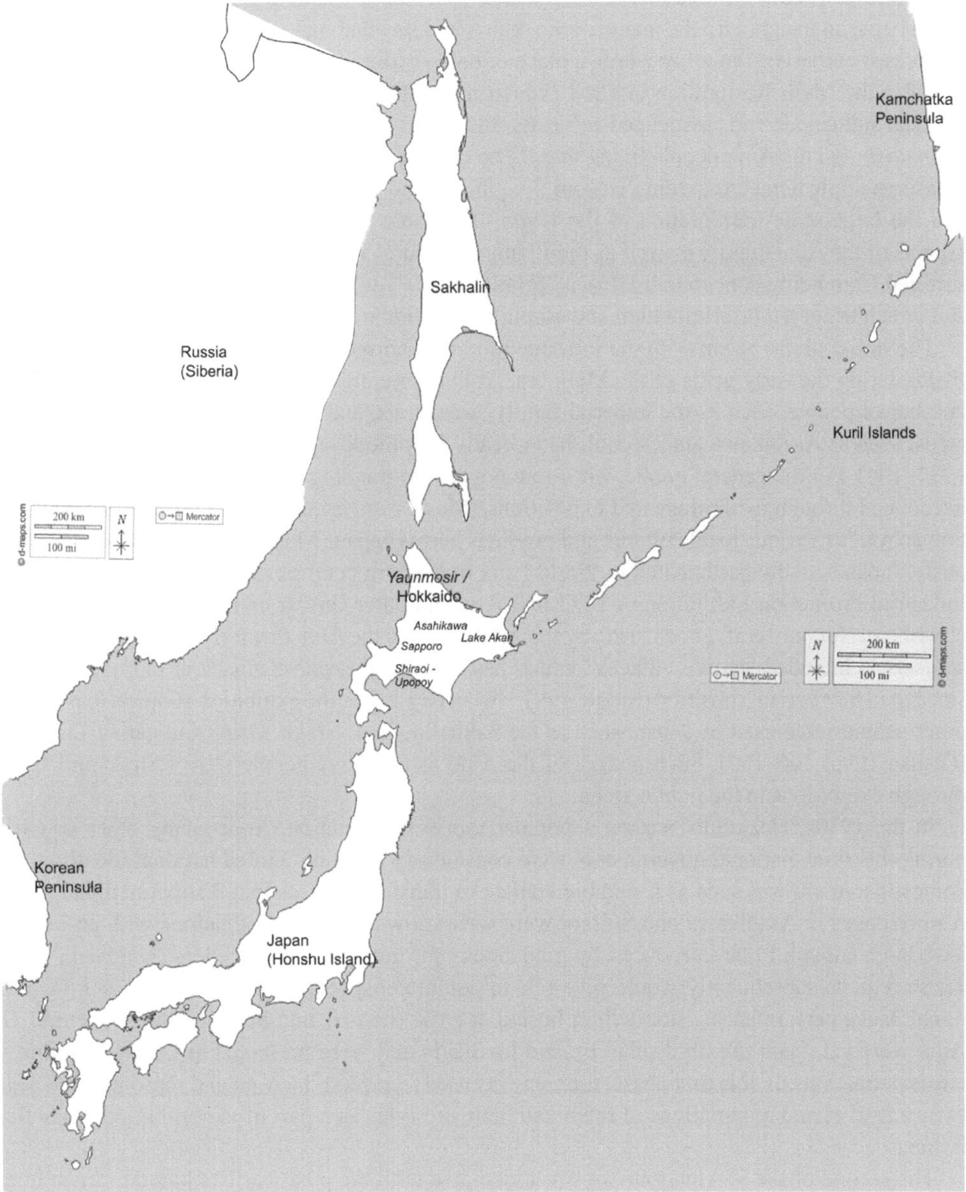

Figure 5.1 The map of *ainumosir*, the original region where the Ainu live and lived, with case locations indicated.

Source: Adapted from www.d-maps.com

known as *Yaunmosir* (Hokkaido Tourism Organization 2019) and Sakhalin, the Kuril Islands, and southern Kamchatka. The Ainu have their own distinct spoken languages and cultures, with regional variations and similarities across these areas.

Before the Meiji era (1868–1912), *Yaunmosir* was beyond the scope of Japan's sovereignty, but considered part of the *Matsumae* domain, land belonging to a feudal clan. Another name for the region was *Ezo*, a name that is still surviving in the names of some animals and plants such as the *Ezo Fox*, a red fox breed native to Hokkaido. The kanji characters that are used to spell Ezo [蝦夷] give an insight into the image of the region and its inhabitants, as a literal translation would be "prawn barbarian", in other words a place of hairy barbarians (Siddle 1997).

After the Meiji Restoration in 1868 (which marks the time of Japanese acceptance of international influences with associated reforms), the Meiji government claimed direct control over *Yaunmosir* and the Ainu people living there. The current name, Hokkaido, was made official. The measures implemented regarding custom, livelihood, and cultural assimilation during this period, and the large-scale immigration of the *Wajin*, the ethnic Japanese, inflicted lasting harm to all aspects of the Ainu culture as well as their natural resource management. Additionally, Ainu have suffered from being conceptualised as a "primitive race", which came into practice in this period and which triggered discrimination and assimilation (Siddle 1997: 5–20, 79–98).

The initial phase began with the introduction of organised or industrial tourism efforts by the *Wajin* during the early years of the Meiji era. At that time, travel was entertainment for a limited number of people, such as the imperial family, aristocrats, and international travellers. Ainu villages, such as Asahikawa and Shiraoi, had already become locations accepting travellers (Siddle 1997: 104). For the general public, who were not able to travel to Hokkaido, the Ainu culture was introduced through exhibitions and expositions, which were established to introduce modernisation as well as various local cultures and products across Japan (Morris-Suzuki 2014). Simultaneously, Japan actively participated in World Fairs in Western countries and held the first Domestic Industrial Promotion Exhibition in 1877 in Tokyo, and later similar exhibitions were carried out elsewhere. Not only craft goods made by the Ainu, but also the Ainu themselves, became exhibition subjects in intended "native villages" which had become popular elements at international fairs since the 1889 Paris Exposition (Siddle 1997: 101–102). Later, the exhibition subjects expanded to other peoples colonised by Japan, such as the Sakhalin Ainu, Nivkh, Orok, and native Taiwanese (Ōtsuka 1996: 106–107). Such images of the Ainu as primitive people were formed and spread through exhibitions in the public arena.

In the 1930s, Hokkaido became a popular tourism destination. As in many other countries during this time, national governments were consciously building unified national identities, and domestic tourism was seen as a suitable vehicle to transform spaces into distinct national places. Ainu villages in Asahikawa and Shiraoi were well-known tourism destinations with an industry base large enough to be introduced by guidebooks for international travellers (Kindaichi 1941). These regions subsequently would give rise to common aspects of Ainu tourism, such as traditional dance performances, storytelling hosted at *chise* (house), and bear carving souvenirs. The Ainu were yet again put on display by and for *Wajin* and were presented as exotic or primitive tourist attractions. In this first phase, tourism was used to exploit the Ainu and reproduce the negative narratives and assumptions of them and their lifestyles as a part of nation-building practices in Japan.

The second phase of Ainu tourism was associated with the post-war Hokkaido tourism boom during the 1960s. As many domestic tourists flocked to Hokkaido, certain traditional aspects of the Ainu culture continued to be highlighted and treated as backward and exotic aspects of tourism in the region. Concurrently, this was also a phase when some Ainu people who engaged in tourism activities were subject to criticism from their own communities. In particular, "Tourism Ainu" (*Kankō* Ainu) was one term used to refer to them. But indignant calls from Ainu activists were also aimed at the ongoing structural inequality between *Wajin* tourism operators and the Ainu (e.g. Arai 1984). In response, some Ainu started utilising tourism as a vehicle for revitalising traditional culture and fostering cultural successors – in other words, consciously training and passing on oral history and traditional skills and knowledge to younger generations. These initiatives slowly gained momentum, and attempts were made at collaboration between the representatives of the Ainu culture and tourism stakeholders in some areas. In 1965, for instance, an Ainu village, originally located within the city of Shiraoi, was relocated to Lake *Poroto* and established as the Shiraoi *Poroto Kotan* by the Shiraoi Tourism Consultant Co., Ltd. In 1976, the Shiraoi Tourism Consultant Co., Ltd., which promoted Ainu culture tourism, was dissolved, and the Foundation for Preservation of Shiraoi Ethnic Culture was established in its place. The aim of the foundation was to contribute to developing the northern culture through the preservation and display of the Ainu heritage in the Shiraoi region. The new foundation supported the protection and promotion of tangible and intangible Ainu heritage through preservation, documentation, and education, partnered with tourism (Ainu Museum 2014). In 1984, based upon these efforts, the foundation opened the Ainu Museum to act as an exhibition space as well as a research facility.

The third phase is characterised by the national Ainu Policy promotion and establishment of *Upopoy,* the National Ainu Museum and Park. This phase began in 2008 when both the Upper and Lower Houses of the Diet of Japan passed a nonbinding resolution to officially recognise the Ainu as Indigenous people of Japan. This provided a springboard for the development of comprehensive policy measures and stimulated discussions highlighting the need to generate income for the Ainu people through cultural revitalisation incorporated with commercial activities. Tourism promotion and branding of Ainu crafts were expected to mobilise the collaborative linkage between cultural heritage and business activities (Advisory Council for Future Ainu Policy 2009: 29–30). For promotion, several measures and campaigns have been implemented such as the "*Irankarapte* (the Ainu greeting) campaign" launched in 2013 as a keyword for hospitality in Hokkaido tourism with industry–academia–government collaboration. As part of this campaign, in 2014, Indigenous artists, collaborating with academic counterparts, installed an artistic monument portraying an Ainu elder's figure, and ritual goods of Ainu, at Japan Railway's Sapporo Station, the main station for Hokkaido. Additionally, the Donan Bus started in-car broadcasts in the Ainu language in the Biratori town area.

Ahead of the official opening of *Upopoy*, the "Council for Excursion Trail of Ainu Culture" was established in 2017, involving Ainu organisations and travel and advertisement agencies. The aim was the creation of a cultural route called the "*Yukar* Road", which would link Shiraoi with Hakodate, Biratori, Sapporo, Asahikawa, and Akan by 2020. This initiative was planned not only to connect tourism destinations related to the Ainu culture with *Upopoy* as a hub, but also to support the development of community-based heritage tourism at each location. These promotions would contribute to greater attention and recognition of the Ainu as Indigenous people. The prefectural government of Hokkaido also developed a regional five-year plan (FY2021–2025) for Ainu policy promotion corresponding to these national measures. New policies expanded upon previous ones for improving living conditions for Ainu communities to include support for industrial development and stable employment. As a result, various forms of tourism featuring the Ainu people and culture are in the process of being planned and promoted across Hokkaido, and this development is the topic of the empirical section of this chapter, which follows the theoretical framework below.

Potentials in Applying "Culturally Sensitive Tourism"

The richness of words such as "culture" and "sustainability" are at the same time both their greatest strength and weakness – if a word means "too much" it ends up meaning next to nothing. For example, examining the official description of *Upopoy*, one can see that it called a place of "ethnic harmony" celebrating "multicultural coexistence", both lofty and worthy descriptions. Though, at political, social, and even academic levels, there is not much discourse to explain what is "harmony" and what makes "coexistence", real for both Ainu and *Wajin*. In our attempt to describe Ainu tourism, we considered different frameworks derived from sustainable or responsible tourism. However, we found that many of these words and theories have already been co-opted into marketing phrases no longer carrying their initial intended meaning. Our aim was to find a motivational framework to analyse Ainu tourism, something that we can hopefully emulate later in our work with the local Ainu tourism stakeholders.

Globally, the coupling of tourism and local indigeneity is broadly considered beneficial for multiple stakeholders, but at the same time fraught with challenges. Benefits that majority populations often happily promote are "economic development", "growth", or opportunities to engage more fully in general societies. Other benefits can be more rooted in peoples' chances to celebrate and take pride in their own traditions, to live and care for the land that they are a part of. The concept this chapter adopts is *culturally sensitive tourism*, as it aims to establish ways in which tourism can be practised in a responsible way in heterogeneous communities.

Culturally sensitive tourism is a relatively recently coined concept; it is the result of a research project (ARCTISEN), which features in other chapters in this book (see Chapters 19 and 2, editors' note), involving local communities, small- and medium-sized enterprises, international researchers, and organisations in the Arctic region. The World Indigenous Tourism Alliance (WINTA) was a core stakeholder in the project. Culturally sensitive tourism highlights that tourism should be created with local communities as the priority and that visitors need to follow the three *R*s of cultural sensitivity, that is, to *Recognise*, *Respect*, and *Reciprocate* the diverse cultural and natural values in existence. A way of distinguishing these practices is by putting them in contrast to cultural insensitivity which includes stereotyping, cultural assimilation, and appropriation (Viken et al. 2021).

Just as all words carry innumerable connotations and denotations, "sensitive" has also its problems as in the past it has been associated with "vulnerability" or "fragility". However, in the context of ARCTISEN, cultural sensitivity is understood as "a subjective orientation towards otherness that simultaneously shapes, and is shaped by, different kinds of social processes, narratives and encounters, including those associated with tourism" (Olsen et al. 2019: 16; Viken et al. 2021: 2). The core is thereby located in self-reflectivity; a distinct understanding of how and why one's own values are shaped as they are; and what practical consequences these values play when dealing with others and when portraying otherness. Following a phenomenological vocabulary, reality is never purely given to us in a fashion of *tabula rasa*: we are always "thrown" into realities with preconceptions that we seldom realise or question until we reflect upon them (Edelheim 2015; Fortin et al. 2021).

A theoretical framework developed by Bennett (1986), describing six stages of cultural sensitivity, forms the basis for that concept: denial, defence, and minimisation (also called ethnocentric); and acceptance, adaption, and integration (also called ethnorelative). In Bennett's original figure, these stages worked as a logical continuum from the left to the right with increased sensitivity for each stage, but we consider this somewhat too optimistic and see the stage rather as parting at the centre, signalling the ever-larger polarisation in society. We claim that even when people are aware of matters, there is a tendency to create dualisms, so the *ethnocentric* and the *ethnorelative* dimensions might in fact not lead into one another, but rather away from one another.

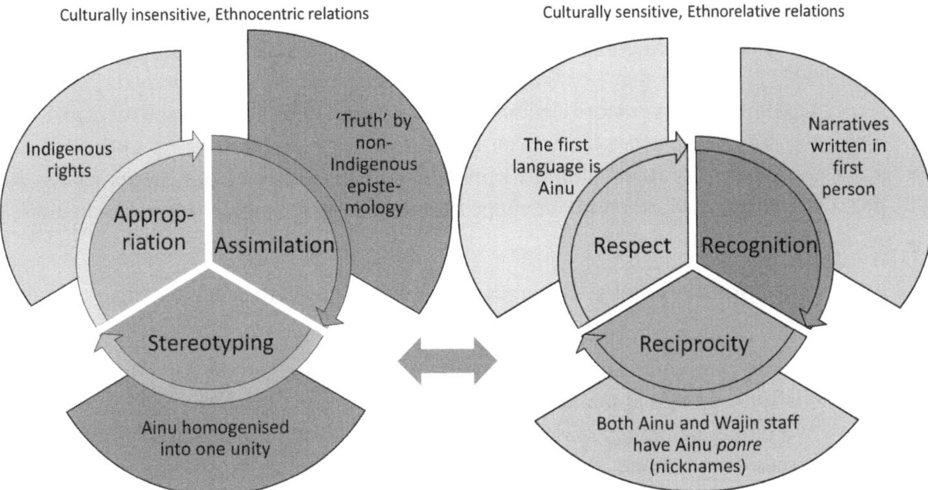

Figure 5.2 A conceptualisation of cultural sensitivity in tourism, with some reflections on Ainu tourism. Image adapted from Viken et al. (2021).

Culturally sensitive tourism has no "one-size-fits-all" as the local is always shaped by its own history and context. However, that same recognition is essential, where "locality" is made central, and the wishes of local communities are always the starting point for any development, rather than "good intentions" introduced by outsiders. A way of fulfilling this is by recognising the different worldviews that different groups inhabit and by respecting these worldviews as equally good, and in cases of decision-making, the local view being preferable to those of outsiders. Examples of culturally sensitive practices in action can be found in different parts of the world. The guidelines and checklists for tourism and human rights, based on the *Larrakia Declaration*, stand as one practical example (PATA 2015). Others include ethical guidelines for Sámi tourism (Viken et al. 2021) and the use of common certifications such as *Duodji* for handicrafts produced by Indigenous peoples (Brattland et al. 2020).

Tourism and hospitality practices in their modern commercial guise are often unnecessarily extractive. Journeys and hospitality have always existed in Indigenous cultures, exchanges and meetings with others have long traditions, and they have always been built on reciprocity, an understanding that favours provided will be reciprocated, maybe not personally, or not even in one person's lifetime, but naturally over time in a cycle of giving and receiving (Mika & Scheyvens 2022). This reciprocity has often been misunderstood by visitors and colonisers, in that it has been perceived as gifts, or trading – even creating the derogatory idiom of being an "Indian giver", when simply reciprocal help is expected (Benson 2006). The respect involved in culturally sensitive tourism comes from going beyond a simplistic economic transfer to personal interactions, including a willingness to learn about others and about oneself (Olsen et al. 2019). This is also a key factor; self-reflection requires time and an openness to challenge taken-for-granted perceptions of one's reality. If Indigenous tourism is seen purely as a product or service bought off the shelf in a detached manner, just like an object in a department store, then there is no self-reflection.

Culturally sensitive tourism practices should therefore provide spaces and experiences to explore not only the "other", but also the self. Figure 5.2, adapted from Viken et al. (2021), shows the relations of culturally insensitive and sensitive tourism as dimensions of ethnocentrism and ethnorelativism, in contrast to Bennett's model discussed earlier. The figure shies away from Bennett's linearity and portrays the relations as cyclical because they relate to self-reflection. Viken et al. (2021) point out that Bennett's highest stages "adaption" and "integration" can lead to practices of assimilation by majority cultures if they are not consciously implemented. The two circular movements are therefore feeding on themselves, but also connecting to the other. It is through this conscious effort that culturally sensitive tourism could lead to better practices, and it is this goal that we aim to work on in our cooperation with local communities in Hokkaido.

New Initiatives and Challenges Arising from Recent Ainu Tourism Development

The establishment of *Upopoy* and new measures and practices based on the Ainu Policy Promotion Act can potentially be an epoch-making in the interaction among the Ainu people, tourism, and host regions. We introduce here empirical cases to consider new initiatives and challenges from recent Ainu tourism development.

Case One: National Ainu Facility Upopoy as a Hub of Cultural Revitalising and Cultural Exchange

Comprised of the National Ainu Museum, National Ainu Park, and National Keeping Place (for the ancestral remains), *Upopoy* assumes the primary role as a national centre for revitalising the Ainu culture and as a base for promoting the understanding of Ainu history and culture among a wider range of people in Japan (Ainu Policy Promotion Headquarters n.d.). *Upopoy*'s functions follow international models and include "1. Research and exhibitions, 2. Continuation of cultural traditions and development of human skills, 3. Exchange of culture and experiences, 4. Dissemination of information, 5. Provision of park facilities, and 6. Fostering of respect for spirituality" (Foundation for Ainu Culture n.d.). To meet these expected functions, some innovations have been made across *Upopoy*. For instance, the Ainu language always comes first at signposts, signboards, and exhibition commentary, as well as in audiovisual captions, followed by Japanese, English, and other foreign languages. Another initiative is that the staff, regardless of whether they are of Ainu or *Wajin* ethnicity, have Ainu business nicknames (*ponre*). All these names have been given under appropriate conditions by an Ainu scholar and *Upopoy* staff; they are meant to create a communal feeling among the staff, but also to protect the staff from showing their actual names, which in the past has led to harassment and trolling in social media. Such a positive and proactive usage of the Ainu language in *Upopoy* contributes not only to create opportunities for those who learn and use the Ainu language, considering that it used to be suppressed, but also to increase the awareness and prestige of the Ainu language among the public (Kitahara 2020: 67–68). Regarding the National Ainu Museum, it is the first and only one dedicated to the history and culture of the Ainu people. Its principle is

> to respect the dignity of the Indigenous Ainu people, promotes the proper recognition and understanding of the history, culture [and more] of the Ainu, both inside and outside Japan, and contributes to the creation and development of a new Ainu culture.
>
> (National Ainu Museum 2020: 2)

To meet this goal, permanent exhibitions aim to be designed from an Ainu viewpoint such as first-person tense in exhibition signages. Also, by avoiding the stereotype of the Ainu culture as inferior and primitive, the exhibition tries to illustrate its variety, continuity, and modernity.

Additionally, but also in contrast, *Upopoy* also carries many hopes of government officials, local municipalities, and host regions based on expectations from Japan's national tourism policies. As stipulated in the pre-COVID-19 Ainu Policy Promotion Act, the facility aims to attract 1 million visitors, and expectations are mounting that it will bring positive benefits for the region's tourism enterprises and also other industries through the development of wide-area tourist routes originating from *Upopoy* such as "the *Yukar* Road" project and the national subsidised tourism promotion project planned by neighbouring municipalities.

Case Two: Community-Based Tourism Management in Lake Akan Ainu Kotan

One of the primary measures based upon the Ainu Policy Promotion Act incorporates new grants as a supporting measure for projects implemented by municipalities. It targets three areas: cultural promotion, tourism and industrial promotion, and community activity assistance (Tsunemoto n.d.: 4). Among these, the regional and industrial promotion assists tourism promotion and branding projects, which also enhance recent Ainu tourism developments and cooperation with businesses areas and municipal tourism policies. A review of subsidised projects based on municipal plans for Ainu policy promotion (from April 2019 to July 2021) found that 28 municipalities in Hokkaido and one outside Hokkaido are implementing projects aimed at tourism promotion and seven municipalities in Hokkaido are working on projects to promote the branding of the local Ainu culture (Japan Cabinet Office 2019). For instance, Kushiro City, which is home to one of the most famous tourism destinations of Ainu tourism, Lake Akan Ainu *Kotan*, has been making municipal plans for Ainu policy promotion focusing on projects for tourism promotion and branding projects with Ainu community-based management. In the financial year 2019, Kushiro City set up nine projects aimed at protecting the intellectual property of the Ainu people, fostering cultural tour guides of Ainu, Ainu tourism promotion, establishing an attractive brand of the Ainu culture through collaboration with craftspeople and artists, as well as an incorporating the Ainu culture with various products and services. Municipalities are supposed to have active collaboration with regional Ainu communities within these municipal plans, and new incorporations in which the Ainu people could be in directory roles have been established in addition to the existing Ainu associations in several regions including Kushiro City.

Akan Ainu *Consulun,* a general incorporated association established by the local Ainu community in 2019, is one of the important partners of Kushiro City. Some observers have raised alarms over the rapid incorporation of the Ainu people/culture and tourism into the general promotion of regional industries and economy, when the public understanding of Ainu history and culture has not deepened nor has the basic structure of tourism industries in Japan largely changed. The Ainu community in Lake Akan Ainu *Kotan* has recognised the importance of autonomic involvements and decision-making rights by the Ainu as good foundations for an equal partnership with *Wajin* enterprises within tourism practices. Akan Ainu *Consulun,* granted by the subsidies, addresses several initiatives including the protection of the intellectual property of the Ainu people (Hirono & Okada 2021).

With the recent increase of interest in the Ainu people and culture, the Ainu language and Ainu patterns are being used on commercial products removed from their original cultural context. The misappropriation of traditional knowledge of the Ainu not only impairs the dignity of the Ainu people but can be a barrier to achieving a balance between the protection of the Ainu culture and economic independence of the Ainu people (Okada 2021; Hirono & Okada 2021). It can also be an obstacle to promoting the appropriate understanding of the Ainu people and culture, which are the essential objectives of the Ainu policy. In response to these issues, by connecting business operators who wish to inject the Ainu culture into their business with local Ainu culture successors and craftspeople, Akan Ainu *Consulun* aims to transmit the authentic Ainu culture and develop branding projects driven by the Ainu themselves. In future, it aims to use revenue from these enterprises to develop Ainu cultural successors and to expand Ainu-community-based cultural promotion at national and international levels (Hirono & Okada 2021). Likewise, there are some initiatives in establishing incorporated associations in Hokkaido aiming at Ainu community involvement, or Ainu-based management for cultural revitalisation and promotion, as well as tourism practices, although this is not at all an overall trend.

Case Three: Challenges Confronting and Arising from Ainu Tourism

Historically, many Ainu who were involved in tourism activities were confronted with discrimination and suffered from trauma (e.g. Muraki 2010; Kitahara 2021), and discrimination against the Ainu involved in tourism is not just an issue of the past but still a major problem at present. A recent article reported a rapid upsurge in online slander and criticism against *Upopoy* shortly after opening (Hokkaido Shimbun 2020). In response to these events, the National Ainu Museum established a special section on its website dedicated to explaining the history and culture of the Ainu people to address misinformation and prejudices about the Ainu (National Ainu Museum n.d.). In addition, Hokkaido Tourism Organization, a public interest incorporated association, published a guidebook for the tourism sector that offers a detailed explanation of the Ainu people, history, and culture, as well as a FAQ section to promote greater understanding. In a revised edition of the guidebook published in 2019, a new section was added entitled "Positive/Negative Checklist for Tour Guides and Interpretations" that showed examples of discriminatory comments made by guides while conducting tours and ways to correct them (Hokkaido Tourism Organization 2019). The publication provides a reflective checklist (comparable to a similar list produced by the Indigenous Tourist Association of Canada, noted in Chapter 31) that includes the following points: (1) "Is your comment based on the perspectives of both Ainu people and *Wajin*?", (2) "Are you relying on stereotype images of Ainu people and culture?", (3) "Are your remarks based on established facts and clear grounds?", and (4) "Are you taking into account the specific characteristics and historical backgrounds of the region?". These questions are important not only for tours but also for the development of a more respectful future for Ainu tourism in Hokkaido.

One of the fundamental issues is the lack of an equitable partnership between the Ainu people and the tourism industries, both in institutional frameworks regarding Indigenous tourism in Japan and more widely of understanding on the part of business operators. This is despite the rapid incorporation of Ainu tourism into the promotional strategies of industries and tourism promotion. For instance, we noted recent misappropriation of the Ainu language and patterns in the previous section. The current legislation in Japan does not have effective mechanisms in place to protect traditional knowledge. Other challenges remain, including a lack of understanding on the part of business operators about the protection of the intellectual property of the Ainu. There is therefore a need to expand these kinds of initiatives into prefecture-wide cooperation with other regions (Hirono & Okada 2021: 119). Equitable Indigenous enterprises and sustainable business practices not only ensure an enhanced economic benefit, but also contribute to protecting cultural and natural resources and intellectual property, fostering community development and improving individual livelihoods.

In order to eliminate discriminations and macroaggressions within tourism practices, current and future stakeholders of Ainu tourism, and even tourists, need to consider once again the fundamental background of the current national Ainu policy. Above all, there must be established an effective mechanism through which the Ainu and *Wajin* can participate in tourism promotion on a mutually beneficial basis.

Discussion

Building on how culturally sensitive tourism is illustrated in Figure 5.2 and incorporating key findings from the empirical cases, we now present ongoing challenges as well as good initiatives for the future improvement in Ainu tourism practice. On the left-hand side of the figure are practices that are still perceived as culturally insensitive, and on the right-hand side are issues that are presented in a culturally sensitive manner.

Appropriation – Indigenous Rights

Indigenous movements are often marred by protracted struggles to seek rights to land, natural resources, hunting and fishing, and language confiscated by the state, and this is true in Hokkaido. The question about Indigenous rights has been slowly gaining attention from the public through past and ongoing lawsuits over the right to enjoy culture, repatriation of ancestral remains, and traditional fishing rights, even though the Ainu Policy Promotion Act does not mention any Indigenous rights. Japan's Ainu policy promotion also does not strongly facilitate measures to protect and promote any of their traditional rights, such as intellectual property in tourism practices. However, as noted earlier, recent tourism developments, including *Upopoy* and Lake Akan Ainu *Kotan*, actively recognise Ainu's intellectual property rights. More discourse on this matter is needed in order to raise awareness of respect for the Ainu people and culture and to establish mechanisms for Ainu communities to be involved and for Ainu-based tourism management.

Assimilation – Non-Indigenous Epistemologies Representing "Truth"

Management of national museums is determined based on academic credentials, and with few Ainu scholars yet reaching high academic positions, all higher management positions are held by *Wajin* (Kitahara 2020: 68). This leads to a dualism in exhibition texts and also in division of roles across *Upopoy*. The written texts accompanying exhibits occasionally use a "neutral" third-person voice, describing "the Ainu", "they", and "them". This lends itself to promoting a non-attached epistemology that gives credence to *Wajin* interpretations of matters, which suits the overtly stated aim where *Upopoy* "contributes to the creation and development of a new Ainu culture". Tourism stakeholders also routinely need to pay attention to this point since tourism industries have the same structure: operating with decision-making led by *Wajin*.

Stereotyping – Ainu Homogenised into One Unity

The Ainu from different regions have distinct traditions, cultures, and dialects; however, in previous cultural representations in tourism and museums, they are generally referred to in a homogenising manner as one. There are naturally benefits to this for *Wajin* visitors as it reduces complexity by creating an image of "one people". The homogenised image also incorporates voluntary and involuntary movements or groups of people, such as modern-day intermarriage between Ainu from different regions or between people of Ainu and *Wajin* descent, and the large-scale evacuation of Japanese nationals, including the Ainu people, from Sakhalin and the Northern Territories that were occupied by the Soviet Union after World War 2, who later were resettled in different parts of Hokkaido. At the same time, some visitors are naturally those who self-identify as Ainu, with their identities varying between conscious and unconscious, overt and covert levels. Professor *Mokottunas* Kitahara, a cultural anthropologist of Sakhalin Ainu descent who supervised Hokkaido Tourism Organization's guidebook (Hokkaido Tourism Organization 2019), indicated, based on his experience, that stereotyping of the Ainu people and culture has excluded interested parties who are not represented there (Kitahara & Okada 2022: 64). In order to improve Ainu tourism to ensure mutual respect with diversity and complexity in identity, tourism stakeholders should pay attention to the multiple layers within the Ainu people and culture as stated within the reflective checklist of the guidebook "Textbook of Ainu culture for the Guide" developed for the tourism sector inf 2019 (Hokkaido Tourism Organization 2019). The challenge in incorporating all Ainu peoples into one entity is that it minimises a layer of everyday realities, including regional differences and rivalries.

Respect – First Language Is Ainu

Through the earlier *Irankarapte* campaign, and now at *Upopoy*, it is positive that signage and presentation greetings are consistently promoted in the Ainu language first. Even if only a minority of visitors speak or understand the language, two contributions are pointed out in terms of helping the revival of a language in danger. One is for those wanting to learn and use the Ainu language, and another is for the general public to appreciate that the Ainu language is a living heritage in daily use, not just a historical remnant (Kitahara 2020: 67). This respect is a tangible reminder of values in play, especially as the official teaching of the Ainu language in schools is still mostly non-existent and the era when children from Ainu families were forbidden to speak their own language on school grounds was not so long ago.

A large number of *Yaunmosir*/Hokkaido geographic names have an Ainu origin; for example, the capital city Sapporo is derived from "Sat-Poro-Pet" referring to a "dry/large/river" (Yamada 1984: 17). However, these names are mostly a palimpsest for residents and visitors alike, seldom reflected upon. It is therefore positive that the Donan Bus and JR, the national railway company servicing Shiraoi, have introduced the Ainu language announcement when approaching the station (together with Japanese). If a similar bilingual announcement could be done at each station, then the language would become a more natural feature of the whole region.

Recognition – Narratives Written in First Person

We commented above that a non-Indigenous "truth" is generally emphasised; there is however a pleasant exception to this found in the descriptive narratives in the permanent exhibition of the National Ainu Museum. These texts are written in the first person by a named person, while identifying which regional dialect each is told in. The narratives are thereafter translated verbatim, giving the text an authentic "insider" character, as if one was being told the account by the person directly. The following quote is an example of such an exhibition text authored by Sanae Ogawa, originally in her Shizunai–Mitsuishi dialect:

> We Ainu have inherited many time-honoured traditions including clothing, necklaces, and earrings; at the same time, we have lost other aspects of our ancestral culture due to government assimilation policies, a source of deep disappointment and regret. Going forward, all of us, not just the Ainu, must work together to avoid repeating the mistakes of the past.

Narratives on history, culture, and environment told by the Ainu are also being initiated through cultural/eco-tours guided by themselves. In Lake Akan Ainu *Kotan*, the Akan Ainu Craft Coop began to run cultural tour programmes guided by the local Ainu. These programmes include a walking tour in the forest and by the lake while listening to the guides' personal narratives, views of history, and self-reflection on tradition and modernity. Cultural tours guided or operated by the Ainu are occurring in Biratori and Sapporo, and this is expected to be an opportunity for the Ainu to bring narratives of their own culture and way of life through tourism.

Reciprocity – New Knowledge and Practice for Interpersonal Relationships

Reciprocity, the giving and receiving of favours, objects, and practices, is always a challenging part of commercial hospitality. Ainu tourism is not always intended to be from the Ainu as hosts towards the non-Ainu as guests. Ainu tourism could be one of the first opportunities to learn about the Ainu people and culture, not only for the majority population, but also for youth of Ainu ethnicity, since no regular school curriculum regarding the Ainu has been developed. The self-reflection that the personal narratives included in Ainu tourism provide can, in this sense, act as reciprocal instruments, not as tangible objects, enabling mutual respect for the other and for the self.

In this regard, Ainu tourism could be part of culturally sensitive tourism contributing to a cultural exchange for the Ainu with multiple realities in self-identification, and Ainu tourism should be transformed from stereotypes, prejudice, and even fantasy (Kitahara & Okada 2022: 48–50). However, as Kitahara points out, Ainu tourism in the past "was aimed at showcasing Ainu culture to *Wajin*, and therefore tended to prioritise the interests of *Wajin* tourists" (2021: 26). Although there are several Ainu-initiated tourism projects, still more is required to expand efforts to establish mechanisms supporting Indigenous governance over tourism. The key to improving Ainu tourism with more reciprocity is to ensure not only an equitable partnership between the Ainu and the tourism industries but also an interpersonal relationship between hosts and guests by all related tourism experts (Viken et al. 2021: 8).

Conclusion

This chapter has followed a historical timeline, investigating the interconnections between the Indigenous peoples of *Yaunmosir*/Hokkaido and tourism. The relationship has never been based on an equal footing, and too much of the current understanding about the Ainu is still based upon stereotypes from the past. These stereotypes are not only damaging to the broader understanding of the people who lived and thrived in the region long before *Wajin* started calling it their home, but also represent a lost opportunity in mutually learning from and respecting one another's lived knowledge. Tourism has in the past been exploitative by essentialising the Ainu as backwards, and traditions have not been seen as cultural expressions and rights, but rather as spectacles to be performed for audiences. Additionally, intangible heritage features of the Ainu, such as designs, folklore, and even the language have been, and are still, appropriated by *Wajin* without learning, or consulting with any representatives of the Ainu, about culturally appropriate use.

We placed the challenges of the past and the present in relation to promising developments such as initiatives to protect the intellectual property of the Ainu people by the Akan Ainu *Consulun* and Hokkaido Tourism Organization's guidebook for Ainu tourism. We also argued for utilising ARC-TISEN's culturally sensitive tourism as a guiding principle for how future Ainu-related tourism developments could be appropriately designed and implemented, and we applied these principles to six examples from current Ainu tourism initiatives.

Culturally sensitive tourism focuses firstly on grounding all initiatives in the local community's wishes and properly listening to how they would like to advance tourism from their own perspectives. Hearing the local voice would be possible by active self-reflection, summed up in the three *R*s of culturally sensitive tourism, *recognition, respect*, and *reciprocity*. Our next steps are to introduce, translate, and adapt internationally available material, such as the *Larrakia Declaration*, the PATA Indigenous tourism human rights checklists, and open courses created by the ARCTISEN project to communities and municipalities in *Yaunmosir*/Hokkaido.

References

Advisory Council for Future Ainu Policy (2009). *Final Report*. Tokyo: Comprehensive Ainu Policy Office, Cabinet Secretariat, Government of Japan. https://www.kantei.go.jp/jp/singi/ainu/dai10/siryou1_en.pdf

Ainu Museum (2014). *Ainu minzoku hakubutsukan kaikan 30 shūnen kinen shi* (*The Ainu Museum in the 30th Anniversary*), Shiraoi: The Ainu Museum. [in Japanese]

Ainu Policy Promotion Headquarters, Cabinet Secretariat "Symbolic Space for Ethnic Harmony." https://www.kantei.go.jp/jp/singi/ainusuishin/index_e.html 1 (last accessed on September 22, 2021; in Japanese)

Arai, G. (1984). *Ainu no sakebi* (*Exclaim of the Ainu*), Sapporo: Hokkaido Publishing Project Centre.Ltd. [in Japanese]

Bennett, M. J. (1986). "A Developmental Approach to Training for Intercultural Sensitivity," *International Journal of Intercultural Relations* 10(2), 179–196.

Benson, M. B. (2006). "Indian Givers: Reterritorializing the South in Contemporary Native American Literature," *The Mississippi Quarterly* 60(1), 101–127.

Brattland, C., Jaeger, K., Olsen, K. O., Dunfjell Oskal, E. M., & Viken, A. (2020). *Cultural Sensitivity and Tourism - Report from Northern Norway*. (Publications of the Multidimensional Tourism Institute, Issue. L. Tourism. http://urn.fi/URN:ISBN:978-952-337-207-8)

Edelheim, J. (2015). *Tourist Attractions - From Object to Narrative*, Bristol: Channel View Publications.

Fortin, K. E., Hurst, C. E., & Grimwood, B. S. R. (2021). "Land, Settler Identity, and Tourism Memories," *Annals of Tourism Research* 91(103299), 1–11. https://doi.org/10.1016/j.annals.2021.103299

Foundation for Ainu Culture (n.d.). *Aboutu Upopoy: Missions.* https://ainu-upopoy.jp/en/about/

Hirono, H., & Okada, M. (2021). "*Kankō wo tsūjite Ainu bunka no keishō wo mezasu: Akanko Ainu kotan no chosen* (Inheritance of Ainu Culture through Tourism: Initiatives by the Akan-lake Ainu Village)," in *Challenges Confronting the Field of Tourism (proceeding for 2020 CATS extension course)*, edited by Okada, M., et al., 107–140. https://eprints.lib.hokudai.ac.jp/dspace/handle/2115/81460 [in Japanese]

Hokkaido Shimbun (2020). *Upsurge in Online Discrimination – Upopoy*. Article from Hokkaido Shimbun Morning Edition of September 2, 2020. [in Japanese]

Hokkaido Tourism Organization (2019). *Ainu bunka gaido kyōhon* (*Textbook of Ainu Culture for the Guide*). https://ainu-guide.visit-hokkaido.jp/guide [in Japanese]

Japan Cabinet Office (2019). FY2019 List of Subsidised Projects to Promote Ainu Policy Measures. https://www8.cao.go.jp/ainu/kouhyou/jigyou_keikaku/reiwa1/reiwa1.html (accessed on September 22, 2021; in Japanese)

Kindaichi, K. (1941). *Ainu Life and Legends*, Tourist Library No.36, Tokyo, Board of Tourist Industry, Japanese Government Railways. (Kōno, Series 1(7)).

Kitahara, M. J. (2020). "*Ainu/ Wajin no tachiba kara kangaeru kyōdō: kokuritsu Ainu minzoku hakubutsukan no torikumi* (Collaboration from the Viewpoint of Ainu and Wajin: Initiatives at the National Ainu Museum)," *For New Historical Science* 296, 66–69. [in Japanese]

Kitahara, M. J. (2021). "*Rekishiteki torauma gainen no Ainu kenkyū heno dōnyū wo saguru* (Examining the Introduction of Historical Trauma Concept into the Ainu Study)," *Journal of Ainu and Indigenous Studies* 1, 7–34. https://eprints.lib.hokudai.ac.jp/dspace/bitstream/2115/80884/4/02_Examining.pdf [in Japanese]

Kitahara, M. J., & Okada, M. (2022). "Ainu minzoku to Bunka koyū to shiteno kankō no shintenkai ni mukete (Seeking New Phase for Ainu People and Tourism as Cultural Exchange)," in *The Proceeding of Tourism Creation Forum 2021*, edited by Yamamura, T., 39–66.

Mika, J. P., & Scheyvens, R. (2022). "Te Awa Tupua: Peace, Justice and Sustainability through Indigenous Tourism," *Journal of Sustainable Tourism* 30(2–3), 637–657. https://doi.org/10.1080/09669582.2021.1912056

Morris-Suzuki, T. (2014). "Tourist, Anthropologists, and Visions of Indigenous Society in Japan," in *Beyond Ainu Studies: Changing Academic and Public Perspectives*, edited by Hudson, M. J., Lewallen, A., and Watson, M. K., Honolulu: University of Hawai'i Press, 45–68.

Muraki, M. (2010). "*Hakubutsukan katsudo to kanko: Ainu minzoku hakubutsukan no jirei kara* (Museum Activities and Tourism: A Case Study of Ainu Museum)," in *Gendai shakai to senjūmin bunka: kanko, geijutsu kara kangaeru (Contemporary Society and Indigenous Culture: Consideration through the Tourism and Art) 1*, edited by Executive Committee of Symposium on Northern Peoples, Hokkaido Museum of Northern People, 25–30. [in Japanese]. Abashiri: Hokkaido Museum of Northern People.

National Ainu Museum (2020). *National Ainu Museum Guidebook*, Shiraoi: National Ainu Museum.

National Ainu Museum (n.d.). *Frequently Asked Questions: Basic Knowledge of Ainu History and Culture.* https://nam.go.jp/inquiry/ (last accessed on September 22, 2021; in Japanese)

Okada, M. (2021). "Prospects and Challenges for Future Development of Ainu Community Based Tourism," in *Proceedings for the 14th Research Meeting on Taiwan's Aboriginal and Japan's Indigenous Peoples*, edited by Centre for Aboriginal Studies, National Chengchi University, 101–104.

Olsen, K. O., Abildgaard, M. S., Brattland, C., Chimirri, D., de Bernardi, C., Edmonds, J., Grimwood, B. S. R., Hurst, C. E., Höckert, E., Jaeger, K., Kugapi, O. R., Lemelin, H., Lüthje, M., Mazzullo, N., Müller, D. K., Ren, C., Saari, R., Ugwuegbula, L., & Viken, A. (2019). *Looking at Arctic Tourism through the Lens of Cultural Sensitivity. ARCTISEN – A Transnational Baseline Report* (Publications of the Multidimensional Tourism Institute, Issue. L. Tourism). https://lauda.ulapland.fi/handle/10024/64069

Ōtsuka, K. (1996). "*Ainu ni okeru kankō no yakuwari: dōka seisaku to kankō seisaku no soukoku* (Role of Tourism for the Ainu: Conflict between Assimilation and Tourism Policies)," in *Kanko no 20 seiki (the 20th Century for Tourism)*, edited by Ishimori, S., Tokyo: Domesu Publishing Inc., 101–122.

PATA (2015). *Indigenous Tourism & Human Rights in Asia & the Pacific Region - Review, Analysis, & Checklists* (PATA Sustainable Practices, Issue. PATA). https://www.humanrights-in-tourism.net/publication/indigenous-tourism-human-rights-asia-pacific-region Siddle, R. M. (1997). *Race, Resistance and the Ainu of Japan*, London: Routledge.

Tsunemoto, T. (n.d.). *Overview of the Ainu Policy Promotion Act of 2019.* https://fpcj.jp/wp/wp-content/uploads/2019/11/b8102b519c7b7c4a4e129763f23ed690.pdf

Viken, A., Höckert, E., & Grimwood, B. S. R. (2021). "Cultural Sensitivity: Engaging Difference in Tourism," *Annals of Tourism Research* 89(103223). https://doi.org/10.1016/j.annals.2021.103223

Yamada, H. (1984). *Geographical Name of Hokkaido: Study on Ainu Geographical Name*, Urayasu: Sōfū sha.

SECTION 2

Indigenous

Over recent decades, Indigenous voices have been much louder in expressing their frustration with undesired and inappropriate tourism development and imagery and the longstanding problem of lack of control over the development and resulting impacts of tourism in their communities. While in the Australian context, the "Welcome to Country" theme has been widely promoted, as Higgins-Desboilles points out, this is often false and forced, and occasionally a truer if less positive emotion is portrayed. It is clear that Indigenous voices need to be expressed, but more importantly, to be heard and listened to, in determining how Indigenous tourism is to be presented to visitors, who should do the presenting, what the content should be, and the rate, scale and locations at which development takes place. In drawing up Codes of Conduct based on local Tibetan opinions, Yang et al. noted local voices arguing for actions such as restricting access to spiritual sites despite external agencies suggesting tourists should visit such locations, as well as clarifying appropriate behaviours of tourists in Indigenous areas in general. Such sentiments were also noted by Curtin et al. in emphasising the ability of learning and knowledge sharing to transform opinions, an issue not limited to Australia in its application. Holder also drew attention to the same problem by pointing out aversions and other negative feelings towards Indigenous tourism by potential visitors.

Utilising traditional Indigenous skills in producing artifacts for tourist consumption can be a sensitive issue and create problems of ensuring authenticity and avoiding cultural appropriation. Boonaboona et al. discuss the role of engaging with tourists at the same time as gaining empowerment, particularly for women and other marginalised community members. Ecotourism development and sales of locally produced goods have assisted in raising the profile of some of the marginalised elements in the communities by enabling them to express themselves in actions and production. Transforming the image and improving the knowledge of non-Indigenous visitors to Indigenous communities through tourism was a theme shared by both Butler and Wilkinson and Jenning in their chapters. They discussed how interpretation and explanation by Indigenous voices can aid reconciliation and respect, key issues to many Indigenous communities, particularly those who have suffered by deliberate actions to eliminate Indigenous languages, beliefs and customs. Expressing clear Indigenous voices is of great importance in the context of reclamation and reconciliation and the need to have accurate portrayals of both

DOI: 10.4324/9781003230335-7

past and present life in Indigenous communities is an issue noted in many of the chapters, along with concern and frustration at presentations by non-Indigenous guides and others. Correcting mistakes and misinterpretations, along with issues such as loss of Indigenous names matter greatly to affected communities. In almost all cases where Indigenous voices have been raised, strong and continuous efforts have been needed to achieve even slight measures of success, suggesting the issue is still far from being resolved despite limited progress having been made in many situations.

6

DECOLONISING INDIGENOUS TOURISM

Reconciliation, Truth-telling, Whiteness and "Welcome to Country" in Australia

Freya Higgins-Desbiolles

Introduction

Whites, it must frankly be said, are not putting in a similar mass effort to re-educate themselves out of their racial ignorance. It is an aspect of a sense of superiority that the white people of America believe they have so little to learn.

- Martin Luther King Jr, 1967

In the aftermath of the COVID-19 pandemic, some tourism analysts anticipated positive transformations in tourism (see Lew 2020). Writing this chapter in 2022, many are looking forward to securing "a new normal" once the worst of this extraordinary global pandemic crisis passes. But from an Indigenous perspective, the pandemic events of 2020–2021 may seem less extraordinary in the light of Indigenous experiences of settler colonialism for more than 250 years bringing invasion, massacres, deaths, removals, controls, losses of many kinds and disease introduction (Allam & Evershed 2022; Mawani 2020). Specifically, many Indigenous Australians have long looked for their "new normal" when their sovereignty might be fully recognised and agreed reparations and restitutions will be made so that they can get on with creating their self-determined futures free of settler colonial violence, disruption and crisis. At this moment of hoping to move beyond this crisis to our new normal, are we able to lift the veil of cultural bias and ignorance to recognise the expressions of Whiteness in our tourism domain? If we could do so, we might recognise our responsibility to support Indigenous pathways to a "new normal" that necessitates a dismantling of ongoing colonising practices, including within tourism, but also more widely (see also Whyte 2020).

There has been promising work on decolonisation in tourism, exploring how we might reconfigure the power equations in tourism and tourism scholarship in the aftermath of settler colonialism and imperialism (e.g. Aquino 2019; Chambers & Buzinde 2015; Wijesinghe 2020). However, as Mowatt (2022) noted, such works too seldom engage with ongoing imperialism and therefore fail to explain why injustices and harms continue in tourism, leisure and recreation. The outcome is that we focus too early on the seemingly positive process of decolonising before we have adequately accounted for the historical and ongoing rapaciousness of imperialism (Higgins-Desbiolles 2022), conducted adequate truth-telling, and estimated and enacted

63

DOI: 10.4324/9781003230335-8

the reparations that are needed to reset relationships (Mowatt 2022). Tourism development in a context of ongoing violence and structural injustice brings exploitation, appropriation, dispossession, commodification and all sorts of injustices (Devine & Ojeda 2017). This critical insight that ongoing structural injustices limit the possibilities for tourism to fulfil its promise is a very important opportunity for our understandings of the possibilities for Indigenous tourism to offer communities, such as those comprising Indigenous Australia, positive, sovereign and self-determining futures as well as reconciliation between Indigenous Australians and settler non-Indigenous Australians.

In this chapter, a critical reading of "Welcome to Country" protocols will be offered through a conceptual, interpretative analysis of a performance piece by Ngarrindjeri/Kaurna/Italian poet Dominic Guerrera performed at Adelaide Writers Week 2021. This performance offered a discomforting moment of truth-telling. The Uluru Statement from the Heart (2017) called for such "truth-telling" about our shared history as a vital part in the path to forming a way of living together in what is now called Australia. This chapter receives this truth-telling with the goal of understanding what light it might shine on the conduct of Indigenous tourism in Australia, including its injustices, inconsistencies and impediments. We find that the structural contexts of injustice limit the possibilities of reconciliation through tourism. Additionally, the continuance of settler colonialism limits the capacities of tourism to support truth-telling, restitution, reparations and agreement-making, which are essential for the pursuit of a decolonising agenda in tourism. Following the advice of Denzin, Lincoln and Smith (2008), this work prioritises Indigenous voices and contributes to the developing dialogue between critical and Indigenous qualitative research and researchers. As the epigraph opening this chapter attests, however, a critical dilemma is whether the "whitestream" majority will be open to hearing, listening and engaging with this truth-telling so that a collaborative practice of rejecting and dismantling colonising agendas can begin.

Indigenous Tourism and Reconciliation

According to Tourism Australia's data, Indigenous tourism is a popular offering. Tourism Australia claimed that in 2019, 1.4 million international tourists (17%) booked an Indigenous tourism experience, a 6% increase each year since 2010; and 1 million domestic tourists engaged in Indigenous tourism in Australia (an increase of 13% each year since 2013) (Tourism Australia: n.d.a). "Australia's Indigenous culture, country, art and history are points of differentiation in today's competitive international tourism market…", and the data show it is particularly international tourists and the "high end" of the tourism market that are most keen on Indigenous Australian tourism experiences (National Indigenous Australians Agency 2019: n.p.).

Higgins-Desbiolles argued: "One positive outcome that Indigenous tourism can offer is opportunities to foster reconciliation. Aboriginal and Torres Strait Islander peoples use tourism to bridge the cultural divides and create better futures by sharing culture, knowledge, and country" (2017: n.p.). Some analysts have supported this idea that Indigenous tourism can foster reconciliation in Australia and have explored the possibilities from the vantage points of tourists, Indigenous tourism operators and "host" communities (Curtin & Bird 2022; Galliford 2010; Higgins-Desbiolles 2003). Recent work by Curtin and Bird (2022) has provided rich insights into the motivations and practices of Indigenous Australian tourism operators who use the tourism opportunity to foster reconciliation with non-Indigenous Australians.

Reconciliation in Australia has been an ongoing process that officially started with a decade-long set of consultations led by the Council for Aboriginal Reconciliation culminating in the celebration through Corroboree 2000. During this time, the former Chair of the Aboriginal and Torres Strait Islander Commission Geoff Clark provided one articulation of "reconciliation" in the Australian context:

> The future of Australia is meshed with the future for the First Peoples. We look back, to find a better way forward. Reconciliation is people being different but finding solutions together. It is about Healing, Justice and Truth. For the future, Australia's heritage must embrace all its peoples and cultures.
>
> (CAR 2000: n.p.)

Since 2001, reconciliation has been supported by the organisation Reconciliation Australia, whose work has tended to focus on education and supporting the development of Reconciliation Action Plans (RAPs) in the business, education and non-governmental organisational sectors (Reconciliation n.d.). However, this has occurred in an environment in which the Commonwealth Government of Australia has consistently worked to control the agenda and push "practical" reconciliation rather than meet the demands of Indigenous Australian leadership for treaty, truth-telling and reconciliation on their terms (see Altman 2016). It has even been described as "coercive reconciliation", in recognition of the fact that the Howard government's Northern Territory intervention in 2007 "for many, … was the final step in implementing a long-held government agenda aimed at re-establishing control over Aboriginal lands" (Marcus 2009: 194; see also Altman & Hinkson 2007). Clearly, reconciliation in Australia is a contested and unresolved space.

In fact, while settler colonial countries such as Australia and Canada have embarked on "reconciliation" processes with great fanfare, these have been criticised as asymmetrical and imposed processes and thereby ineffective from Indigenous perspectives. Corntassel, Chaw-win-is and T'lakwadzi explained how efforts are designed to limit claims made about injustices through historicising the concerns:

> Such a convenient framing of the issue allows political leaders and settler populations to deal with residual guilt on their own terms, which often follows all too familiar scripts of "forgiving and forgetting," "moving on from the past," and "unifying as a country," all the while brushing aside any deeper discussions of restitution or justice. Reconciliation becomes a way for the dominant culture to reinscribe the status quo rather than to make amends for previous injustices.
>
> (2009: 144)

There has been greater interest in issues of truth-telling as one vital step in the process towards reconciliation. Maori scholar Linda Smith explained that Indigenous communities continuously emphasise this need to address truths in settler colonial histories: "'The talk' about the colonial past is embedded in our political discourses, our humour, poetry, music, storytelling, and other common-sense ways of passing on both a narrative of history and an attitude about history" (1999: 19). This truth-telling addresses the historical injustices and subsequent consequences of violent settler colonialism. In his analysis of using study tours to indigenise the curriculum, Jakobi offered the category of "truth-telling dark tours" (2019). He used this term to describe his experience of leading largely "whitestream" students and colleagues on fieldtrips to learn from the "pain narratives" of Indigenous Australians as part of institutional commitments to indigenise university curricula and tell the real history of settler colonial Australia. But situated in the managerial university and subject to unequal power configurations, Jakobi's analysis:

…explores whitestream institutional expectations of the Aboriginal teacher educator tour guide as, increasingly, the pre-paid "student experience" dictates the curriculum and pedagogical destinations of these tours and poor "trip reviews" may trigger surveillance. In this context the only purpose and function of the Aboriginal academic tour guide is to produce pain narratives that authenticate social justice programs that work to include but enclose Aboriginal difference and exert demands upon the tour guide to be less provocative, and more performative.

(2019: 103)

Jakobi's analysis reminds us that the receivers of Indigenous truth-telling may not be open to the message as it implicates their very existential legitimacy for their being on and benefiting from stolen Indigenous lands. It is also important to underscore these educational tours comprise a significant, if unmeasured, facet of the Indigenous Australian tourism market (the Northern Territory of Australia offers a rare exception in having a campaign to foster this promising niche; see Tourism NT 2014). As Curtin and Bird found in their in-depth interviews with Western Australian Indigenous tourism operators, they identified themselves as at the forefront of reconciliation – "we are reconciliators" – and their role as educators is a key facet: "This relationship between hosts as teachers and visitors as learners creates a space where truth telling leads to understanding: 'making people understand is how you do it'" (2022: 472).

It is clear that facilitating reconciliation through Indigenous tourism requires considered critical thought. There is a clear disjuncture between the non-Indigenous tourists who decide to participate in an Indigenous tourism experience and those Indigenous leaders guiding such tours who recount confronting truths of settler colonial histories and the subsequent, ongoing damages. The development of "Welcome to Country" and "Acknowledgement of Country" protocols has been greeted as a positive indication of the evolution of settler colonial mindsets and one step on the path towards reconciliation.

Welcome to Country

In recent years in Australia, contemporary practices of "Welcome to Country" and "Acknowledgement of Country" have been developed and practised around the country (see Kowal 2015: 174). Performer and Yamatji man Ernie Dingo together with musician and Noongar/Yamatji man Richard Walley claims that they were the first to perform Welcome to Country for non-Indigenous Australians in 1976 (Tan 2016). However, welcome ceremonies and protocols between Indigenous Australian nations go back millennia.

As Indigenous scholars Peters and Lambert explained:

A formal Welcome to Country is only performed by a member (usually an Elder) of the Traditional Owners of the land on which it is performed… An Acknowledgment of Country is, as Emma Kowal asserts, a "twin ritual" to the Welcome to Country (Kowal, 2015). Acknowledgments are performed by non-Indigenous people, or Indigenous people who are not Traditional Owners of the land on which it is performed.

(2022: 31–32)

These protocols for opening events express a recognition of Indigenous Australia's millennia of custodianship of the Country, marking their prior habitation before settler colonial invasion. The typical words a settler might make in an Acknowledgement of Country are: "I begin today by acknowledging the Traditional Custodians of the land on which we <gather/meet> today, and pay my respects to their Elders past and present. I extend that respect to Aboriginal and Torres Strait Islander peoples here today" (Indigenous.gov.au n.d).

There has been some high-profile criticism of these developments. For example, former Prime Minister Tony Abbott (when he was Leader of the Opposition in 2010) criticised politicians for acknowledging traditional owners at official events, labelling it "tokenistic" and politically correct (Rodgers 2010). Non-Indigenous newspaper columnist Andrew Bolt declared them "divisive" and "racist" and claimed that they "make intruders of non-Aboriginal people" even those that have been born in Australia and whose roots may go back generations (Bolt 2010). Peters and Lambert acknowledged these critical views but explained that they themselves instead champion the "... ability [of Welcomes and Acknowledgements] to connect peoples, cultures and histories in contemporary settler-colonial settings. Emerging from the pre-colonial traditions of connecting with spirts and land, these ceremonies highlight the human place in this context" (2022: 33). This is a noticeable disjuncture in views, with critics refusing to listen to the intended purposes of these protocols and wanting to obscure issues of settler colonialism through an imposed assimilation in Australia's one (and only) nation.

Australia's tourism authorities have accepted the value of these protocols and recognise Indigenous Australian culture as a marketable asset with its longest enduring human custodianship. Tourism Australia has developed a "Welcome to Country Guide" (n.d.b). While it explains what are "Welcome to Country" and "Acknowledgement of Country"; ways to arrange these; and the need to remunerate Elders for performing such Welcomes, it is silent on the need for more active, meaningful, ongoing and impactful relationships that could be catalysed from the observance of these protocols. In July 2021, Tourism Australia released a campaign called "Welcome to my Country" to coincide with the annual National Aborigines' and Islanders' Day Observance Committee (NAIDOC) celebrations. This was described as:

> More than a film, it is a personal welcome to step onto country, meet its ancestors and become part of their family and their dreaming…By engaging with respected Elders in their country and by choosing their voices to guide viewers, for the first time we are welcomed to their country to meet their ancestors.
>
> (Murphy 2021)

But despite Tourism Australia's recent embrace of Welcomes and Indigenous Australian Elders, it is not as unproblematic as it would seem. Welcomes and Acknowledgements may be viewed as acts of "socialising tourism", as explained by Higgins-Desbiolles, Doering and Bigby (2022) who provide a number of definitions and ideas of "socialising tourism", including that it "…means engaging with the difficulties of the times and finding ways to fit tourism in the societies and ecologies in which it is occurring" (2022: 17). But they also note the very problematic and contested grounds of tourism in settler colonial contexts such as Australia:

> Using a socialisation lens, we might understand settler-colonialism, colonialism and other forms of dispossession of First Nations peoples as the most violent and damaging form of violation of host-guest protocols. As a result, the modern meeting grounds between

Indigenous hosts and non-Indigenous tourists too often symbolises a good deal of what is wrong in marketised forms of tourism.

(2022: 4)

Such a comment provides an insight that settler colonial injustices that remain unaddressed in countries such as Australia lay precarious foundations for Indigenous tourism.

Everingham, Peters and Higgins-Desbiolles (2021: 2) discussed the "coloniality of power embedded in tourism development" at Uluru, Central Australia. In their analysis of the dynamics of the closing of the climbing of Uluru, they drew attention to tourism as an expression of dominance and an attempt to suppress the rights and narrative of the Anangu custodians by constructing Uluru as "Ayers Rock" and declaring that it belongs to all Australians:

It is these constructions, embedded within the colonial mapmaking and conquest of Australia, that underpin much of the tourism consumption at Uluru and consequently, the tensions that arose in the Australian public in relation to the closure of the climb.

(Everingham, Peters & Higgins-Desbiolles 2021: 2)

Arguably, understanding these dynamics is aided with an exploration of Whiteness and White privilege in such settler colonial contexts (Figure 6.1).

Figure 6.1 A Black Lives Matter protest held in Tarntanyangga (Victoria Square), Adelaide, in 2021, during the COVID-19 pandemic era. Deaths in custody were a key concern of this gathering.

Whiteness

While race is well recognised as a socially constructed phenomenon and therefore designations such as "Black" and "White" have no inherent meaning, race thinking has serious material impacts in creating racialised Others that are acted upon in settler colonial and other societies. Benjamin and Dillette argued: "Although there is much attention focused around whiteness within leisure studies…, there is a dearth within tourism studies - acknowledging whiteness within tourism scholarship is a distinctive epistemological standpoint" (2021: 3). Whiteness refers to "a set of locations that are historically, socially, politically, and culturally produced and moreover, are intrinsically linked with unfolding relations of dominance" (Frankenburg 1997: 6). In their analysis of Indigenous Australian enterprise development, Banerjee and Tedmanson argued:

> There is … we argue, an overarching discourse of "whiteness" - the practice of white privilege which disempowers Indigenous communities - informing the practices and politics of representation and governance. It is the political economy of whiteness in the context of Indigenous economic development….
>
> (2010: 152)

It is this assertion of dominance through Whiteness and expressions of White Supremacy which is yet to be unpacked sufficiently in studies of Indigenous tourism (however, critical reflective work on settler positionality is a topic of recent focus, see Stinson, Grimwood & Caton 2021).

Professor Aileen Moreton-Robinson (2015), a Goenpul woman, has written about "white possessive logics" and argued it is accompanied by the denial of Indigenous sovereignty in order to take Indigenous lands and all of the wealth of these lands. Houtekamer argued that Moreton-Robinson's approach "allows for the concept of 'indigenous sovereignty' instead of the concept of 'indigenous Otherness' to be central to the social construction of whiteness, which leaves significant room for an indigenous perspective…" (2016: 78).

We might ask: Is there a "white possessive logic" in Indigenous tourism? It is visible when tourism authorities claim Indigenous Australians for branding in marketing campaigns overseas, selling Australia by leveraging this opportunity for differentiation. It is also apparent when non-Indigenous tourism operators interpret and thereby profit from Indigenous Australian knowledge and cultures, a source of some complaint among Indigenous Australian tourism operators. It is evident when government authorities, consultants and trainers try to discipline Indigenous Australian tourism operators to market imperatives (Higgins-Desbiolles, Trevorrow & Sparrow 2014) and only include inbound, market-ready businesses in their distribution systems.

There is a famous speech that was spoken by Rosalie Kunoth-Monks, an Arrernte Anmatjere woman, on Australian national television in 2014. She said:

> … My language, in spite of Whiteness trying to penetrate into my brain by assimilationists, I am alive, I am here and now and I speak my language. I practice my cultural essence … Don't try and suppress me and don't call me a problem. I am NOT the problem I have never left my country, nor have I ceded any part of it.
>
> (Kunoth-Monks 2014)

Indigenous Australians are aware of Whiteness and the tendencies to dominance and speak of it often, as suggested by Smith above. Non-Indigenous Australians are made uncomfortable with such words as "Whiteness" and "whitestream", some even suggesting such words are "racist".

But if Jakobi designates his work in indigenising the curriculum through education tours as "truth-telling dark tourism" performing "pain narratives" for the "whitestream", it calls for us to listen, reflect and perhaps act. It would seem that a critical question to pose, in concerns for utilising Indigenous tourism as a conduit to truth-telling on historical and ongoing acts of settler violence and abuse occurring in Australia, is: How do we pierce through settler deafness and indifference to even make possible any notions of "reconciliation"? This chapter suggests art provocations, such as Guerrera's performance, may offer one pathway.

A "Welcome to Country" as Challenge

This chapter proposes to listen to and engage with Dominic Guerrera's (2021) performance piece entitled "Welcome to Country" presented at Adelaide Writers Week's First Words event in 2021. Guerrera describes himself as a Ngarrindjeri, Kaurna and Italian person who lives on Kaurna Yerta [land], a poet and writer. His performance also takes heed of Denzin's work on "new interpretative approaches to inquiry" in which he responded to Indigenous calls to privilege Indigenous voices and act collaboratively "committed to the principles of performance, resistance, and political integrity" (2007: 455). Through this spoken word performance, Guerrera engages in an act of "truth-telling" that conflicts with preconceived ideas of "welcome". Taking Guerrera's challenge seriously, this chapter will engage with sections of his work and reflect on how these words might ask us to dig more critically into tourism and societal dynamics in Australia and interrogate the notion that reconciliation through tourism is possible or a priority. The spoken word performance was transcribed by the author and analysed through a critical, interpretative lens.

It would have been preferable to present Guerrera's words transcribed in their entirety in order to avoid disfiguring or diminishing the work's impact. This is important as a committed act of listening to Indigenous Australians' voices in full without violent interventions for whatever purposes (including limited word counts and academic conventions). However, this was not possible due to Guerrera's publication plans for the work (pers. comm., 22 September 2021). Therefore, this analysis will present selected sections of the performance (designated in italics) and then relate these to the conduct of Indigenous tourism in Australia, a settler colonial society built on unceded, stolen lands and presenting unresolved conflicts in dire need of considered truth-telling.

> Thank you for gathering unlawfully on my land. It is often said that we as Aboriginal people live within two worlds. That is not true. We live in a state of colonisation.

This addresses the point of stolen land and how little land Indigenous Australians hold in Australia. A total of 134 million hectares of land in Australia (17%) is Indigenous-owned (Jacobsen, Howell & Read 2020: 1). How is it possible to hold a "Welcome to Country" address when in fact Indigenous Australians hold little such Country in their grasp? It has been long recognised that the lack of land ownership and particular types of land ownership structures have been major inhibiting factors in the development and thriving of Indigenous Australian tourism (Schmiechen & Boyle 2007: 63). This also names Australia as a settler colonial nation that is defined by ongoing colonising processes that establish the context of structural injustices under which current tourism operations occur.

> I was invited by a white person to come here today and welcome you to my Country. But that won't be happening because to be frank, you are not welcome. You've overstayed your self-imposed welcome and we are fed up with you and your colony. We are fed up with your

destructive and murdering ways, we have had enough. It has gone on for too long that even the land is rejecting you. If our lives matter and they do, then your presence is no longer needed.

This section implies the falseness and fragility underpinning many Welcome to Country performances. It is too frequently not a heartfelt, authentic welcome but rather a white-invited component to open a whitestream event. It also indicates how the acts of first contact and subsequent interactions in the colonising process between colonisers and Indigenous Australians violated all conventions of host–guest protocols, e.g. the guests return to their homes in a timely fashion. This wording also indicates that, even at this late date in 2021, colonisation is not accepted by Indigenous Australians and any goals of assimilation have failed and will continue to fail.

Because no matter what scenario of reconciliation or cohabitation is proposed, we are always the ones that pay the heavy costs and usually it's with our lives.

Reconciliation processes in Australia have arguably attempted to reconcile Indigenous Australia to the conquest and colonisation of the Country; this is implied in even the name of the former organisation leading the reconciliation process from 1991 to 2001, the Council for Aboriginal Reconciliation. It extends to current processes where corporations and institutions adopt RAPs that fail to address the ongoing structural injustices perpetuated against Indigenous Australians and the Country (see Lloyd 2019). This enables ongoing exploitation and abuse in the spaces of tourism, including the appropriation of native foods, racism in tourism spaces, destruction of sacred sites and ongoing removals and alienation of people from the Country. Recent research has explained how ongoing racism limits the success of Indigenous Australian tourism businesses (Ruhanen & Whitford 2018).

How about instead of country we welcome you to the poverty and the disparity … the instability that echoes through our lives because of your disruption.

This commentary addresses the often discussed "indicators of disadvantage" and the "gap" in Indigenous Australian quality of life *vis a vis* settlers. It speaks, also, to the ongoing injustices that many settlers remain ignorant of or indifferent to, including: ongoing child removals, the 2020 destruction of Juukan Gorge by the mining corporation Rio Tinto (who had a RAP), the ongoing high rates of Indigenous Australian incarceration and associated deaths in custody (McGrath 2020). There was also the unceremonious and immediate rejection of The Uluru Statement from the Heart by then Prime Minister Malcolm Turnbull in 2017, which represented an Indigenous Australian-led articulation of a pathway to *rapprochement* including a process of Makarrata, a Yolngu word describing when "two parties coming together after a struggle, heal… the divisions of the past" (Noel Pearson cited in McKay 2017). It suggests the underlying hypocrisy of official Welcomes conducted for the whitestream because colonising injustices are clearly not limited to the historical past; they are continuing and compounding relentlessly.

Acknowledge the privilege of being able to gather in public spaces uninterrupted by the colony or its forces. Because when white people drink in parklands, they call it a festival but when black people drink in parklands, they call the cops. The disruption of our gatherings is a deliberate attempt to sever our connection to land and with each other. But it is time to imagine and bring forth a future that does not centre this continent on white lies. And until then you are unwelcome.

This passage refers to particular occurrences in Guerrera's location of Adelaide, where, for example, the traditional meeting place of Tarntanyangga (Victoria Square) has seen settler authorities allow settler consumption of alcohol during festivals and events (such as the "Tasting Australia" food and wine festival) while simultaneously driving out Indigenous Australians as "problem drinkers" and public nuisances in such public spaces (see Hall 2004). But Adelaide is not the only site of such dynamics of marginalisation, exclusion and racism. For instance, Graham explained:

> An ABC [Australian Broadcasting Corporation] investigation revealed that a major hotel chain in Alice Springs was segregating guests, providing Aboriginal people from "out bush" with dirty, inferior rooms, while giving everyone else standard, clean rooms. The story barely rated in Australia, although it made big news overseas, with media giants BBC and CNN both weighing in, along with New Zealand radio and a major news service in Asia.
>
> (2019: n.p.)

Clearly, racism is a widescale if unacknowledged problem in settler colonial Australia, and tourism facilities are sites for its practice. Guerrera underscores the need for "truth-telling" in his final sentence of this performance, suggesting at the heart of settler Australia rests a volume of "white lies". More typical Welcomes to Country are in fact most often part of this web of white lies and take us further from addressing the structural injustices that are embedded in settler colonial Australia. These facts supporting the performance, understood in conjunction with Ruhanen and Whitford's (2018) research demonstrating racism as a barrier to Indigenous Australian tourism, invite our critical reflection and reconsideration of practice in efforts to use Indigenous tourism as a pathway to reconciliation in Australia.

Discussion

Relations remain unreconciled in settler colonial Australia. Uses of the tour interface by Indigenous tourism for truth-telling and artistic and cultural performances, such as Guerrera's, are acts of resistance that make concerted attempts to cut through settler deafness and indifference. Settler colonial nations and nations that have emerged from conflict may use tourism for reconciliation and conflict reduction, but insufficient change is likely unless underlying structural injustices are addressed (see Higgins-Desbiolles, Blanchard & Urbain 2022). While recent research confirms Indigenous Australian tourism operators see themselves as "reconciliators" and they use their tour interpretations and experiences for truth-telling and sharing Indigenous Australia's enduring cultures despite colonisation (Curtin & Bird 2022), it cannot be argued that non-Indigenous Australians are doing their part to understand and act to remedy the violent injustices of invasion and settlement that continue to reverberate still.

Professor Megan Davis, a Cobble Cobble woman, explained: "One of the sources of disgruntlement and frustration is how rarely the justice requirements – what does repair look like? – follow the truth-telling, and how little changes in power relations" (2021: n.p.). It is such "disgruntlement and frustration" that we hear in Guerrera's "Welcome". Davis' articulation of the problem raises important questions for the conduct of Indigenous tourism in Australia. When Indigenous Australian tourism operators conduct truth-telling of colonial history, how are settler tourists to respond? Is it with poor tour evaluations and rejection of Indigenous Australian tourism experiences, or is it with a mindset of reconciliation and openness to listening, learning and then acting? Will educational and tourism policy spaces converge to support the thriving of Indigenous Australian tourism businesses and the development of critical consciousness of young Australians so that support for just reconciliation might grow, following the example of NT Tourism (2014)?

Such work can be done from a positive base of love, integrity and responsibility. Higgins-Desbiolles, Blanchard and Urbain (2022) cite Dr. Emma Lee's work, a Trawlwulwuy woman from Tebrakunna country, referring to "love-bombing" to describe her strategy in decolonising relationships with government agencies that are interfacing with her community (e.g. Lee & Hamilton 2016). They stated:

> Lee's works suggest that using Indigenous epistemologies, values and practices can reset long-standing violent and negative relationships in settler-colonial contexts. This "love-bombing", with its paradoxical juxtapositions, describes overturning power relationships by inviting non-Indigenous others to transform their relationships to place and First Nations, to the benefit of all. Such a vision moves on from reconciliation as mutual tolerance to reconciliation as mutual thriving through relatedness and reciprocity.
>
> (Higgins-Desbiolles, Blanchard & Urbain 2022: 346)

Guerrera's artistic provocation presses uncomfortable questions on settlers in Australia. Can "Welcome to Country" be performed by Indigenous Australians when they may indeed not have rights to any of the Country in question? Can tourism bodies and tourists contribute to the repatriation of Indigenous lands to Indigenous Australians? Can tourism discourses (in interpretative materials, in all media communications, in tourism curricula) "unsettle the invasive settler logic of elimination?" (Jakobi 2019: 98). Finally, in what ways can tourism and other sectors substantially contribute to truth-telling, reparations and restitutions when clearly they continue to benefit from the proceeds of the stolen Country, enslaved peoples and appropriated resources, both tangible and intangible? Until Australians, non-Indigenous and Indigenous, come to legitimate agreements on these things, Welcomes and Acknowledgements are likely to lack the rich meaning and value they have the potential to offer and reconciliation will remain elusive.

Conclusion

Tourism has been accused of white-washing in order to avoid thorny issues of justice (see Butler 2008; Mowatt 2022). Corntassel, Chaw-win-is and T'lakwadzi (2009: 137) claimed Indigenous story-telling and truths are essential to the cultural and political resurgence of Indigenous nations. If we view Indigenous tourism as one possible pathway to resolving outstanding injustices of settler colonialism and its effects in Australia, it is essential that we move beyond superficial engagements such as token Welcomes and constrained pain narratives. This chapter has posed a number of questions, after laying out facets relevant to assessing the value of Indigenous tourism for the achievement of reconciliation in Australia, including reconciliation through tourism, the adoption of Welcomes and Acknowledgements and considerations of the impact of Whiteness. This was followed by a considered analysis of Guerrera's provocative Welcome in order to listen to and understand Indigenous Australian challenges calling for justice.

It is paradoxically possible that this chapter written by a settler may be viewed as just another case of settler exploitation as Guerrera's words have been used extensively to develop this academic argument. This indeed may be accurate, while simultaneously true that this work offers a response of deep engagement with the provocation of Guerrera's performance. For indeed, settlers such as this author continuously appropriate Indigenous resources by their very act of being on the Country, on still unacknowledged stolen lands, in the first place. We need to understand this enduring fact of settler colonisation and not recoil or refute this. The critical question to pose though is: How are we to follow acts of truth-telling with actions in solidarity that might properly constitute responses of restitution, reparations and settlement making in a place so violently colonised? And what roles might tourism play in such agendas?

This chapter has attempted to engage with these issues in consideration of Indigenous tourism as one tool we might use. As Whyte (2020) argued, we need to evolve epistemologies of coordination rather than crisis epistemologies as we struggle to build safe futures. For if Australia is to have a positive future, it must first recognise that it unjustly wrought devastating crises on the legitimate peoples of the Country and make amends for this. Historical injustices of the past loudly reverberate in the present, poisoning Australia's future. Guerrera's unwelcome presents a gift; like good guests, we should receive it with appreciation and energy to join in coordinating epistemologies that can see us unweave the colony and colonising mindset, allowing us to move on together. Indigenous tourism conducted in such a decolonised context then would become a life-enhancing activity of significant value.

Acknowledgements

This chapter is a revised version of a conference paper presented at the CAUTHE 2022 Shaping the Next Normal in Tourism, Hospitality and Events conference, convened online on 7–9 February by Griffith University, Australia.

References

Allam, L. and Evershed, N. (2022, 16 March) "Almost half the massacres of Aboriginal people were by police or other government forces, research finds", *The Guardian* (Online), retrieved 16 March 2022, from https://www.theguardian.com/australia-news/2022/mar/16/almost-half-the-massacres-of-aboriginal-people-were-by-police-research-finds

Altman, J. (2016) "Reconciliation and the quest for economic sameness". In S. Maddison, T. Clark & R. de Costa (Eds.) *The Limits of Settler Colonial Reconciliation* (pp. 213–230), Singapore: Springer, https://doi.org/10.1007/978-981-10-2654-6_13

Altman, J. and Hinkson, M. (Eds.) (2007) *Coercive Reconciliation: Stabilise, Normalise, Exit Aboriginal Australia*, Melbourne: Arena.

Aquino, R.S. (2019) "Towards decolonising tourism and hospitality research in the Philippines", *Tourism Management Perspectives* 31, 72–84, https://doi.org/10.1016/j.tmp.2019.03.014

Banerjee, S. and Tedmanson, D. (2010) "Grass burning under our feet: Indigenous enterprise development in a political economy of whiteness", *Management Learning* 41(2), 147–165, https://doi.org/10.1177/1350507609357391

Benjamin, S. and Dillette, A.K. (2021) "Black travel movement: Systemic racism informing tourism", *Annals of Tourism Research* 88, 103169, https://doi.org/10.1016/j.annals.2021.103169

Bolt, A. (2010) "I need no welcome to my own land", *Daily Telegraph* (Online), retrieved 28 August 2021, from https://www.dailytelegraph.com.au/blogs/andrew-bolt/column--i-need-no-welcome-to-my-own-land/news-story/fc4732aa3ec0c234fe64947e05137b9f

Butler, D.L. (2008) "Whitewashing plantations", *International Journal of Hospitality & Tourism Administration* 2(3–4), 163–175, https://doi.org/10.1300/J149v02n03_07

Chambers, D. and Buzinde, C. (2015) "Tourism and decolonisation: Locating research and self", *Annals of Tourism Research* 51(1), 1–16.

Corntassel, J., Chaw-win-is and T'lakwadzi (2009) "Indigenous storytelling, truthtelling, and community approaches to reconciliation", *ESC: English Studies in Canada* 35(1), https://ojs.lib.uwo.ca/index.php/esc/article/view/9788/7888

Council for Aboriginal Reconciliation (CAR) (2000) "Some reflections by Council members", retrieved 28 September 2021, from http://www.austlii.edu.au/au/other/IndigLRes/car/2000/16/appendices05.htm

Curtin, N. and Bird, S. (2022) "We are reconciliators": When Indigenous tourism begins with agency', *Journal of Sustainable Tourism* 30(2–3), 461–481, https://doi.org/10.1080/09669582.2021.1903908

Davis, M. (2021) "The truth about truth-telling", *The Monthly* (Online), retrieved from https://www.themonthly.com.au/issue/2021/december/1638277200/megan-davis/truth-about-truth-telling

Denzin, N.K. (2007) "Grounded theory and the politics of interpretation". In A. Bryant & K. Charmaz (Eds.) *The Sage Handbook of Grounded Theory* (pp. 454–471), London: Sage.

Denzin, N.K., Lincoln, Y.S. andSmith, L.T. (2008) *Handbook of Critical and Indigenous Methodologies*, London: Sage.

Devine, J. and Ojeda, D. (2017) "Violence and dispossession in tourism development: A critical geographical approach", *Journal of Sustainable Tourism* 25(5), 605–617, https://doi.org/10.1080/09669582.2017.1293401

Everingham, P., Peters, A. and Higgins-Desbiolles, F. (2021) "The (im)possibilities of doing tourism otherwise: The case of settler colonial Australia and the closure of the climb at Uluru", *Annals of Tourism Research* 88, 103178, https://doi.org/10.1016/j.annals.2021.103178

Frankenburg, R. (1997) *Displacing Whiteness*, London: Duke.

Galliford, M. (2010) "Touring 'Country', sharing 'home': Aboriginal tourism, Australian tourists and the possibilities for cultural transversality", *Tourist Studies* 10(3), 227–244, https://doi.org/10.1177/1468797611407759

Graham, C. (2019) "Clayton's apartheid: The racist things Alice Springs does when it doesn't want to appear racist", *New Matilda*, retrieved 3 September 2021, from https://newmatilda.com/2019/03/09/claytons-apartheid-racist-things-alice-springs-doesnt-want-appear-racist/

Guerrera, D. (2021) "First words", retrieved 28 August 2021, from https://vimeo.com/561736318

Hall, L.-A. (2004) "Sitting down in the square: Indigenous presence in an Australian city", *Humanities Research* 11(1), 54–77.

Higgins-Desbiolles, F. (2003) "Reconciliation tourism: Tourism healing divided societies?" *Tourism Recreation Research* 28(3), 35–44.

Higgins-Desbiolles, F. (2017) "How Indigenous tourism can help bring about reconciliation in Australia", *The Conversation* (Online), retrieved 22 September 2021, from https://theconversation.com/how-indigenous-tourism-can-help-bring-about-reconciliation-in-australia-78344

Higgins-Desbiolles, F. (2022) "The ongoingness of imperialism: The problem of tourism dependency and the promise of radical equality", *Annals of Tourism Research* 94, 103382, https://doi.org/10.1016/j.annals.2022.103382

Higgins-Desbiolles, F., Blanchard, L.-A. and Urbain, Y. (2022) "Peace through tourism: Critical reflections on the intersections between peace, justice, sustainable development and tourism", *Journal of Sustainable Tourism* 30(2–3), 335–351, https://doi.org/10.1080/09669582.2021.1952420

Higgins-Desbiolles, F., Doering, A. and Chew Bigby, B. (Eds.) (2022) *Socialising Tourism: Rethinking Tourism for Social and Ecological Justice*, London: Routledge.

Higgins-Desbiolles, F., Trevorrow, G. andSparrow, S. (2014) "The Coorong Wilderness Lodge: A case study of planning failures in Indigenous Tourism", *Tourism Management* 44, 46–57.

Houtekamer, T. (2016) "Review: The White Possessive: Property, power, and Indigenous sovereignty", *Junctions: Graduate Journal of the Humanities* 1(1), 77–79, http://doi.org/10.33391/jgjh.21

Indigenous.gov.au (n.d.) "Welcome to Country or Acknowledgement of Country", retrieved 17 March 2022, from https://www.indigenous.gov.au/contact-us/welcome_acknowledgement-country#:~:text=The%20words%20are%3A

Jacobsen, R., Howell, C. and Read, S. (2020) "Australia's Indigenous land and forest estate", retrieved 23 September 2021, from https://www.agriculture.gov.au/sites/default/files/documents/ABARES_Aus_Indig_land_forest_estate-Separate_reporting_FINAL_02Dec20.pdf

Jakobi, M. (2019) "Truth-telling the dark tourism of Australian teacher education". In D. Bottrell & C. Manathunga (Eds.) *Resisting Neoliberalism in Higher Education*, Vol. I, Cham: Palgrave Critical University Studies.

Kowal, E. (2015) "Welcome to Country: Acknowledgement, belonging and White anti-racism", *Cultural Studies Review* 21, 173–204.

Kunoth-Monks, R. (2014, 10 June) "I am not the problem", ABC Q&A (TV), retrieved 3 March 2022, from https://www.youtube.com/watch?v=birnA3_tm5E

Lee, E. and Hamilton, F. (2016) "Tasmania – After a long journey, World Heritage Area delivers Indigenous rights", retrieved 10 August 2020, from https://www.iccaconsortium.org/index.php/2016/09/25/tasmania-after-a-long-journey-world-heritage-area-delivers-indigenous-rights/

Lew, A. (2020) "Reset: Visions of travel and tourism after the global COVID-19 transformation of 2020", retrieved 3 July 2021, from https://www.tgjournal.com/transformation.html (accessed 10 May 2020).

Lloyd, C. (2019) "Managing indigeneity, cultivating citizens: Reconciliation action plans in Australian organizations", Doctoral dissertation, Harvard University, Graduate School of Arts & Sciences.

Marcus, J. (2009) "Ruptured reconciliation", *Cultural Studies Review* 15(2), 194–198.

Mawani, R. (2020) "A historical account of the pandemic: Health, colonialism and racism in Canada", retrieved 22 September 2021, from https://rsc-src.ca/en/covid-19/impact-covid-19-in-racialized-communities/historical-account-pandemic-health-colonialism

McGrath, A. (2020) "The reverberations of deep time", retrieved 22 March 2022, from https://re.anu.edu.au/the-reverberations-of-deep-time

McKay, D. (2017) "Uluru statement: A quick guide", retrieved 5 May 2022, from https://www.aph.gov.au/About_Parliament/Parliamentary_Departments/Parliamentary_Library/pubs/rp/rp1617/Quick_Guides/UluruStatement

Moreton-Robinson, A. (2015) *The White Possessive: Property, Power and Indigenous Sovereignty*, Minneapolis: University of Minnesota.

Mowatt, R. (2022) "The Dylann Roof Road Trip: A report on the banality of evil". In F. Higgins-Desbiolles, A. Doering & B.C. Bigby (Eds.) *Socialising Tourism: Rethinking Tourism for Social and Ecological Justice* (pp. 109–128), London: Routledge.

Murphy (2021) "Indigenous Elders give personal welcome to Country in Tourism Australia campaign", *AdNews*, retrieved 23 September 2021, from https://www.adnews.com.au/news/indigenous-elders-give-personal-welcome-to-country-in-tourism-australia-campaign

National Indigenous Australians Agency (2019) "Indigenous tourism fund: Discussion paper", retrieved 3 September 2021, from https://www.niaa.gov.au/sites/default/files/publications/indigenous-tourism-fund-discussion-paper.pdf

Peters, A. and Lambert, S. (2022) "'Wominjeka' / 'Haere Mai': The role of Indigenous ceremony in socialising tourism". In F. Higgins-Desbiolles, A. Doering & B.C. Bigby (Eds.) *Socialising Tourism: Rethinking Tourism for Social and Ecological Justice* (pp. 25–39), London: Routledge.

Reconciliation Australia (n.d.) "Our work", retrieved 3 February 2022, from https://www.reconciliation.org.au/our-work/

Rodgers, E. (2010, 15 March) "Aboriginal recognition a farce: Tuckey", *ABC News* (Online), retrieved 22 September 2021, from https://www.abc.net.au/news/2010-03-15/aboriginal-recognition-a-farce-tuckey/364450

Ruhanen, L. and Whitford, M. (2018) "Racism as an inhibitor to the organisational legitimacy of Indigenous tourism businesses in Australia", *Current Issues in Tourism* 21(15), 1728–1742, http://doi.org/10.1080/13683500.2016.1225698

Schmiechen, J. and Boyle, A. (2007) "Aboriginal tourism research in Australia". In R. Butler & T. Hinch (Eds.) *Tourism and Indigenous Peoples* (pp. 58–72), Amsterdam: Elsevier.

Smith, L.T. (1999) *Decolonizing Methodologies: Research and Indigenous Peoples*, New York: Zed Books.

Stinson, M.J., Grimwood, B.S.R. and Caton, K. (2021) "Becoming common plantain: Metaphor, settler responsibility, and decolonizing tourism", *Journal of Sustainable Tourism* 29(2–3), 234–252, http://doi.org/10.1080/09669582.2020.1734605

Tan, M. (2016, 23 February) "Ernie Dingo and Richard Walley on the 40th year of their welcome to country", *The Guardian* (Online), retrieved 18 March 2022, from https://www.theguardian.com/australia-news/2016/feb/23/ernie-dingo-and-richard-walley-on-the-40th-year-of-their-welcome-to-country

Tourism Australia (n.d. a) "Aboriginal and Torres Strait Islander Tourism", retrieved 22 September 2021, from https://www.tourism.australia.com/en/events-and-tools/industry-resources/resources-for-industry/aboriginal-torres-strait-islander-tourism.html

Tourism Australia (n.d. b) "Welcome to Country guide", retrieved 23 September 2021, from https://www.tourism.australia.com/en/events-and-tools/industry-resources/resources-for-industry/aboriginal-torres-strait-islander-tourism.html

Tourism NT (2014) "Australian School Education Tourism Activation Plan", retrieved 4 March 2022, from https://www.tourismnt.com.au/system/files/uploads/files/2020/australian-school-education-tourism-activation-plan.pdf

"The Uluru Statement from the Heart" (2017) Retrieved 23 September 2021, from https://ulurustatement.org/the-statement

Whyte, K. (2020) "Against crisis epistemology". In A. Moreton-Robinson, L.T. Smith, C. Andersen & S. Larkin (Eds.) *Handbook of Critical Indigenous Studies* (pp. 52–64), London: Routledge.

Wijesinghe, S.N.R. (2020) "Researching coloniality: A reflection on identity", *Annals of Tourism Research* 82, 102901, https://doi.org/10.1016/j.annals.2020.102901

7

LEARN, TEACH, HEAL

Indigenous Tourism as a Site for Reclaiming and Becoming

Helen Jennings

Introduction

Tourism is often seen as a shallow, commercial and artificial activity, yet such a view risks speaking over the various reasons why hosts choose to engage in the industry. This research foregrounds the voices and experiences of: six First Nations individuals, all engaged in tourism for their own reasons, but with a common goal, to use tourism to learn, teach and heal, both for themselves and for their guests. Healing is gained through having a space to learn, teach, and to restore pride to the communities by taking control of the narratives. Indigenous Tourism is being used by these six people as sites of 'becoming' and 'reclaiming' in ways that put decolonization into practice.

> Tourism is really important to our people as it is a form of reconciliation'..[yet]..
> our community were very very hesitant in doing tourism, but then one Elder stood up and said you know our youth are starting to lose their way, their culture and their traditions and we don't know another way to bring that back, other than tourism.

This comment comes from Sierra Hall of the Xai'xais Nation in a video entitled 'Why Tourism Matters?' produced by *Indigenous Tourism British Columbia* (ITBC, 2020). It resonates with popular local narratives that Indigenous tourism offers a form of reconciliation, whilst also highlighting the need for cultural recovery and reclamation. Indigenous Tourism is booming in British Columbia and has been partly facilitated by the political movement of Truth and Reconciliation. This chapter draws on recent ethnographic research in British Columbia to show how Indigenous-owned and operated tourist sites can serve as a resource for guides, create jobs for Indigenous people, and open new avenues for communication with each other, with guests and with national and global audiences (Jennings, 2023). These jobs are providing opportunities to learn, create and share practices – practices that might have been previously prohibited or stigmatized – which are now through tourism incentivized and facilitated. What emerged was how the guides were using tourism to 'learn, teach and heal' – both for themselves and their guests. Learning how to be guided, their languages, traditional practices, histories and politics, they were able to explore with tourists' aspects of their indigeneity, illustrate the diversity of peoples and practices, and share values and hopes for the future. To illustrate how Indigenous tourism can work as a site of 'becoming and reclaiming', this chapter will discuss various contexts, situate the research within relevant scholarship, and use six accounts to show how and why guides are using tourism to put decolonization into practice.

Context

Indigenous Tourism in British Columbia has grown rapidly: some 400 businesses have generated an estimated $705 million GDP and created about 7,400 full-time jobs (DestinationsBC, 2020). A regulatory and marketing body called *Indigenous Tourism British Columbia* aims to promote a 'sustainable and culturally rich Indigenous tourism industry'; it liaises with business entrepreneurs and government agencies to ensure 'quality experiences' and promote tourism that is in the ownership and control of Indigenous communities (ITBC, 2023). This is an important contextual point, as there are many cases around the world where Indigenous people have little or no control as to what is on offer for tourists. In British Columbia, it seems at least, that engagement in tourism is an active choice. The main tagline of the tourist site 'HelloBC' is 'Super, Natural British Columbia'. The beauty of the province contributes to the success of its tourism, as nature facilitates many of the experiences on offer, and this fits a widely held perception that Indigenous people have a deep connection to the land and nature.

The focus on Indigenous tourism in the province fits many of the 'calls to action' that came out of the work of the Truth and Reconciliation Commission (TRC) completed in 2015 (TRC, 2015). Many of those calls spoke about the importance of having a space to learn and engage with different Indigenous languages and cultures, as well as the ability to partake effectively in the economy. This narrative has been explicitly embraced by *Indigenous Tourism British Columbia (ITBC)*. The *Indigenous Tourism Association of Canada* proclaimed that 'Indigenous Tourism has the power to change perspectives, preserve culture, language and community and provide our relatives with a platform to be the leading voice in reclaiming our space in history – both ancient and modern' (ITAC, 2022).

The Truth and Reconciliation program stems from the fact that Canada is a settler, colonial state and since the 1800s state-sanctioned assimilation policies controlled who does and does not count as Indigenous, contributed to a dramatic decline in fluent Indigenous language speakers, restricted Indigenous people to reserves, imposed colonial governance on these areas, and prohibited traditional land-use practices. From the mid-19th century until almost the end of the 20th century many Indigenous children were forcibly removed from their families and placed in residential schools. These schools were often far away from their own communities, where they were forced to speak English and prohibited from speaking their mother tongues; such policies were designed to alienate children from their languages and cultures and assimilate them into the Canadian Nation. The Indian Act of 1876 sought to regulate Indigenous governance, education, economy, and religion, and included a ban on a variety of spiritual practices; in British Columbia, this most notably included a ban on Potlatch ceremonies. The impact of the policies, attitudes, and the ongoing colonial realities have – and continue to have – major effects on the health and well-being of Indigenous peoples and communities.

Whilst reconciliation features strongly in marketing, the guides I spoke with suggested that this was often framed historically, rather than as a contemporary struggle; several of them expressed frustration with slow progress regarding land rights. Similar critiques have been voiced by scholars, notably the failure to address land rights, matters of self-determination, and the implied sovereignty of the settler colonial state. As Hargreaves and Jefferess (2015; 207) argue, people are being asked to accept a linear history of Canada, that begins with discovery and ends with reconciliation, and in between has a 'series of regrettable colonial harms.' Taiaike Alfred (2008; 182:3), similarly argues that reconciliation constitutes a pacifying discourse demanding that Indigenous peoples become 'reconciled with imperialism.' Others have raised concerns about the onus placed on Indigenous peoples to heal for the benefit of the whole nation (Coulthard, 2014; 109). As Hargreaves and Jefferess (2015; 201) put it, the official discourses of reconciliation in Canada seek to provide closure, which fits into Canada's national mythology of progress and inclusion. They suggest that it might be more useful to see reconciliation as a starting point from which to rethink relationships and responsibilities.

Decolonisation is a wider discourse that can incorporate elements of reconciliation but is less tied to local and national political agendas and entails ideas of wider cultural/social change. Both terms have diverse meanings and implications in practice. Tuck and Yang (2012; 36) note that, unlike reconciliation, 'decolonization is accountable to Indigenous sovereignty and futurity.' For them, reconciliation is mostly concerned with the life and well-being of the settler. In contrast, decolonization is predominantly about repatriating land and life to Indigenous peoples. In British Columbia, decolonization is playing out in diverse ways that include improving education on colonial histories and realities in the province; the use of museums, art galleries, and public spaces to celebrate the work of Indigenous artists, art, languages, and practices; the involvement of universities in rethinking curricula, performing land acknowledgments, and challenging certain established norms within the academy. Across Vancouver Island, road names have been changed, and land disputes have been taken to court. In 2019, BC became the first province to sign up to the UN Declaration of Indigenous Rights which has already had an impact on land claims. Decolonisation is a massive – and longstanding – process that entails large-scale political, social, and cultural change.

Tourism has been dismissed by many academics who point to how invasive it can be, how it can be used to smooth over issues of colonial occupation, and thus work against goals towards Indigenous sovereignty (Williams & Gonzalez, 2017). Some suggest that it works to uphold neo-colonialist attitudes, with communities being subjected to protectionist attitudes by people who believe they know better, and that the products and businesses that are developed prioritize the interest of consumers (Coronado, 2014). Others argue that tourism shapes, commodifies and (re)constructs Indigenous culture to suit visitor expectations, often leading to past practices and ways of life being foregrounded and modern realities side-lined. Indigenous peoples who meet such expectations have been accused of being inauthentic and commercialized (Johnston, 2013).

Tourism studies need to give more consideration to ways in which colonization has affected debates such as questions of authenticity, who is considered Indigenous, and within that, 'right ways' to be Indigenous. Binaries – such as tradition versus modern, nature versus capitalism, sacred versus profane, private versus public, genuine versus staged, we versus them – work to feed and maintain these narratives. Assumptions embedded within this thinking are that Indigenous people are victims, passive onlookers to dominant forces like capitalism, that they all think and experience the world in the same way, and when they do engage in 'modern economies', they somehow cease to be Indigenous.

Indigenous Tourism can highlight important political power relations and historical contexts. Ann Laura Stoler (2016; 13), for example, invites scholars working in Tourism Studies to 'reconsider today's imperial continuities and to identify existing colonial dynamics shaping nowadays multiple spaces, practices and imaginaries.' Dimitrios Theodossopoulos (2019; 99) argues for the distinctive dynamic of Indigenous tourism and notes that it provides new avenues for Indigenous people 'to reach out to the world, in search of new allies and supportive connections.' In comparison with other types of tourism, 'Indigenous tourism more closely addresses issues of Indigenous representations, with indigeneity being simultaneously an attraction and a vehicle for escaping economic and political peripheralization' (Theodossopoulos, 2019; 99).

It is also arguable that Indigenous tourism is a strategy for Indigenous people to participate in the global economy ethically and appropriately. Alexis Bunten (2010; 285) has suggested this with reference to 'Indigenous Capitalism' noting how Indigenous tourism has the potential to increase political power and provide cultural and educational reinvestment into communities. She claims that Indigenous guides, through encounters with their guests, work to dispel negative stereotypes and address historical inaccuracies whilst building understanding across cultural divides. Likewise, Laura Peers (2007) argues that historical sites can be arenas in which guides challenge stories, myths, histories, and stereotypes. James Clifford (1997) has suggested that tourism can replace narratives of cultural disappearance and salvage stories of revival, remembrance, and struggle.

My own research in this field has concentrated on encounters with six Indigenous guides working in British Columbia in 2018. My approach was informed by the scholars noted above and by Graham Harvey's notion of 'Guesthood', which recognizes 'the powerful priority, sovereignty and intellectual property rights of hosts' (2003; 142). The six guides, three women and three men of different ages lived in different parts of Vancouver Island and the Haida Gwaii and were engaged in activities including working in art and cultural centers, at festivals, and offering guided tours in nature. This represents the full range of experiences advertised by *ITBC*. I highlight these encounters because personal details of why Indigenous guides choose to engage in tourism are often neglected in academic tourism literature. I accept the advice that 'academic researchers gain a more comprehensive understanding of Indigenous tourism from the perspective of Indigenous stakeholders' (Carr, Ruhanen & Whitford (2016; 1080).

Indigenous Voices

Tana Thomas

On a 'cultural canoe tour' off Vancouver Island our guide for the day was a young woman named Tana, who introduced herself as a member of the Ahousaht First Nations - part of the Nuu-chah-nulth speaking peoples. Tana started by telling us that she was in the process of building her own canoe, learning to do so in the traditional manner from a local master carver and elder in her community. The process involved several rituals and prayers and was controversial in that carving a traditional canoe is typically considered a man's game; Tana was warned that she would be unable to use the tools correctly or adhere to the strict rules involved. She spoke at length about her dreams for her canoe and how the process of making it was healing and transformative; she had grown up away from this territory, and this task was part of her journey literally to reconnect with the land and culture that she considered to be her home. A year later she completed her task and named her canoe *Šaahyačistup*, which means "it'll bring healing wherever it goes" (Jennings, 2023; 75). Tana said, "my canoe is a big part of who I am and what I represent, we come as one!" (Jennings, 2023; 75).

On the tour, Tana sang us a song, telling us that she had only recently plucked up the courage to sing on these tours. She said she finds it easier to sing in such tourist settings than in the drum circles and ceremonies that take place in her community; for they often involve gender and power structures, and she feels unsure about where and how to fit in. Through tourism, Tana gets a practice run, where she can build up her confidence ready for community and ceremonial settings. Having grown up very shy and quiet – always feeling out of place – she felt that involvement with tourism was helping her to find her voice!

In explaining her involvement in tourism, Tana noted that:

We cannot thrive if we are not healthy people, we need to be learning from our elders and teaching the youth. We need a culture we know well and are proud to share. Communities have been so broken and displaced, […] the older generations went through the residential school system they learnt not to talk about their teachings and ceremonies, and this is deep rooted in them. They were taught to be ashamed, and some people do not realise that they are passing that sense of shame on by keeping it all so hidden. We need a change; I think we need to share; we need to stop keeping things secret.

(Jennings, 2023; 69)

Tourism is providing Tana with the time and opportunity to work, learn and connect with her land and culture, build confidence and practice using her voice.

Andy Everson

I met Andy Everson at an 'Indigenous cultural festival' in Victoria in 2018 when he was offering a tour of the totem poles that stand outside the Royal BC Museum. He was adorned in regalia made from cedar, painted on his face, and wearing a top in the design of a chilkat blanket. He introduced himself in Kwak'wala and then again in English. He confessed to often feeling reluctant to do these tours, because he does not feel he has the right to tell a lot of the stories that the different poles illustrate. What he can do is explain the different types of poles, and their different purposes and point out some of the images represented. He told us that it is important to talk about culture, for this is what informs monumental art like totem poles. His main aim with the tours is to convey that he is a working, living culture. Totem poles are a tourist symbol of indigeneity on the Northwest coast and Andy's explanations ranged over the nuances, uses and protocols involved in the production of these carvings.

Andy explained that in the 1880s the government of Canada made it illegal for his people to potlatch. He described the potlatch as a ceremony where people from other villages were brought together to witness and share in celebrations such as marriages, baptisms, raising of poles, and new chiefs; it also served as a kind of banking system. He said,

The potlatch was an essential part of our society, at a potlatch you would give almost everything you owned away, and by doing so, you were symbolising your wealth, and that this was the most important thing a Chief could do.

(Jennings, 2023; 52)

Andy is best known for his contemporary art and more recently he enjoys working with pop culture figures and feels that it is a useful tool to attract people to traditional art forms. He noted that he "likes to make things people want to hang on their walls" (Jennings, 2023; 43). Partly for that reason, he does not do a lot of carving: " In our culture and our teachings we don't hang our carved art like masks on our walls" (ibid.). These, he said, are carefully stored away ready to be used in ceremonies. Carving is for traditional use; flat designs are for sale. This is how he "mediates between those worlds" (ibid.). Andy explained that he was uncomfortable with the colonial notions of some tourists who buy masks to display like trophies. He said that making and selling limited edition prints feels like a more honest encounter.

Andy hopes that his images help to bridge the gap for people approaching his art and

> another part of it is that traditionally our artwork, is really in many respects, a mnemonic device, it triggers memory etc. About different stories and events because as we are an oral based tradition, we did not write down our histories and legends in the past.
>
> (Jennings, 2023; 45)

He hopes that his art can act like a 'gateway drug' and give:

> me space to talk about colonisation, treaties, my concerns, negotiations, settlements and the current relationship between Indigenous peoples and the colonial government. So, it's loaded with so many different things, not only does it draw people in visually, but it also enables me to hack it with way more meaning than I could, if I just drew a thunderbird for example.
>
> (Ibid.)

Andy's art and tour challenges perceptions of what art produced by an Indigenous person should look like, and how contemporary, relevant and relatable these images can be.

Tsimka Martin

Tsimka Martin was the part-owner of the '*T'ashii Paddle School*' and described herself as a musician, knowledge seeker and language learner. On a cultural canoe tour, she introduced herself first in Nuu-chah-nulth and then in English. She explained that we were currently in Tla-o-qui-aht territory and told us about the industrial logging company McMillian Bloedel that was operating on unceded lands; with pride, she spoke about the protests that followed, and as a result, how much of the territory is now protected. Tsimka spoke passionately about an initiative of which she is part called *Tla-o-qui-aht Tribal Parks,* which began in response to logging threats. She explained that, unlike *National Parks Canada, Tribal Parks* are about working out what land should be used and for what purposes, where tourism should take place for example, and what activities might be appropriate (see Chapter 18 this volume, editors' comment). This community perspective is what distinguishes this initiative from Canada Parks. Tsimka remarked that:

> They see this area like a wilderness, and an animal habitat. But the term wilderness – people think it's untouched by man, but actually you can see the selective harvests and places that have been gardened. We think of it as ancestral gardens. Some areas don'tvsee as many humans as others, but you can always see the stewardship if you know how to look for it.
>
> (Jennings, 2023; 112)

Tsimka was working on a project with *Tourism Tofino* to help tourists visiting the area. She has worked on a brochure to be distributed about Tla-o-qui-aht values, "to give a welcome from our people. At the moment, people can come here and not even know what nation or what area they are in. So, I think it is important to offer some understanding and core principles" (Jennings, 2023; 113). Tsimka emphasized the importance of having a physical presence on the land and said that: "from March to October, that is a good amount of time to have eyes on land and water. It is so important to have that presence on the land. It is a sharp contrast from residential school times" (Jennings, 2023; 108).

Tsimka spoke about the three residential schools that were in the area and directed our attention to a totem pole that stood on Meares Island near where the Christie residential school had once stood. Totem poles "work as histories and as clues to the past" (Jennings, 2023; 104). Tsimka noted, " You just have to know how to read them; they help keep our stories alive" (Jennings, 2023; 105). Tsimka spoke about the history of residential schools which were mostly run by Catholics. She described the abuse that occurred in the institutions and the destructive impact that has had on successive generations. She added that at least her people were able to remain in their traditional territory and "there is strength in that; healing can happen because of that. That's more than some Indigenous people around the world today" (Jennings, 2023; 110). She wanted to talk about this history "because context matters and affects everything. Tourism therefore is not just good or bad, it's context dependent" (ibid.).

> I talk about this on our tours, because there is such a disconnect to people who have never heard about it, and we cannot reconcile if we do not know the stories of Indigenous peoples in a place!
>
> (Jennings, 2023; 111)

Tsimka is passionate about giving young people the opportunity to develop important skills that come from guiding and described the work she does as a guide as holistic, "because when you are speaking and sharing history and culture you have to be constantly learning and growing to avoid sounding like it is just an 'autospiel'" (Jennings 2023; 116).

Roy Henry Vickers

Roy Henry Vickers is an artist, storyteller, and Chief, who owns a gallery in Tofino at which he runs storytelling sessions. Roy told us how he became a storyteller: back in the 1990s, on a day out fishing with his son, an eagle came down and began circling him. Roy made a noise, like a birdcall, and the eagle spun off to fly away; as it did, the eagle released a feather and with some nifty maneuvringof the boat, Roy caught it. This was a profound moment for him. That evening, whilst putting his son to bed, he had become overwhelmed with emotion and had a conversation with God about what it all meant. Through that conversation, he realized two things: first, relief on becoming the father he had always hoped he could be, and second, that it was time for him to use his voice and his story and to become a teacher.

With pride, Roy said that 12 million people have been through his gallery and read his books. Roy told me that "all our education prior to colonization was through storytelling, stories in songs, dances, in art, and storytellers like myself, tell the same stories over and over again, why, to teach people" (Jennings, 2023; 85). Just the day before, he had met with representatives of the Anglican Diocese of Vancouver Island to give a speech. He saw this as an opportunity to speak to colonisers – to the Church which had systematically broken down the culture of his ancestors – to show and tell them what they had done, not to guilt them, but so that they could understand the legacy of abuse. He was not angry, for he no longer saw the point. Roy was baptized in an Anglican Church and grew up as a Christian, but the more he learned, the more he hated the Church, to the point where he said that it was destroying him. On realizing that, he said: "Ok, I need to begin a healing journey" (ibid.). That journey has moved from a personal one to include his teaching of others – emphasizing the responsibility people hold for one another.

Roy spoke of the fears he had in speaking about the Canadian settler side of his identity, because of how and on what grounds an Indigenous identity can be claimed. Roy says he is always conscious of his art teacher's advice to create from his 'true self.' Because his mother was "white with mixed European heritage", it was important to acknowledge that side of him or he would not be able to create his whole self (Jennings, 2023; 92). In response, he decided to push the boundaries and move away from the rigid iconography of his ancestors. He told me how worried he had been about what people would think, but in the end, he knew he had to be true to himself.

Many of the pictures in his gallery are derived from local landscapes. Roy argues for the 'spatial and visionary connection' that is made when people communicate through art and says that this connection comes from being open to the images that emerge from that shared encounter. He stresses that that is how it was done prior to colonization, and that is a big part of why he is passionate about sharing his knowledge and beliefs in this format. Through his art and storytelling, Roy grapples with aspects of the past – notably colonialism – his identity as a Canadian First Nations man, and the connections he holds for people and places.

K'odi Nelson

K'odi Nelson is the lead cultural guide for *Sea Wolf Adventures* and the Executive Director of the *Nawalakw Healing Society and Lodge*. I met K'odi in Yalis (Alert Bay) where he introduced himself first in Kwak'wala and then in English; he added that he has only recently learned to speak Kwak'wala. We set off in the boat in search of whales. After a few minutes, we paused as K'odi pointed out the Namgis century-old burial ground, which has, amongst the stone graves, some large Memorial Totem Poles, to commemorate deceased members of the Kwakwaka'wakw, typically the Chiefs. As a sign of respect, we were asked not to take pictures and informed that no tourists are allowed on the site as it is a sacred space reserved for the community and family members. Some of the poles were lying on the ground; when a pole falls, some see it as a sign that it has done its work. This area is a well-known whale-watching hotspot and we were told that whales play an important role in Kwakwaka'wakw culture.

Throughout the trip, K'odi made efforts to teach his son – who looked approximately ten years old – how to use the boat and be a guide. He explained how important it was to him that his son feels that he has a future here, adding that it is hard to find people to do this job, and that although he could find people willing to drive the boat, those same people were often unwilling to be a guide or vice versa. Although there are lots of different origin stories from the many different Kwakwaka'wakw Nations, all of them show that the Kwakwaka'wakw were created right here. That is why, K'odi said, they fight so hard for their land, water and lives.

After a couple of hours on the water, we traveled back to Alert Bay and walked to the U'mista Cultural Centre, a beautiful building right on the shore. Before we entered, K'odi directed us to the field right behind, where we stood by a plaque that commemorates St Michael's Residential School. The school was built in 1929 and was one of the last residential schools to close in 1974; it was the largest Anglican run residential school in Canada. The building was turned over to the Namgis First Nations and demolished in 2015, at which time the Namgis held a ceremony to bless the land. The Umista center is a UNESCO World Heritage site and a thriving community space. There were many exhibitions, including one on the Potlatch ceremony and the historic potlatch ban. K'odi said the potlatch ban was enforced because it involved giving away wealth, which was the opposite of "the West and so they were frightened" (Jennings, 2023; 137). They celebrate important events that way, and they were in the process of relearning many old dances. A potlatch helps you to grieve. K'odi said that this is a way to say an official goodbye, and you burn things that the departed will need on their journey, like food and clothes. Whenever food falls from the table or a cupboard, it is a sign that they need to hold one of these ceremonies.

K'odi spoke at length about the benefits of tourism, remarking that "obviously the economics of tourism is important in that it's creating jobs for our people", adding if he or other members of his community will not build it, then someone else will, and he feels that they are the ones that need to benefit from their traditional land (Jennings, 2023; 140). K'odi noted that his work in tourism had given him:

a sense of pride in being First Nations, because we have clients that are asking us specific questions about who we are and where we come from. It creates an environment that starts to value that. […] It's also good that we are out on our territory. We become the eyes and ears of the territory, because lots of people don't have boats anymore, so we need to be those eyes and ears. So, if there's fish farms etc., we know and tell the communities.

(Ibid.)

K'odi described tourism as "dancing with the devil" and acknowledged the need to be careful with it (Jennings, 2023; 144). His father had wanted Kodi to become a teacher "to help my people revive our culture", and he noted, "it's difficult for kids to wrap their heads around the relocation of our people. The social spin-offs that are so negative" (Jennings, 2023; 142). He added that:

I truly believe that we need to instil pride back into our people and I'm convinced that they will blossom. We are very connected to our environment, and it's mostly about educating people so they respect the land. The tides are slowly starting to change, it's medicine for our elders to see the young full of pride.

(Ibid.)

S*G̲aana G̲aahlandaay (Alix Goetzinger)*

In the old-growth rainforests in K'yah G̲awG̲a (Windy Bay) SG̲aana G̲aahlandaay (Alex) introduced herself first in English and then in Haida. Before we landed, James (the skipper) called ahead and asked the 'watchmen' for permission to come ashore. The watchmen program is based on a traditional governance system once used to alert the community to the approach of an enemy; their mandate today is to safeguard 'Gwaii Haanas National Park Reserve', to welcome and teach visitors about the natural and cultural heritage of the area, and to ensure that the tourists show respect and 'leave no trace'. The vision is that each nation can manage and safeguard the land and waters in their territories according to their own laws.

We visited a 42 ft Legacy Pole which was raised in 2013 to commemorate the twentieth anniversary of the agreement between the Canadian Government and the Haida Nation regarding the co-management of the region when it was designated a National Park Reserve. This legacy pole is the first totem pole to be raised in Gwaii Haanas in 130 years and Alix described the pole and all its carvings in great detail. This was Alix's fourth year as a "cultural ambassador, tour guide or cultural interpreter, whatever you want to call it" (Jennings, 2023; 157). When she began:

I was terrified of public speaking, I was very much learning the culture. […] But I hadn't realised how much there is to know and learn and what knowledge gaps to fill. I felt huge internal pressure to learn about my own history, collect as much information as I can, not just for the sake of tourism, and to be able to get the information across properly, so guests leave with correct information, but for my own sake as well, so I can pass that on

(Jennings, 2023; 158)

On first being asked to give tours Alix was apprehensive but:

On my first tour I looked at the totem poles and realised that just from growing up here and seeing these figures so often, I can do this, I know this, and I was like yeah actually it is all right there, that's just what I need to know. When you're talking about your own history and culture there's no way to feel embarrassed or to feel anxious to say something because, when you're passionate – no matter how introverted you feel – that will push you to speak…that was my big starting point.

(Ibid.)

She remarked how much she enjoyed interacting with people, making new connections with different cultures whilst sharing aspects of her own:

I think that's the most powerful part of tourism, that there is a way we can all get to the same level of understanding and compare our own cultures. There are always similarities, and I think that really moves us forward in Truth and Reconciliation, just the ability to understand each other is really important.

(Ibid.)

She added that "every day I make little mental notes like this is something I need to know more about. I often tell myself, research this, ask for something that I need – to fill this gap" (ibid.).

The hard part about tourism for Alix, was being stereotyped, tourists sometimes asked harsh and ignorant questions about her identity: "saying, " You don't look like you're native", "oh do they allow non-native people to work here?" (Jennings, 2023; 159). As she said: "I am young, Indigenous and a woman and these three things kind of work against me" (Jennings, 2023; 160). This had become so common that Alix had worked on a stock response about how everyone looks and sounds different, warning people not to judge. Responses apparently varied, with some people being apologetic, acknowledging their ignorance. Alix would feel good if she felt people had learned something. For the most part, she said, " The people that visit the Haida Gwaii, it's such an effort to make it here when they arrive, they are willing to be educated" (Jennings, 2023; 160). Alix commented : "We are teased for it, but the Haida are really proud people, but I think when you are fighting you need to be loud and proud. I think tourism has helped that" (Jennings, 2023; 163).

Conclusion

In talking with these six guides, I became aware that tourism was being presented not only as a valuable source of employment but as something being used to 'learn, teach, heal'. This characterization captures the complexity of their work in Indigenous tourism. It includes aspects of their personal development, the local Indigenous communities to which they belong, and the long-term project of reconciliation and decolonization. Through tourism they are learning and teaching about their past, their identities and languages, and reconnecting with their land, practices, spiritualities, communities, and environment.

Tourism is providing a space to learn and practice specific languages, languages that were once banned under colonialism. Learning a language and performing that process for tourists is a practical step that invites tourists to witness and support the processes of reclaiming and decolonizing. 'Tradition' is an established marker of indigeneity and learning about such practices is one of the unique selling points for Indigenous tourism. Many of the guides talked about needing to re-learn traditions long denigrated under colonialism. Working in tourism was often described as 'in keeping with' traditional practices, including how knowledge has been passed on, using totem poles, ceremonies, dance, drums and singing, through to storytelling. Traditions offer tools by which people can orientate, differentiate, position, and make claims about themselves to others. Practices articulated as traditions were not presented as fixed. Totem poles that now stand outside museums, for example, once stood on the grounds as artifacts with labels that rendered them part of a vanishing culture. They are currently open to a variety of interpretations, including continuity with, and reclaiming of the past, and rejection of the colonial regime that removed them. Joyce Green (2007) notes that although embracing the symbols of one's culture and traditions – particularly against the imposition of colonialism – can be a strategy for reclaiming primacy in the world, the power relations embedded must be considered for who decides what tradition is and for whom. There can never be a single cultural version of tradition.

Tourism has facilitated a presence, having 'eyes and ears' on the land. The land is central to colonialism and decolonization, and as historian Cole Harris (2004; 167) has put it: 'the experienced materiality of colonialism is grounded in the dispossessions and repossessions of land.' The issue of land claims featured strongly in the tours, performances, and interviews. The guides talked about the land that was once theirs, about past and current protests and evinced a strong sense of stewardship. One discourse that pervades imaginaries around Indigenous peoples is that they inhabit a wilderness. The idea of wilderness is rooted in and perpetuated through colonial narratives. Nature plays a huge role in the Canadian narrative and has helped forge an 'imagined community'; ideas of wilderness are highly valued in the national rhetoric. Challenging notions of wilderness open questions about how to conceptualize a place and its history, correct perceptions, and teach guests new ways of thinking.

Indigenous Tourism tells people that there are Indigenous people here. By providing face-to-face communication, the tours can place stories in the 'here and now', offer personal context and meaning, and ground references to past colonial oppression. The histories told within Indigenous tourism generally tend to be 'therapeutic histories', (Niezen, 2009) where positive elements of the past and present are highlighted, and negative aspects excoriated. The dominant discourses are used as a reference point with which to interact. One of the most important aspects of 'therapeutic history' is its potential to change for the better the way members of a community feel about themselves. This positive feeling, Niezen (2009) argues, is the main criterion for determining what should be included in the histories. Choosing which narratives to tell is part of a process of self-discovery and affirming collective identity. By highlighting the achievements of one's ancestors, a sense of responsibility and vision for the future is provided. Tourism provides a setting for both guides and visitors to discuss history.

Tourists often come expecting to learn, hopeful of reinforcing positive ideas and images of Indigenous peoples. The guides make use of this space by meeting those expectations, with some critical discussion of past oppression. The process of guides choosing what to say and how to say it is 'actually part of a creative process of becoming.' Tourism can in this sense serve to heighten visibility and assist in the reformulation of a peoples' history. Indigenous Tourism is providing spaces for the guides to write that script (Jennings 2023: 188).

'Learn, teach, heal' emphasizes agency, process, and change and so challenges notions of indigeneity as fixed in the past and lacking a future. Learning signals change and development; teaching denotes knowledge, rather than merely entertainment. Teaching was a core motivation for all the guides. The teaching was designed to disrupt, historicize, and inform. Healing comes from the learning and teaching, from the pride that comes with sharing, and having something special to share, from being heard – breaking past silences. For the younger guides, tourism offered a space in which to deepen their knowledge of the land and associated traditions, and a practice ground free from the rules and hierarchies that might exist in their communities. For the older guides tourism offered a way to share their knowledge, experiment with traditional matters, and develop 'gateways', channels through which tourists can be invited in, learn more, and join them in their causes. 'To teach' signals control in these contexts: over what to share, with whom and in which circumstances. It signals competence, authority, and responsibility. The notion of 'Learn, Teach, Heal' connects a variety of interwoven processes. That healing comes from learning and teaching is important for the processes of decolonization; it provides a focus on shared wounds and a collective process of reclaiming and becoming.

References

Alfred, T. (2008). "Opening Words." *Lighting the Eighth Fire: The Liberation, Resurgence and Protection of Indigenous Nations*, ed. by L. Simpson, 9–11. Winnipeg: ARP BOOKS.

Bunten, A. (2010). "More like Ourselves: Indigenous Capitalism through Tourism." *The American Indian Quarterly* 34 (3) pp. 285–311.

Bunten, A. (2015). *So, How Long Have You Been Native? Life as an Alaska Native Tour Guide.* Lincoln: University of Nebraska Press.

Carr, A., Ruhanen, L. and Whitford, M. (2016). "Indigenous Peoples and Tourism: The Challenges and Opportunities for Sustainable Tourism." *Journal of Sustainable Tourism: Sustainable Tourism and Indigenous Peoples* 24 (88) pp. 1–13.

Clifford, J. (2013). *Returns: Becoming Indigenous in the Twenty-First Century.* Cambridge, MA: Harvard University Press.

Coronado, G. (2014). "Selling Culture? Between Commoditisation and Cultural Control in Indigenous Alternative Tourism." *PASOS Revista de Turismo y Patrimonio Cultural* 12 (1) pp. 11–28.

Coulthard, G.S. (2014). *Red Skin, White masks: Rejecting the Colonial Politics of Recognition.* Minneapolis: University of Minnesota Press.

Green, J. (2007). *Making Space for Indigenous Feminism.* New York, NY: Bloomsbury Academic.

Hargreaves, A. and Jefferess, D. (2015). "Always Beginning: Imagining Reconciliation beyond Inclusion or Loss." *The Land We Are: Artists and Writers Unsettle the Politics of Reconciliation*, ed. by G. L'Hirondelle and S. McCall, 200–210. Winnipeg: ARP Books.

Harris, C. (2004). "How Did Colonialism Dispossess? Comments from an Edge of an Empire." *Annals of the Association of the American Geographers* 94 (1), pp. 165–182.

Harvey, G. (2003). "Guesthood as Ethical Decolonising Research Method." *International Review for the History of Religions* 50 (2) pp. 125–146.

Jennings, H. (2023). *Learn, Teach, Heal: Articulations of Indigeneity and Spirituality in Indigenous Tourism in British Columbia, Canada.* https://munin.uit.no/handle/10037/27718 (Accessed March 2023).

Johnston, A. M. (2013). *Is the Scared for Sale: Tourism and Indigenous Peoples.* Oxford: Earthscan.

Niezen, R. (2009). *The Rediscovered Self: Indigenous Identity and Cultural Justice.* Montreal and Kingston, QC: McGill–Queen's University Press.

Peers, L. (2007). *Playing Ourselves: Interpreting Native Histories at Historical Reconstructions.* Lanham, MD: Altamira Press.

Stoler A. L. (2016). *Duress: Imperial Durabilities in Our Times.* Durham, NC and London: Duke University Press.

Theodossopoulos, D. (2019). "Indigenous Tourism as a Transformative Process: The Case of the Embera in Panama." *Indigenous Tourism Movements*, ed. by A. C. Bunten and N. Graburn, 99–116. Toronto, ON: University of Toronto Press.

Tuck, E. and Yang, K. W. (2012). "Decolonisation Is Not a Metaphor." *Decolonisation: Indigeneity, Education & Society* 1 (1) pp. 1–40.
Williams, l. Z. and Gonzalez, V. V. (2017). "Indigeneity, Sovereignty, Sustainability and Cultural Tourism: Hosts and hostages at 'Iolani Palace." *Hawai'i Journal of Sustainable Tourism* 25 (3) 668–683.

Websites

DestinationsBC (2020) (https://www.destinationbc.ca/research-insights/type/industry-performance/ (Accessed March 2023).
Indigenous Tourism Association of Canada (ITAC) (2022) https://indigenoustourism.ca/plans-reports/indigenous-tourism-is-reconciliation-in-action-2022-2023-action-plan/ (Accessed March 2023).
Indigenous Tourism British Columbia (ITBC) (2020) https://www.indigenousbc.com/corporate/news/why-indigneous-tourism-matters-tourism-week-2020/ (Accessed March 2023).
Indigenous Tourism British Columbia (ITBC) (2023) https://www.indigenousbc.com/corporate (Accessed March 2023).
Truth and Reconciliation commission (TRC) (2015) https://www.rcaanc-cirnac.gc.ca/eng/1450124405592/1529106060525 (Accessed March 2023).

8

"COME AND KNOW A LITTLE ABOUT YOUR OWN BACKYARD"

Transformative Learning Potentials through Knowledge Sharing in Indigenous Tourism in Australia

Nicole Curtin, Tracy Woodroffe and Clinton Walker

Introduction

For people who completed their schooling in Australia, it is likely that little was learnt about Aboriginal and Torres Strait Islander histories and cultures. The little that was taught was likely prejudiced, even racist, with Indigenous Australians portrayed as uncivilised or unintelligent. As a society, how can Australians learn about true Australian history and unlearn the prejudices that are widely held? And how can the rhetoric that frames Indigenous matters and Indigenous people as a "problem" to be solved, rather than as holders of their own solutions, be dispelled?

Education is the starting point for a society where people are confident in engaging and working with Indigenous peoples and communities. For progress towards a broader awareness of Indigenous peoples, cultures, and histories, the visibility and recognition of Indigenous peoples as knowledge holders need to be increased. Co-author Warumungu Luritja woman Dr Tracy Woodroffe (2021) explains: "My strength, and the strength of Indigenous people, is in who we are at the core. The core is our Indigenous knowledge. It is the foundation for our strength of character and our Indigenous perspectives are a uniting force" (p. 74).

Recognising Indigenous tourism operators as educators can be part of the solution. Within the context of cross-cultural communication and learning, we consider Indigenous tourism as a site of Indigenous knowledge sharing where operators are choosing what, how much, and how to share their cultural knowledge with visitors (Curtin & Bird, 2022). The sharing of local knowledge and prioritisation of the needs and wants of locals above those of tourists represents a less exploitative style of tourism, aligned with calls to localise tourism by grounding operations in "social and ecological contexts for greater justice and sustainability" (Higgins-Desbiolles & Bigby, 2021, p. 3).

The work of shifting stereotypes and biases requires both individual and collective transformation. In this chapter, we explore learning potentials that represent opportunities for visitors to establish and deepen their respect for the sovereignty of Indigenous tourism hosts who are sharing their knowledge and culture with visitors. We begin this chapter by discussing deficits in the Australian education system and considering Indigenous tourism as a site of transformational

DOI: 10.4324/9781003230335-10

learning. We then propose a model of transformative learning potentials and discuss implications for cultural competency and systemic change.

Background

Across Australia, Indigenous and non-Indigenous students alike are not consistently taught accurate, truthful, and full stories of Australian history that portray Indigenous peoples and cultures in a positive light (Woodroffe, 2020). Chronic deficits in the education system have led to a broad lack of understanding of true local and national histories, peoples, and cultures across the Australian population.

The outcome of these deficits is a widespread lack of understanding, and frequent misunderstandings, as pervasive negative stereotypes of Indigenous Australians are left unchallenged (Reconciliation Australia, 2021). As a result, motivated people are seeking to learn about Indigenous histories and cultures elsewhere. Education is thus implicated as part of the problem of a lack of social cohesion stemming from a lack of understanding and misunderstanding of Indigenous peoples, cultures, and histories. Education is also implicated as a solution for achieving greater understanding and improved race relations.

Indigenous tourism operators are sharing their cultural knowledge with visitors with the aim of sustaining culture and practicing reconciliation (Curtin & Bird, 2022). This leads us to consider tourism as an informal adult education activity in which Indigenous operators are knowledge holders and visitors are learners. Informal education is often unstructured and is a kind of learning that results from daily life: learning *through* experience and learning *as* experience. Through Indigenous tourism engagement, visitors are likely to come across information that is new to them and that challenges their preconceptions. Tourism interactions thereby present opportunities for critical self-reflection and transformational learning.

As a social force, tourism possesses transformative possibilities that deserve considered analysis, as "when Indigenous people share their Indigenous ontologies with non-Indigenous people through [tourism], transformations in non-Indigenous consciousness can occur to profound effect" (Higgins-Desbiolles, 2009, p. 146). Indigenous tourism embodies transformative potentials through the presentation of realities that challenge hegemonic narratives. Through the sharing of stories and worldviews, tourism engagement can accelerate reflexive dialogue necessary for transformative learning (Ormond & Vietti, 2022). Individuals and groups can consider the experiences of others in relation to their assumptions and pre-conceived perspectives of the world (Mezirow, 2012). Indigenous tourism engagement can contribute to transforming collectively held worldviews, expressed as cultural paradigms, such as pervasive white ignorance in Australian society (Taylor & Habibis, 2020).

A defining human condition is the need to understand and make sense of our experiences and to integrate new experiences with what we already know about the world. Transformational learning occurs through a process of becoming aware of, questioning, and revising our perceptions of the world due to new information, or an alternative perspective, challenging our frames of reference. Frames of reference are the assumptions that shape experience, expectations, perceptions, and feelings (Mezirow, 2012).

When newly encountered experiences or information do not fit with our frames of reference, two outcomes are possible. If we are unable to understand new information or experiences, we may turn to uncritically assimilated perspectives of the world, drawing on established traditions and authority figures, or adopt psychological mechanisms such as schemas, biases, prejudices, or stereotypes to create coherent meaning (Mezirow, 2012). Alternatively, we may become critically aware of our assumptions and expectations, leading to perspective transformation (Mezirow, 2012).

As transformative change is disruptive, some level of discomfort and uncertainty are unavoidable. Visitors experience these unsettling emotions in Indigenous tourism settings when they encounter stories that challenge the ways that they understand Australian histories, peoples, and cultures. Shifting biases requires metacognition: visitors must reflect upon discrepancies between information shared by Indigenous tourism operators and their own assumptions and preconceptions. Through transformative learning, meaning structures become more open and permeable, and autonomous thinking is facilitated, whereby a person is able to critically reflect on their assumptions rather than uncritically acting on problematic frames of reference, such as stereotypes (Mezirow, 2012).

The transformation of biases is required for the necessary cultural shift for improvements towards Australia's *Closing the Gap* targets[1]. New ways of thinking and engaging require discomfort. For disruption to occur, a better understanding of social disadvantage, how disadvantage can be seen across a population, and the conditions that have led to experiences of social disadvantage for Indigenous Australians is crucial. To transcend the practice of reconciliation beyond awareness raising, visitors must reflect on their assumptions. By doing so, transformative learning potentials can arise. In this chapter, we explore the nature of these learning potentials in Indigenous tourism.

Research Design

We adopted critical Indigenous methodology and constructivist grounded theory as culturally appropriate research approaches to explore the experiences of Indigenous tourism stakeholders. Working ethically with Indigenous participants in an Indigenous research context required us to be responsive to the concerns of participants while involving participants in knowledge creation and dissemination (Denzin & Lincoln, 2008). Constructivist grounded theory (Charmaz, 2014) provided protocols for developing a ground-up theory by privileging the lived experiences of the people who participated in our research.

We undertook theoretical sampling and data analysis simultaneously until theoretical sufficiency was reached in an all is dataapproach[2] (Charmaz, 2014). Field work activities (26) were carried out by the first author, a non-Indigenous woman, across the Northern Territory (NT) and Western Australia (WA), with a focus on spending time with, and listening carefully to operators and visitors. We conducted interviews with 22 participants: 6 Aboriginal tourism operators, 5 non-Indigenous tourism hosts, 1 government representative, and 10 non-Indigenous tourism visitors.

The participating Aboriginal tourism operators[3] chose to be identified and to have their quotes attributed to their names

- Bart Pigram, Yawuru owner/operator of Narlijia Experiences Broome (Broome, WA);
- Clinton Walker (co-author), Ngarluma and Yindjibarndi owner/operator of Ngurrangga Tours (Karratha, WA);
- Kerry-Ann Winmar, Whadjuk Nyungar owner/operator of Nyungar Tours (Perth, WA);
- Marissa Verma, Noongar owner/operator of Bindi Bindi Dreaming (Perth, WA);
- Mick Hayden, Njaki Njaki Nyoongar owner/operator of Njaki Njaki Aboriginal Cultural Tours (Merredin, WA);
- and Roland Burrunali, Bininj tour guide with Injalak Arts and Craft (Gunbalanya, NT) (Figure 8.1).

Figure 8.1   The locations of participating Aboriginal tourism operations.

Findings

Transformative Learning Potentials: "They're Actually Challenging Themselves"

Indigenous tourism operators are sharing cultural knowledge with the aim of offering another perspective to visitors. Bart explains that learning is the desired outcome: "We are creating a platform for people to engage that are interested in learning and understanding our culture". Through Indigenous tourism, opportunities for unlearning prejudices also arise, as Kerry-Ann illustrates: "There are a lot of misconceptions about Aboriginal people. ... It's time that [Aboriginal people] are recognised and seen".

Visitor engagement in Indigenous tourism presents opportunities for visitors to critically reflect on their frames of reference. Bart says that visitors are *"actually challenging themselves"*. These transformative learning potentials can be categorised as respecting Indigenous sovereignty: *"our shared history"*; valuing Indigenous knowledges: *"belonging to Australia together"*; and supporting Indigenous affairs: *"go and sit and learn"*.

Respecting Indigenous Sovereignty: "Our Shared History"

Through listening to stories shared by Indigenous operators, visitors can come to an understanding of Indigenous sovereignty and the myth of *terra nullius*.[4] Visitors gain a new appreciation of the diversity of Indigenous peoples and cultures as they come to understand truths about Australian history for the first time: "I've really understood the connection that Aboriginal people have to their Country which I hadn't really understood before. ... 200 years [of colonisation] in their history of over 50,000 years is quite recent still" (visitor).

Operators emphasise the importance of facilitating an understanding of the diversity of Aboriginal peoples and cultures to challenge pervasive negative stereotypes. Kerry-Ann explains: "It gives [visitors] the understanding that Aboriginal people are still existing. They're still here. They still have their language. They still have their traditions. It's time that [Aboriginal people] are recognised and seen". Through contact with Aboriginal tourism operators, beliefs about pan-Indigeneity are challenged.

Connecting with Aboriginal tourism operators can break down barriers and facilitate future contact. Operators provide basic guidance on how to respectfully interact with Aboriginal people, as Marissa describes: "The most asked question I get all the time is 'how do I talk to an Aboriginal person?' ... And I say, 'well, how would you talk to anybody?'" Here, visitors can reflect upon dehumanising stereotypes that portray Aboriginal Australians as so radically different from themselves that visitors may have presumed that radically different ways of interacting are required.

A vital starting point for respecting Aboriginal Australians is for visitors to understand the disruptive nature of colonisation on Aboriginal communities. Bart emphasises the importance of tracing social disadvantage to colonial legacies:

The government may blanket us with all the same issues, but this and every single community has a very broad spectrum within it of the histories of the families. There are reasons why people are in certain ways, why some are successful and some aren't. ... To break that down for [visitors] is really important. ... You can't respect or take value in anything if you don't understand it.

Through listening deeply, visitors can understand how the impacts of colonisation are continuing to the present day. Clinton recounts a commonly held misconception:

A lot of people think that we get things for free. But when they listen to me and they hear my story, they realise that I'm just like them, except that as an Aboriginal person, I've actually had to come out of an impoverished society, rather than grow up in middle class. Everything that I've got I've built from nothing. And it's not just, when we say nothing, because a lot of people grew up with nothing, but your entire people have nothing. So, you've got no-one to aspire to be like. And so, you have to be the change.

Through the sharing of personal and family stories, visitors can make links between the past, present, and ongoing impacts of colonisation. One visitor describes that it is important to "put context around why people do the things they do or why things are the way things have been, or are, now".

Once the basic facts of sovereignty are established, there is room for seeing nuance and holding multiple, and often conflicting, realities simultaneously, which can then make way for flexibility and openness in thinking. Tourism engagements whereby visitors encounter people going about their daily lives present opportunities for seeing beyond the commercialisable aspects of culture to understand the complexity of unmet needs faced by Indigenous peoples and communities: "these tourists aren't flying in to walk on a beach. They're flying in, the environment is part of it, but getting a snapshot of life in community and the culture of the community, is another thing. And that's a contested space at the best of times" (non-Indigenous host).

Through being open and having a willingness to learn, visitors can come to see and think about places differently, transforming their way of living in their local space. Kerry-Ann says that

a lot of [visitors] that I've come across, they don't know anything about Aboriginal people. [Visitors] always apologise because they say 'we never knew that'. ... So, they come now to learn about it on a tour. ... They never knew these things. Yet, they drive past it every day.

Ultimately, understanding Indigenous cultures and histories can facilitate empathy: "I think that it enables people to have more empathy towards their First Nation counterparts" (Clinton).

Respecting Indigenous sovereignty provides visitors with opportunities for self-reflection. Visitors can situate their personal stories within a larger national narrative by considering *our shared history*:

> [The host] spoke a little bit about the history of the place, and the massacres that happened there. But [they] also said that that's '*our* history'. We shouldn't be thinking in terms of Indigenous history and our, white people's, history. It's *our shared history*. … [In schools] it's very 'Indigenous and non-Indigenous history', whereas it's really *our* history.

Viewing history as shared transcends the pervasive historical denial of harms inflicted upon Indigenous peoples and is seen as a starting point for moving forward together.

Valuing Indigenous Knowledges: "Belonging to Australia Together"

Through respecting Indigenous sovereignty, visitors are able to value Indigenous knowledges by appreciating the importance, worth, and utility of the Indigenous knowledges and sustainability values that operators are sharing. Kerry-Ann explains that visitors "want to know how to pay respects, or to respect the land, or even how to work alongside, or work with, Aboriginal people". Through seeking this insight from their hosts, visitors can come to know "how important it is for Aboriginal culture to be integrated into Australian society and ways of identifying as a nation" (visitor). Bart explains that

> we strive to ensure that [a tourism experience] resonates with them. … We truly strive to penetrate into, we say, *liyan*. *Liyan* is that understanding, or that spiritual essence beating in your heart. We try to penetrate into there and resonate with them so that they take it away and change their lives in a sense.

This resonance, or self-reflection, can change the way that visitors understand the world and themselves: "I am starting to see [local Aboriginal culture] as something that is becoming a part of me. I've lived here most of my life, but I never knew those stories of Country, and the depth of knowledge that the Noongar people have has created a connection to Country as a non-Indigenous person" (visitor). Self-reflection can lead to visitors considering their own family history through coming to understand the depth of an Indigenous host's connection to Country and culture, as Bart describes:

> I actually explain my whole family going back five generations. And I sort of do that to justify my heritage, but it also challenges the people on the tour, because they go, 'what am I? I don't know my fifth generation back'. So, they go away becoming their own family researcher.

Hosts extend the invitation for visitors to "come and know a little about your own backyard" (Mick) so that they can know more about their national or local 'backyard' through an appreciation of Indigenous knowledges. Reflecting on true Australian and local history can enhance a sense of belonging. Visitors can come to appreciate how they see themselves in relation to *our shared history*. Kerry-Ann explains: "If [visitors] learn something from about where they are, they belong there too". Visitors frequently acknowledge the value and importance of local knowledges: "The best knowledge that we have locally is the [knowledge of] local Indigenous operators" (visitor). Valuing Indigenous knowledges is important for achieving unity: "I think there's a lot that can be learnt from Indigenous people and an extra depth that you can have in belonging to Australia and *belonging to Australia together*" (visitor).

Visitors described negotiating mixed messages from various sources regarding how they can integrate a connection to Aboriginal culture in their everyday lives. Visitors described grappling with knowing when to act as opposed to when not to act, such as when to refer to Elders for authorisation of tasks, as is often suggested, as opposed to being told not to "waste the time of Elders". Visitors also spoke about navigating who is aligned with who, identifying the right person to speak with and seek authorisation from, especially where there is contested authority such as being aligned with certain family groups, and negotiating the impacts of family feuding. These mixed messages were considered to contribute to a feeling of paralysis, whereby fear of the consequences of "getting it wrong" often resulted in visitors not acting at all. One non-Indigenous host summarised this paralysis:

> You can go too far and you can not go far enough. It's a double-edged sword of criticism. Because either it's 'just a white person doing something', or it's 'why aren't you doing more?' I just don't know what you want me to do?

Another dynamic that visitors described when considering how they can value Indigenous knowledges in their own lives was negotiating the line between cultural appreciation and cultural appropriation. Some visitors were more confident than others in knowing whether they could share what they had learnt about Aboriginal culture, as opposed to not having the authority to share culture or speak on behalf of Aboriginal people and communities. Navigating these dynamics requires critical reflection and ongoing negotiation. In general, Aboriginal tourism operators were generous towards visitors making sense of how to value Indigenous knowledges in their own lives.

Supporting Indigenous Affairs: "Go and Sit and Learn"

Through respecting Indigenous sovereignty and valuing Indigenous knowledges, visitors described thinking about what they can "do" after the tourism experience. Visitors asked questions about how to translate what they have learnt and experienced into their own lives. Visitors describe working out how to support Indigenous affairs and considering individual actions in relation to systemic and political progress. Momentum for political change was seen to be built through grass-root social action. When considering individual action, visitors often expressed uncertainty: "I don't really know where to start".

Visitors spoke about feeling a sense of responsibility for advocating for Indigenous affairs. This responsibility also requires ongoing negotiation:

> Sometimes it can be difficult to try to share the message about what I've learnt and what reconciliation is about, and it's also acknowledging the limits of what I can talk about. It's actually up to the Indigenous Nations of Australia to say what they want, not for me to say what they want or need.

Here, a visitor reflects on support for Indigenous affairs needing a base understanding of the need for self-determination.

Visitors often spoke about "wanting more" or seeking a sense of certainty from their tourism experience but were not easily able to express what they were expecting or hoping to experience. Operators speak of not imposing suggested actions on visitors but helping them come to their own understandings. Bart describes that visitors often ask him how they can support Indigenous affairs, with his common response being:

> Don't ask me how to fix, or how to help, the people [from other places]. .... It's a different language, it's a different landscape, it's a different culture, different history, different impacts, different socially. ... This is what Australia and every single Australian needs to be aware of. It's very complex.

From a place of empathy and understanding, visitors can then consider how to interact with Aboriginal people within their own communities and consider tangible actions they can take. Building a sense of confidence in taking action requires risk through being prepared to make mistakes, and correcting oneself where necessary.

Operators emphasise the need for visitors to commit to ongoing learning as an intentionally active and personal, rather than passive and objective, process. Bart comments that:

> There have been times when I've finished a tour and [a visitor] said, how can I go back and help the Aboriginal people in my community? And my straight answer is ... go and learn. *Go and sit and learn.* Start understanding first. Don't go into a community and say I want to save you. These are human beings. We have a social structure, we have history, we have culture. Understand all that and then you'll find your own answer on how to help.

Deep listening is a vital part of the reciprocity required, as Mick explains: "old Noongars, we talk and talk and still we're blue in the face and still no-one hears us". Learning is more than a one-way process and is an ongoing action and an outcome: "I learnt a few things, I'm forever constantly learning new things about Indigenous culture" (visitor).

Discussion

A Model of Transformative Learning Potentials

The thematic headings from the findings are presented below in Figure 8.2 in a model of transformative learning potentials in Indigenous tourism. Visitors often come across information that does not fit with their pre-existing frames of reference. Through critical reflection, transformative learning potentials arise. Here, learning can be considered an action: "*they're actually challenging themselves*". By respecting Indigenous sovereignty, visitors can come to appreciate "*our shared history*". By valuing Indigenous knowledges, visitors can negotiate "*belonging to Australia together*". And in supporting Indigenous affairs, learning is considered an ongoing action: "*go and sit and learn*".Learning as action positions Indigenous operators as educators and knowledge holders, with opportunities for learning and unlearning continuing beyond the tourism experience. The need for truth-telling in establishing respect for Indigenous sovereignty and the valuing of Indigenous knowledges is well recognised (Reconciliation Australia, 2021). Although there may be a base level of information required for visitors to understand the causal contributors to social disadvantage and intergenerational trauma, and the magnitude of the past, present, and ongoing

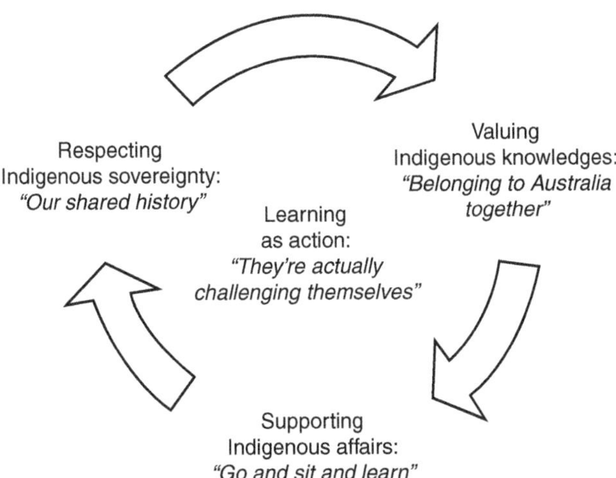

Figure 8.2 A model of transformative learning potential in Indigenous tourism.

impacts of colonisation, there is no ceiling to the learning potentials of respecting Indigenous sovereignty, valuing Indigenous knowledges, and supporting Indigenous affairs.

Although hierarchical in nature, as each category scaffolds the next, these potentials also exist concurrently (Figure 8.2). This is why the model of transformative learning potentials is cyclical, as transformative learning potentials exist across these domains. Here, learning is seen as an action and as a key pathway to supporting Indigenous affairs, which in turn can lead to visitors gaining deeper respect and value for Indigenous sovereignty and knowledges.

Co-author Clinton Walker is a Ngarluma and Yindjibarndi man, and Traditional Owner of the West Pilbara region of WA, and owner/operator of Ngurrangga Tours. Through his tours, Clinton showcases the unique and vast landscape, and strong Aboriginal communities, languages, and cultures, of the region. Clinton and his guides share the beauty of their Country with visitors and challenge the reputation of the region as a mining dust bowl (Figure 8.3).

Clinton is in a unique position as a cultural authority who provides cultural awareness training and tours for the same corporations that are threatening his Country through industrialisation (Curtin et al., 2022). Clinton has learnt how to work with and alongside industry. Visitors come on tours with Ngurrangga Tours to learn about ancient history, rock art, bush tucker, and local Aboriginal connection to Country. Clinton also talks about threats to rock art, sacred sites, and Indigenous sovereignty at Murujuga with his visitors because they see these threats firsthand:

> Being on Murujuga is a disorienting experience. Look in one direction and you'll see ancient art, towering red-rock hills, wildflowers, kangaroos and the glistening ocean. Look the other way and you'll see chemical plants, pipelines, and flares, emitting toxic chemicals and humming with industrial noise.
>
> (Mazza, 2022, para. 3)

Visitors commonly ask Clinton, "How is this possible? How was this industrialisation given approval?" Visitors express outrage, confusion, and disgust at the environmental and cultural devastation facing Murujuga. Clinton sees educating visitors so that they respect Indigenous

Figure 8.3 Traditional Owner and co-author Clinton Walker with rock art.
Source: City of Karratha.

sovereignty and value Indigenous knowledges as one way to encourage a shared responsibility for the environment.

Tourism is an important mechanism for raising awareness of the impacts of environmental and social injustice. Transformative learning potentials arising through Ngurrangga Tours highlight pathways for raising community awareness and fuelling social momentum to exert pressure on decision makers to protect environmental and cultural heritage at Murujuga. Clinton sees ongoing learning as necessary for supporting Indigenous affairs:

> I get the question 'what can I do to help?' all the time. Wherever you go, when you're travelling, and in your day to day life, just learn about Aboriginal culture and history. Do it in your own time and don't wait for events like NAIDOC.[5]

A Quandary on the Theme of Understanding

Problematic frames of reference about Aboriginal people, places and cultures are held broadly across the Australian population. Pervasive and damaging stereotypes and biases arise from the chronic deficit within the education system. Visitors are engaging in Indigenous tourism as a way to learn about Indigenous peoples, cultures, and histories. Operators express that they "want people to understand". Visitors often ask operators "what can I do to help?" to which a common response is "go and learn". This presents a quandary for visitors. They may think "but I'm here learning and I still don't know what to do" or "what do you want me to understand? And what then?".

Individual and community actions exist within the need for broader systemic action, meaning that social and political reconciliation processes are inextricably linked. Awareness raising efforts can fuel social momentum, with individual and collaborative efforts exerting pressure on decision makers to implement change. The findings of this research support the notion that reconciliation needs to move beyond the current period of awareness raising towards action-oriented social change. Here, we see that learning and understanding have tangible utility. Reaching critical mass for political change requires the voicing and consideration of many different perspectives simultaneously to navigate a course of action that will serve as many people as possible. Reconciliation paralysis can be seen on individual and political levels, with Reconciliation Australia (2021) asserting that brave action is required for progress. The question is, what does this brave action look like?

Within Indigenous tourism interactions, a one-way delivery and consumption of information can be transcended to an informed approach to working together when visitors are also open to sharing. Visitors can engage openly and share their own insights and self-reflection. This sharing from visitors might include common feelings such as guilt or shame around their own lack of understanding or an awareness of their preconceptions, or feelings of uncertainty around individual action. This represents two-way knowledge-sharing. Brave action can begin here.

Difficult emotions reflect engagement with a difficult process, but the experience of difficult emotions by visitors is not an end in itself (Stinson et al., 2022). Indeed, Indigenous tourism operators are also undertaking emotional labour (Graham & Dadd, 2021). They are explaining, again and again, the basic facts of Indigenous sovereignty, guiding visitors as they grapple with what it means to value Indigenous knowledges, and patiently modelling learning as both a process and an outcome as visitors consider how they can support Indigenous affairs.

Although not every tourism experience is transformative, every experience *embodies* transformative potentials, even if these appear relatively minor. Learning requires moving beyond what we already know about the world. Transformative learning cannot be accurately designed for, as each person holds their own unique biases. However, listening deeply can encourage visitor self-reflection: "The world is not divided up into those people who have bias and those who don't. It is divided up, though, into those people who recognise they have bias and those people who think they have none" (Kandola, 2013, para. 7).

Learning as Action

Through acknowledging the need for deep listening within Indigenous tourism, the question arises, what does reciprocity look like beyond these tourism encounters? Here, we come full circle. Learning is an aspiration of hosts and a motivation of visitors, and it is a process and outcome within the Indigenous tourism arena. One of the most common questions that visitors ask of operators is, "what can I do?" The operators suggest: engage and learn, and continue engaging and learning. Ongoing learning is thus part of the required action for moving forward.

Stories shared by operators of individual, cultural, community, and environmental impacts of colonisation can evoke empathy, which can enhance social relationships (Reconciliation Australia, 2021). Constant negotiation is required to address the void of cultural understanding from the deficient education system and support white capacity building (Taylor & Habibis, 2020). Learning therefore requires commitment.

Learning is active, not passive, and it is both powerful and simple. Unlearning bias is deep, personal work, as critical reflection is necessary for both individual and collective transformation (Mezirow, 2012). Transformation requires awareness raising to acknowledge the facts of Indigenous sovereignty, working to find a shared national identity with shared values, and a commitment to constantly learning and negotiating mutually beneficial outcomes that support Indigenous rights and self determination. Transformative learning within a tourism context can lead to more sustained involvement in activism outside of the tourism experience by facilitating visitors' "introspection about their perceived role in being an agent of change" (Soulard & McGehee, 2023,

p. 16). This leads us to consider how lifelong learning beyond the tourism arena can be promoted and how this model can be used to inform practice.

Firstly, this model emphasises the holistic nature of the learning journey. An awareness of these learning potentials can help tourism operators conceptualise how they can share knowledge to maximise impact and adjust their practice if necessary to cater best to each new audience. The ability to differentiate is attributed to expert educators. The model presents learning as a deeply personal process, highlighting the importance of connection and two-way dialogue (Ormond & Vietti, 2022) where visitors think about themselves in relation to a shared history and their own connection to Indigenous peoples, histories, and cultures. This is in line with Indigenous relationality where "everyone is someone" (Bart Pigram) and "All persons matter. All of us belong" (Ungunmerr-Baumann, 2002, para. 3). When reflecting upon what valuing Indigenous knowledges means for them in their own lives, visitors were often conflicted and confused about what genuine allyship entails. From the findings, we can see how a commitment to ongoing learning means that the more that visitors understand, the more they can advocate for Indigenous rights and position their individual actions within a broader sociopolitical context.

Next, when considering the processes and goals of reconciliation, we need to keep in mind that the outcome of a transformative journey is not easily predictable (Reisinger, 2013). As educators, Indigenous operators face specific challenges as they attempt to "position themselves in relation to material of an emotive and uncertain quality" (Garvey, 2015, p. 349). Garvey proposes that a praxis of uncertainty can facilitate educators to engage imperfectly with learners while modelling ways of addressing uncertainty: "The role of the [educator] in this circumstance need not be as counsellor nor solution provider. It may instead focus on supporting the questions and acknowledging the different standpoints that [learners] bring with them to the educational encounter" (p. 350). As taking action is never risk-free, a degree of uncertainty is unavoidable. Autonomous thinking developed through transformative learning can assist visitors to manage this uncertainty.

Finally, the importance of local knowledges cannot be overemphasised. As Indigenous populations are not homogenous, locally negotiated solutions that represent people and place are vital. Concepts of connectedness are central, as visitors "come to know a little about their own backyard" (Mick). Clinton explains that in Ngarluma language, there are two categories of people, those local to the place, *ngurrara*, and those from outside the area, *manyjangu*. This distinction between local custodians and visitors dissolves the Indigenous-non-Indigenous divide and is consistent with the "old practices of welcoming people to our Country" (Bart). Maybe this gives us clues to achieving more equal terms of engagement?

The privileging of local Indigenous perspectives can assist with transcending difference. Although visitorsexperience uncertainty about the best course of individual and collective action to support Indigenous affairs they also resonate with the stories shared by hosts and can discover a sense of congruence and relief in arriving at an understanding of *"our shared history"* and a sense of *"belonging to Australia together"*. To *"go and sit and learn"* about local wisdom not only presents opportunities for social cohesion but also proven benefits for sustainable land and sea management (Indigenous Land and Sea Corporation, 2021). Visitors may genuinely want to experience and learn about Aboriginal cultures however, they are only able to take on the knowledge that they are ready for at that time. This is why ongoing exposure to Indigenous peoples and cultures is important.

Operators are sharing knowledge in generous and positive ways, meaning that the "opportunity to meet Aboriginal people and share some degree of intimacy with them was for many tourists of far more significance than simply experiencing the cultural aspects which the tours have on offer" (Galliford, 2010, p. 230). Positive and connected experiences mean that visitors will be more inclined to continue to engage and reflect with a willingness to *"actually challenge themselves"*, which is one way to think about brave action.

Implications for Cultural Competency

These learning potentials open doors for a more cohesive national narrative founded upon truthful and accurate stories of Australian history and a national identity that views culture and difference as strengths. Cultural competency, although a growing requirement and trend, is often seen as a once-off "tick box" training. It is often incorporated into Reconciliation Action Plans and taught in university degrees as ancillary, not core, business. This tokenistic approach does not embed knowledge into practice and can do more harm than good.

These findings lend support for moving towards "cultural agility" (Garvey, 2015) or "culturally responsive" (Morrison et al., 2019) models, which view cultural competency as an ongoing action rather than an endpoint. Indeed, the diversity of individual understandings and connections to culture and cultural identity means that cultural competency cannot be a state to be achieved, but is under constant negotiation, and enacted through ongoing reflection. To embed cultural reflexivity as an ongoing practice would equip workforces with the skills to work and engage with Indigenous peoples and communities in meaningful ways to deliver truthful and accurate accounts of Australian history and culture in inclusive and culturally respectful ways. Culturally responsive models address the dehumanising and damaging discourses that disempower and disenfranchise Indigenous Australians (Woodroffe, 2020). Embedding cultural competency requires listening to the needs and concerns of Aboriginal Australians and increasing the visibility and recognition of Indigenous people as knowledge-holders. Indeed, Indigenous tourism operators are on the ground, day after day, contributing to building the capacity of white Australia. Indigenous tourism operators are modelling reconciled interactions to teach for change.

Conclusion

Engaging in Indigenous tourism presents transformative learning potentials whereby visitors' pre-existing ways of understanding the world can be challenged and enhanced. Respecting Indigenous sovereignty can assist visitors to understand Australia's colonial legacy and have empathy for the resulting social disadvantage and the urgent need to address *Closing the Gap* targets. Valuing Indigenous knowledges means doing things in culturally appropriate ways and points to the need for reciprocal and mutually beneficial relationships where Indigenous tourism operators get something other than monetary gain back from sharing their knowledge and culture so generously and patiently with visitors. Supporting Indigenous affairs is an ongoing learning process, requiring individual and collective commitment to stay the course to find new ways of working together, even when difficulties and challenges arise. Visitors must think about Indigenous histories, cultures, and knowledges in relation to their own lives to address biases. They must also be willing to *"actually challenge themselves"* and understand that negotiations and tensions will continue to be ongoing.

The model of transformative learning potentials has implications for viewing Indigenous tourism as a cultural competency activity, with culturally responsive approaches requiring reflexivity and agility. Deep listening presents opportunities for visitors to reflect on their assumptions and stereotypes (Ungunmerr-Baumann, 2002). Critical self-reflection is necessary for transformative learning and is a way that we can understand the experiences of discomfort and uncertainty that visitors describe. A commitment to ongoing learning means that the more that visitors understand, the more they can advocate for Indigenous rights in their own personal and professional lives. Here, feelings of uncertainty and guilt can be replaced with a more informed approach to reconciliation. Risk is never totally avoidable, which is why brave action is required. This bravery is not solely individual, as social and political reconciliation processes need to be understood not as neatly separate but as inextricably linked.

By considering the visitor learning journey as both holistic and ongoing, we see the need for thinking more broadly about Indigenous affairs. Indigenous affairs are not siloed and compartmentalised. Indigenous voices must be integrated across industries and systems rather than relegated to Reconciliation Week, one week of the year, where reconciliation agendas are displayed as corporate badges of honour, where Indigenous people are treated like "silverware dragged out for special occasions and then shoved back in the drawer for the rest of the year" (L. Watene, personal communication, November 11, 2021), and where the responsibility is placed on Indigenous Australians to teach, rather than acknowledging the responsibility for non-Indigenous Australians to *"go and sit and learn"*.

As a social justice activity, Indigenous tourism provides a valuable contribution to Australian society. We need to begin from a place of actually challenging ourselves and respecting Indigenous sovereignty, valuing Indigenous knowledges, and supporting Indigenous affairs to start the real work of more nuanced discussions that embed a diversity of ways of being, doing, and knowing across systems and levels. Indigenous tourism engagement presents transformative opportunities, not as a token buzzword, but as a viable and pragmatic approach to ensuring that the future is radically different from what has come before.

Acknowledgements

We acknowledge the Whadjuk Noongar, Larrakia and Ngarluma peoples as the Traditional Owners of Country where the authors live and work. We thank Mick Hayden for allowing us to quote his words in the title of this chapter. We extend our thanks to Bart Pigram, Kerry-Ann Winmar, Marissa Verma, Mick Hayden and Roland Burrunali for their generosity and for allowing us to share their stories. We are also grateful to the non-Indigenous hosts and tourism visitors who have shared their stories with us. We acknowledge the guidance and input of Professor Ruth Wallace and Professor Steven Bird on this project. This research was supported by a Commonwealth Research Training Programme scholarship for the first author. This project has approval from the CDU Human Research Ethics Committee.

Notes

1 *Closing the Gap* is an Australian government strategy to improve outcomes for Indigenous Australians to reduce entrenched disadvantage.
2 Sampled data includes semi-structured interviews and yarning, field notes and observational data, and other relevant data such as promotional and website material, media and social media data, and presentations.
3 Operators who are located in WA are members of Western Australian Indigenous Tourism Operator's Council (WAITOC) with the tour guide from the NT employed as an Aboriginal staff member with Injalak Arts, a non-profit Aboriginal Corporation.
4 *Terra nullius*, which translates to "nobody's land", was the doctrine used by British colonists to justify the occupation of so-called Australia.
5 NAIDOC is the acronym for National Aborigines and Islanders Day Observance Committee. National NAIDOC Week in July celebrates the history, culture, and achievements of Aboriginal and Torres Strait Islander peoples.

References

Charmaz, K. (2014). *Constructing grounded theory* (2nd ed.). London: Sage Publications Ltd.

Curtin, N., & Bird, S. (2022). "We are reconciliators": When Indigenous tourism begins with agency. *Journal of Sustainable Tourism, 30*(2–3), 461–481. https://doi.org/10.1080/09669582.2021.1903908

Curtin, N., Walker, C., & Woodroffe, T. (2022, August 26). Sacred Aboriginal sites are yet again at risk in the Pilbara. But tourism can help protect Australia's rich cultural heritage. *The Conversation.* https://theconversation.com/sacred-aboriginal-sites-are-yet-again-at-risk-in-the-pilbarabut-tourism-can-help-protect-australias-rich-cultural-heritage-188524

Denzin, N. K., & Lincoln, Y. S. (2008). Critical methodologies and indigenous inquiry. In N. K. Denzin, Y. S. Lincoln & L. T. Smith (Eds.), *Handbook of critical and indigenous methodologies* (pp. 1–20). London: Sage Publications Ltd. https://doi.org/10.4135/9781483385686

Galliford, M. (2010). Touring 'country', sharing 'home': Aboriginal tourism, Australian tourists and the possibilities for cultural transversality. *Tourist Studies, 10*(3), 227–244. https://doi.org/10.1177/1468797611407759

Garvey, D. C. (2015). *A causal layered analysis of movement, paralysis and liminality in the contested arena of Indigenous mental health* (Doctoral dissertation, Curtin University).

Graham, M., & Dadd, U. L. (2021). Deep-colonising narratives and emotional labour: Indigenous tourism in a deeply-colonised place. *Tourist Studies, 21*(3), 444–463. https://doi.org/10.1177/1468797620987688

Higgins-Desbiolles, F. (2009). Indigenous ecotourism's role in transforming ecological consciousness. *Journal of Ecotourism, 8*(2), 144–160. https://doi.org/10.1080/14724040802696031

Higgins-Desbiolles, F., & Bigby, B. C. (2021). A local turn in tourism studies. *Annals of Tourism Research, 92.* https://doi.org/10.1016/j.annals.2021.103291

Indigenous Land and Sea Corporation. (2021). *Annual report 2020-21.* Indigenous Land and Sea Corporation. https://www.ilsc.gov.au/wp-content/uploads/2021/10/ILSC-AR-Web-s.pdf

Kandola, B. (2013). *Diffusing bias: Reimagining self and other* [Presentation]. Facing History and Ourselves. https://www.facinghistory.org/resource-library/video/day-learning-2013-binna-kandola-diffusing-bias

Mazza, G. (2022, September 23). *What you need to know about the Burrup Hub.* Australian Greens. https://greens.org.au/magazine/what-you-need-know-about-burrup-hub

Mezirow, J. (2012). Learning to think like an adult: Core concepts of transformation theory. In E. W. Taylor & P. Cranton (Eds.), *The handbook of transformative learning: Theory, research, and practice.* San Francisco, CA: Jossey-Bass, 73–95.

Morrison, A., Rigney, L. I., Hattam, R., & Diplock, A. (2019). *Toward an Australian culturally responsive pedagogy: A narrative review of the literature.* Adelaide: University of South Australia.

Ormond, M., & Vietti, F. (2022). Beyond multicultural "tolerance": Guided tours and guidebooks as transformative tools for civic learning. *Journal of Sustainable Tourism, 30*(2–3), 533–549. https://doi.org/10.1080/09669582.2021.1901908

Reconciliation Australia. (2021). *2021 State of reconciliation in Australia report: Moving from safe to brave.* https://www.reconciliation.org.au/wp-content/uploads/2021/02/State-of-Reconciliation-2021-Full-Report_web.pdf. Accessed August 12, 2022.

Reisinger, Y. (Ed.). (2013). *Transformational tourism: Tourist perspectives.* Wallingford: CABI.

Soulard, J., & McGehee, N. (2023). Transformative travel and external activism: Framing external activism outcomes within the travelers' discourse of perceived transformation. *Journal of Sustainable Tourism, 31*(3), 821–839. https://doi.org/10.1080/09669582.2022.2029871

Stinson, M. J., Hurst, C. E., & Grimwood, B. S. R. (2022). Tracing the materiality of reconciliation in tourism. *Annals of Tourism Research, 94,* 1–12. https://doi.org/10.1016/j.annals.2022.103380

Taylor, P. S., & Habibis, D. (2020). Widening the gap: White ignorance, race relations and the consequences for Aboriginal people in Australia. *Australian Journal of Social Issues, 55*(3), 354–371. https://doi.org/10.1002/ajs4.106

Ungunmerr-Baumann, M. (2002). *Dadirri: Inner deep listening and quiet still awareness. A reflection by Miriam-Rose Ungunmerr-Baumann.* Thornleigh, NSW: Emmaus Productions.

Woodroffe, T. (2020). Improving Indigenous student outcomes through improved teacher education: The views of Indigenous educators. *AlterNative: An International Journal of Indigenous Peoples, 16*(2), 146–152. https://doi.org/10.1177/1177180120929688

Woodroffe, T. (2021). Presentation feedback as an Indigenous methodology. *Australian Journal of Education, 65*(1), 73–83. https://doi.org/10.1177/0004944120969987

9

TOURISM, TRUTH-TELLING AND SOVEREIGNTY

A Gitxaała Perspective

Caroline Butler and Bruce Watkinson

Introduction

Indigenous Tourism has been an economic strategy for many communities and First Nations in Canada for decades. The performance and exhibition of culture for visitors can also support community-based cultural revitalization goals. As Indigenous governments and entrepreneurs expand their role in the tourism industry, the opportunities to increasingly link political and social goals to tourism development should be explored. The Gitxaała Nation on the north coast of what is now British Columbia, Canada, looks to tourism as supporting economic and cultural outcomes, but also as a tool to educate the Canadian public and international visitors and to advance 'reconciliation' with the governments of British Columbia and Canada. This paper explores the ways in which intentional and strategic tourism development can potentially support the political actions and aspirations of a small Indigenous Nation.

This chapter is co-written by a Gitxaała citizen and a collaborative researcher, both former employees of the Gitxaała government. Bruce Watkinson is a member of the House of Gitnagun'aks of the Gispudwada (Killerwhale) clan of the Gitxaała holds the hereditary name Wudimes. Trained as a biologist, he is the former Director of the Gitxaała Territorial Management Agency is the current Executive Director of the North Coast Skeena First Nations Stewardship Society. Caroline Butler is non-Indigenous and trained as a cultural anthropologist. She worked with the Nation for 22 years and was formerly the Research Program Manager for the Language and Culture Department. The paper is written from a Gitxaała perspective and structured by Gitxaała cultural concepts and principles. While linked to the academic literature on Indigenous tourism and related work, this chapter strives to remain grounded in territory and community and focuses on Gitxaała's priorities and considerations. This is a community-based case study of strategically integrating political goals of decolonization and reconciliation into economic development planning. Tourism, for Gitxaała, is part of a comprehensive and coordinated approach to reclaiming rights and territory and rebuilding community vitality.

This exploration of tourism goals is structured to reflect the three structuring principles of Gitxaała society in the Sm'algyax language: *adawx* (true tellings, oral record), *ayaawx* (laws) and *gugxwilya'ansk* (inheritance). Indigenous tourism experiences are frequently linked to the history, protocols and cultural practices that are, in Gitxaała society, grounded in these principles. What

DOI: 10.4324/9781003230335-11

106

we intend to argue in this paper, is that a fourth principle, *malsk* (history) needs to be integrated into contemporary Indigenous tourism to promote greater awareness, understanding and political support.

The adawx are the "true tellings" or lineage histories that document the origins of Gitxaała clans and hereditary names, the ownership of territory, and major events, and "references places, events, people and privileges (crests, songs, stories etc.), things (tanglible and intangible) that form property, and rights within Gitxaała society" (Menzies 2016: 71). The adawx section will provide an introduction to Gitxaała territory, culture, and history. The *ayaawx* section will describe the ways in which Gitxaała laws are practiced and protected. Gitxaała ayaawx is a collection of precepts that range from broad principles to specific protocols. Ayaawx governs relations between individuals, between communities, between humans and the natural world, and the spiritual world. There are ayaawx about inheritance, rights of passage, territorial governance, justice, and inter-nation conflict resolution.

Translated literally, *gugwilxya'ansk* means 'to pass down through generations.' It can also be understood as birthright or inheritance; it is the process, structure, and tradition of passing Gitxaała names and territory to future generations. The *gugxwilya'ansk* section will outline current efforts to protect, and reinvigorate Gitxaała birthrights for future generations, including the repatriation of Gitxaała cultural treasures to support cultural tourism and cultural education.

Tourism development is part of this project of rebuilding for the future. An intentional approach to tourism that we outline in this chapter can be understood by adding the concept of *malsk*, the telling of recent history, to the cultural practices that are the more expected forms of Gitxaała tourism. Tourism experiences in Gitxaała territory should be about hearing the truth and building understanding to enhance support for Gitxaała political action and recognition of Gitxaała rights.

Ayaawx – Gitxaała Society and Governance

Gitxaała people are born into a walp (house group or lineage) of one of four exogamous matrilineal pteex (clan or tribe): G̱anhada (Raven) Gisputdwada (Killerwhale), Lasgiik (Eagle) and Laxgibuu (Wolf). Walps are led by Smgigyet (house chiefs) who own chiefly names and territories and will pass them to a matrilineal nephew, as identified by the Sigyidm hana'ax (matriarchs). Ranked hereditary names, and the other forms of property and rights associated with these positions, are passed down through the feasting system (more widely known as the potlatch).

The Canadian government established elected councils in most Indigenous communities as part of the implementation of the federal Indian Act, which also outlawed, for almost seven decades, the feasting events through which Gitxaała hereditary governance is practiced. Today, the Gitxaała Nation is governed collaboratively by Hereditary and Elected leadership. Gitxaała territories and society are governed by a Hereditary Table of Smgigyet (chiefs) lik'agigyet (head men) and Sigyidm hana'ax (matriarchs). Each chief is traditionally responsible for the health and well-being of the members of their walp (house group/matrilineage) and for the management and enhancement of the lands held by the walp (see Menzies and Butler 2007). Smgigyet advises on the harvesting of resources in their territories and guides the elected Gitxaała council on Nation-level territorial issues. The Hereditary Table meets quarterly to be updated by the elected council, department managers, and to determine key mandates regarding territorial and cultural issues.

While the ayaawx have persisted and have continued to shape Gitxaała society despite colonial efforts to undermine this legal system, they are increasingly integrated into Gitxaała Nation structures, processes and protocols and school curriculum.

The Elected Council oversees the administration of the Nation, manages the community programs and departments, and enacts the mandates of the Hereditary Table. There are seven elected positions which are held for a three-year term. The Gitxaała government employs over 75 people, most of whom are Gitxaała citizens. The economic development initiatives of the Nation are run by Gitxaała Enterprises which invests in a variety of businesses and joint ventures, including tourism enterprises. Recent business strategies include the purchase of a share in an eco-tourism operation and a hotel in Prince Rupert.

Adawx – *Gitxaała Oral History*

Gitxaała adawx is the origin stories that link the names and territories held by Gitxaała leaders today to the deep past and to the supernatural beings, the nax nox, that inhabit the territory, including worlds in the sky and in the depths of the ocean. These narratives tell people who they are and link to the ayaawx in that they illustrate how Gitxaała people should behave to each other, to outsiders, and other living creatures. The adawx are part of the property of Gitxaała houses and many of the totem poles, masks and other treasures created by Gitxaała artists are visual representations of these narratives.

The adawx tells us that Gitxaała Nation is one of the most ancient Indigenous societies on the North Pacific coast. Following the matrilineal system of inheritance, hereditary leaders manage specific marine and terrestrial territories that have been used and enhanced for many millennia. Following the population loss caused by contact with European explorers and settlers, Gitxaała people chose to leave their many villages stretching from Kaien Island to Aristazabel Island and settled together in the ancient winter village of Lach Klan (Kitkatla BC) on Dolphin Island. Their population had dropped by almost 90%. Today there are 2008 Gitxaała citizens, approximately a quarter of whom live in the village of Lax Klan (Kitkatla BC), which, previously a Gitxaała winter village, has been inhabited for over 9,000 years. Lax Klan is 60 km southwest of Prince Rupert, the hub of the North Coast with a population of 12,500. The village is accessible by ferry or floatplane. While linguists, anthropologists and colonial governments have classified the Gitxaała as 1 of the 14 'tribes' that make up the 'Tsimshian' in British Columbia, the Gitxaała highlight their distinct governance system, oral histories, language dialect and contemporary history from Tsm'syen people.

Gugwilx'ya'ansk: *Cultural Continuity and Protection*

Translated literally this means "to pass down through generations". It can also be understood as birthright or inheritance; it refers to the process, structure, and tradition of passing on Gitxaała names and territory. We often use this concept to frame discussions of territorial protection and efforts to assert Gitxaała rights and sovereignty. The protection of the Gitxaała language, culture and way of life is part of the contemporary commitment to *gugwilx'ya'ansk*.

The Gitxaała Territorial Management Agency engages in a variety of initiatives and projects to protect the resources and territory of the Nation (see Butler, Watkinson and Witzke 2021). Marine spatial planning occurs on several scales and in collaboration with other governments to identify protected areas and oil spill response strategies. Fisheries co-management structures ensure that Gitxaała fisheries interests are protected and prioritized. Fisheries monitoring and water quality sampling are some of the scientific research that supports territorial health. The early 2000s saw a flurry of new industrial development proposals in the region including an expansion of the container port, marine and terrestrial windfarms, liquid natural gas export terminals and an oil pipeline and tanker route. The regional economy has shifted from fishing and forestry to a port-based focus on marine shipping and resource export. GTMA participates in environmental assessments on industrial development proposals to protect harvesting and other cultural rights.

The Language and Culture Department supports Sm'algyax revitalization through language documentation, community classes and curriculum development. The repatriation program is arranging the return of Gitxaała belongings from museums throughout the world, to develop a Gitxaała community museum.

Lach Klan School integrates Sm'algyax language instruction and land-based learning in efforts to provide a culturally relevant and appropriate curriculum and support the strengthening of Gitxaała culture while achieving other learning outcomes.

All of these initiatives to maintain, strengthen and protect Gitxaała's cultural and natural resources are part of the ethic of *gugwilx'ya'ansk.* Tourism development will also be shaped by this principle of protecting and passing down a rich heritage to future generations of Gitxaała.

Malsk

Many Indigenous tourism experiences focus on what in Gitxaała society are the stories, practices and visual representation of the adawx, and gugxwilya'ansk. These tend to be representations of pre-contact ceremonial practice. We feel Indigenous tourism needs to deliberately improve Canadian and international understandings of First Nations society through engagement with the true stories of colonialism, cultural resilience and representations of contemporary Indigenous life. In Sm'alygax, the telling of more recent history or stories - 'malsk' - and an emphasis on ayaawx and governance can support awareness of Indigenous issues and rights. These more recent stories provide another component of the Gitxaała oral history. The adawx and malsk are complementary forms of history upon which Gitxaała people are drawn on to understand their past and their contemporary identity. It is this holistic historical perspective that must shape and inform the presentation of Gitxaała culture and society to visitors. It is *malsk* that provides an understanding of the last 250 years of Gitxaała history and the Nation's interactions with colonial intrusions and settler culture. It is *ayaawx* that contextualizes the continuity of Gitxaała governance and commitment to the protection of territory and culture.

K̲'amksiw̲a̲hs (translated as ghost people) arrived in Gitxaała territory in the late 18th century, on fur trading ships. Over the next century, significant changes were wrought on the territory and the people. Population loss through epidemics was as high as 90% in some villages. Gitxaała territory was declared Crown Land, except for 21 small reserves. Before the imposition of the reserve system, Gitxaała house groups moved between various harvesting camps and winter villages within their territories. Gitxaała people continue to access over 100 marine and terrestrial food resources. When the industrial fishery was developed on the North Coast, Gitxaała people integrated this new market into their economy, participating in both fishing and cannery work (see Menzies and Butler 2008). In the winter months, Gitxaała families worked their traplines for furs and engaged in small-scale coastal hand logging in their house territories. This weaving of the traditional and colonial economic systems allowed the Gitxaała people to continue to monitor and use their territories and resources. The commercial fishery declined drastically in the latter decades of the 20th century, and government restructuring policies significantly impacted coastal Indigenous communities, including Gitxaała, unsettling the fragile combination of industrial and traditional harvesting. The increasing restriction of access to the commercial fishery from the 1960s onwards has resulted in significant poverty for the Gitxaała people.

The economic disruptions of colonialism however were more easily absorbed than many of the social disruptions. Missionaries arrived in 19th century, encouraging a shift to single-family dwellings and outlawing traditional cultural practices. These policies were supported by the federal government banning the potlatch system of governance for over 67 years. Many Gitxaała children were taken to residential schools in the south, where they suffered diverse forms of abuse. The impacts of colonialism are evident in Gitxaała society, like many other Indigenous communities, with high rates of unemployment and poverty, poor health outcomes and the social issues associated with trauma.

However, the Gitxaała have strategically resisted colonial intrusions throughout and are currently striving to reinvigorate their society and economy. Gitxaała harvesters continue to feed their families and community, with many households continuing to harvest up to 90% of their protein from the territory. Feasts are held regularly in the winter to mark the passing of names within a lineage, and cultural and spiritual practices continue. Sm'algyax is taught in the village school and in community classes for adult learners. The first longhouse to stand in the village in over a century is beginning construction in summer 2023. This longhouse will be the center of Gitxaała tourism development.

Tourism and Reconciliation

Indigenous tourism can engage visitors in cultural learning. Bringing people into Gitxaała territory provides an opportunity for a broader learning experience, including the history of oppression and genocide, and land annexation, and the contemporary forms of colonialism that still impact our people, in addition to a celebration of our ancient and resilient culture.

In June 2021, the unmarked graves of 215 children were found on the grounds of the former Indian residential school in Kamloops. This saddened but did not shock Indigenous people. Between 1867 and 1996, over 150,000 Indigenous children were forced to attend residential school at 1of 140 federal institutions in Canada. Many did not come home, and countless children suffered incredible abuse and neglect. Between 2007 and 2015, 6,500 witnesses told their stories to the Truth and Reconciliation Commission of Canada which was established to investigate this genocidal project. The Commission's report included 94 Calls to Action, several of which spoke to the need to investigate residential school grounds for grave sites. Federal funding did not materialize for this work and the Secwempemc Nation's self-funded investigation blew the lid off one of Canada's best-kept secrets. Residential school histories have only recently been integrated into the school curriculum. Many generations of non-Indigenous Canadians were legitimately shocked by the number of unmarked graves when the story broke. Indigenous leaders warned it was just the beginning. At the time of writing, more than 2,000 graves have been located.

The discovery of these lost children has initiated meaningful public discussion about the impacts and legacies of colonialism in Canada, which is much needed at the current juncture in Canadian history. Over the last three decades, there have been significant advances in the legal recognition of Indigenous rights in Canada. Key court cases have drastically altered the ways in which federal and provincial governments engage with Indigenous governments and communities. This is good news for Indigenous communities in many ways but is also resulting in new potential sources of division between Indigenous and non-Indigenous Canadians.

In British Columbia, federal and provincial recognition of Indigenous land titles and other rights has resulted in significant changes in the federal and provincial government relationships with First Nations, altering the way industrial development can proceed, how commercial fisheries and forestry are structured, and the protocols associated with public events. Indigenous leaders are present at many events, land acknowledgments are now a standard part of the opening of any meeting or event, and Indigenous culture and art are increasingly profiled. Reconciliation, in these new forms, can be somewhat jarring for non-Indigenous Canadians. This is a somewhat rapid socio-political change that is driven by legal rulings on rights and titles. While the achievement of meaningful change in decision-making power and economic inequity is in reality very slow, the optics of change are fast.

Especially in northern resource-dependent communities, there is some frustration with and resistance to the shift in the recognition of Indigenous rights. Land claims in coastal British Columbia invariably involve fisheries measures, which can often directly impact commercial fish harvesters. This is a major flaw in the way Canada and BC approach the settlement of claims; local resource industries often bear the brunt of the settlements more than other Canadians. In this way, advancements in social and economic justice are leading to a polarization of public opinion.

Furthermore, the understanding of Indigenous Nations as having special rights is counter-intuitive to the average person's understanding of equality. The legal basis for these rights in Canadian law is poorly understood by most citizens. While it may feel like new liberal laws are being created, in fact, the recognition of Indigenous rights and title was built into British and then Canadian law; Canada has been breaking its own laws for centuries when it comes to Indigenous land.

Gitxaała and other Indigenous Nations have argued for social justice based on "special rights". The federal and provincial governments have been forced by the courts to recognize these rights. One of the main challenges is to build an understanding of these rights in the general public. The current moment is one with a high risk of alienation. The ability of people to hear and actually listen to the atrocities Indigenous peoples have experienced is limited and potentially shrinking. There are new stories every day about unmarked graves, about Murdered and Missing Indigenous Women and Girls, and communities without clean drinking water. Indigenous issues are in the news, but we worry that they are becoming ubiquitous.

Reconciliation can't be anonymous. It needs to be grounded and personal. That is what can be achieved through tourism experiences. We can weave a more digestible way of telling and teaching into ecotours and cultural experiences. Tourists are looking for an "authentic" experience. An authentic experience of Gitxaała is to hear about our challenges and our resilience. It is to hear our language but to understand how few fluent speakers we have left, and why. It is to see our repatriated totem pole but to engage with the pressure from missionaries to cut it down, the duress under which it left the community, and the challenges faced bringing it and other treasures home.

Tourists are paying to hear our story. We need to take the opportunity to tell the whole story and all the stories. Ideally, when people have seen our territory and heard our stories, the next time we litigate against the province, veto a pipeline, or remove a treasure from an external museum, they will understand why and support us or at the very least, respect our position. In this way, tourism can combat stereotypes and misconceptions and contribute to the growth of public support for Indigenous political action and initiatives.

In our ongoing battle for sovereignty, there is a place for protests, litigation, and other direct action. But we also need a softer approach that rallies support gently. While the courts can advance some of our interests and rights, political power is determined by votes and public opinion. We need to slowly build support for our rights in the court of public opinion. Canadians need to understand that we are different and that our difference is dependent on our territory and our right to use it. Public understanding of how Gitxaała governance and law functions and continues to guide our practice and decision-making can increase public support for Gitxaała rights. Tourism is an opportunity to share our culture but also our governance structures, our rights, and our challenges. People must understand that Gitxaała hereditary governance is still practiced and that the rights and responsibilities of the hereditary chiefs shape our Nation's policies and mandates. This goal is greater than reconciliation; it is about creating allies and broader support for our political initiatives.

It is important to note that as a small Nation, far from urban centers and large populations, our ability to sway the public through tourism will be limited. Given our location and limited infrastructure, we can expect only a small number of visitors a year to Gitxaała tourist attractions. It will take us a long time to engage with enough visitors to make meaningful changes. In this, as in many arenas, decolonization is a team sport. We hope that our neighbors and other First Nations in BC will also be telling their true stories to tourists. We can build up the truth and grow understanding across the province while showcasing our cultural practices and arts.

We need to update the public understanding of our history from a focus on pre-contact to contemporary life and a celebration of resilience and revitalization that inspires respect, empathy, understanding and political support. The time of us highlighting longhouses and children in buckskin is gone. Indigenous tourism must include the "modern" Indigenous peoples affected by all the impacts of colonialism.

While a part of the motivation is garnering political support, the Canadian and international public deserves and needs to know an accurate history and the realities of contemporary Indigenous life. Many non-Indigenous people in this region grew up knowing nothing of the richness of our cultures or the deliberate dismantling of those ways of life. They still don't understand our ongoing experiences of being treated differently in stores and hospitals. The outcomes of a tourism program in our village can be multi-level. We can push our political agenda but also break down local stereotypes. The recognition of our fisheries rights might land better when the public understands our continued dependence on local marine resources. Our legal battles against pipelines and gold mines might have more external support when people see the land we are trying to protect. The recognition of our leaders beside elected officials will make sense to them when they have heard how our ayaawx is still practiced through our hereditary system of governance. We can dismantle myths held by many Canadians about our rights and build understanding in local populations about what was here before them and what remains.

Gitxaała Intentional Tourism in Context

Gitxaała's goals for an integrated and intentional approach to tourism as a social justice lever align with what Higgins-Desbiolles (2003) has named "reconciliation tourism" in Aboriginal Australian communities. Reconciliation tourism can include some or all of the following attributes: contact between Indigenous and non-Indigenous peoples, experience to foster social justice, understanding or healing, and experiences that create "bridges" between communities, or that educate settlers (Higgins-Desbiolles 2003).

While some tourism experiences employ a direct call to action, the more subtle effects of connection can build the foundation of reconciliation. Community guides at Goombaragin implement "hospitality, compassion, respect and understanding" to create a bond with visitors and to foster dialogue (Jacobsen 2018: 29).

Tourism can thus be a tool to "chip away" at the barriers between Indigenous peoples and the newcomers to their territories (Higgins-Desbiolles 2006a: 143). While gentle, it is not necessarily comfortable and requires the tourist to "have their ideas challenged to affect the societal changes that reconciliation requires" (2006a: 148). Higgins-Desbiolles frames tourism as a "social force", not just an industry (2006b: 1192), indicating its potential for creating social and ultimately political, change. Similarly, Jacobsen (2018: 22) understands that reconciliation tourism can "stimulate genuine dialogue that compels visitors towards transformative change".

Kramvig and Førde (2020: 39) describe a Sami story-telling as "enacting" reconciliation, by bringing the colonial past into the present and creating connections and "relation-weaving" between Saami, locals and visitors. Indigenous tourism can thus be considered an embodiment of reconciliation where the interaction between two cultures and shared celebration of Indigenous resilience both is and fosters reconciliation. Understanding Indigenous tourism as storytelling and thereby an act of re-empowerment supports the integration of contemporary experiences and issues into representations of the deep past and ceremonial present. The Gitxaała tourism strategy of integrating *Malsk* along with *adawx, ayaawx* and *gugwilx'ya'ansk* into the tourism experience weaves recent history and experience into each part of the tour, stories of the land and water, of loss and genocide, of repatriation and resurgence.

Weaving social justice and healing into tourism is also an intentional component of Civil Rights tourism and the history of slavery in the American South that hopes to enhance public education around racism and the African American experience (see for example Loewe 2014). For some attractions, racial reconciliation is a "by-product" with an emphasis on economic development and historic preservation (Barton and Leonard 2010), but others focus on public education targeting a tourism audience and addressing the root causes of contemporary inequality (Skipper 2016). Skipper (2016: 525) emphasizes that tourism initiatives that have an education focus rather than profit-focus can leave more room to "strategize around interpretation and education than returns" and that "Tourism strategies that prioritize community wellbeing counteract the structural silencing of racially underrepresented groups" (ibid.). This is in contrast to many historic sites in the US South that tend to minimize slavery (Butler 2001) and the general critique that Timothy and Boyd describe many heritage attractions providing a "sanitized and idealized" picture of the past (2003).

Gerberich (2005) relates the educational component of an Indigenous tourism experience directly to the cultural sustainability of the enterprise. Indicators of cultural sustainability include educational activities, guided tours, and the availability of information. This suggests a direct link between tourism as a public education strategy and community cultural experience. Similarly, Galliford (2010: 242) asserts that tourism experiences of Aboriginal country in Australia can result in a new way of thinking about Aboriginality and Aboriginal issues From a Gitxaała perspective, the inclusion of contemporary history as an aspect of the educational component only serves to strengthen the cultural sustainability of tourism. An intentional and holistic approach to cultural revitalization and presentation upholds Gitxaała *ayaawx* and celebrates the resilience of Gitxaała culture and society. This approach also supports an authentic representation of the Gitxaała experience.

The question of authenticity in Maori tourism was linked directly to contemporary Maori life by McIntosh and Johnson who suggest that authenticity is "an interactive experience and an experience of a present (as opposed to a past) reality" (2005: 49). The quest for authenticity in Indigenous tourism has sometimes been linked to issues of ownership and local production; a more significant issue is whether tourists are engaging with the truth of Indigenous experience and the legacy of colonialism and joining Indigenous communities in the celebration of survival and resurgence.

An Integrated and Intentional Gitxaała Tourism Experience

Tourism is both part of a larger economic strategy, and part of a larger political strategy. Tourism is a gentler approach than direct action and litigation for creating a positive awareness around our battle for sovereignty and feeds into a larger process of education and advancing social issues. What gets people into a natural history museum? A big dinosaur. However, while there, they can learn about evolution and smaller, less iconic species that require conservation and protection. We know that it is a very small niche market that will come to a "remote" north coast village to learn about colonialism. We need to appeal to "mainstream" tourists but improve upon the outcomes of mainstream Indigenous tourism. If we offer sports fishing and cultural performances, we can also provide a dose of history from our perspective. We need a main attraction to draw people in, but the value-added pieces are educational. Tourists can order the regular fare: drumming totem poles and wildlife, but it needs to come with a side of truth and reconciliation.

We can build on the established local opportunities and indigenize them, or more specifically, Gitxaała-ize them. One of the most important aspects of a community-based tourism program is to build awareness and understanding of the differences between Indigenous cultures and communities. While we share a language, a way of life, and cultural structures with our neighboring Nations, each village and each government are unique.

Whale watching and wildlife tours are a key attraction for Prince Rupert and Gitxaała has recently entered into a partnership with the largest ecotourism operator on the North Coast. Through this partnership, whale watching and wildlife tours can move further south into Gitxaała territory and can directly integrate information about our relationships with these species. The special relationship of the Gisputdwada (Killerwhale clan) with the orca can be part of a whale watching tour, resulting in a new understanding of our clan-based social structure and the cultural practices associated with it. A visit to the sea lion rookery near Wil hał psah on the northwestern edge of Gitxaała territory can integrate information about how we harvest and process t'iibn (sea lion), with the opportunity to explain how the seasonal arrival of tsgah (herring) and Wah (oolichan. forage fish) impact the regional harvesting cycle and Indigenous trade.

A day tour to Gitxaała will potentially include the following experiences:

A ferry ride out to the village will include information about the territory, the laws that govern our management and stewardship of the land. When we see whales we can discuss our clan system and our tribal crests. We can also introduce our ongoing efforts through marine conservation planning to create protected areas for whales and other species and to establish marine transportation routes and speed limits to reduce marine mammal strikes. If we visit a rookery or a seal haul out, we can talk about contemporary harvesting and our continued interdependence with our natural resources.

When the tourists reach the village of Lax Klan, we won't take them directly to our cultural center as we don't intend to hide our reality but highlighting it. A bus tour of the village will open their eyes to both the beauty of our way of life and the continued disadvantages we face. Our world is illustrated very well in the village. The paths to the houses are made of crushed clam and cockle shells, indicating our harvesting culture, which can be contrasted to the statistics on our unemployment and poverty. We can highlight our water treatment plant which allows us to enjoy clean water, unlike the dozens of Indigenous communities across the country who have lived with boil water advisories for decades. We can celebrate our building and renovation projects but speak to the housing shortage that results in overcrowding and prevents some of our people from moving home.

Our cultural center will include a Gitxaała timeline exhibit that will include all the main events and transitions in our world. The period after the arrival of K'amksiwah̲s will be honest; there have been wins and losses. Walking through this exhibit on the way to the feast hall for a dance performance is an important way of providing context and tone. The experience cannot be an à la carte menu of experiences. Visitors will need to pass through our contemporary history and learn about the removal of our land and our children before they get to the drumming.

In the feast hall, will be displayed our treasures; this will include contemporary art and ancestral pieces, both repatriated belongings and items that remained in the community, protected from collectors and missionaries. Research has identified numerous Gitxaała belongings in museums across North America and Europe. Several have already been repatriated to the Nation, including a totem pole from the Peabody Museum at Harvard University in 2023. The Gitxaała living museum will grow as more belongings are returned and provenance research links more items to the territory. There are thousands of treasures from the north coast that are poorly catalogued and simply labelled as "Tsimshian" in dozens of Western museums; the search for Gitxaała belongings in these large unattributed collections will be long and arduous. The stories of theft and duress that resulted in the exhibit of these beautiful treasures in museums thousands of miles from their home will be part of the exhibit. The stories of their journeys home and a celebration of their Gitxaała origin and meaning will be highlighted. The persistence of these arts despite the colonial disruption and the continuity of practice will be showcased by presenting ancient and contemporary pieces together, emphasizing our living culture.

Our drum and dance group, *Gulaplip* (thunder), will perform in the feast hall. The protocols surrounding our songs will be explained with an emphasis on the clan-based ownership of particular songs and the stories they tell. The crests our dancer wear will be explained, another opportunity to share our *ayaawx* that continue to govern our society. Our performative traditions will be contextualized in our governance and living society.

Gitxaała people are renowned for their seafood and our visitors will be able to sample our local foods. Understanding our ongoing reliance on these resources, our science and monitoring programs to protect fish and crab stocks, and our traditional harvesting regulations will be woven into the enjoyment of fried seaweed, smoked fish, and salmonberries.

Our gift shop will stock history books, wildlife guides, Indigenous literature and children's books, to support the educational component of our tourism mandate. There is a burgeoning literature of children's stories about residential school, language loss and other issues that provide a way to learn about these issues gently. These books are important for Indigenous children to see themselves and their cultures in books, but they are equally important to other Canadian children, to have these beautiful stories of survival woven into their young consciousnesses. The works of Gitxaała artists and artisans will be for sale, to support the revitalization of our arts and our economy.

Target Markets

Our goals for tourism are multiple, including economic development, cultural revitalization and public education, meaning that our target markets are also diverse.

A key market is local including school groups and families. Reconciliation and understanding is most needed at a local level, to heal the relationships with non-Indigenous people with whom we interact on a regular basis. This could create more cooperative, mutually beneficial relationships in many areas – an important aspect of reconciliation. If we reach enough school groups, we can integrate the Gitxaała story with the important Indigenous education initiatives in our regional school districts. Many children in Prince Rupert, even those of Indigenous ancestry, do not regularly visit a reserve community. Witnessing life in a village like Lax Klan, which is remote from even a small urban center, is important for understanding our relationship with our territory and the importance of our social connections. We hope the steady growth of the Gitxaała repatriation exhibit and rotating loans from partnering museums will encourage repeat visits.

We hope that our very public repatriation efforts will encourage domestic tourism from other parts of British Columbia and Canada. We hope to leave contemporary Gitxaała art in the museums that return our ancestral belongings, encouraging people to visit the repatriated pieces in their original context. We will collaborate with other Indigenous community-based museums and cultural centers to market an Indigenous cultural tour route of the region. Ksan Historical Village and Museum in Gitxsan territory, Hli Goothl Wilp-Adokshl Nisga'a (Nisga'a Museum) and the Haida Heritage Centre on Haida Gwaii showcase unique cultures and the Museum of Northern British Columbia in Prince Rupert celebrate multiple Indigenous and settler societies. These museums could be combined with ecotourism, sports fishing and other activities in a self-guided regional tour. Eventually, we hope to provide Gitxaała versions of all the main regional attractions including whale watching and fishing charters to independent tourists as well as the growing number of cruise ship passengers visiting Prince Rupert. Day trips from the cruise ships could provide an additional significant source of revenue and reconciliation reach.

Conclusion: The Outcomes of Truth-Telling Tourism

There are many benefits of an integrated and intentional approach to tourism. The goal is to develop sustainable tourism enterprises that provide local employment and revenue, with the added benefit of internal cultural awareness and education. Tourism will support intergenerational and intercultural knowledge transfer. The training of guides, dancers, and artists all results in the passing down of Gitxaała knowledge to youth, supporting the reinvigoration of cultural traditions and the Sm'algyax language, and the strengthening of cultural pride. The visitor experience of these practices and performances, and the hearing of Gitxaała stories enhance local, national and international understanding of the ancient and recent history of Gitxaała people, their values, governance structures, and territorial rights.

The goal of truth-telling in Gitxaała tourism will not be to traumatize visitors or make them feel guilty. The truth of Gitxaała history is the strength of the laws and practices that have survived disruption and a way of life rooted in the territory. Tourism performances, exhibits and activities can be a celebration of contemporary Gitxaała life rather than a representation of the pre-colonial past. Weaving adawx and malsk to present the full history and full truth of Gitxaała life can educate and motivate visitors. We are seeking to gain respect and understanding to create allies who support our efforts to achieve sovereignty over our territories and the repatriation of our treasures. We want Canadians to know who we are and engage with the issues that face us and other First Nations. Given the remote location of our village, tourism is not going to provide us with significant economic development opportunities. At best, tourism will be a seasonal income for a handful of village residents. What it can do is contribute to the education of visitors and their understanding of Indigenous issues. Truth-telling through tourism is one way to chip away at the barriers to true reconciliation and decolonization in Canada. Our culture is a culture of respect, and the goal of hosting visitors is to gain their respect for our people and our way of life.

References

Barton, A. W., & Leonard, S. J. (2010). Incorporating social justice in tourism planning: Racial reconciliation and sustainable community development in the Deep South. *Community Development, 41*, 98–322.

Butler, C., Watkinson, B., & Witzke, J. (2021). The immovable object: Mitigation as Indigenous conservation. *Collaborative Anthropologies, 13*(2), 1–28.

Butler, D. L. (2001). Whitewashing plantations: The commodification of a slave-free antebellum South. *International Journal of Hospitality & Tourism Administration, 2*, 163–175.

Galliford, M. (2010). Touring 'country', sharing 'home': Aboriginal tourism, Australian tourists and the possibilities for cultural transversality. *Tourist Studies, 10*(3), 227–244.

Gerberich, V. L. (2005). An evaluation of sustainable American Indian tourism. In C. Ryan & M. Aiken (Eds.), *Indigenous tourism: The commodification and management of culture* (pp. 75–86). Amsterdam: Elsevier.

Higgins-Desbiolles, F. (2003). Reconciliation tourism: Tourism healing divided societies. *Tourism Recreation Research, 28*(3), 35–44.

Higgins-Desbiolles, F. (2005). Reconciliation tourism: Challenging the constraints of economic rationalism. In C. Ryan & M. Aiken (Eds.), *Indigenous tourism: The commodification and management of culture* (pp. 223–246). Amsterdam: Elsevier.

Higgins-Desbiolles, F. (2006a). Reconciliation tourism: On crossing bridges and funding ferries. In P. Burns & M. Novelli (Eds.), *Tourism and social identities: Global frameworks and local realities.* Amsterdam: Elsevier.

Higgins-Desbiolles, F. (2006b). More than an "industry": The forgotten power of tourism as a social force. *Tourism Management, 27*(6), 1192–1208.

Jacobsen, D. (2018). Aboriginal domestic tourism leadership towards reconciliation in Australia. *Journal of Linguistics and Education Research*, 1(1), 19–31.

Kramvig, B., & Førde, A. (2020). Stories of reconciliation enacted in the everyday lives of Sámi tourism entrepreneurs. *Acta Borealia, 37*(1–2), 27–42.

Loewe, R. (2014). Civil rights tourism in Mississippi: Openings, closures, redemption and remuneration. *Sociology Mind, 4*(1), 84–92.

McIntosh, A. J., & Johnson, H. (2005). Understanding the nature of the marae experience: Views from hosts and visitors at the Ng Hau E wha National marae, Christchurch, New Zealand. In C. Ryan & M. Aiken (Eds.), *Indigenous tourism: The commodification and management of culture. The commodification and management of culture* (pp. 35–50). Amsterdam: Elsevier.

Menzies, C. R. (2016). *People of the saltwater: An ethnography of Git lax m'oon.* Lincoln: University of Nebraska Press.

Menzies, C. R., & Butler, C. F. (2007). Returning to selective fishing through Indigenous fisheries knowledge: The example of K'moda, Gitxaała territory. *The American Indian Quarterly, 31*(3), 441–464.

Menzies, C. R., & Butler, C. F. (2008). The Indigenous foundation of the resource economy of BC's North Coast. *Labour*, 61, 131–149.

Skipper, J. (2016). Community development through reconciliation tourism: The behind the Big House Program in Holly Springs, Mississippi. *Community Development, 47*(4), 514–529.

Timothy, D. J., & Boyd, S. W. (2003). *Heritage tourism.* Harlow: Prentice Hall.

10
YOU CAN'T ASK THAT! PROJECTIVE TECHNIQUES UNEARTH SOCIO-CULTURAL AVERSIONS TOWARDS INDIGENOUS TOURISM

Afiya Holder

Introduction

Are tourists racist? Are their choices of a tourism experience just a matter of preference, safety, comfort, or mere hedonism? Ongoing scholarly debates concerning tourist socio-cultural aversions have accelerated as socio-cultural tensions heighten on the heels of global trends such as socio-demographic movements (e.g., #BlackLivesMatter, #metoo), socio-political transitions (e.g., Palestine/Iran, Russia/Ukraine wars) (Benjamin & Dillette, 2021; Higgins-Desbiolles, 2006) and socio-economic and technological shifts (e.g., redistributed capitalism – racial and gender financial inclusion and accessibility to economic opportunities) (Bunten, 2010; Cave & Dredge, 2020). For instance, researchers have documented how tourism at times can be a contributor to social inequalities, injustices, and tensions due to the economic and power imbalances between tourists and tourism service providers (Chien & Ritchie, 2018; Graham & Dadd, 2021; Ruhanen & Whitford, 2016). This chapter adds to these ongoing debates, more specifically by building upon contextual and empirical work that shows socio-cultural aversions do exist and can predict tourist behaviour towards Indigenous tourism in Australia (Holder, Ruhanen, Mkono, & Walters, 2021).

This study sought to unearth tourists' sentiments towards Indigenous tourism marketing stimuli. In doing so, it reveals a critical shift in tourism research on how socio-cultural aversions are manifested through the third-person projections. The qualitative methodological approach applied consists of photo-elicitation and in-depth projective interviews with forty-one (41) domestic (Australian) participants. The temporal context of this research occurred during the COVID-19 pandemic between 2020 and 2021. During this period, the global tourism and hospitality sector, especially in Australia, faced major challenges due to extreme border control and COVID-19 restriction policies (Luo & Zhai, 2017; Zhong, Sun, Law, & Li, 2021). Notwithstanding these factors, tourists' appeal and affinity towards Indigenous tourism does exist, despite it being mostly from niche segments consisting of serious and intentional cultural tourists (Fan, Zhang, Jenkins, & Lau, 2020; McKercher, 2002) and in some cases incidental tourists (Moscardo & Pearce, 1999). These considerations underpin the chapter's aim to understand those tourists who are averse to having an Indigenous tourism experience in order to enlighten Indigenous tourism stakeholders who are positioned to reap socio-economic and cultural benefits such as self-determination, cultural revitalisation, and community pride.

First, a review of the literature on the context and ongoing challenges is highlighted. Second, a presentation of the key findings is provided in the case of how tourist socio-cultural aversions are manifested towards Indigenous tourism in Australia. Finally, the chapter discusses the key themes and draws conclusions as to how these aversions influence tourist behaviour towards Indigenous tourism, as well as strategies to limit or address potential aversions towards Indigenous tourism.

Tourist socio-cultural aversions

Tourist socio-cultural aversions was proposed by Holder et al. (2021, p. 443) as:

> avoidance associated with an ingrained dislike for, and distancing from, a tourism product, experience, or interconnected system of consumerism, representative of a particular social or cultural group's identity (implicit or explicit; assumed or real).

Despite the seminal works in consumer behaviour studies that have explored multiple emotive, psychological, and behavioural reactions such as dislike, avoidance, and rejection of products/brands associated with marginalised groups (Lee, Motion, & Conroy, 2009), there have been only limited, albeit growing, investigations of this research phenomenon. For instance, Kock et al. (2019) apply unitary variable tourist xenophobia to explore tourists' anxiety towards trying local foods within a tourism context generally. However, the holistic conceptual framing by Holder et al. (2021) demonstrates that aversions of a socio-cultural nature can be implicit, explicit, passive, or aggressive, and can also evolve based on situational and temporal factors. Those authors also highlight the relevance of exploring aversions within contexts where tourist demand is inhibited by marginalised and contested factors such as social, cultural, political, religious or gender. Applying a holistic approach to explore tourist socio-cultural aversions and how they are manifested towards Indigenous tourism in Australia is essential for the complex nature of the study within this chapter.

Direct reference to the Tourist Socio-Cultural Aversion Typology conceptually demarcates the degree/intensity of aversions ranging from passive to aggressive (vertical) and levels of consciousness ranging from implicit to explicit (horizontal) (Holder et al., 2021). In Quadrant 1, the *micro-aggressor* tends to react in subtle yet aggressive forms. These can be displayed as minimisation of one's race, passive/aggressive insults, or invalidations. Existing research highlights some evidence of micro-invalidations and negative reviews towards Indigenous businesses, labelling them unprofessional, incapable, archaic, and tourist traps based on westernised ideologies of service excellence (Foley, 2006; Holder & Ruhanen, 2017; Mkono, 2016). Quadrant 2, the *aggressor*, tends to display both aggressive and explicit behaviours, intentionally seeking to hurt, destroy and/or in extreme situations, actively deter others from the brand/business. Here, participants express or tend to engage in behaviours such as boycotting and brand sabotage. Vorobjovas-Pinta (2018) shares an indication of deliberate anti-consumption and social activism towards tourism establishments known to disregard the rights of tourists based on their religious beliefs or sexual orientation.

Quadrant 3, the *avoider* tends to express a more passive form of avoidance, mostly linked to one's social and self-identity. Kannisto (2018) found this to be true in the investigation of tourists who tend to distance themselves and disidentify with tourism experiences considered to be different from their social group and the stereotyped values, worldviews, and practices of that (social/cultural) group. Those that tend to experience ambivalence and states of passive-implicit feelings or behaviours are mostly situated in Quadrant 4 and are the least averse and most persuadable. Yet this group is the most difficult to identify and can be the easiest to shift from mixed feelings to hate, especially if they develop feelings of distrust and discomfort. For example, Holder and Ruhanen (2017) explain how incidental tourists' tend to disregard and ignore any messaging or marketing about Indigenous tourism experiences due to discomfort in learning about historic and colonial narratives that form part of some tours. Lastly, the *discriminator* intersects multiple quadrants and tends to reveal different forms and degrees of passive-aggressive/implicit-explicit feelings and behaviours. This is the most challenging segment to identify and address, often triggered by socio-cultural inter-group tensions such as animosity, racism, and xenophobia developed due to ignorance, fear, and other socio-cultural biases (Chien & Ritchie, 2018). For instance, Siyami-yan Gorji, Almeida Garcia, and Mercadé-Melé (2022) found that American tourists developed ingrained cultural and social animosity underpinned by historic, religious, military, and political tensions. They found tourists expressed both cognitive and emotional animosity towards the destination image, as well as behavioural intentions to boycott the destination.

It is evident that the nature of this body of scholarly work is complex, multifaceted, and therefore often overlooked. Yet, the results of these studies collectively highlight the need for further exploratory research, especially in the realm of Indigenous tourism with its ongoing demand challenges.

Indigenous Tourism and Ongoing Challenges

Indigenous peoples, tourism businesses and their representations are one of the most marginalised and misunderstood in Australia (Frederick & Foley, 2006; Higgins-Desbiolles, 2022; Ruhanen & Whitford, 2016). However, these challenges are not isolated within the tourism sector. Historically, Indigenous peoples in Australia have been made to feel inferior and have faced discrimination in sports, education, law enforcement, and the criminal justice system, as well as within the economic, financial and entrepreneurial systems. For instance, existing research reveals deep-rooted suspicion and enlightened racism towards athletes, which enables the continuation of structural barriers and overt discrimination towards Aboriginal and Torres Strait Islander athletes despite their high levels of performance (Coram & Hallinan, 2017; Hippolite & Bruce, 2010). Similarly, research also highlights racial profiling, over-surveillance, under- and over-policing in Indigenous communities, the overrepresentation of Indigenous people in the criminal justice system, prison deaths, and the existence of systemic bias, discrimination, and institutional racism in the law enforcement and criminal justice system, especially amongst Indigenous youth (Blagg, Morgan, Cunneen, & Ferrante, 2005; Cunneen, 2020). Other studies found inequality in Indigenous employment rates and support for Indigenous entrepreneurs, which are especially evident in their dealings with government agencies, financiers, and banks (Foley, 2006; Mika, Warren, Foley, & Palmer, 2017).

Although research suggests Indigenous tourism stakeholders may face similar biases, there remains a lack of substantive empirical evidence (Espinosa Abascal, Fluker, & Jiang, 2015; Everingham, Peters, & Higgins-Desbiolles, 2021; Ruhanen & Whitford, 2016). Even more so when acknowledging that the tourism eco-system is complex and includes the involvement and engagement of numerous stakeholders on different levels, performing different roles such as governance, trade, supply, and demand. These tourism players often tend to have different business practices and agendas utilising the same limited resources (ecological, socio-cultural, and economic). Everingham et al. (2021), for instance, raise the argument that existing Indigenous tourism practice engaged mostly colonial market logic that embodies westernised capitalism, privileging the eco-centric tourist gaze and denying local Aboriginal custodians and their country agency, respect, or care. Such practices are documented in other research that spotlights how discriminatory colonial and heritage practices and policies present hindrances for Indigenous peoples from assertion of control, eviction from native lands to missionary land or reserves, and suppression of traditional practices to assimilation (Bell, 2014; Frederick & Foley, 2006; Schaper, 2007). More specific to its impacts on the tourism sector, these challenges have impacted both the Indigenous supply-side and the tourist demand-side.

Key tourism reports claim that tourism contributes to community poverty alleviation, cultural pride, sovereignty, and connection to a country/land (Tourism Research Australia, 2011; UNWTO, 2012). Yet, critical tourism scholars question the continuous failed efforts to maintain responsible and sustainable operations (Chambers & Buzinde, 2015; Graham & Dadd, 2021; Higgins-Desbiolles, 2022; Stinson, Grimwood, & Caton, 2021). For instance, Graham and Dadd (2021) address the traumatising encounters Indigenous tour operators experience, often exposed to offensive, insensitive, and ignorant stereotypes and being made to feel guilt, shame, embarrassment, and displacement from the country by government officials, the travel trade, the media, and tourists. Compounding these socio-cultural challenges are other supply-side hindrances, including limited access and support for financial resources, training and skill-development (Foley, 2012), access to social capital, partnerships, and support or uptake of Indigenous products by travel trade (Foley & O'Connor, 2013; Hollinshead, 2007; UNWTO, 2017).

Demand challenges and the relatively low appeal and interest by tourists have also been attributed in part to the impacts of Indigenous supply challenges, especially the misrepresentation and stereotypical ideologies perpetuated in Australia's colonial heritage, systems, and media (Holder et al., 2021; Ryan & Huyton, 2002). It has been found that tourists often perceive Indigenous tourism as boring, socially confronting, too far, inaccessible, and located in environments that are too hot, uncomfortable, and unsafe (Holder & Ruhanen, 2018). The low appeal and unwillingness to support Indigenous tourism have also been connected to tourists' perceptions of Indigenous tourism as being inauthentic 'tourist traps' (Holder & Ruhanen, 2018) and tourism operators being unprofessional and untrustworthy (Mkono, 2016; Whitford & Ruhanen, 2014).

The review of the literature demonstrates the potential for a deeper exploration of socio-cultural aversions as an explanatory research area to better understand the potential challenges facing the Indigenous tourism sector in Australia, more specifically from the demand side. While previous studies have explored these issues using conventional methodological tools, the combined historical and accelerated trending challenges associated with socio-cultural factors are a clear indication of cultural sensitivities and tensions. More importantly, the hidden and conspicuous nature of these aversions calls for more sophisticated investigative tools to understand tourist aversions.

Methods

Underpinned by a pragmatist paradigm, this study offers practicality for investigating issues in research, a valuable ontological tool that acknowledges emergent intangible factors that include culture, subjectivity, and values, offering research and reasoned inquiry towards larger truths and insights (Morgan, 2014). This study does not aspire to suggest another hindrance impacting Indigenous tourism but rather, it aims to unearth the potential impacts of larger truths and insights and potentially tourist socio-cultural aversions. Past research notes the challenge of undertaking such a study due to social desirability bias (Curtin & Bird, 2022; Ruhanen & Whitford, 2016) However, a combination of projective techniques were utilised that include choice-ordering, photo-elicitation, and third-person projective questioning. Nine (9) images were selected from a bank of destination marketing images from Tourism Australia (2019), the national tourism marketing organisation. The selected images represented water-based, land-based, culinary-based, and city-based Indigenous tourism experiences. Holistically, these techniques and tools were used to examine choice, projective reactions, and narratives that reveal participants' perceptions, emotions, values, motives, and subconscious sense-making, as well as hidden meanings, thus limiting social-desirability bias (Hindley & Font, 2017).

Data collection occurred between August 2020 and March 2021. A Qualtrics research panel meeting Tourism Research Australia (2011) criteria of targeted Australian tourists was recruited via a convenience sampling process, and a snowballing approach was used to identify the 41 participants. The interview ranged from 30 to 90 minutes in length and was conducted utilising Zoom video. The interviews were recorded, and responses were transcribed. The sample included a range of ages (age ranges: 18–29, 30–44, 45–64, 65–79), genders (females: 63%, males: 37%), ethnicity (Europeans: 63%, Asians: 24%, mixed ethnicity: 7%, African: 2%), education (high school: 7%, undergraduate: 51%, postgraduate: 41%), geographic location (New South Wales: 24%, Victoria: 27%, Queensland: 49%), and income (average: AU$120–190k).

Firstly, the process started with choice-ordering, where participants were asked to view the photos and select their favourable and unfavourable choices for an Indigenous tourism experience. Participants were then asked to hypothetically place themselves in the contexts or scenarios of the images (relate in the third person). Participants were then prompted to project their thoughts and feelings about the characters in both their most favourable and unfavourable image choices (e.g., imagine you are the tourist character in that image; can you describe your thoughts during this experience? Describe how you are feeling in this moment? What do you think the tourist and tour guide characters are saying/asking?). During the third-person projective questioning, participants were probed to elicit their deep thoughts, feelings, and reactions. Participants' contextualising the images from their worldviews helped trigger deeper discussions of personal experiences, existing feelings and recall of their own memories from engaging with Indigenous peoples in tourism, service, and causal settings to reveal deep and hidden projections. Utilising NVivo 12 Pro software, a critical content analysis of the verbatim transcriptions underwent an iterative process and multiple rounds of coding, thematizing and analysis to address the research aims to explore potential tourist socio-cultural aversions towards Indigenous tourism (Braun & Clarke, 2006).

Findings and Discussion

Through the analysis of the data and inferences from the multi-purpose projective techniques, the study generated insightful results on the preferences, associations, and projections that characterise tourist socio-cultural aversions, according to Holder et al. (2021). The analysis yields novel and significant findings based on tourist aversions towards Indigenous tourism experiences in an Australian context, as discussed below.

Table 10.1 summarises the most favourable and unfavourable choices, associations, and other salient factors arising from interviews. Emerging from the choice ordering and projective questioning, these factors reveal significant influences on participants' reactions, feelings, and worldviews.

Factors Related to Mixed and Positive Projections

Mixed feelings. Interestingly, participants made associations and projections with negatively loaded expressions for both their most favourable and unfavourable image choices. For instance, one participant asserts:

> They look enthusiastic…. but although this image is really glorious and spectacular, I still have a bit of a question mark, like what is the actual community living standards outside of this image? … that would be quite a big culture shock and what if the accommodation in the area is

Table 10.1 Summary of participant preferences, associations and projections

Type of Indigenous tourism experience	Associations	Projections and other salient factors of influence	Perceived target market
Most favourable image: **Outback bush experience (image 5):** key characteristics include European couple sitting listening to Indigenous storytelling with Indigenous elders in front campfire	Cultural exchange; authenticity; traditional; adventurous; learning; Indigenous connection; exclusive; curiosity; novelty; landmark; bushlands; remote; unsafe; dry; hot; far; inconvenience; inferior; expensive	**Feelings:** Enjoyment; memorable; discomfort; empathy; hope. **Other salient factors:** Adorable elders but confronting stories; Indigenous stereotypes; actors ill-matched to the experience; out-numbered guides to tourists; concern for safety; weary of not seeing other tourists; Indigenous elders need English interpreter	Young adventure-seekers Age: 20s–30s Country of origin: Germans, Canadian, Americans, Australians Grey nomads Age: 50s–60s Country of origin: Australians
Most unfavourable image: City tour experience **(image 7):** key characteristics include European middle-aged woman engaging with tour guide who provides historic interpretations of Sydney areas- the Rocks, Sydney Harbour and Sydney Opera House	Fake; avoidance; cultural distancing; inauthentic; cultural sensitivities; unprofessional; inappropriate; victimisation; condescending; cultural commodification; misappropriation.	**Feelings:** Racial bias, boredom; fear; guilt; hate; unrelatable; angry; indifferent; irreverent; miserable; discomfort; distracted; confused; confronted. Other salient factors: Fake and forced tourism; false representation; them vs us; non-traditional; preference for an audio self-guided tour; tourist trap	Mass bus tour groups All ages Country of origin: Chinese, Indian, Japanese

inferior and low grade and located in an unsafe area. So, what it is like for the tourist or outsider beyond this stunning photography. I mean you know the state of most of these remote areas.

(P26)

These associations reveal mixed emotions and careful risk assessments in participants' stated opinions. However, these included both intrinsic (e.g., feelings, preferences) and extrinsic (e.g., perceived quality) associations. These findings are aligned with the findings of Kock et al. (2019), who observed that some xenophobic tourists still had intentions to visit a destination, but under consideration of risk management factors such as planning and travelling in large groups.

Another participant showcases ambivalent attitudes towards Indigenous culture:

I also feel to some extent that living in Australia, that I have some obligation to engage with that a little bit more to gain some deeper understanding of that culture so that I can, when the opportunity presents itself, engage with indigenous issues with the kind of respect that I would want and that sort of thing. There is that desire somewhere, but whether that desire it would actually cause me to book a holiday one day, it hasn't gotten to that so far.

(P14)

This finding substantiates existing behavioural studies that make accounts of ambivalence as states of incongruence or mixed feelings and beliefs, which can lead to states of discomfort, emotional uncertainty, avoidance or delayed decision-making (Bee & Madrigal, 2013; Penz & Hogg, 2011).

Positive and transformative experience. Negative associations and projections were made more frequently than positive associations in each of the interviews. However, participants who made positive associations mostly projected a sense of empathy and hope in their comments relating to cultural exchange, learning, and novelty-seeking. As one participant eloquently expresses:

I'd want the hands-on stuff, I don't know if that would be included here, but I would want to get in there and touch those little berries. And if they are cooking, I'd want to have my own bowl that I can mix up those spices or whatever they're doing. I think that hands-on element ... that's the way I learn best. I think it's that multi-sensory thing it's not just watching and listening, but it's also seeing and smelling and feeling and getting to know them. And that to me is all part of a rich experience.

(P29)

Tourism scholars who explore positive and transformative tourism do highlight similar factors of cultural exchange and learning in their findings on cultural tourism (Mkono, 2019; Walker & Moscardo, 2016). For instance, Walker and Moscardo (2016) suggest that Indigenous interpretation experiences that provide positive learning experiences were positively received by tourists, who also reported a heightened sense of appreciation and awareness of the environmental and social issues in the community.

Factors Related to Negative Projections

Discriminatory bias. In this study, broader sociocultural aversions included biases and negative attitudes towards experiences involving Indigenous peoples as salient factors, mostly projecting the link to Indigenous phenotypes and perceptions of Indigenous culture. Together, these factors contribute to both explicit and implicit aversive tourist behaviour and a greater predisposition for disinterest and avoidance based on underlying biases and preconceived stereotypes (Hill & Paphitis, 2011; Lee et al., 2009). For instance, one participant explicitly highlights:

Some might expect ... Aboriginal tour guides not to be dressed in pants and shirts and hats .... we expect them to be dressed in what we know as traditional, like a lap and things. Also, another challenge is the skin colour, you know, we expect all Aboriginals to have skin colour, a lot darker than these fellas.

(P25)

Another participant projects feelings of discomfort towards Indigenous cultural content:

...there are so many things that are a mystification of the whole culture ... you don't know what you should be able to get out from it, like what's appropriate, I would say. It's too much headache, and uncomfortable.

(P04)

Authenticity. In tourism literature, authenticity is a contested concept with many classifications and variations. Scholarly work tends to ascribe characteristics such as *real, true* or *unique* to an experience, object, event, people or place in a way that suggests an undisclosed objective, constructed, or personal bias (Cohen, 1988; Jamal & Hill, 2004; MacCannell, 1973; Pearce & Moscardo, 1986). According to research, cultural and heritage places also define place authenticity or sense of place through meaning-making activities, people, and events, that historically, politically, or culturally define the location (Jamal & Hill, 2004). Participants also made strong assertions on place authenticity to express their explicit disinterest:

> I think the location matters a lot. If you're looking for really nice architect work, go to this place, enjoy it. Enjoy the opera house, enjoy the water, look at the bridge. Amazing. But you're not there for that experience ... then if not here go to a different place that is not so busy and crowded. Nothing about this feels authentic.
>
> (P25)

Social servicescape. Presented with their image choices, participants tended to focus on multiple factors such as the actors, the location, and the social servicescape, making associations and projections based on their current expectations, past experiences, memories, and influences from media and socialisation within Australia. Noteworthy, there was also much mention of extrinsic factors beyond the control of Indigenous providers, like elements related to the weather, location, and place. Consumer behaviour research on social servicescape also indicates similar findings where consumers tend to feel uneasy when they are outnumbered by members of their ingroups, whether it be based on ethnicity, nationality, status, language or sexual orientation (Malik & Paswan, 2022; Olson & Park, 2019; Rosenbaum & Montoya, 2007). In the context of this study, however, it was mostly common for participants to recall past experiences of being in an outback setting and describe feelings of distress related to the prominent presence of Indigenous peoples in the streets:

> ...they were just fighting on the streets and drinking and passed out. And my kids were quite frightened in Coober Pedy. So they had that very negative experience about it and they're just like, oh my God, it was really frightening. They remember when the Aboriginals were fighting in the street and yelling at each other and throwing things.
>
> (P21)

Low awareness. Brand awareness and popularity of the tourism offering were also upheld as salient factors in their choice preference:

> I think (Indigenous tourism) is under promoted, seemingly inaccessible and sort of a little bit difficult to get to, difficult to get the real thing.
>
> (P18)

> I think that there'll be gaps because of the level of promotion it's only local.... They do need state government support with that sort of stuff. Then generally, I think state governments are fairly good at promoting regions, particularly post-COVID. Obviously, the Victorian government is trying to get people to travel to different parts of Victoria and they've offered all these incentive programs to doing that. Again, there's not a particular focus on indigenous

cultures, so that's a gap. Again, it points back to my view on the fact that our governments don't have enough understanding or knowledge of indigenous culture and where it fits in contemporary society to be confidently promoting it the way that they should. If that makes sense?

(P14)

These assertions are similar to the findings of Frederick and Foley (2006), who claim Indigenous service providers face distributive injustices from the unfair distribution of resources by government officials and recognition by the media. Holder and Ruhanen (2018) also found stereotypical representations, misrepresentations, and an overall lack of promotion and marketing accounted for low market appeal and perceptions of Indigenous tourism being inaccessible.

Perceived target markets. In most cases, initial participant reactions to the marketing images focused on the actors revealing the target market based on their perceived accounts. Participants made explicit comments about the mismatch between the actors and the Indigenous tourism experience, which preceded their criticism and projections of the experience in the third person. According to Hindley and Font (2017) such reactions are common and serve as the main purpose of utilising projective techniques to reveal deep-seated emotions and limit social desirability biases. Relating to the Outback bush experience (image 5), participants felt this experience was more suited to two main types of tourists: (1) young adventure-seekers from North America, Europe or Australia; or (2) Australian retirees as evidenced by these indicative excerpts:

Getting these experiences and what looks like quite a private experience here, to me this is for senior people with plenty money and time. There might be a few others, but I don't imagine there'd be groups or anything like that. I don't think you're getting a Greyhound bus into these places.

(P13, reference to image 5)

If that's their market audience, it probably might work in a certain way. But me as a consumer is not something that will attract me. It's for a younger and more adventurous generation than mine that maybe willing to waste money on anything really.

(P6, reference to image 5)

Yeah, hmm she sort of looks like an angry German and a bit older. To be honest she could come from anywhere in Europe, but she is obviously not interested in what's going on. Those Chinese groups may more appreciate this sort of thing… A quick guided tour across Sydney, take some pics and hop back on the bus.

(P9, reference to image 7)

Affective, Cognitive, Behavioural and Sensory Responses

Notably, all participants took different approaches to expressing their feelings and about of their chosen Indigenous tourism experience, which showcased a variety of perceptual pathways to reaction, inclusive of affective, cognitive, behavioural, and sensory (see Table 10.2). These findings provide deeper insights into the relationship between the representation of an Indigenous tourism experience and one's reactive and sensory responses. Both positive and negative associations and projections were present, however, negative were more dominant.

Table 10.2 Multiple types of projections and responses

Responses	Indicative quotes
Affective	I'm not saying I wouldn't do it, but it's just least appealing. Plus for my kids, they're young. So for my kids to sit down and do something like this is difficult. They would feel very bored doing that.
	How people acted in all times and all that comes from those backgrounds. I think normally I wouldn't have no issues with the word indigenous or whatever, but because it's treated as such a sensitive topic
Cognitive	I think it has a long history, but it has a different history, a different perspective of their stories and how they perceive the world and history in itself… I think there often many misunderstandings between the cultures and I think sometimes that makes Australians not want to experience these local experiences. (P11)
	I'm so curious about those things like making some handcrafts they usually have. (P27)
Behavioural	I like my freedom, I don't want anything that where I feel that I'm being forced into, let's say, a very rigid routine where I'm stuck for three hours with someone, and that sort of thing. (P09)
Sensory	My major concern would be ending up in a restaurant with an Indigenous dance troop … I find it cringe-worthy. I don't want to see half-naked people shaking their bodies while I eat. (P13)

Table 10.2 summarises indicative quotes that reveal a plethora of responses. Affective reactions develop implicitly or explicitly from one's mood or feelings (Fishbein & Ajzen, 1977). Affect played a major role in participants responses to experiences where they felt bored or confronted by the content. Cognitive reactions form based on organic or induced awareness, knowledge, or beliefs (Pike & Ryan, 2004).

In this study, what made participants uncomfortable, or on a more positive note, curious, were associations linked to history, culture, and the arts. This was especially prominent with references to traditional experiences in the Outback, demonstrating a clear distinction between thoughts about urban and contemporary experiences. Behaviours linked to preference, desire, intention, or future motives to participate according to Chen, Lai, Petrick, and Lin (2016), were expressed explicitly in most instances and indifferently to a lesser extent. Interestingly, conscious associations to the visual and gustatory triggers of chosen images accounted for participants' explicit intentions to engage in similar Indigenous tourism experiences. This finding extends sensory marketing research that proposes marketing and advertisement copy can enhance multi-sensory stimulation and shape consumer's evaluation and behaviour (Kirillova, Fu, Lehto, & Cai, 2014; Krishna, Cian, & Sokolova, 2016). One participant responded excitingly to an image about food and foraging (image 2):

Wow look at that… It's so cool, we'd love to go there… What does it taste like? How can I use it? Where do you find it? …What else can you get from the land? … I think of it more of my mom who would be like, How can I use that to cook? What does it taste like? Where can I get it? This is going to be amazing. Teach me how to do it…. Take me there!

(P01)

Sensory reactions tend to engage consumers' senses and affect their perception, judgment, and behaviour (Krishna et al., 2016). The aforementioned reference draws on how visual and gustatory sensory projections and judgements can inform future motives and intentions, confirming the common colloquial adage 'you taste with your eyes'. Sensory projections were also used by participants to personify emotional triggers associated with their chosen marketing images. For instance, participants used personifying adjectives to describe disgust or hate towards the imagery of a particular experience. Deeper probing revealed in several instances that positive projections of the senses were associated with familiarity or curiosity, for instance, the example provided above recalls memories of their mother and asks questions that showcase curiosity. On the other hand, negative projections were mostly associated with ignorance or preconceived stereotypes, as illustrated in the following example:

> Seeing that makes me feel a bit yuck…It just seems tacky nothing I've ever seen or experienced before …I don't really know how else to describe it… I don't even know if that's something that all indigenous groups do… that kind of dot painting, or if that's just one thing that's taken by everyone.

(P29)

Conclusions

This chapter has discussed empirical evidence of tourist socio-cultural aversions and tourist affinity for Indigenous tourism experiences using projective techniques. New insights reveal not only the key elements of Indigenous tourism that incite tourist aversions, but also, those elements that influence tourist appeal, whilst both explicit and implicit aversions were identified. Explicit aversions were often associated with Indigenous peoples, their identity, and references to their culture and do suggest tourists can be averse towards Indigenous tourism, peoples, and their representations. Implicit aversions were more subtle and associated with intrinsic elements such as participants' indifference or mixed emotions, underpinned by biases and negative stereotypes towards Indigenous peoples and their culture.

Overall, the use of these projective techniques provided more detailed insights into how tourist's affinity or aversion towards a particular Indigenous tourism experience can influence participants' affective, cognitive, behavioural, and sensory projections and associations, as well as their intentions to participate. Such revelations were unexplored in previous studies because it was deemed politically incorrect to explicitly question one's implicit and explicit biases and aversions. Interestingly, the link between the influence of socio-cultural factors, tourist aversions, and these multiple reactions has not yet been made in existing scholarly work, despite suggestions implied by researchers such as Ruhanen and Whitford (2016).

These findings fill a gap in the literature that accounts for target markets based on quantitative statistics, highlighting singular characteristics such as age, country of origin, and income. For instance, Ruhanen, Whitford, and McLennan (2015, p. 30) stated their understanding of Chinese tourists' low interest in Indigenous tourism to be somewhat 'scrambled' based on their findings that contrast other reports claiming high interest (e.g., Tourism Research Australia, 2011). They suggest this low interest was due to lack of awareness and Chinese travel patterns of avoiding acclaimed remote Indigenous destinations. However, in this study, alternative findings reveal Chinese tourists to be an appropriate target market if a packaged bus tour is offered with opportunities for photography. A major finding in this study also supports other research that suggests low brand awareness and poor media representations and associations of Indigenous culture and peoples account for low tourist appeal (Holder & Ruhanen, 2018; Ruhanen et al., 2015; Taylor, Carson, Carson, & Brokensha, 2015). The study results add more context and reveal that participants who had low awareness of or negative associations with Indigenous tourism tended to be more critical and negative in their responses.

Furthermore, the results also highlight participants' affective, cognitive, behavioural, and sensory responses towards the marketing stimuli of Indigenous tourism experiences. Interesting links were discovered between multisensory triggers and participants' intentions to participate in Indigenous tourism, with visual stimulations taking precedence in most cases. However, caution should be taken as the types of projective tools used could be an influencing factor based on the methods and context of this study. Nonetheless, consumer behaviour research does confirm that visual senses can have a dominant effect on one's judgements, perceptions, cognition, memory, and behaviours (Kirillova et al., 2014; Russell, 2002). Indigenous tourism providers could consider the findings of this study to inform the effective marketing and development of Indigenous tourism experiences, in particular, to test and utilise visually stimulating marketing imagery that can limit potential aversions and evoke positive reactions and by extension willingness to participate. Participants' suggestions of appropriate target markets based on activity and place may also provide valuable insights for tourism marketers to test. In addition, these findings can also benefit destination managers and tourism operators in other contexts challenged by socio-cultural tensions caused by differences such as nationality, religion, culture, or beliefs.

Ultimately, unearthing socio-cultural aversions in their many forms can be challenging for both Indigenous tourism stakeholders and researchers, particularly in a culturally sensitive society where truth-telling is limited. More appropriate research methods are needed to productively deconstruct these challenges and ask uncomfortable questions to forge reciprocity and respectful relations over division and discrimination amongst Indigenous peoples and tourists.

References

Bee, C. C., & Madrigal, R. (2013). Consumer uncertainty: The influence of anticipatory emotions on ambivalence, attitudes, and intentions. *Journal of Consumer Behaviour, 12*(5), 370–381.

Bell, A. (2014). *Relating Indigenous and settler identities: Beyond domination.* Palgrave Macmillan UK.

Benjamin, S., & Dillette, A. K. (2021). Black travel movement: Systemic racism informing tourism. *Annals of Tourism Research, 88*, 103169.

Blagg, H., Morgan, N., Cunneen, C., & Ferrante, A. (2005). Systemic racism as a factor in the overrepresentation of Aboriginal people in the Victorian criminal justice system. *Report to the Equal Opportunity Commission of Victoria.*

Braun, V., & Clarke, V. (2006). Using thematic analysis in psychology. *Qualitative Research in Psychology, 3*(2), 77–101.

Bunten, A. C. (2010). More like ourselves: Indigenous capitalism through tourism. *The American Indian Quarterly, 34*(3), 285–311.

Cave, J., & Dredge, D. (2020). Regenerative tourism needs diverse economic practices. *Tourism Geographies, 22*(3), 503–513.

Chambers, D., & Buzinde, C. (2015). Tourism and decolonisation: Locating research and self. *Annals of Tourism Research, 51*, 1–16.

Chen, C.-C., Lai, Y.-H. R., Petrick, J. F., & Lin, Y.-H. (2016). Tourism between divided nations: An examination of stereotyping on destination image. *Tourism Management, 55*, 25–36.

Chien, P. M., & Ritchie, B. W. (2018). Understanding intergroup conflicts in tourism. *Annals of Tourism Research, 72*, 177–179.

Cohen, E. (1988). Authenticity and commoditization in tourism. *Annals of Tourism Research, 15*(3), 371–386.

Coram, S., & Hallinan, C. (2017). Critical race theory and the orthodoxy of race neutrality: Examining the denigration of Adam Goodes. *Australian Aboriginal Studies (Canberra)* (1), 99–111. Retrieved from https://search.informit.org/doi/10.3316/informit.907032851637853

Cunneen, C. (2020). *Conflict, politics and crime: Aboriginal communities and the police.* Routledge.

Curtin, N., & Bird, S. (2022). "We are reconciliators": When Indigenous tourism begins with agency. *Journal of Sustainable Tourism, 30*(2–3), 461–481.

DFAT. (2018). Indigenous tourism surge. *Business Envoy*. Retrieved from https://www.dfat.gov.au/about-us/publications/trade-investment/business-envoy/Pages/january-2019/indigenous-tourism-surge

Espinosa Abascal, T., Fluker, M., & Jiang, M. (2015). Domestic demand for Indigenous tourism in Australia: Understanding motivations, barriers, and implications for future development. *Journal of Heritage Tourism, 10*(1), 1–20.

Everingham, P., Peters, A., & Higgins-Desbiolles, F. (2021). The (im) possibilities of doing tourism otherwise: The case of settler colonial Australia and the closure of the climb at Uluru. *Annals of Tourism Research, 88*, 103178.

Fan, D., Zhang, H., Jenkins, C., & Lau, C. (2020). Towards a better tourist-host relationship: The role of social contact between tourists' perceived cultural distance and travel attitude. *Journal of Sustainable Tourism, 31*(1), 1–25.

Fishbein, M., & Ajzen, I. (1977). *Belief, attitude, intention, and behavior: An introduction to theory and research*. Reading, MA: Addison-Wesley.

Foley, D. (2006). *Indigenous Australian entrepreneurs: Not all community organisations, not all in the outback*. CAEPR.

Foley, D. (2012). Teaching entrepreneurship to Indigenous and other minorities: Towards a strong sense of self, tangible skills and active participation within society. *The Journal of Business Diversity, 12*(2), 59.

Foley, D., & O'Connor, A. (2013). Social capital and the networking practices of indigenous entrepreneurs. *Journal of Small Business Management, 51*(2), 276–296.

Frederick, H., & Foley, D. (2006). Indigenous populations as disadvantaged entrepreneurs in Australia and New Zealand. *International Indigenous Journal of Entrepreneurship, Advancement, Strategy and Education, 2*(2), 34.

Graham, M., & Dadd, U. L. (2021). Deep-colonising narratives and emotional labour: Indigenous tourism in a deeply-colonised place. *Tourist Studies, 21*(3), 444–463.

Higgins-Desbiolles, F. (2006). More than an "industry": The forgotten power of tourism as a social force. *Tourism Management, 27*(6), 1192–1208.

Higgins-Desbiolles, F. (2022). The ongoingness of imperialism: The problem of tourism dependency and the promise of radical equality. *Annals of Tourism Research, 94*, 103382.

Hill, S., & Paphitis, K. (2011). Can consumers be racist? An investigation of the effect of consumer racism on product purchase. *Asia Pacific Journal of Marketing and Logistics, 23*(1), 57–71.

Hindley, A., & Font, X. (2017). The use of projective techniques to circumvent socially desirable responses or reveal the subconscious. In *Handbook of research methods for tourism and hospitality management* (pp. 201–211). Edward Elgar Publishing. https://doi.org/10.4337/9781785366284.00023

Hippolite, H., & Bruce, T. (2010). Speaking the unspoken: Racism, sport and Maori. *Cosmopolitan Civil Societies: An Interdisciplinary Journal, 2*(2), 23–45. Retrieved from https://search.informit.org/doi/10.3316/informit.072475016229860

Holder, A., & Ruhanen, L. (2017). Identifying the relative importance of culture in Indigenous tourism experiences: netnographic evidence from Australia. *Tourism Recreation Research, 42*(3), 1–11.

Holder, A., & Ruhanen, L. (2018). Exploring the market appeal of Indigenous tourism: A netnographic perspective. *Journal of Vacation Marketing*. https://doi.org/10.1177/135676671775

Holder, A., Ruhanen, L., Mkono, M., & Walters, G. (2021). Tourist socio-cultural aversions: A holistic conceptual framework. *Journal of Hospitality and Tourism Management, 49*, 439–450.

Hollinshead, K. (2007). Indigenous Australia in the bittersweet world: The power of tourism in the projection of 'old' and 'fresh' visions of Aboriginality. In R. Butler & T. Hinch (Eds.), *Tourism and indigenous peoples: Issues and implications* (pp. 281–304). Elsevier.

Jamal, T., & Hill, S. (2004). Developing a framework for indicators of authenticity: The place and space of cultural and heritage tourism. *Asia Pacific Journal of Tourism Research, 9*(4), 353–372.

Kannisto, P. (2018). Travelling like locals: Market resistance in long-term travel. *Tourism management, 67*, 297–306.

Kirillova, K., Fu, X., Lehto, X., & Cai, L. (2014). What makes a destination beautiful? Dimensions of tourist aesthetic judgment. *Tourism Management, 42*, 282–293.

Kock, F., Josiassen, A., & Assaf, A. G. (2019). The xenophobic tourist. *Annals of Tourism Research, 74*, 155–166.

Krishna, A., Cian, L., & Sokolova, T. (2016). The power of sensory marketing in advertising. *Current Opinion in Psychology, 10*, 142–147.

Lee, M., Motion, J., & Conroy, D. (2009). Anti-consumption and brand avoidance. *Journal of Business Research, 62*(2), 169–180.

Luo, Q., & Zhai, X. (2017). "I will never go to Hong Kong again!" How the secondary crisis communication of "Occupy Central" on Weibo shifted to a tourism boycott. *Tourism Management, 62*, 159–172.

MacCannell, D. (1973). Staged authenticity: Arrangements of social space in tourist settings. *American Journal of Sociology, 79*(3), 589–603.

Malik, A. Z., & Paswan, A. (2022). Linguistic racism in inter-culture service encounter. *Journal of Consumer Marketing* (ahead-of-print).

McKercher, B. (2002). Towards a classification of cultural tourists. *International Journal of Tourism Research, 4*(1), 29–38.

Mika, J., Warren, L., Foley, D., & Palmer, F. (2017). Perspectives on indigenous entrepreneurship, innovation and enterprise. *Journal of Management & Organization, 23*(6), 767–773.

Mkono, M. (2016). Sustainability and Indigenous tourism insights from social media: worldview differences, cultural friction and negotiation. *Journal of Sustainable Tourism, 24*(8–9), 1315–1330.

Mkono, M. (2019). *Positive tourism in Africa* (1st ed.). Taylor & Francis.

Morgan, D. (2014). Pragmatism as a paradigm for social research. *Qualitative Inquiry, 20*(8), 1045–1053.

Moscardo, G., & Pearce, P. (1999). Understanding ethnic tourists. *Annals of Tourism Research, 26*, 416–434.

NIAA. (2019). Indigenous Tourism Fund: Discussion Paper. Retrieved from https://www.niaa.gov.au/sites/default/files/publications/indigenous-tourism-fund-discussion-paper.pdf

Olson, E., & Park, H. J. (2019). The impact of age on gay consumers' reaction to the physical and social servicescape in gay bars. *International Journal of Contemporary Hospitality Management. 31*(9), 3683–3701

Pearce, P. L., & Moscardo, G. M. (1986). The concept of authenticity in tourist experiences. *The Australian and New Zealand Journal of Sociology, 22*(1), 121–132.

Penz, E., & Hogg, M. K. (2011). The role of mixed emotions in consumer behaviour: Investigating ambivalence in consumers' experiences of approach-avoidance conflicts in online and offline settings. *European Journal of Marketing, 45*(1/2), 104–132.

Pike, S., & Ryan, C. (2004). Destination positioning analysis through a comparison of cognitive, affective, and conative perceptions. *Journal of Travel Research, 42*(4), 333–342.

Rosenbaum, M., & Montoya, D. (2007). Am I welcome here? Exploring how ethnic consumers assess their place identity. *Journal of Business Research, 60*(3), 206–214.

Ruhanen, L., & Whitford, M. (2016). Racism as an inhibitor to the organisational legitimacy of Indigenous tourism businesses in Australia. *Current Issues in Tourism, 21*(15), 1–15.

Ruhanen, L., Whitford, M., & McLennan, C. (2015). Exploring Chinese visitor demand for Australia's indigenous tourism experiences. *Journal of Hospitality and Tourism Management, 24*, 25–34.

Russell, C. A. (2002). Investigating the effectiveness of product placements in television shows: The role of modality and plot connection congruence on brand memory and attitude. *Journal of Consumer Research, 29*(3), 306–318.

Ryan, C., & Huyton, J. (2002). Tourists and aboriginal people. *Annals of Tourism Research, 29*(3), 631–647.

Schaper, M. (2007). Aboriginal and Torres Strait Islander entrepreneurship in Australia: Looking forward, looking back. In L.-P. Dana & R. B. Anderson (Eds.), *International handbook of research on indigenous entrepreneurship* (pp. 526–535). Edward Elgar.

Siyamiyan Gorji, A., Almeida Garcia, F., & Mercadé-Melé, P. (2022). How tourists' animosity leads to travel boycott during a tumultuous relationship. *Tourism Recreation Research*, 1–18.

Stinson, M. J., Grimwood, B. S. R., & Caton, K. (2021). Becoming common plantain: Metaphor, settler responsibility, and decolonizing tourism. *Journal of Sustainable Tourism, 29*(2–3), 234–252.

Taylor, A., Carson, D. B., Carson, D. A., & Brokensha, H. (2015). 'Walkabout' tourism: The indigenous tourism market for outback Australia. *Journal of Hospitality and Tourism Management, 24*, 9–17.

Tourism Australia. (2019). Discover Aboriginal experiences portfolio. Retrieved from https://www.tourism.australia.com/content/dam/assets/document/1/7/5/4/0/2012832.pdf

Tourism Research Australia. (2011). Snapshots 2011: Indigenous tourism visitors in Australia. Retrieved from http://tra.gov.au/documents/ snapshots/Snapshots2011Indigenous.pdf

UNWTO. (2012). Larrakia declaration on the development of Indigenous tourism. Retrieved from http://www.winta.org/wp-content/uploads/2014/12/Larrakia-Declaration-endorsed-by-UNWTO.pdf

UNWTO. (2017). UNWTO panel on Indigenous tourism: Promoting equitable partnerships. Retrieved from http://ethics.unwto.org/event/unwto-panel-indigenous-tourism-promoting-equitable-partnerships

Vorobjovas-Pinta, O. (2018). Gay neo-tribes: Exploration of travel behaviour and space. *Annals of Tourism Research, 72*, 1–10.

Walker, K., & Moscardo, G. (2016). Moving beyond sense of place to care of place: The role of Indigenous values and interpretation in promoting transformative change in tourists' place images and personal values. *Journal of Sustainable Tourism, 24*(8–9), 1243–1261.

Whitford, M., & Ruhanen, L. (2014). Indigenous tourism businesses: An exploratory study of business owners' perceptions of drivers and inhibitors. *Tourism Recreation Research, 39*(2), 149–168.

Zhong, L., Sun, S., Law, R., & Li, X. (2021). Tourism crisis management: Evidence from COVID-19. *Current Issues in Tourism, 24*(19), 2671–2682.

11

AN INDIGENOUS COMMUNITY CODE OF CONDUCT FOR TOURIST BEHAVIOURS VOICES FROM THE TIBET AUTONOMOUS REGION, CHINA

Xiaotao Yang, Heather Mair and Bryan Grimwood

Introduction

Scholars involved in Indigenous studies argue that it is important for researchers to prioritise community benefits from their research. One strategy for centering community priorities is to engage with Indigenous participants to determine research questions that resonate with them. The purpose of this chapter is to report on the process and outcome of developing, in collaboration with Tibetans, a list of codes of conduct for visitors to Tibet Autonomous Region (TAR), China. The chapter draws on four months of fieldwork in TAR, which focused primarily on the travel- and tourism-related issues that Tibetans consider important for further discussion and examination. Consultations with 19 participants led to the decision to document codes of conduct that represent tourists behaviors and attitudes that are welcomed. Tibetan collaborators sought to align these codes with religious requirements and to address increasing tourist numbers and inappropriate tourist behaviours. The codes of conduct that were identified focused on Buddhist items, photography, and environmental pollution. The necessity of documenting and informing tourists about codes of conduct ultimately aims to support and inform three main community-identified priorities: the respect needed at religious sites, the abundance of religious sites across TAR, and the need to gain some degree of local autonomy in tourism management.

Chilisa (2012), Smith (2012) and Wilson (2008) strongly believed that the purpose of conducting research should be to promote love and harmony, a view that highlights the relationships between researchers and participants, as well as non-human beings, and the cosmos. We may call this participant or community-centered research, which is different from the current conventional gap-oriented research that flourishes under the post-positivism paradigm. Such research is generally driven by the urge to identify and meet knowledge gaps that contribute to and reinforce norms within cycles of academic knowledge production. Pedri-Spade (2016) argued that this research logic is based on competition, significantly different from the Indigenous focus on respectful and reciprocal relationships. The latter emphasises not results, but the mutual learning and sharing

133　　　　DOI: 10.4324/9781003230335-13

process. Ignoring community benefits from research is one of the critiques of hopeful tourism by Higgins-Desbiolles and Whyte (2013).

As a response to community-centered research, this paper priorities Tibetan community needs and focuses on an issue of concern for some Tibetans—namely that some visitors to the TAR do not respect local culture. Consequently, Tibetan participants in our study developed codes of conduct for visitors to the TAR. The chapter first introduces literature on Community-Based Participatory Research (CBPR) and codes of conduct, followed by the process of generating the codes and their details, and related discussions.

Indigenous methodologies are closely aligned with existing CBPR approaches that have been widely applied in work with marginalised or oppressed groups (Holmes 2015; Victor et al. 2016), such as Indigenous people and third-world women as listed by Chilisa (2012, p.21). As its name suggests, CBPR engages participants throughout the research process, from research design to presentation and knowledge dissemination. The degree of participation depends on negotiation between researchers and participants. Victor et al. (2016) highlighted that CBPR is closely aligned with their Nehinuw (Cree) worldview, which emphasises interconnection and interactivity among members. de Leeuw, Cameron, and Greenwood (2012) on the one hand, have recognised the wide application of CBPR in Indigenous studies; but they also warn of legitimizing CBPR as 'best practices' and thus possibly devaluing non-participatory methods.

Informed by CBPR and Indigenous methodologies, we conducted our participatory research in the Tibetan context. Tibet can refer to both the geographical area of the Tibetan Plateau and the administrative area of the TAR. Only half of Tibetans live in the TAR, whereas the other half is spread throughout the Tibetan autonomous prefectures or counties in different Chinese provinces. According to the statistics of the Sixth National Census, 3 million people live in the TAR, among whom 2.71 million are Tibetans (90.48%) with the remainder composed of other ethnic groups, including Han Chinese, Hui, Monpa, Lhoba and Nashi (National Bureau of Statistics of the PRC, 2012). I follow Tibetan scholar Gaerrang's (2017) application of the term 'Indigenous people' in the Tibetan context. Gaerrang (2017) explained that some scholars refer to the term 'Indigenous' from a political perspective, whereas he uses this term to address the epistemological difference between Tibetans and mainstream society. Other scholars, such as Chen et al. (2016) and Wu (2012), also apply the term 'Indigenous people' to describe Tibetans in China. All participants and I agreed with the current Chinese government's policy that TAR is an inseparable part of China and subsequently developed our discussion under this premise. Although there are contesting debates on the Tibet relationship with China, especially in the Western media, all the Tibetan people contacted shared the consensus that TAR is an inseparable part of China. In this chapter, we hereafter use the term 'Tibetans' to refer to the Tibetan people living in the TAR, unless otherwise specified.

Research Process and Topic Selection

In this study that centered on Tibetan people's knowledge of their own travel experiences as tourists, the first research question asked was: what travel and tourism-related issues do Tibetan participants consider important? We purposively left this research question open to engage with Tibetan participants to determine research questions that resonate with them. Unstructured interviews were used because the overall design was broad and flexible. We tried to de-center the researchers' voice in order to give privilege to the Tibetan view. 'Unstructured interviewing' offers 'maximum flexibility to pursue information in whichever direction appears to be appropriate, depending on what emerges from observing a particular setting or from talking to one or more individuals in that setting' (Patton 2015, p.462). The strength of unstructured interviews lies in their flexibility and responsiveness to specific individuals and situations. Possible weaknesses of unstructured interviews might include the fact that this form of interview is time-consuming, requiring several rounds to clarify any researcher confusion; and that researchers must have the skill to ensure that impromptu questions are quickly and smoothly formulated (Patton 2015). We used unstructured interviews to invite participants to share issues important to them. The four months' fieldwork and follow-up via second and third interviews, in person or via WeChat (like Facebook) messages helped address these weaknesses. We now elaborate on the research process and the literature on codes of conduct, followed by the findings from our participants. Situating our discussions with existing literature was a response following Hollinshead's (2012) suggestion of cultivating dialogue space between Indigenous and non-Indigenous knowledge.

Due to its participatory nature, it is necessary to describe the collaborative process that led to determining the research topic on codes of conduct. The section will first introduce the overview of fieldwork and research topics proposed by participants, followed by their concerns on codes of conduct. Four-months of fieldwork, from July to December 2019, were conducted in Lhasa and Shigatse City in the TAR, China. The first author had worked as an officer in a Chinese Government tourism bureau in TAR for 2.5 years. TAR was selected as the study case because the tourism literature and general media lack information on Tibetan people as tourists and Tibet's unique culture. The typical discourse on tourism in TAR is always about welcoming hosts and drawing thousands of tourists to travel around 'the roof of the world'; however, the voices of Tibetans as travelers themselves are missing. Gaerrang (2017) emphasised that the Tibetan culture presents a unique-alternative development path, but one that is engulfed in the 'modernity' and 'market' discourse.

During the fieldwork, the first author followed a snowballing approach to invite participants, beginning with seven Tibetans known to her for nearly ten years ago. Those friends introduced others who were also willing to discuss their travel experiences. In total, 35 informants shared their travel stories, with respondents aged from 20s to 60s. Thirty-four of them were of Tibetan ethnicity and one was of Monpa ethinicity. All participants are considered minority ethnic peoples in China, as the Chinese government officially uses the term 'minority ethnic peoples' rather than 'Indigenous', although Bunten and Graburn (2018) argue that China's 55 minority groups are Indigenous people. One-third of participants held a university degree, and occupations included construction workers, small business owners, temporary workers, and hotel and restaurant workers.

In beginning the fieldwork, the first author asked Tibetan participants: what travel and tourism-related issues they considered important, and also indicated that the topic selected could be developed into an academic study to bring their voices into academic tourism studies circles. To inform participants better on the roles of research question, the first author also explained that through conducting a study on the research topic, researchers would collect views from various Tibetan participants to learn whether they shared similar concerns over tourism. Respondents listed several topics they felt important, including identifying generational and time differences among Tibetan travelers, differences between urban and rural residents in travel patterns, psychological changes between pilgrimage and daily lives, causes of generational differences on pilgrimages and behaviour during pilgrimage trips.

Since large numbers of non-Tibetans visit TAR, one interview question asked Tibetans about non-Tibetan visitors' impacts on Tibetan residents. Our participants listed many examples, such as visitors donating clothes, food, and books to local people; helping local farmers to sell their products; collecting litter on their trip; following local rules and rituals; and showing familiarity with Tibetan culture. All participants also shared their concerns about inappropriate tourist behaviours.

'Nowadays, there are more and more tourists, but I think there are more and more tourists who don't know about tourism and about respecting local culture.' was a comment made by one participant, and this was reinforced by another participant who made a nearly identical comment in a different interview during our discussions. The first author was surprised by this overlap which demonstrated Tibetan's concerns about tourists' behavior in their province.

Initially this was regarded as a possible topic for future research, given our research was designed to focus on Tibetans' own travel knowledge, rather than appropriate behaviour of tourists visiting Tibet, and there had been working on the same topic in a Canadian context (see Holmes et al. 2016). However, the importance of focusing on codes of conduct was emphasised to the first author while they were transcribing an interview. During this interview, the participant pointed directly at the first author's notebook and said:

> Have you written it down? You really should help us to advertise these concerns. We cannot blame those visitors since they don't know about our religion. But if two parties quarrel over these concerns, it could exacerbate conflicts between two ethnic groups.

The strength of his concern resulted in a decision to focus the study on developing codes of conduct. This participant was the 35th our last, a point at which we had decided to stop interviewing any new participants due to the large amount of rich and insightful information we already had. The first author then summarised concerns expressed, and was able to re-connect with 19 of the participants to check the accuracy of reporting. She also consulted an additional five Tibetans (in this case monks), via WeChat (a Chinese social media messaging app) messages to seek their comments on the new focus, as they were knowledgeable on Tibetan Buddhism and potential issues relating to visits.

The determination of our research topic is clearly aligned with the principle of CBPR and Indigenous methodologies, which encourages participants to decide on research topics. The decision on developing codes of conduct was based on all 35 participants' concerns on tourism impacts and their contribution to the specific contents, and 19 out of 35 participants were involved in finalizing the list.

The necessity of documenting and informing tourists about codes of conduct in our study was raised by participants. This concern echoes existing publications about codes of conduct and religious tourism. In this section, we first review the literature of codes of conduct, followed by the necessity of documenting codes of conduct in TAR.

Literature of Codes of Conduct

Cultural tourism is a popular form of travel and many tourists visit different cultural communities that practice different behaviour to their own. While it is argued that people within the same culture share the same system of language and symbols (Crotty 1998), enormous barriers in language and symbols exist in cross-cultural communication situations. Cohen (1972) referred to an 'environmental bubble' to depict various degrees of tourists' reflection of their own cultural background in viewing destination culture. Clearly, misunderstanding and conflicts are possible if tourists interpret hosts' behaviours from their own viewpoint, rather than the hosts'. Thus, the informative roles of appropriate conduct could help minimise perceived negative effects of tourism and encourage desirable and beneficial behaviour (Mason 2007; Ell 2015).

Codes of conduct are practical and pragmatic guidelines that are relevant to ethical principles and philosophies (Radin 2004; Fennell 2014; Ell 2015). The main function of codes of conduct is to educate tourists and influence their behaviours to fit the local sociocultural context. Such instruction can reduce negative visitor impacts, enhance tourist confidence, raise cultural awareness, lower misunderstanding, reduce offensive behaviours, and prevent conflicts between stakeholders, particularly between hosts and tourists (Cole 2007; Fennell 2014; Ell 2015). A search on Google Scholar and Science Direct, indicates that most studies on codes of conduct originated between the 1990s and 2010s.

Many codes of conduct are culturally contextualised and locally oriented, containing a mixture of social, cultural, environmental, and in some cases, Codes of conduct not only provide necessary information about local communities but also address ethical concerns, which are culturally contextualised and individually verified. Those ethical concerns are normally expressed by encouraging desired conduct and discouraging undesirable conduct and are also named as 'deontology' and 'teleology' codes respectively (Mason 2007; Fennell 2014). Some codes of conduct are created mainly to solve ethical dilemmas as the likelihood of conflict may be caused by different players in tourism who have diverse value positions and use different languages and symbols (Fennell 2000; 2014). Even within the same region or context, codes of conduct developed by different organizations may differ. For instance, Ong et al. (2013) conducted qualitative content analysis and tag clouds to identify themes of guiding principles of 12 individual volunteer tourism organizations. The emerging themes included: local needs, participant-focused, consideration toward continuity, community, exploration of issues, interaction and organizational goals. They further compared these themes with two global codes in volunteerism and tourism and found that most themes matched the two international codes, except for organizational goals (e.g. religious goals or industry leadership goals). Their study suggested that even within the same topic of volunteer tourism, different developers of guiding principles would shape specific items differently (Ong et al. 2013). Therefore, being aware of 'who' developed the codes is important.

The tourism industry or government agencies/departments are the parties that most frequently construct such codes and tend to favour a top-down approach. Few existing codes of conduct were developed through a bottom-up approach that enabled local communities to provide input. Holmes et al. (2016) reminded us to scrutinise these codes and to consider the question 'What values become authorised in such circumstances, and who benefits, how, and at whose expense?' (p.1180). Accordingly, collaborating with the Lutsel K'e Dene First Nation, Holmes and her colleagues created an indigenised visitors' code of conduct. Other scholars, such as Buzinde (2019) and Higgins-Desbiolles et al. (2019), proclaimed that: 'tourism must be redefined and redesigned to acknowledge, prioritise and place the rights of local communities above the rights of tourists for holidays and the rights of tourism corporates to make profits' (Higgins-Desbiolles et al. 2019, p.1927). Constructing and disseminating an indigenous-community-determined code of conduct can therefore move toward this community-centric direction.

Engaging multiple stakeholders in the formulating process of codes of conduct will affect its implementation. The effectiveness of codes of conduct will be enhanced if the intended user groups are involved in creating the code, not only the content but also the tone and format, as well as in providing the rationale behind the instructions (Mason 2007). Fennell (2014) echoed the point that better understanding of the codes with detailed explanation is related to actual changes in behaviors. Sethi and Williams (2000) suggested that these codes should be mutable, dynamic, capable of adjust, ament and relate to the evolving societal context, instead of being static codes of conduct. Codes of conduct need to respond to issues and situations that arise during the implementation of the code.

Cole (2007) attempted to implement codes of conduct for tourists and evaluated their effectiveness. Contrary to the common notion that it was an easy, quick, and cheap way to manage tourists, the development of an appropriate code of conduct requires time, patience, and extensive research. On the basis of Cole's ethnographic fieldwork and discussion with local stakeholders, a code of conduct that was specific to the studied region was drafted. After its two-year implementation, the draft was tested on tourists to see whether they changed their behaviours or attitudes and how they responded to the codes of conduct, and a revision was made consequently. In order to make sure that the new version was agreeable to both local villagers and guides, it was modified according to their comments and reactions. Cole also suggested that the effective display and distribution of codes of conduct relied on official endorsement.

Constructing visitors' codes of conduct or guidelines is not a new practice. However, incorporating indigenous voices into the development of such codes should be a priority for many researchers and practitioners (Ell 2015; Holmes et al. 2016). Such a move would offer local people a degree of autonomy in deciding how tourism activities in their communities should function. Ell (2015) underscored that the mitigation of negative misunderstanding about Indigenous practices via codes of conduct could help preserve Indigenous cultural ways of life from erosion. Moreover, it shifts the power away from Western models to practices deeply rooted in local communities. Blowfield (1999) suggested that codes of conduct that are local, emic, and case by case, work more appropriately than general and universal principles. Therefore, although various studies have worked on developing codes of conduct in various contexts, in this Tibetan example, producing codes of conduct based on the context of the TAR is necessary.

The Necessity of Documenting Codes of Conduct in TAR

The necessity of documenting codes of conduct in TAR comes from two main considerations: the need for respect by visitors at religious sites and the sheer abundance of religious sites across Tibet.

Government and tourism officials increasingly promote religious sites and destinations to encourage tourist visits, thereby greatly boosting the number of visitors, regardless of their motivations or religious affiliation (Olsen 2019). Blending religious and tourist spaces has long been an ongoing challenge for many religious sites. Religious heritage tourists, those visiting religious sites for educational and leisure purposes, might lack sufficient knowledge of the appropriate ways to behave at such sites, causing conflicts with religious tourists (Olsen & Timothy 2006, p.27). As a result, the mixture of religious and recreational visitors has led to considerable management pressure to mitigate any potential conflicts among these two groups of visitors, as well as between one religious group and another (Olsen & Timothy 2006; Timothy & Conover 2006). Timothy and Conover (2006) reported conflicts between New Age believers and Native Americans because the two groups hold different beliefs about the same sites. With the number of tourists involved, it can be challenging for some religious sites to maintain a spiritual nature (Yang et al. 2019) because the impression of sanctity tends to disappear with crowds and commercial enterprises. In fact, balancing sanctity and commercialisation of religious sites has been a key area in religious tourism studies.

In addition to religious concerns, locations such as sky burial sites are also problematic places for tourists to visit. Scholars drew attention to a local community's perceptions when dark tourism took place (Sharpley & Stone 2009; Wright & Sharpley 2018). They were sensitive about and fragile in the face of the outsiders' activities because of their emotional connections with their deceased loved ones. Respecting the deceased person has been identified as an important rule for visitors. In Wang et al.'s (2020) study, local villagers expressed welcome to visitors because most visitors to the site of interest respected and worshiped their deceased family members. Sky burial sites in TAR are both religious and dark places that attract Tibetan and non-Tibetan visitors.

Accordingly, educating non-religious visitors on the meanings behind sacred sites has been proposed to help those involved move from conflicts to solutions (Olsen & Timothy 2006). Wang (2016) emphasised educating non-Tibetan visitors on respecting deity mountains and lakes as a way of respecting local Tibetan culture in order to avoid conflicts between different ethnic groups. However, few studies have explored in detail how to achieve this goal, what educational content to include, and whose voices should be heard. It is not surprising that visitors are sometimes in direct conflict with pilgrims at religious sites but regulating policies are normally developed by the government or specific management at sites, rather than allowing the pilgrims to have a voice in creating such policies. This chapter aims to amplify the voices of local pilgrims through codes of conduct.

Two Tibet-specific factors might lead participants to be concerned about visitors' behaviours: the large number of visitors and the lack of information/education about behavioural conduct in Tibet for these visitors. First, because of Tibetans' long-standing dedication to Buddhism, there are large numbers of religious sites across TAR, open to both believers and tourists: around 1,700 temples and monasteries, with 40 Buddhist-related festivals at least (Yu et al. 2016). Moreover, many mountains, rivers, caves, and lakes are sanctified as well. These sacred sites are commonly viewed as rich touristic resources.

Figure 11.1 highlights the sharp increase in visitors to TAR in the past decade. In 2019, there were 40 million tourists, nearly 12 times the local population. As a result, many sacred sites are visited by large numbers of tourists and local believers, a factor that might lead local participants to be very concerned about tourists' behaviour at Buddhist sites. Even facing so many visitors, it is notable that most participants involved in our study were not hostile towards tourists, as demonstrated from this sympathetic quote: '*We are aware of that visitors to Tibet might experience High*

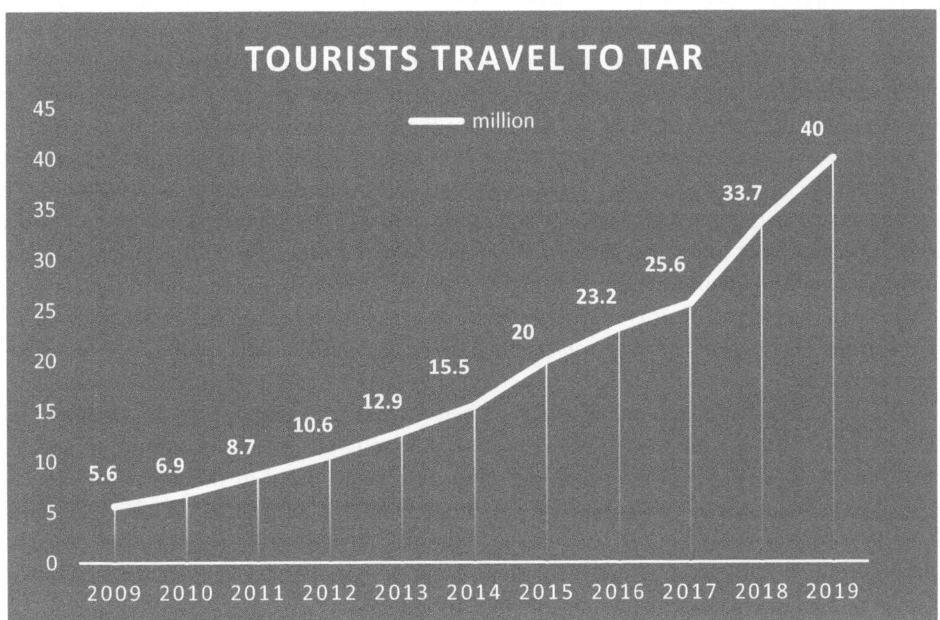

Figure 11.1   Tourist travel to TAR between 2009 and 2019.

Source: Tibet Youth International Travel Service (2022).

Altitude Sickness, language barriers, and different culture'. However, they felt that tourists could demonstrate more respect and be careful in their behaviour during their visits.

Second, while tourists are typically welcomed by governments for economic stimulation, informing visitors of appropriate conduct is uncommon in current tourism practices. For cross-culture travelling, suggestions and tips could inform tourists about local cultural features or contexts. However, a search on the Google and Baidu (the major search platform in China) with keywords of 'suggestions/tips/must-known for visiting Tibet revealed that these must-know tips were either about protecting the tourists or reflected an industry-driven perspective. For instance, many tips listed detailed information on UV protection, such as sunscreen, sunglasses, and clothing, and on mitigating high mountain sickness, such as getting enough sleep, avoiding showers, taking medicines in advance, and reducing intense movement. Almost no tips mentioned suggestions for appropriate behaviour when visiting religious sites. The only ones the first author found were in an attachment to a publication of a travel agency intended for its customers. It briefly mentioned respecting local customs but without explaining specifically how to do so. Also, it stated that 'tourists are not allowed to visit the sky burial sites', which is not true, as our field work showed, and Tibetan culture encourages people to visit these sites as a way of learning. These problems of omission and errors therefore, mean it would be both useful and practical to have a code of conduct designed by local residents to reflect appropriate behaviour and actual situations.

Codes of Conduct for Tourists' Behaviour

The following recommendations in Table 11.1 were based on the participants' concerns. The words in italics are the original words of participants. The word 'we' refers to the Tibetan participants and 'you' refers to tourists who visit TAR.

Most Tibetans are dedicated Buddhists. We are aware that visitors to Tibet might experience High Altitude Sickness, language barriers, and different culture. We welcome any visitors, both Buddhists and non-Buddhists, who are interested in and respect our cultural and religious practices. Even though a few simple words cannot explain the true meaning of Buddhism, some fundamental principles include, but are not limited to: *the mind of bodhicitta – the great compassion and the great love; accumulations of merits; not to commit any sin, to do good, and to purify one's mind, that is the teaching of (all) the Awakened Ones; the law of cause and effect; and vicissitudes of life in cyclic existence.* [Note: These principles are based on participants' summaries. There are various interpretations of the essentials of Buddhism. The renowned Buddhist Master—Dzongsar Jamyang Khyentse—reminded that: 'Although essentially very simple, Buddhism cannot be easily explained', (2007, p.2)]. *It is always better to learn about this land before stepping on to it. The following is a brief list of inappropriate behaviors that might conflict with our faith, make us angry, or hurt us:*

Table 11.1 documents commonly reported behaviours, both appropriate and inappropriate, noted by respondents, who also informed us about the religious reasons for doing or not doing certain things. In total, 13 issues were identified. Six were related to Buddhist items, such as prayer flags, mani stones, or deity lakes. Four were related to photography, which suggested that taking photos, a common practice for tourists, may easily conflict with local customs if not done appropriately and had in fact become an issue concern to local Tibetans. Other concerns included Tibetan dress culture, garbage/pollution and hitchhiking.

Tibetan Buddhism is a unique branch of Buddhism, with a set of quite special practices. Items such prayer flags or mani stones are among its unique features. Even Buddhists who follow other branches of Buddhism might make mistakes without realising. The first author had lived

Table 11.1 Codes of conduct for welcomed tourist behaviours in Tibet Autonomous Region (TAR), China

Items	Inappropriate behaviour	Appropriate behaviour
Buddha images	Print out on cards, T-shirts, beds, tattoos, even shoes	*Buddha is the highest and most holy status in Buddhist world. Thus, as his follower, we demonstrate our most venerated respects to Buddha and his associates, such as images. We print out Buddha' image and carefully put them on clean and high places for worship. We never print it on any inappropriate context, such as cloth or beds. We never put Buddha related stuff, such as prayer beads, below our waist, to show our respect. Thus, if we see someone wearing Buddha images on shoes, we are extremely hurt.*
Prayer flags / Wind-horse/ Lungta	Cross over or step on Lungta while taking photos	*For you, they are colorful flags that attract your eyes for photo taking (we understand such curiosity); For us, they are carrying devoted prayers. We call it, 'Lungta', literally meaning 'Wind-horse'. Divinity sutras have been printed on them. We hang them up on high mountains or near clean water to pray for all the living beings. If you have to go cross the flags, lift them up and walk under/below them, rather than cross over/above them.*
Mani stone piles	Step on or sit on Mani stone piles	*Mani stones are named because the stones are engraved with the six-syllable prayer word 'om mani padme hum', which send out our praying message to Buddha and demonstrate compassion. We pile them up in mountain-passes and on river banks. It is inappropriate to step on or sit on them.*
Buddhist Pagodas	Climb or step on pagodas	*There are many pagodas that contain deity materials, such as relics or sutras; thus, we worship and circle pagodas, rather than climb them. For example, there is a pagodas at the foot of Potala Palace. We frequently see some tourists step on it for play or taking photos. Such behaviour is inappropriate.*
Sanctified mountains, rivers, caves, and lakes	Inappropriate behaviors at these sites	*In addition to monasteries, many mountains, rivers, caves, and lakes are sanctified as well. Certain visiting rules are required thought it is hard to list all the rules. A simple tip is to pay attention to how the locals respond to them. Two commonly seen inappropriate behaviors are: circulate mountains or monasteries in reverse clockwise; step into deity lakes. Tibetan Buddhism normally circulate religious sites, such as mountains or monasteries in clockwise. Some lakes are sanctified, such as the two famous tourists' lake, Namtso and Yamdrok Lake. We Tibetans worship these deity lakes rather than step into it for fun.*
Tibetan dressing culture	Dress shorts while visiting monasteries	*Tibetan people normally do not wear shorts in public space. It is better to dress long sleves, especially during the visit at scared places, like the monasteries.*

(*Continued*)

Table 11.1 (Continued)

Items	Inappropriate behaviour	Appropriate behaviour
Elements that you cannot understand	Laugh, surprise	*Some religious elements, including images, status, words, or behaviors, might beyond your understanding. Please pay attention to how the locals respond to them and do not laugh at or surprised about them. Maybe try to search for the actual meanings of these elements.*
Photographing in the temple	Shoot photos at temples where such activities are forbidden	*Some temples allow for photo shooting inside of the temple/hall; however, many temples do not allow photo shooting inside for either religious or protection reasons. Therefore, check carefully, particular at the entrance of the building or by Buddha statues, for any specific regulation. Follow any signs, such as 'no photographs allowed'.*
Photographing someone with Buddha statues	Take photo with Buddha statues with postures, such as hand in hand, or shoulder by shoulder	*Buddha holds the highest position, whereas we humans are in a lower position than him. Obviously, it is inappropriate to have any posture that indicates parallel status with Buddha. You are suggested to not take photos with Buddha statues. Buddhism believes that all sentient beings carry Buddhist seeds, a metaphor which suggests everyone has the potential to be fully enlightened— become a Buddha—the ultimate goal. Please respect Buddha statues.*
Photographing at the sky burial sites	Take photo or video of the site or the ritual, and even upload them online	*As part of Buddhism teaching, we encourage people to visit the sky burial sites to get the direct view of how we people come to the world with a bare body and leave the world with bare body as well. Meanwhile, donation is a practice that we carry out throughout our life so that we donate our bodies to other living beings, such as vultures and other birds. That's a meaningful thing that we could do after death. You are allowed to visit these sites and sometime might encounter some rituals; however, you should respect the dead bodies and families and no photographs or videos are allowed, not to mention uploading such files to the Internet.*
Photographing at people	Give local people 5 or 10 Yuan, or none, and take photo directly even when the people act shy, turning their faces away, or covering them	*How would you feel if someone took photos of you and uploaded them online? It would be better to ask for permission of anyone involved, get their permission, and inform them about how these photos will be used and disseminated. If they do not give permission, do not take photos.*
Garbage	Throw away garbage anywhere	*We only have one mother earth* who has generously supported our human development; thus, it is our responsibility to keep it clean. You are enjoying the magnificent scenery, so please *save the same view for OUR kids. No tourists are allowed to visit Mt. Everest Base Camp anymore since 2019 due to pollution concerns. Look at this, who threw trashes there? What a pity!* The Only Cure for Litter is YOU!

and worked in TAR for two and half years, but some of the features were new to her, such as not stepping into deity lakes. Many non-Tibetans would normally step into a sea or lake just as they would outside of TAR. However, deity landscapes, such as mountains and lakes, must be respected. Another example, relates to the Tibetan belief in rebirth and reincarnation, emphasizing good deeds leading to higher levels of reincarnation. For example, Tibetans normally do not ask for compensation from a driver at fault in an accident, instead relying on karma. Compensation is believed to create an enormous burden for the innocent driver. One phenomenon that many Tibetans (participants and non-participants) talked about is the fact that non-Tibetans ask for a large amount of compensation for accidents, which is in conflict with Buddhist belief so that some of them were not willing to get involved with non-Tibetan hitchhikers. Clearly, these differences come from different belief systems or cultural backgrounds.

Conclusion

Analysis of the Tibetan participants' interviews identified 13 key themes or issues around visitors' behaviours in the TAR. Six were related to Buddhist items, one for Tibetan dressing culture, and six for tourists' activities, such as taking photos, throwing garbage, and hitchhiking. Tibetan participants also explained related reasons to help visitors better understanding these cultural specific items, as documented on the third volume in Table 11.1. Their explanations reflect their expectation that the tourists visiting TAR could respect their culture. Overall, our study intertwines with existing literature from the following four aspects. First, our findings reflected Holmes et al.'s (2016) study in the Lutsel K'e Dene First Nation in Canada and Cole's (2007) study in Ngadha, Flores, Indonesia: where local communities generally welcomed tourists. Thus, the goal of developing codes of conduct is to shape tourism toward a moral, just, and culturally sensitive development (Holmes et al. 2016). Despite the codes described some inappropriate behaviors of visitors, hosts' welcoming attitude is important because they are not rejecting cross-cultural communication, but expecting for more culturally sensitive ways. Developing such codes aims for smoothing such communication.

Second, although some items may vary, respecting local culture is the fundamental rule of such codes. According to Holmes et al. (2016), respecting local culture refers to understanding and respecting Indigenous world views. Similarly, in our study, the local culture is deeply rooted in Tibetan Buddhism. Although engaging in Indigenous world views sounds challenging, Tibetan participants provide a very useful tips: for difficult understanding elements, visitors could pay attention to how the locals respond to them. Echoing Cole's (2007) finding, our study reported that local communities observe conservative dressing.

Third, Holmes et al. (2016) reported issues on environmental pollution and reminders to visitors to 'keep the land and water clean,' a concern shared by our Tibetan participants. Tourist attractions are the homeland of local communities, so environmental pollution can be a constant concern for the hosts.

Fourth, all three studies mentioned that tourists should ask permission first before taking photos of locals. Taking photos is the most frequently activities for tourists, but ask for permission could serve as a general rule for tourists. Dark tourism literature highlights the necessity to be sensitive on such sites (Sharpley & Stone 2009; Wright & Sharpley 2018; Wang et al. 2020). Similarly, participants shared their specific experiences of the restriction of no photography allowed at the sky burial sites.

This chapter has first elaborated Tibetan participants' request that we develop guidelines for tourists who visit Tibet, followed by a detailed list of items that could inform a code of conduct.

We then contextualised the idea of tourist codes within the existing literature, which aligns with Buddhist religious requirements, reflecting the local culture of Tibet. Methodologically, it demonstrated how the research question emerged from the fieldwork and resonated with Tibetan participants' concerns. Theoretically, it contributes to the move towards a moral and culturally sensitive forms of tourism by adding the Tibetan perspective. Practically, it aims to inform visitors who are visiting or plan to visit TAR towards responsible behaviors. In addition to disseminating the findings in this chapter, the authors will disseminate the codes via non-academic channels because of its significance for the participants. Current activities include converting the codes into short animation videos to improve accessibility and trying to include these codes in the travel contracts of travel agencies for visitors.

References

Blowfield, M. (1999) "Ethical Trade: A Review of Developments and Issues," *Third World Quarterly* 20, 753–770.

Bunten, A. C., & Graburn, N. H. (Eds.) (2018) *Indigenous Tourism Movements*, Toronto: University of Toronto Press.

Buzinde, C. (2019) "Tourism and Indigenous Communities: Linking Reterritorialisation and Decolonisation in North Africa," in D. J. Timothy (ed.) *Routledge Handbook on Tourism in the Middle East and North Africa*, London: Routledge, 83–94.

Chen, K., Li, Y., & Wang, J. (2016) "The Railway Connects Plateau Tourism Sites". Available at: http://www.vtibet.cn/tbch/2016zt/pdxz_7887/cs7/201606/t20160629_408024.html

Chilisa, B. (2012) *Indigenous Research Methodologies*, Thousand Oaks: Sage Publications.

Cohen, E. (1972) "Toward a Sociology of International Tourism," *Social Research* 39(1), 164–182.

Cole, S. (2007) "Implementing and Evaluating a Code of Conduct for Visitors," *Tourism Management* 28, 443–451.

Crotty, M. (1998) *The Foundations of Social Research: Meaning and Perspective in the Research Process*, Thousand Oaks: Sage.

de Leeuw, S. D., Cameron, E. S., & Greenwood, M. L. (2012) "Participatory and Community-based Research, Indigenous Geographies, and the Spaces of Friendship: A Critical Engagement," *The Canadian Geographer/Le Géographe Canadien* 56(2), 180–194.

Dzongsar Jamyang Khyentse. (2007) *What Makes You Not a Buddhist*, Boston: Shambhala Publications, Inc.

Ell, L. (2015) *Codes of Conduct for Indigenous-Inspired Spa and Wellness Tourism* [Master thesis, Royal Roads University, Canada]. Available at: https://central.bac-lac.gc.ca/.item?id=TC-BRC-827&op=pdf&app=Library&oclc_number=1032922480

Gaerrang, K. (2017) "Contested Understandings of Yaks on the Eastern Tibetan Plateau: Market Logic, Tibetan Buddhism and Indigenous Knowledge," *Area* 49(4), 526–532.

Fennell, D. (2000) "Tourism and Applied Ethics," *Tourism Recreation Research* 25, 59–69.

Fennell, D. (2014) *Ecotourism*, London: Routledge.

Higgins-Desbiolles, F., & Whyte, K. P. (2013) "No High Hopes for Hopeful Tourism: A Critical Comment," *Annals of Tourism Research* 40, 428–433.

Higgins-Desbiolles, F., Carnicelli, S., Krolikowski, C., Wijesinghe, G., & Boluk, K. (2019) "Degrowing Tourism: Rethinking Tourism," *Journal of Sustainable Tourism* 27(12), 1926–1944.

Holmes, A. P. (2015) *'This Sacred Land Is Our Home: Respectful Visitors Welcome': An Indigenized Code of Conduct for Visitors to the Traditional Lands of the Lutsel K'e Dene First Nation* [Master thesis, University of Waterloo]. UW Space.

Holmes, A. P., Grimwood, B. S., King, L. J., & Lutsel K'e Dene First Nation. (2016) "Creating an Indigenized Visitor Code of Conduct: The Development of Denesoline Self-determination for Sustainable Tourism," *Journal of Sustainable Tourism* 24(8–9), 1177–1193.

Hollinshead, K. (2012) "Tourism and Indigenous Reverence: The Possibilities for Recovery of Land and Revitalisation of Life," in A. Holden & D. Fennell (eds.) *The Routledge Handbook of Tourism and the Environment*, New York: Routledge, 43–54.

Mason, P. (2007) "'No Better Than a Band-Aid for a Bullet Wound!': The Effectiveness of Tourism Codes of Conduct," in R. Black & A. Crabtree (eds.) *Quality Assurance and Certification in Ecotourism*, Wallingford: CABI, 46–64.

National Bureau of Statistics of the PRC. (2012) *Statistics of the Six National Census of the Tibet Autonomous Region*. Available at: http://www.stats.gov.cn/tjsj/tjgb/rkpcgb/dfrkpcgb/201202/t20120228_30406.html

Olsen, D.H. (2019) "Religion, Pilgrimage, and Tourism in the Middle East," in D. J. Timothy (ed.) *Routledge Handbook on Tourism in the Middle East and North Africa*, London: Routledge, 109–124.

Olsen, D. H., & Timothy, D. J. (2006) "Tourism and Religious Journeys," in D. J. Timothy & D. H. Olsen (eds.) *Tourism, Religion and Spiritual Journeys*, London and New York: Routledge, 1–21.

Ong, F., Pearlman, M., Lockstone-Binney, L. & King, B. (2013) "Virtuous Volunteer Tourism: Towards a Uniform Code of Conduct," *Annals of Leisure Research* 16(1), 72–86.

Radin, T.J. (2004) "The Effectiveness of Global Codes of Conduct: Role Models That Make Sense," *Business and Society Review* 109(4), 415–447.

Patton, M. Q. (2015) *Qualitative Research and Evaluation Methods: Integrating Theory and Practice (Fourth Edition)*, Thousand Oaks: SAGE.

Pedri-Spade, C. (2016) "'The Drum is Your Document': Decolonizing Research through Anishinabe Song and Story," *International Review of Qualitative Research* 9(4), 385–406.

Sethi, S.P., & Williams, O.F. (2000) "Creating and Implementing Global Codes of Conduct: An Assessment of the Sullivan Principles as a Role Model for Developing International Codes of Conduct - Lessons Learned and Unlearned," *Business and Society Review* 105(2), 169–200.

Smith, L.T. (2012) *Decolonizing Methodologies: Research and Indigenous Peoples*, London: Zed Books Ltd.

Sharpley, R., & Stone, P. R. (2009) *The Darker Side of Travel: The Theory and Practice of Dark Tourism*, Bristol: Channel View Publications.

Tibet Youth International Travel Service. (2022) "Number and Income of Tourism Reception in TAR Over the Years," Available at: https://www.qnly.com/guide/15782.html

Timothy, D. J., & Conover, P. J. (2006) "Nature Religion, Self-spirituality and New Age Tourism," in D. J. Timothy & D. H. Olsen (eds.) *Tourism, Religion and Spiritual Journeys*, London and New York: Routledge, 139–155.

Victor, J. M., Goulet, L. M., Schmidt, K., Linds, W., Episkenew, J. A., & Goulet, K. (2016) "Like Braiding Sweetgrass: Nurturing Relationships and Alliances in Indigenous Community-based Research," *International Review of Qualitative Research* 9(4), 423–445.

Wang, F. (2016) "Analyzing the Ecological Impacts of Tibetan Pilgrimage," [藏传佛教生态朝圣旅游试析] *Journal of Yunnan Socialist College* 1, 76–83.

Wang, J., Xie, L., & Zhang, S. (2020) "Residents' Perceptions and Participation in Dark Tourism on Natural Disaster Site—A Case Study of Jina Qiang Stockaded Village, China," *Tourism Tribune*, 36(1), 1–18.

Wilson, S. (2008) *Research Is Ceremony: Indigenous Research Methods*, Black Point: Fernwood Publishing.

Wright, D., & Sharpley, R. (2018) "Local Community Perceptions of Disaster Tourism: The Case of L'Aquila, Italy," *Current Issues in Tourism* 21(14), 1569–1585.

Wu, Mao-Ying. (2012) *Tourism at the Roof of the World: Young Hosts Assess Tourism Community Futures in Lhasa, Tibet* [Doctoral dissertation, James Cook University]. Available at: https://researchonline.jcu.edu.au/24990/

Yang, X., Hung, K., Huang, W. J., & Tseng, Y. P. (2019) "Tourism Representation by DMOs at Religious Sites: A Case of Shaolin Temple, China," *Tourism Management* 75, 569–581.

Yu, Z., Tian, X., & Zhu, P. (2016) "Present Development Situation and Countermeasures of Religious Tourism in Xizang," *Resource Development & Market* 32(3), 369–373.

12

CONFRONTING MARGINALISATION? GENDER DYNAMICS AND BATWA ENGAGEMENT WITH TOURISM IN UGANDA

Brenda Boonabaana, Amos Ochieng and Christine Ampumuza

Introduction

In many developing countries, tourism has been widely touted as a tool for addressing poverty by creating employment opportunities for the local people (Ahebwa, Sandbrook, and Ochieng, 2018). In Uganda, tourism is considered amongst the top industry sectors to leverage socio-economic development in the country. Several attempts are being made to further develop tourism, to enable the country realise social-economic benefits – with a trickle-down effect to the local communities living in areas adjacent to the tourist attractions. The country's policy framework such as the *Uganda Vision 2040*, the tourism sector development plan 2020/2021–2024/2025 and the *National Development Plan* (NDPIII) recognize the crucial contribution of tourism to Uganda's economy.

Before the COVID-19 pandemic and associated lockdown restrictions, Uganda registered over 1.5 million tourist arrivals. This contributed to about US$ 1.6 billion in foreign exchange with an approximate contribution of 7.7% of the country's gross domestic product (GDP) (MoTWA, 2020). These figures, however, drastically dropped by more than 1 million foreign tourist arrivals in the year 2020 and 2021 following global reductions in travel. As the country gradually eases its lockdown restrictions including opening borders and the removal of mandatory COVID testing on arrival, there is hope that tourism will pick up again. Although Uganda's tourism is diverse, including culture and heritage, religion, food, entertainment. MICE – the country's tourism sector is largely nature-based. There are currently ten national parks, eight wildlife reserves, four wildlife sanctuaries and nine Ramsar sites which make Uganda one of the most biodiversity-rich nations in the Eastern African region. The top tourist attractions in Uganda are the primates, including the mountain gorilla (*Gorilla beringei beringei*) in the Bwindi Impenetrable National Park (BINP) and Mgahinga National Park (MNP) in South-western Uganda and Chimpanzees in Kibale National Park and other conservation areas in the country (Ochieng, Ahebwa, and Twinomuhangi, forthcoming). The BINP and MNP host about 60% of the 1,000 endangered mountain gorilla population in the world, with the BINP alone hosting about 43% of the individuals (Weber, Kalema-Zikusoka, and Stevens, 2020). The country is also endowed with diverse birds, reptiles, mammals, fish and butterflies alongside a rich cultural heritage. The broad range of biodiversity has boosted ecotourism which has in turn created employment opportunities, foreign exchange and improved livelihoods. Ecotourism has therefore become an empowering tool for marginalised communities living adjacent to protected areas in Uganda (Ochieng, Ahebwa, and Twinomuhangi, forthcoming).

This chapter focuses on the Batwa, an Indigenous people who were forcibly evicted from their ancestral home (Bwindi Forest) in 1992, to pave the way for gorilla ecotourism. For millennia, the Batwa had lived in this forest and practised their hunter-gatherer lifestyle and depended on the forest for shelter, foods such as honey, for medicinal herbs and spiritual and religious connection (Grant, 2018; Tumusiime and Svarstad, 2011). However, since the eviction, the Batwa have faced challenges associated with land compensation and survival and have now engaged in tourism to survive in the new environment (Ampumuza, 2020; Tumusiime and Vedeld, 2015). Local and Indigenous peoples in developing countries often experience threats, abuse and oppression in the wake of their governments' pursuit of tourism and conservation goals and commoditisation of nature (Laudati, 2010). Guided by Foucault's (1982) notion of power and agency, this chapter interrogates how the Batwa men and women are engaged in tourism initiatives while confronting social marginalisation and vulnerability, as they carve out space for themselves as key actors in Uganda's tourism sector.

Theoretical Framing

Foucauldian Notion of Power and Agency

Foucault (1982) has been widely cited for his scholarly work on power and agency. His seminal work on power has been widely theorised, for example as a form of social control (Gallagher, 2008). Power, as widely used in the Foucauldian view is considered as a construct that is ubiquitous (Brown, 2006; Foucault, 1982) and can be used to enable others to control their fellows or subdue others to subjugation. In this chapter, we analyse the circumstances that led to the eviction of the Batwa from the forests and the subsequent declaration of illegality, having them lose access to resources they once owned and freely accessed for survival (Kagumba, 2021). Similarly, the bio-power concept (Bio-power is about the subjugation of bodies and control of particular populations in general (Clegg, 1998)) provides important clues in understanding how the Batwa population has been able to adapt and survive in their 'new' environment. In all these circumstances, it is worthwhile noting the role of human agency and the power it holds in enabling and constraining human actions.

The concept of agency, as used in the social sciences, is one of the dimensions of power – which is widely used in the study of human existence and society. An individual's agency involves defining own goals and acting upon them (Kabeer, 1999). According to Kabeer (1999, p.438), agency takes '"the form of bargaining and negotiation, deception and manipulation, subversion and resistance, as well as more intangible, cognitive processes of reflection and analysis'. From a gender perspective, 'agency' is founded on the assumption that women are not passive victims of oppression but are active participants who make choices and participate in structuring their identities (Weedon, 1997). We consider the actions of Batwa men and women in the context of tourism, since eviction from the forest and how their actions are challenging their long-held marginalisation.

Indigenous People and Tourism

According to Butler and Hinch (2007), indigenous tourism is a form of tourism where Indigenous people have direct control of the delivery and interpretation of their eco-cultural heritage to the tourists. Worldwide, Indigenous people suffer marginalisation and threats as governments pursue conservation goals (Nielsen and Wilson, 2012). As biodiversity-rich countries embrace the idea to commodify nature as a tradable good driven by the laws of supply and demand (Laudati, 2010), the plight of Indigenous peoples has often been ignored or paid little attention. Instead, they have been represented for the tourist gaze (Urry and Larsen, 2011), which further reinforces their individual and cultural marginalisation. Even so, Indigenous and local communities have had a long relationship with natural resources in which they act as both custodians and at the same time beneficiaries from the use of nature (Dawson et al., 2021), which has enabled many indigenous and local communities to accumulate a wealth of knowledge and experiences in dealing with their surrounding natural resource base.

Community Involvement in Tourism Initiatives – A Gender Perspective

Globally, women make up a larger proportion of the tourism workforce than any other sector (World Tourism Organisation, 2019), thereby contributing to two key sustainable development goals – gender equality and women empowerment (Goal 5) and decent work and economic development (Goal 8). However, the literature on tourism employment and gender demonstrates that unequal gender relations limit women's opportunities to access, participate in and benefit from tourism developments (Boonabaana, 2014; Tucker and Boonabaana, 2012). Women's tourism work remains largely confined to low-ranking and lesser remunerated positions relative to the men's, where they work as chambermaids, cleaners, receptionists, telephone controllers, housekeepers, handcraft vendors and restaurant waitresses, while men become tourism entrepreneurs, managers, guides, transporters, security personnel and porters (Ferguson, 2010; Moodley et al., 2016; Nyaruwata and Nyaruwata, 2013; Tucker and Boonabaana, 2012). In Kenya, men are perceived to be more appropriate to lead expeditions than women and have ended up dominating the safari guiding roles (World Bank, 2013). More evidence shows that, women are more likely to work as handicraft producers and cultural dancers (Babb, 2012; De, 2013; Ferguson, 2009; Tucker, 2007; World Bank, 2013). Indigenous women are often viewed as the most culturally authentic members of their communities which gives them cultural capital and some economic advantages within tourism (Babb, 2012; Taylor, 2017). Trau and Bushell (2008) have argued that indigenous people in Australia are motivated to engage in tourism because of its economic linkage to protection and transfer of their traditional cultural custom and knowledge. Despite the prevailing opportunities for local and Indigenous men and women in tourism, African women often miss out on tourism opportunities due to the patriarchal system that limits their mobility and free participation in the public space. Women often have to seek permission from their husbands to work in tourism to avoid negative backlash including violence and community gossip (Boonabaana, 2014).

As argued by Ferguson (2007, 2009, 2010), tourism-based gender benefits would have more impact if the constraints that disadvantage women to join the tourism space were understood and addressed. This aligns with Scheyvens' (2002) argument that both mainstream and alternative tourism initiatives that do not have a specific mandate to be sensitive to the needs and interests of women actively run the risk of disadvantaging and marginalising local women. Indigenous people are usually more affected by conservation than other groups. For instance, the Batwa communities living around protected resources in Uganda and the Democratic Republic of Congo have experienced dispossession after forced displacement, oppression of their cultural identity, exclusion from jobs and inability to pay for formal schooling (Laudati, 2010; Ramsay, 2010; Simpson and Geenen, 2021). Batwa women face additional challenges because they are discriminated against for being Batwa, whilst experiencing gender discrimination as the majority of African women do (Ramsay, 2010).

Methodology

The results presented in this chapter are based on two PhD studies that were completed by the first and third authors in 2012 and 2021 respectively. The first author utilised a case study research design and ethnographic fieldwork that utilised 44 in-depth interviews (29 women and 15 men), participant observations, informal interactions and key informant interviews in Mukono Parish, South-western Uganda. Men and women participants were purposively drawn from four villages of Nkwenda, Buhoma, Mukono and Kanyashande where most tourism businesses and activities are concentrated.

The second PhD study also utilised ethnographic participant observations involving daily visits to Mukono Parish villages through joining Batwa meetings, workshops, training, as well as cultural tours. Observation data was triangulated with qualitative unstructured interviews (23) and group interviews with Batwa men and women (two with women alone and one with a mixed-sex group), as well as 37 key informant interviews from the Batwa community and tourism stakeholders around Bwindi. Participants were selected from five villages (Sanuriro, Kalehe, Ruhija, Kanyamahene, and Nkuringo). Additional data were obtained by studying various documents such as the NGO reports, tourist blogs, social media posts (Facebook) and websites that promoted the Batwa experience, reviews of the Batwa experience posted on TripAdvisor between 2016 and 2018 and local newspaper articles. Generally, participants were identified by preliminary informal interactions as we moved around the villages, trading centres and businesses while observing and noting down relevant issues. The first participants then led us to other relevant participants, following the snowball sampling method.

Throughout the fieldwork period, a local dialect – *Rukiga*, which is widely spoken by both the Bakiga and Batwa, was used during interviews and translation made into English. During fieldwork, the authors applied the concept of reflexivity, by continuously remaining aware of their personal attributes as researchers and women who hail from the same district as the participants, while focusing on participant meanings by reflecting, discovering themes and linking up emerging ideas. Data were analysed for themes arising from recurrent patterns of ideas. The above similarities in focus and methodological processes by the two authors provide a rich account of Batwa's experience with tourism in Bwindi National Park.

Findings

Batwa and Tourism Development

Gazetted as a National Park in 1991, BINP in South-western Uganda, was declared a UNESCO Natural Heritage site in 1994 and is Uganda's most sought-after destination due to gorilla tourism (Tumusiime and Vedeld, 2015). The park is located on the eastern edge of the Albertine Rift covering 321 square kilometres and borders the Democratic Republic of Congo to the west and Rwanda to the south (Ampumuza, 2021). The Batwa first inhabited the Bwindi area between 32,000 and 47,000 years ago, as hunter-gatherers who depended on the Bwindi forest resources for survival (Ampumuza, 2021). During the interviews, the Batwa tell a story of their origin as follows:

> The first family to settle in central Africa comprised of father - Gihanga, who had three sons - Gatwa, Gahutu and Gatutsi. Gihanga put his three sons to a stewardship test. He gave each one a jar full of milk. Gatwa and Gahutu took a sip of milk, but Gatutsi drank half of

his milk and spilt the rest as he dozed off during the night. In the morning, the two sympathized with Gatutsi and decided to pour some milk into his jar thus almost filling Gahutu's jar while theirs remained half full. As they went about the day's chores, Gatwa decided to quench his thirst with part of his remaining milk reducing it considerably. At sunset, their father called them up to account for their jars of milk. Each one of them presented their jars and accounted for the milk. Their father then rewarded his sons, Gahutu and Gatutsi, with hoes and cows respectively to survive. Gatwa received bows, spears and arrows for hunting and gathering plants and fruits from the natural forest.

(Mukasa, 2014)

As a result of the rewards, the Batwa occupied and entirely depended on the central African forests and adopted a hunting and gathering lifestyle. Oral histories from elderly Batwa indicate that these forests were known as the "domain of the bells," named after the bells in Batwa dogs 'collars' (Van der Duim, Ampumuza, and Ahebwa, 2014, p.59). However, in 1991, the Batwa who lived in the Echuya, Bwindi and Mgahinga forests in South-western Uganda were evicted from the forests to create room for conservation and tourism developments (Kagumba, 2021). This occurrence, unfortunately, turned the Batwa into 'conservation refugees' (Harper, 2012). The Batwa and the neighbouring Bakiga became landless, homeless, jobless and faced food insecurity which was exacerbated by increased crop raiding by park animals (Tumusiime and Vedeld, 2015). Over the years, some of the local Bakiga farmers, who were also evicted, have already been resettled and are currently farming on parcels of land they possess outside the forest, as opposed to the Batwa who still lack these options (Tumusiime and Vedeld, 2015). Also, while the Bakiga people, traditionally known for farming, have now ventured into tourism work (Uganda Wildlife Authority, 2018), the minority indigenous Batwa are still surviving on casual farm labour, illegal hunting and providing entertainment and handcraft sales to tourists (Ampumuza, 2021; Boonabaana, 2012). Entertainments are either provided at wedding parties in the villages in exchange for food or at lodges, as part of tourist package offered by lodges (see Figure 12.1), tour companies or as part of the Batwa forest experience at Bwindi and Batwa trails in Mgahinga and recently, Echuya forest. Crafts are usually sold in the scheduled community markets or during the forest experience and trails as souvenirs and Batwa run craft shops established with support from the Kellerman Foundation at Buhoma and the International Gorilla Conservation Program [IGCP], at Sanuriro.

Batwa Tribe Experience

Learn about and interact with the local Batwa People. This full-day activity involves a three-hour walk followed by about an hour spent with the Batwa. The walk back to camp is two hours. The visit includes a music performance, visitation by the goddess, hunting and gathering demonstration and an opportunity to see how they live. The Twa tribe who are also known as the Batwa are the oldest inhabitants of the Great Lakes region of Central Africa…

Source: https://www.sanctuaryretreats.com/uganda-camps-gorilla-forest-activities

---

Despite some support from the Uganda Wildlife Authority through tourism revenue sharing, the Batwa people remain vulnerable, facing limited access to health care, education, clean water,

Figure 12.1   A high end lodge (Sanctuary Gorilla Forest Camp) marketing Batwa experience online.

employment and adequate clothing, food and security (Harper, 2012). They suffer stereotyping, sexual and physical brutality perpetuated by the non-Batwa communities (Kagumba, 2021) and are sometimes described by the local Bakiga and NGOs as primitive, being landless and unable to save food, money and other resources for future use and therefore expected to be dependent on handouts (Ampumuza, Duineveld, and Van der Duim, 2020). Conversely, tourism promotional materials portray their 'exotic images' as an authentic lifestyle and the Batwa experience as a 'must do' activity after gorilla viewing as reflected in Box 1 above.

Batwa Involvement in Tourism Initiatives

Batwa as Tourist Guides

Our combined research findings revealed that the Batwa have a rich ecological knowledge about various plant and animal species which they are able to interpret to tourists for a fee. Batwa traditional ecological knowledge is now being used to position them to survive in their new livelihood environment, out of the forest. The Batwa women eloquently explain the various plants, their habitats and dyes that can be extracted from each plant to make crafts (Ampumuza, Duineveld, and Van der Duim, 2020). This enables them to position themselves as knowledgeable about ecology as opposed to being viewed as powerless actors. While women do not join the hunting expeditions, they pray to their gods to protect the men and for successful hunting, whilst collecting fruits and firewood. As soon as the men approach the home carrying their hunt, the women start the fire and chant praises to gods and their men. The men skin the animal and the women, with the men, cook a feast. This is followed by entertainment involving dancing and singing, led by all men, women and children. This enhances their income generated through the production of arts, crafts and the trade in medicinal plants.

Buhoma Village and Cultural Walk: A Site for Exercising Batwa Agency

The Batwa are currently owning and running tourism enterprises, either as individuals, groups or in partnerships with other people. The enterprises include Buhoma-Mukono cultural walk, Batwa forest experience, Batwa craft shops and a Batwa homestay accommodation (Ampumuza, Duineveld, and Van der Duim, 2020; Boonabaana, 2012). The Buhoma village walk that officially opened for operation in 2002 was initiated by the community under their umbrella organisation – Buhoma-Mukono community Development Association (BMCDA), with support from Uganda Wildlife Authority – the Mgahinga and Bwindi Forest Conservation Trust (MBIFCT), with the aim of: providing tourists with alternative attractions outside the park that enabled the locals to diversify their livelihoods. The village walk lasts about three hours and visits six designated sites (see Table 12.1 below);

Specifically at the last site (*Batwa*/pygmy group site at Kalehe), the males re-enact their former forest life by demonstrating the houses they used to construct, the safe tree houses for children, while the females display the crafts for sale as souvenirs. Other activities include demonstrations of traditional fire making, cooking and birthing/child delivery mainly led by women, while demonstrations on hunting and survival skills living in a cave are led by males. These are mainly performed in the forested land at Kalehe village in Buhoma, bought for the Batwa by the Mgahinga and Bwindi Impenetrable Forest Conservation Trust (MBFICT) that resettled more than 403 Batwa since 1992 in the three districts of Kabale, Kanungu and Kisoro close to Bwindi and Mgahinga National Parks and former home for Batwa communities. MBFICT was initially funded by the World Bank through the Global Environmental Facility (GEF) in May 1991, to the tune of US 4.3 million, to establish a trust fund in Uganda to cater for the needs of the Batwa by

Table 12.1 Tourist village walk sites and activities in Buhoma community

#	Sites	Activities
i.	Women's craft business centre	• Display and sale of women's handcraft designs and products, that are produced using local materials • Demonstrating how to weave baskets
ii.	Local tea and coffee plantations	• Display and demonstration of traditional cashcrop/income source for locals • Demonstrating coffee and tea harvesting methods
iii	Traditional healer	• Demonstration of how traditional medicinal plants/herbs are prepared • Citing the different diseases, misfortunes and love relations the herds handle • How the herbs are grown and conserved at the healer's home and the different roles of family members (husband, wife and children) in managing herbs and healing processes
iv.	Two brewing homesteads	• showcasing the brewing of local gin (*waragi*), juice (*eshande*) and beer (*tonto*) • Showcasing a local banana plantation with different banana species used to make these local drinks and tasting of the drinks
v.	Community primary and secondary schools	• Interacting with students and teachers • Explaining the school history, current enrolments, development opportunities and challenges
vi.	*Batwa*/pygmy group site at Kalehe	• Cultural entertainment and craft market • Demonstration of fire making out of sticks • Demonstration of life in the forest using a leafy structure at the site

Source: Boonabaana (2012).

compensating them for land lost and supporting them in maintaining access to the forest for herbs and food (Zaninka, 2001).

The Batwa also collaborate with some owners of accommodation facilities such as Silverback Backpackers and Gorilla-close-up-lodges to deliver tourist experiences at a fee. The Batwa experience alone costs US$70 per person (2022 rate): http://www.buhomacommunityug.com/dt_portfolios/meet-the-batwa-pigmies/. The payments are shared among site owners and the community management organisation that oversees the activities (30% for guides; 10% to BMCDA; 40% to site owners; and 20% to community projects). Since 2002, the Batwa have been recognised as key players in the tourism processes of the locality, depicting both relevance and opportunity. The money earned from these activities enable the Batwa to meet household needs including paying school fees and buying scholastic materials.

Batwa as Craft Makers and Sellers

In addition, Batwa is increasingly becoming renowned for making traditional crafts using natural materials. Their unique crafts are preferred in the market that is otherwise dominated by products out of synthetic dyes and materials. One of the female Batwa elders explained as follows:

> Our crafts are preferred to those of the non-Batwa because we use natural materials and dyes unlike the Bairu (used to mean non-Batwa groups) who use synthetic fibre to produce them. Moreover, our finishing is neater because we do it with utmost precision. That is one characteristion of Batwa you know.
>
> (Interview with Female Batwa elder, Sanuriro, 2018)

Although some of them revealed that they never made baskets, mats and other products while in the forest, they clarified that their exposure to tourism granted them an opportunity to use their vast knowledge of plants such as the different plants and how they could change colour with natural treatments such as soaking them in water-wood ash solution or rubbing them at different intensities to produce unique souvenirs. At Sanuriro village, south of Bwindi, the female village guide teaches tourists how to make their own souvenirs. In all these activities, there are clear roles between men, women, children and the elderly. For example, the men focus on collecting materials used by both men and women and make wooden crafts while the women predominantly do the weaving and teach tourists how to weave.

The Batwa have also opened a craft shop in the village and constructed a shelter where tourists can pitch their tents for guests who prefer to spend a night. They also operate a kiosk where they sell food and drinks to visitors, community members and passers-by needing food when they make a stopover at their village. This shelter operates on a semblance of time-share arrangement in the sense that when there are no tourists, the shelter is used for village meetings and other functions such as weddings. We interpret this integration of tourism in the Batwa lifestyle as a way of confronting marginalisation, asserting collective power and surviving within a new and challenging livelihood environment.

Building Actor Relationships

Through the NGO, Batwa have managed to establish relationships and networks with other actors (donors, anthropologists and other researchers) to secure scholarships for Batwa students, purchase land for the Batwa, establish livelihood projects, as well as in negotiating policy changes and representing fellow Batwa in national and international meetings (Group Interview, Kalehe, 2017). These relationships illustrate Batwa's self-organising capabilities. Our observations of the Batwa's negotiations with other partners revealed that the Batwa valued honesty, togetherness and sustainability. Some Batwa are engaged in activities such as mining, teaching, tourist guiding, stone quarrying and agriculture in addition to daily home chores and leisure activities. In all conversations, most of the Batwa constantly associated Batwa-ness with hard work, dynamism, precision, creativity and intelligence. Batwa has been and are still involved in negotiations, advocacy campaigns and other engagements with politicians on restoring their access rights to the forests, affirmative action to eliminate their marginalisation and deliberate efforts to ensure their political representation in parliament and other key decision-making positions (Ampumuza, Duineveld, and Van der Duim, 2020).

The Batwa Development programme (BDP) was started in 2001 by an American missionary couple, Dr. and Mrs. Scott Kellerman, under the Kellerman foundation www.kellermann foundation.org). It initially aimed at empowering the Batwa with education, farming land and associated inputs, homes, health care, water and sanitation, income-generating initiatives, as well as encouraging indigenous rights. This NGO runs the Batwa cultural centre aimed at preserving Batwa's rich culture, currently situated on 100 acres of land adjacent to Bwindi Impenetrable Park.

The programme also strengthens generational cultural interactions among the Batwa, whereby the culturally knowledgeable (older) Batwa are encouraged to pass along their cultural values, beliefs and practices to their children and visitors. These include ftraditions such as telling and showcasing traditional practices like disease treatment by medicinal plants and making fire out of sticks. They also pass on Batwa legends and songs to their children. These experiences are open to tourists at a cost of US$ 70 per person per day. The collected funds are meant to sustain the cultural centre (Batwa cultural experience brochure, n.d; www.kellermannfoundation.org) and 10% of the revenues are given to BMCDA, to support community developments and payment of community guides while the rest supports the foundation activities. This programme also runs a craft enterprise and cafe whose income supports Batwa families to acquire basic needs (see Figure 12.2). Batwa youth are trained to run the business to enable them to take over the management roles in future. At the time of fieldwork, the manager was non-Mutwa (singular for Batwa) training two young Batwa girls in all the business activities. However, she cited low interest by the girls who often abandon the training before completion, to join their Batwa kin to 'loiter around' the villages in groups. This was attributed to Batwa being less accustomed to settled lives than the dominant non-Batwa communities (Bakiga), with aspects of income generation, education and health-seeking behaviours still strange for them, irrespective of these services being mostly free to eligible Batwa.

Figure 12.2   Batwa Development Café at the entrance of Bwindi Impenetrable National Part.

Photo credit: Brenda Boonabaana.

The Thorny Part of Batwa Engagement in Tourism

Limited Representation on Project Committees

Founded in 1992, BMCDA, is an umbrella development organisation that brings together all Mukono locals (about 7,000 members) to work together for the common causes of development, conservation of natural resources and improved livelihoods. It started through collaborative work between the locals, International Gorilla Conservation Programme (ICGP) staff and the American Peace Corps. Its aims were: promoting community development in the Parish through local group and organisation assistance; provision of local infrastructure; initiation of tourism-supported employment for the locals; and provision of tourist accommodation. All locals are required to pay a small membership fee of UGX 500 (USD 20 Cents) as a form of commitment towards community development efforts. However, analysis shows that management positions were mainly occupied by non-Batwa men (Bakiga). The positions of Chairperson, Vice Chairperson and Treasurer were held by men. Out of 12 committee positions, only two were held by women, one occupied by a woman representing Bakiga women and the other representing Batwa women. The advisory committee positions had seven men and one woman. Out of 22 members of the governing council, only three were women, two of them representing women groups and the elderly. The leadership structure indicates the limited representation of Batwa men and women. With village walk guiding, while there was equal distribution of employment slots for men and women, the Bakiga (non-Batwa) occupied these positions across all the village sites, including the Batwa site. Findings further depict that the management of community development projects were male dominated with no Batwa representation, due to the historical and continuous marginalisation of Batwa by non-Batwa communities. For instance, the vegetable project had only Bakiga male managers; the water project's monitoring committee had one woman (non-Mutwa) out of ten members while the micro-finance initiative was mainly managed by non-Batwa women.

Marginalisation in Accommodation Job Opportunities

Fieldwork observations and interviews with management of foreign-owned lodges by the first author indicate that only two Batwa women worked at the position of housekeeping while the remaining staff members, mainly the Bakiga men, occupied positions of manager, assistant manager/bar manager, treasurer, driver, table waiting staff and chef. This was because women were seen to maybe not cope well with tourism work demands and may not be as efficient as the male employees. Generally, except for Batwa-focused organisations and programmes, the general tourism employment patterns in the area demonstrates the exclusion of Batwa men and women, as well as the non-Batwa women, with opportunities skewed to young Bakiga men that are relatively better literate. Bakiga men dominate management and most jobs of the community projects while women occupy the cultural entertainment, craft businesses and gardening roles, where the Batwa men and women belonged. The jobs occupied by Bakiga men were also better remunerated and attracted social respect.

Cultural and Gender-Based Marginalisation

Analysis of the cultural entertainment and handicraft activities revealed that these were women-focused and divided along ethnic lines. While both Batwa and non-Batwa women make traditional baskets, necklaces and offer cultural dances and songs, they operate and deliver their products and services along ethnic lines due to the historical marginalisation of Batwa who are perceived by non-Batwa

Figure 12.3 Handicrafts in Nkenda trading centre, next to Bwindi.

Photo credit: Brenda Boonabaana.

(Bakiga) as 'backward' and ineligible to mix and work with. Women involved in these activities are predominantly elderly with very little or no education. For Batwa, they mainly operate in mixed-sex groups of different ages. Overall, most souvenir and handicraft businesses are owned and run by non-Batwa local male youths who mostly produce and sell gorilla wood carvings and other crafts in Nkwenda trading centre, the closest village to the park (see Figure 12.3).

Discussion

Ecotourism has demonstrated potential to positively impact on neighbouring communities including the historically marginalised social groups like the Batwa and other traditionally marginalised groups such as women and the youth. While gorilla ecotourism in BINP has displaced the Batwa communities from their forest homeland, it has also opened space for them to reconstruct their identities and livelihoods and assert their power in a socially diverse terrain. This has been made possible through their engagement in cultural tourism activities and the attraction of development opportunities for the wider community. Further, ecotourism has become a platform for challenging ethnic-oriented power dynamics, whereby, as Batwa men and women become active participants in cultural tourism, they are confronting power dynamics that have traditionally marginalised them and their ethnic roots.

Batwa has demonstrated abilities with forest ecology knowledge, tourism entrepreneurship, organisational capacity and political activism (Ampumuza, Duineveld, and Van der Duim, 2020). Together with Civil Society Organisations, Batwa has demanded restitution of their rights of ownership and access to their traditional forest homes through the United Organisation for Batwa Development in Uganda [UOBDU] (Ampumuza, 2021). For example, on 13th February 2013, the UOBDU petitioned the Constitutional Court of Uganda seeking recognition of their rights and redress over long-time marginalisation and are still engaged in these legal processes (Ampumuza, 2020). Several approaches and strategies are being deployed to restore Batwa agency, dignity and overcome their marginalisation.

Because Batwa cultural knowledge is highly sought after by conservationists and tourists alike, the Batwa are using their strengths to remain connected to their ecological and cultural beliefs, values and practices and passing over the same values to their younger generations. As noted by Kagumba (2021), by maintaining access to their ancestral forest through the cultural trail, the Batwa have been able to facilitate their younger generations to gain knowledge about their hunter-gatherer livelihood. The Batwa cultural values have therefore become core to the tourism supply and demand value chain that blends gorilla ecotourism and community-based cultural tourism, while enabling the restoration of their cultural dignity. Dignity restoration through tourism has been identified as critical for the wellbeing and sustainability of Indigenous communities (Camargo, Winchenbach, and Vázquez-Maguirre, 2022).

While some authors have raised the challenge of cultural commodification through tourism (Laudati, 2010), we argue that for the case of the Batwa being included in the community village trail/walk product has pushed boundaries and redefined experiences around local cultural power relations. The dominant ethnic community members (Bakiga) who previously undermined the Batwa are beginning to partner and work with them to create 'authentic' tourism experiences for tourists. Tourism is therefore playing a critical bridging role for the two ethnic groups, while elevating Batwa's presence and visibility. In relation to another Ugandan gorilla tourism site [Mgahinga National Park, Kisoro District], Kagumba (2021) has argued that tourism has become a temporal space in which the Batwa overturn local power dynamics, assume a more dominant position of host and in so doing, retaliate by liberating themselves from the chains of marginalisation through performance.

Although the Batwa cultural site could be perceived as a symbol of cultural commodification, as portrayed in promotional materials, their presence legitimises their emerging relevance in the conservation and tourism processes. Importantly, Batwa have been able to tell their own stories about their relationship with the forest, tourism and current forms of livelihoods they have experienced. While the level of authenticity is compromised due to showcasing their experiences outside the forest environment, by reconstructing their own experiences, Batwa cultural experiences have elicited sympathy and legitimised philanthropic projects and donations (Ampumuza, Duineveld, and Van der Duim, 2020) that support them and the wider community.

As well, the Batwa tourism activities are delivered through collective efforts for collective purpose and gains, which resonates with what other scholars have emphasised in other Indigenous contexts (Kagumba, 2021; Scheyvens et al., 2021; Trau and Bushell, 2008). In the context of Kisoro District, Uganda, Kagumba (2021) has articulated how the Batwa use their tourism trail to demonstrate their culture through songs that portray resistance to marginalisation by the non-Batwa communities that often undermines them and demand for their rights. Trau and Bushell (2008) have alluded to the role of tourism in keeping indigenous culture alive and the ability to link the economic incentive to protection and transfer of traditional cultural custom and knowledge.

Batwa Development Organisation itself employs Batwa as guides, markets and sells their cultural products and acts as a unifying development platform for Batwa. Other than the cultural dividend, there are several opportunities being enjoyed by the local communities because of Batwa's conservation connections. The case in point is the sole hospital, the Bwindi Community Hospital, that was constructed to support the Batwa and has since expanded services to other local ethnic communities and beyond.

Conclusions

Nevertheless, while we argue that Batwa women and men have claimed a place in the conservation and tourism space, there are still problematical discriminatory perceptions and practices that constrain their agency. For instance, they are poorly represented in the community development organisation, which is dominated by Bakiga men. The community development projects that provide cultural, education, agritourism, water and microfinance services have very few or no Batwa in management positions. Having Batwa-only organisations and groups symbolises continuities of local exclusion and marginalisation. Much needs to be done to embrace the Batwa men and women as active agents of tourism development across the different opportunity structures of the community.

From the gender point of view, while both Batwa men and women are underrepresented in management roles of tourism-supported ventures, the non-Batwa women (Bakiga women) are also underrepresented, except within the cultural enterprises. Bakiga men remain dominant in tourism management, decision making and other lucrative opportunities, depicting gender inequality in the wider community. Generally, low education levels, cultural and ethnic power relations play out to disadvantage Batwa men and women and non-Batwa women, to meaningfully engage in different tourism spheres in the Bwindi area. Nevertheless, Batwa men and women are creating collective structures to counter such marginalisation.

The role of external NGOs including faith-based agencies, government's conservation agencies and local leadership in enabling Batwa's visibility and agency cannot be underestimated. The financial, training, marketing and health support provided by national and local NGOs, community-based organisations and high-end tourist lodges have become key supportive resources for them to exercise their agency within the tourism space. As Kabeer (1999) has argued, it is indeed difficult for people to exercise their agency in the absence of supportive resources and environment to do so.

We conclude that community-based cultural tourism in BINP has provided space for Batwa to participate in and contribute to their community's wellbeing, while confronting some of the traditionally held stereotypes, ethnic power relations and marginalisation associated with being a Mutwa. The Batwa men and women's ability to work collectively has also narrowed their gender divides and enabled them to harness each other's cultural capital to tap into ecotourism opportunities. Secondly, both Batwa men and women are using their vulnerability and commodification identities to gain visibility, solicit support and reconnect to their traditional cultural beliefs, values and practices, while reinventing new identities through a commercial lens. Finally, Batwa identities are being redefined outside the forest through a complex process that integrates the Batwa themselves, tourists, local non-Batwa, government and development agencies, depicting different actors, interests and expectations that Batwa have to navigate to successfully compete within the ecotourism space.

References

Ahebwa, W, M, Sandbrook, C, & Ochieng, A, (2018), 'Parks, people, and partnerships: Experiments in the governance of nature-based tourism in Uganda'. In C.J. Cavanagh, C. Sandbrook & D. M. Tumusiime (eds.) *Conservation and development in Uganda*, pp. 148–170, London: Earthscan.

Ampumuza, C, Duineveld, M, and van der Duim, R, (2020), The most marginalized people in Uganda? Alternative realities of Batwa at Bwindi Impenetrable National Park. *World Development Perspectives* 20, 100267.

Ampumuza, C, (2021), *Batwa, gorillas and the Ruhija road: A relational perspective on controversies at Bwindi Impenetrable National Park, Uganda*, Doctoral Dissertation, Wageningen University and Research.

Ampumuza, C, Duineveld, M, & Van der Duim, V, R, (2020), The most marginalised people in Uganda? Alternative realities of Batwa at Bwindi Impenetrable National Park, *World Development Perspectives*, 20, https://doi.org/10.1016/j.wdp.2020.100267

Babb, F, E, (2012), Theorising gender, race, and cultural tourism in Latin America: A view from Peru and Mexico, *Latin American Perspectives*, Issue 187, 39(6), 36–50.

Boonabaana, B, (2012), *Community-based tourism and gender relations in Uganda*, Doctoral Dissertation, University of Otago, Dunedin, New Zealand.

Boonabaana, B, (2014), Negotiating gender and tourism work: Women's lived experiences in Uganda, *Journal of Tourism and Hospitality Research*, 14(1–2), 27–36.

Brown, W, (2006), 'Power after Foucault'. In J. Dryzek, B. Honig, and A. Phillips, A. (eds.) *The Oxford Handbook of Political Theory*, pp. 65–84. Oxford: Oxford University Press.

Butler, R, & Hinch, T, (2007), *Tourism and Indigenous Peoples: issues and implications*, Amsterdam: Butterworth-Heinemann.

Camargo, B, A, Winchenbach, A, & Vázquez-Maguirre, M, (2022), Restoring the dignity of indigenous people: Perspectives on tourism employment, *Tourism Management Perspectives*, 41, 100946.

Clegg, S, (1998), 'Foucault, power and organizations'. In A. McKinlay, and K.P. Starkey (eds.) *Foucault, management and organization theory*, pp. 29–48. Newbury Park, CA: Sage.

Dawson, N, M, Coolsaet, B, Sterling, E, J, Loveridge, R, Gross-Camp, N, D, Wongbusarakum, S et al., (2021), The role of Indigenous peoples and local communities in effective and equitable conservation, *Ecology and Society*, 26(3), 19. https://doi.org/10.5751/ES-12625-260319

De, U, K, (2013), Sustainable nature-based tourism, involvement of indigenous women and development: A case of North-East India, *Tourism Recreation Research*, 38(3), 311–324. https://doi.org/10.1080/0250 8281.2013.11081756

Ferguson, L, (2007), *Reinforcing inequality: Service sector activities and the new entrepreneurial model of development in Central America* (CIP Working Paper Series, No.26). University of Manchester.

Ferguson, L, (2009), *Analysing the gender dimensions of tourism as a development strategy* (No. pp03/09): ICEI Working Paper, Universidad Complutense de Madrid.

Ferguson, L, (2010), Tourism development and the restructuring of social reproduction in Central America, *Review of International Political Economy*, 17(5), 860–888.

Foucault, M, (1982), 'The subject and power'. In H. Dreyfus & P. Rabinow (eds.) *Beyond structuralism and hermeneutics*. Chicago: University of Chicago Press.

Gallagher, M, (2008), Foucault, power and participation, *International Journal of Children's Rights*, 16(3), 395–406.

Grant, P, (ed.) (2018), *Minority and Indigenous trends: Focus on migration and displacement*, London: Minority Rights Group International.

Harper, S, (2012), *Social determinants of health for Uganda's Indigenous Batwa population*, Africa Portal Backgrounder No. 32, project of the Africa Initiative. https://doi.org/10.1080/09669582.2016.1217871

Kabeer, N, (1999), Resources, agency, achievements: Reflections on the measurement of women's empowerment, *Development and Change*, 30(3), 435–464.

Kagumba, A, K, (2021), The Batwa Trail: Developing agency and cultural self-determination in Uganda through Indigenous tourism and cultural performance, *AlterNative*, 17(4), 514–523.

Laudati, A, (2010), Ecotourism: The modern predator? Implications of gorilla tourism on local livelihoods in Bwindi Impenetrable National Park, Uganda, *Environment and Planning D: Society and Space*, 28, 726–743.

Ministry of Tourism Wildlife and Antiquities [MoTWA], (2020), *The impact of COVID-19 on the tourism sector in Uganda*, Government of Uganda.

Moodley, L, Holt, T, Leke, A, & Desvaux, G, (2016), *Women matter in Africa*, New York: McKinsey and Company.

Mukasa, N, (2014), The Batwa Indigenous people of Uganda and their traditional forest land: Eviction, non-collaboration and unfulfilled needs, *Indigenous Policy Journal*, XXIV(4).

Nielsen, N, & Wilson, E, (2012), From invisible to Indigenous-driven: A critical typology of research in Indigenous tourism, *Journal of Hospitality and Tourism Management*, 19, 1–9.

Nyaruwata, S, & Nyaruwata, L, T, (2013), Gender equity and executive management in tourism: Challenges in the Southern African Development Community (SADC) region, *African Journal of Business Management*, 7(21), 2059–2070.

Ochieng, A, Ahebwa, W, M, & Twinomuhangi, R, (forthcoming), Nexus between biodiversity, climate change and tourism: implications for sustainable tourism in Uganda.

Ramsay, K, (2010), *Uncounted: the hidden lives of Batwa women*, London: Minority Rights Group International.

Scheyvens, R, (2002), *Tourism for development: Empowering communities*, Essex: Pearson Education Limited.

Scheyvens, R, Carr, A, Movono, A, Hughes, E, Higgins-Desbiolles, F, & Mika, J, P, (2021), Indigenous tourism and the sustainable development goals, *Annals of Tourism Research*, 90, 103260. https://doi. org/10.1016/j.annals.2021.10326.

Simpson, F, O, & Geenen, S, (2021), Batwa return to their Eden? Intricacies of violence and resistance in eastern DRCongo's Kahuzi-Biega National Park, *The Journal of Peasant Studies*. https://doi.org/10.1080/ 03066150.2021.1970539

Taylor, R, S, (2017), Issues in measuring success in community-based Indigenous tourism: Elites, kin groups, social capital, gender dynamics and income flows, *Journal of Sustainable Tourism*, 25(3), 433–449. https://doi.org/10.1080/09669582.2016.1217871

Trau, A, & Bushell, R, (2008), 'Tourism and Indigenous peoples'. In S. F. McCool & R. N. Moisey (eds.) *Tourism, recreation and sustainability*, Linking culture and the environment, 2nd edn, pp. 260–282, Wallingford: CAB International.

Tucker, H, (2007), Undoing shame: Tourism and women's work in Turkey, *Journal of Tourism and Cultural Change*, 5(2), 87–105.

Tucker, H, & Boonabaana, B, (2012), A critical analysis of tourism, gender and poverty reduction, *Journal of Sustainable Tourism*, 20(3), 437–455.

Tumusiime, D, M, & Svarstad, H, (2011), A local counter-narrative on the conservation of mountain gorillas, *Forum for Development Studies*, 38(3), 239–265.

Tumusiime, D, M, & Vedeld, P, (2015), Can biodiversity conservation benefit local people? Costs and benefits at a strict protected area in Uganda, *Journal of Sustainable Forestry*, 34, 761–786.

Uganda Wildlife Authority, (2018), Bwindi Impenetrable National Park. https://ugandawildlife.org/wp-content/uploads/2022/01/Bwindi-cc-2018-2.pdf

Urry, J, & Larsen, J, (2011), *The Tourist Gaze* 3.0, London: Sage.

Van der Duim, V, R, Ampumuza, C, & Ahebwa, W, M, (2014), Gorilla tourism in Bwindi Impenetrable National Park, Uganda: An actor-network perspective, *Society & Natural Resources*, 27(6), 588–601.

Weber, A, Kalema-Zikusoka, G, & Stevens, N, J, (2020), Lack of rule-adherence during mountain gorilla tourism encounters in Bwindi Impenetrable National Park, Uganda, places gorillas at risk from human disease, *Frontiers in Public Health*, 8, 1.

Weedon, C, (1997), *Feminist practice and poststructuralist theory*, Malden, MA: Blackwell.

World Bank, (2013), 'Case studies of the horticulture, tourism, and call centre industries'. In C. Staritz & J. G. Reis (eds.) *Global value chains, economic upgrading, and gender*, Washington, DC. The World Bank Group.

World Tourism Organisation, (2019), *Global report on women in tourism* (2nd Edition), Madrid: UNWTO. https://doi.org/10.18111/9789284420384

Zaninka, P, (2001), The impact of (forest) nature conservation on indigenous peoples: The Batwa of South-Western Uganda – A case study of the Mgahinga and Bwindi Impenetrable Forest Conservation Trust. https://www.forestpeoples.org/sites/fpp/files/publication/2010/10/ugandaeng.pdf

SECTION 3

Indigenous Alternatives

While it is clear that Indigenous tourism is not a single form of tourism, but varies widely in scale, nature and focus from one community to another, it is equally clear that as a generic form of tourism it is quite distinctive and has far-reaching implications and needs. One such need is to educate and train Indigenous community members to participate in tourism in more than just relatively low-paid and low prestige positions, and to consider taking control of tourism development in Indigenous territories. Berno et al. explain how a completely different approach is needed to education and training of Indigenous peoples in tourism and associated industries, as well as the skills necessary to become leaders rather than just followers in the provision of tourism opportunities in their communities.

Part of this process as McGinnis notes is visibly claiming and "Indigenising" the landscape, both physically and metaphorically so that tourists are aware they are visiting something that is different with a different history and culture to other places. Similarly, Puriri and McIntosh, Carr et al. and Mika et al. vividly illustrate how incorporating Indigenous values and practices with the principles of sustainable and appropriate development can result in highly distinctive and important examples of Indigenous-led development. All of the developments they discuss truly reflect the cultures that have developed those enterprises and their relationships with their natural environments and their ways of life. In so doing, their examples show how important such developments are in increasing suitable opportunities for Indigenous youth and helping preserve specific elements of their way of life.

The implications of Indigenous alternatives are profound and numerous, in that they offer new ways to preserve and present Indigenous culture to visitors, to remind visitors of the long and often painful histories of the Indigenous communities, to demonstrate how their traditional activities have shaped the land and settlement being visited, and how Indigenous communities can create,

DOI: 10.4324/9781003230335-15

lead, and control attractive and successful tourism opportunities on their own terms. Running throughout these examples of uniquely indigenous approaches to tourism development in their communities is the theme of development with rather than against the natural environment and the traditional beliefs in the symbolic importance of respect for the living landscape, processes and other inhabitants.

13

DOING IT THE 'PACIFIC WAY'

Indigenous Education and Training in the Pacific Islands

Tracy Berno, Rerekura Teaurere and Dawn Gibson

Introduction

Tourism presents an opportunity for Indigenous Pacific people to develop skills and utilize traditional hospitality practices while also providing tourists with exceptional, grounded-in-the-Pacific hospitality experiences. For the most part however, Pacific education in hospitality and tourism has adopted Western models which are limited in considering traditional practices of hospitality. Pacific notions of hospitality in tourism education in a region subject to colonization and contemporary change limit the use of traditional practices that have the potential to enhance tourists' experiences. This also has implications for Indigenous tourism and hospitality students. Education in tourism and hospitality for Indigenous students has created a dichotomized perspective of Western ways of hospitality for tourists and traditional ways of hospitality in Pacific Island communities. Ultimately, the enhanced benefits of integrating Indigenous hospitality into education models are overlooked. This has consequences limiting positive tourists' experiences, and Indigenous empowerment and the growth of positive socio-cultural exchange.

In 2006, Berno (p. 29) asked in relation to tourism education in the Pacific

> ...does this [current educational] approach address the needs of sustainability, empowerment and self-determination of Indigenous people? Is tourism education (both in terms of content and delivery) an etic (universal) field of knowledge or does there need to be consideration given to the socio-cultural context in which it is delivered?

As tourism has continued to grow both in volume and importance over the past 15 years in the Pacific, how has the education landscape responded?

These questions highlight the importance of understanding the educational needs of communities in the Pacific region in ways that enable Pacific ways of hospitality to be at the forefront of tourists' experiences. This may require the need for policy infrastructure changes to acknowledge distinctive Pacific ways of hospitality. Pacific islanders have unique ways of displaying hospitality, which in a tourism context can differentiate the Pacific tourism product from other sea, sand, and sun destinations, ultimately contributing to sustainable tourism growth. The need to understand the benefits of acknowledging and embedding Indigenous hospitality into educational policy

163

DOI: 10.4324/9781003230335-16

infrastructure is needed at a regional level. Moreover, there is a need for a regional collaborative approach to tourism and hospitality education across the Pacific, given the movement of staff within the Pacific region, although a collaborative approach to tourism education works against the individualistic Western models of tourism education that have been embedded into education systems across the Pacific. Therein lies a significant issue. There is a need to transition away from the Western models of education that have become embedded in the Pacific systems as a result of intense colonization to a contextually relevant Pacific educational model that acknowledges the benefits of Indigenous hospitality, recognizing its value and normalizing its use in the tourism industries across the Pacific.

This chapter will revisit these critical issues raised in 2006 in relation to tourism development and Indigenous people by considering the Indigenous voice in relation to tourism and hospitality education in the Pacific region in the current context. This will be achieved through the consideration of the following: Indigenous tourism education in general; the Pacific cultural and geographical context; how tourism training and education is currently structured in the region; characteristics of policy infrastructure, (HRD and tourism education that are particular to the region; and the cumulative effect of these characteristics in providing a context from which a Pacific Way for tourism education can be shaped.

Tourism Education in an Indigenous Context

To fulfill the human resource needs resulting from the growth of tourism demand, many developing countries have established tourism and hospitality training programs and institutes. In many instances, Western models of education have been imported from Europe or North America (Craig-Smith, 2005; Lewis, 2004; Theuns & Go, 1992; Tight, 2022). This has led to programs that are oriented towards the management of Western-style tourism and hospitality facilities but applied to an Indigenous context. We, however, argue that tourism educational programs must be placed in the context of their societies' needs and be interconnected with well-conceived development plans (see for example Nabobo-Baba, 2012; Thaman, 2013). The onus is therefore placed on the hospitality and tourism education programs to demonstrate that knowledge benefits not only the individual who acquires it, but also the society of which they are a part (Hegarty, 1990, p. 41, quoted in Lewis, 2004, p. 5).

Although this chapter focuses on the tourism education context in the Pacific, the issues we raise are not unique to the Indigenous people of this region. A growing literature recognizes that "[a] pedagogy of tourism informed by Indigenous approaches" (Higgins-Desbiolles, 2017, p. 439) is much needed to address power differentials in our education systems, to decolonize the curriculum, and to create salient benefits for Indigenous people through acknowledging the legitimacy of Indigenous ways of knowing. To reduce Indigenous knowledge to imperialistic assumptions of being lesser than scientific, sitting at the bottom tier of the hierarchy of knowledge legitimacy, and thereby privileging Western ways of knowing is what Spivak (1988) describes as "epistemic violence" (p. 281). We need to acknowledge that the traditional (Western) view on tourism and tourism education may be entirely different to that of Indigenous peoples, and navigating this space, especially for Western educators unfamiliar with Indigenous ways of knowing, can be fraught with tensions.

Tourism as a form of commodified leisure travel is a Western concept (Berno, 1999) and thereby inherently colonialist in its educational application. Recognizing this, tourism scholars need to re-examine tourism education through a critical lens that respects Indigenous perspectives, rights, and self-determination (Higgins-Desbiolles, 2017). In essence, academics need to engage in "a broad process encompassing the iterative de-privileging of dominant colonial discourses" (Young & Maguire, 2017, p. 458). To do so not only benefits Indigenous peoples but invites both Indigenous and non-Indigenous academics and students to reflect critically on the role of tourism in society and their roles within it (Dredge et al., 2013). As Hollinshead (2013, cited in Chambers & Buzinde, 2015) has pointed out,

> First [tourism education] tends to be the product of the disciplines that have traditionally influenced thinking on tourism…Second, our conceptualizations of tourism tend to be overly informed by the cosmologies of the societies from which tourism scholars have traditionally hailed – what we might clumsily call 'western' ways of understanding the world.
>
> (no page)

In considering this push for Indigenous tourism education, we place emphasis on the need to include tourism education as part of a holistic and integrative planning approach. It is also suggested that within this framework, tourism education must be compatible with culturally informed pedagogies, as well as balancing content and level of delivery to achieve a type of Indigenous tourism education that supports the aspirations of Indigenous people. Moreover, although modernization and development agendas reflective of colonization and the West dominate the socio-economic and political landscape of the Pacific, we argue that the embedding of Indigenous entrepreneurship in Pacific tourism education can contribute to empowering marginalized Indigenous communities as a movement toward self-efficacy and self-determination.

The Pacific Context

The island nations of the Pacific are spread across 33 million square kilometers of the Pacific Ocean and include countries from all three of the major communities of interest: Polynesia, Micronesia and Melanesia. The close to 2 million inhabitants of the region represent a broad range of cultural, ethnic and linguistic backgrounds. These island states and nations also represent a cross-section of types and levels of development, colonial heritage, population types and densities, and physical geographies. In many ways, their only common feature is that they reside within the tropical belt of the Pacific Ocean.

The state of HRD for the tourism sector in the region has been reviewed on several occasions (see for example Burns, 1999; Dowse, 1994; King, 1994, 1996; Medlik, 1989; SPTO, 2002, 2020; Tourism Council of the South Pacific (TCSP, 1995, 1999). Continuing interest in HRD needs in the region is indicative of the inherent challenges in trying to implement tourism training and education throughout such a large and diverse cultural and geographical area. The Pacific region requires a system of training and education that addresses the needs of the current and potential future employees in the travel and tourism industry, as well as the future needs of tourism. In addition to the demand for entrepreneurship, business skills, and labor, consideration must also be given to the large range of diversity in the region and the challenges of delivering the training across huge geographical distances.

Due to the vast distances between countries, limited transportation between and within some countries and at times limited access to communication technology outside of the main centers, strong connections between the public and private sectors can be difficult to establish and maintain on a regional basis. Given the scope and scale of the region, no single organization can afford to employ enough educators to meet all the disparate needs of all areas of the tourism industry and associated communities. Additionally, it cannot be forgotten that regionality itself is potentially problematic. On one hand, it is desirable to work collaboratively and collectively to optimize the use of training and education resources, but on the other hand, there is a sense that destinations are competing against each other for tourists. Notwithstanding any inherent concern about competition, given the particularities of the regional context, a cooperative approach to tourism training and education is required (Berno, 2001, 2007; SPTO, 2002, 2020).

Current Tourism Training and Education in the Pacific

Despite the importance of tourism to the region and the past (pre-COVID) demand for human resources, there exists a relatively limited public awareness of the current and potential benefits of tourism. The quality of the tourism and hospitality facilities and training across the region varies. Much of the tourism training in the region is undertaken at the pre-degree level in hospitality and catering schools and by vocational providers. This training focuses on the acquisition of vocational and trade skills, rather than a critical social scientific perspective or management concepts applied to the context of tourism. This 'front line' does little to address the serious shortage of regional entrepreneurs and management staff and maintains the current status quo — a reliance on expatriates to fill these positions, with the majority of sub-managerial positions being filled by Pacific Islanders or migrant workers.

Tourism has been recognized as an area of importance in the Pacific since the 1990s and there are now numerous providers of tourism education and training in the region. The University of the South Pacific (USP) and the Fiji National University (FNU), both based in Fiji, are the only regional university-level educational providers in the Pacific. Both institutions deliver pre-degree, degree and postgraduate qualifications in tourism. The University of Papua New Guinea offers a three-year business and management degree with a major in tourism and hospitality and Divine Word University (also in Papua New Guinea) offers a Bachelor of Tourism and Hospitality Management. Another significant regional educational provider is the Asia Pacific Training Coalition (APTC). An Australian government initiative, the APTC was established in 2007, initially as the Australia-Pacific Technical College. The APTC has campuses in Fiji, Papua New Guinea, Samoa, the Solomon Islands and Vanuatu, and offers a range of sub-degree vocational qualifications in several areas, including tourism, hospitality and cookery. The National University of Samoa also offers sub-degree qualifications in tourism and hospitality.

Most national tourism organizations (NTOs) also offer some form of tourism training, and many major resorts offer in-house training to staff. Several countries in the region also have national training councils (NTCs) that offer short-course vocational training for the tourism industry. Additionally, several national non-governmental organizations (NGOs) and regional organizations such as the South Pacific Regional Environmental Programme (SPREP) and the Pacific Asia Travel Association (PATA), have offered short courses and workshops for the tourism industry. Finally, bilateral aid agencies (such as NZAid and AusAID) also play a role in the provision of community-based tourism training in the region (Cheer et al., 2018; SPTO, 2002, 2020).

The quality of the facilities and training across the region varies. Although there has been some improvement since a (TCSP, now the South Pacific Tourism Organization - SPTO) study on HRD needs was undertaken in 1995, many countries still lack formal forms of tourism education, other than visits in the past from the SPTO mobile team (SPTO, 2002) and remote study through the University of the South Pacific. Additionally, training providers have been criticized for their lack of industry standard equipment and facilities, poor quality teaching resources and materials, outdated technologies, and failing to expose students to current industry practices and standards (Cheer et al., 2017; Morris, 2015). Tourism plays a significant role in the economic development of Pacific countries, foreign exchange earnings, and employment. 2019 saw the arrival of 2.2 million visitors to the Pacific region, which generated over US$4 billion in receipts and a contribution of 8% to Gross Domestic Product (GDP) and employed over 90,000 people (IDEEA Group, SPTO, 2021). However, since March 2020 tourism in the region has been seriously impacted by the COVID-19 pandemic and travel lockdowns were implemented in most countries in the Pacific region. Tourist arrivals in 2019 in the region reached 2,264,791, but dropped to 391,303 by the end of 2020 (SPTO, 2021). While some countries such as Fiji, have recently opened many countries as of writing, they have still not reopened their borders. In consideration of the projected growth of tourism in the Pacific region and the importance of tourism to the region, regional tourism businesses should be gaining a highly skilled workforce. However, the reality is that expatriates still dominate the managerial tourism industry positions, leaving Pacific Islanders filling the lower end positions. This highlights the need to determine how Pacific Islanders want tourism to look and develop in their respective islands, thereby shaping what is taught in regional tourism education with a movement toward Indigenous entrepreneurship.

Characteristics of HRD and Tourism Education in the Pacific

Irrespective of the level of delivery, or the agency by which it is delivered, there are particular contextual issues in the Pacific that need to be considered in any discussion of tourism education and training. On the surface, the relevance of these issues to tourism education is not always readily apparent. However, they are issues that are integral to creating opportunities and equity for Indigenous people in the tourism industry in the Pacific. The nature of landownership in the Pacific is one such issue, as is the collectivist nature of Pacific society.

Land ownership in the Pacific is mostly based on a collective model. In respect to tourism development, the land tenure system may be of benefit to Indigenous people. Unless a rare freehold property is obtained, lands are leased from the collective ownership, with leases ranging from a few years to more than 75 years. Leases are encumbered with a range of conditions that vary depending on the particular country and the aspirations of the Land-Owning Community (LOC). One of the areas that may be affected by this form of land tenure is employment, which is related to training and education. National law or local conventions may result in the situation where staff for tourism operations on leased land are recruited from the local LOC. Leases often have job rights attachedguaranteeing the LOC jobs at the resorts and/or operations.

Although these contractual arrangements facilitate employment opportunities for Indigenous people in the tourism industry, they may also act as barriers to entrepreneurship and employment at higher levels. For example, as jobs are available by right, motivation to attend pre-employment and in-service training can be negatively affected (SPTO, 2002). Movono et al. (2015) and Gibson (2019) found that guaranteed employment meant little incentive for youth to continue their education. This lack of training can result in Indigenous staff being concentrated at lower levels of the employment hierarchy. Additionally, even if staff are motivated to undertake pre-employment training, the remoteness of villages or offshore islands (where many tourism operations are situated) restricts access to training opportunities. Potentially, these contextual issues limit labor mobility and could impact who will benefit from education and training. Differing authority structures at the workplace and in the villages may lead to tensions. As a result, employees may also be reluctant to undertake additional training to enable them to move on to supervisory positions, as it may challenge traditional village lines of authority (TCSP, 1992). In some cases, employees may choose to opt out of the responsibility that comes with promotion to supervisory positions (Gibson, 2019).

To fully understand tourism development and its potential progress and benefits to Indigenous people in the Pacific, Britton (1982) suggested that it was important to situate it in the historical political environment in which it emerged. Of particular importance is understanding the impacts of foreign ownership of tourism on the socio-political, economic, cultural, and for the purposes of this topic in particular, the educational landscape in the Pacific Islands.

Foreign interests established themselves early in Pacific tourism development due to the lack of investment capital from the Indigenous people. In modern tourism environments, the implications of this include the limited ability of small- scale Indigenous -owned businesses to compete and establish themselves in the tourism market and the fact that those foreign investors first in the market continue to have a competitive advantage (Apostolopoulos et al., 2013). This domination has had impacts on regional tourism education. Rather than education being the impetus for self-determination and self-efficacy of Indigenous people, tourism education has reinforced imperialism in the Pacific by training Indigenous people in tourism service rather than in the dominant positions of entrepreneurship and tourism management, reinforcing and maintaining the imperialistic status quo. This has also reinforced dependency, not only on the direction of development, but dependency of developing countries from the West for tourists and tourism development (Apostolopoulos et al., 2013). Tourism development reinforces class differences and widens inequalities in communities as tourism expands the divide between the haves (early investors in tourism being either local elites or foreign investors) and the have- nots (Bianchi, 2018).

However, given the land tenure system across much of the Pacific that determines land ownership by Indigenous people, there is the opportunity for Indigenous people to turn tourism ownership and control around from dominant foreign ownership and subordinate tourism employment to Indigenous ownership and empowerment through Indigenous entrepreneurship. The land tenure system also provides a greater chance for Indigenous people to compete with foreign interests in tourism. A potential advantage for Indigenous people to compete with foreign investors is the use of cultural capital and unique Indigenous and cultural forms of hospitality which may provide a competitive advantage. However, this success would require the focus of Indigenous entrepreneurship in education.

The cultures of the Pacific are, for the most part, collectivist, rather than individualistic in nature. These collectivist characteristics can exert additional constraints on opportunities for Indigenous people in the tourism industry. For those in paid employment, there is often significant social pressure to share wealth with the extended family, or in some cultures, to make financial contributions to the church. These demands grow proportionally with increases in income. This pressure to provide financial support may at times act as a disincentive for Indigenous people to take up paid employment in the tourism industry or to undertake additional training and education to support job advancement (TCSP, 1992). On the other hand, Trupp et al. (2021) discuss collectivism in Indigenous entrepreneurship as having social capital benefits. Although the economics of entrepreneurship and its associate benefits cannot be disregarded, it is important to acknowledge other advantages which move away from the economic dimensions associated with business.

The changes to tourism as a result of the COVID context also contribute to the complexities of understanding tourism education needs in the Pacific. In their research in Fiji, Movono et al. (2017) found that as *iTaukei* (Indigenous Fijians) became more dependent on resorts for employment, they became distant from traditional activities including crafts, performance and traditional food production systems. This in turn increased dependency on tourism, vulnerability to systemic change and diminished resilience. However, Movono and Scheyvens (2022) found that the significant COVID-related downturn in tourism in the Pacific heralded a period of deep reflection with an emerging sense of self-determination,

> Pacific peoples are thus trying to improve their situation in the face of COVID-19, and do not see themselves as victims…they are ruminating deeply on the past, appreciating more about their own people and cultures, and thinking about what they want in the future….
>
> (p. 139)

Movono and Scheyvens go on to suggest that these skills and resources had been largely ignored or overlooked for some time because of people's focus and dependence on tourism as their primary livelihood source. This is reflected in the Cook Islands' Mana Tiaki campaign, launched in May 2020 in which Cook Islanders, as guardians of their lands and waters, re-vision a tourism that is more culturally focused and based on traditional Indigenous systems (see for example the Mana Tiaki YouTube campaign: https://www.youtube.com/watch?v=iHlkYr80dNo). Movono and Scheyvens (2022) suggest that this type of COVID-driven adaptation to tourism has strengthened Pacific Islanders' social cohesion and engagement with their cultural and ecological environment. They further suggest that these community capacities, local cultural systems, customary resources and connection to their land are useful tools that can be embraced as a means to "increase resilience and add value to tourism management practice for the future" (p. 143). This has implications for Indigenous tourism education in the Pacific and should be acknowledged in a movement to establish the importance of Indigenous entrepreneurship in tourism education.

If Western models of tourism education and HRD are adopted without adequate consideration of these socio-cultural characteristics and the cultural context (an 'imposed etic') (Berno, 1996; Tight, 2022), there is a danger that Indigenous people will fail to reach their full potential and achieve their aspirations. However, if the local context in which the tourism education is to be delivered is considered, tourism training and education, and importantly Indigenous entrepreneurship in tourism education can become a means for the empowerment of Indigenous people, as well as a means to contribute to self- determination and support the overall resilience and sustainability of tourism in the region. Considering this, it speaks to the importance of cultural capacity and competency that can be incorporated in tourism education by Indigenous academics. This can be seen in some Pacific institutes, such as the National University of Samoa (NUS) (NUS, 2022).

Tourism Education and 'The Pacific Way'

As discussed above, the need for HRD and capacity building in Pacific tourism has been the subject of discussion and research for some time now. In response to these HRD needs of the industry, a range of providers exist throughout the region who offer tourism and hospitality education. Although this supply of educational programs helps to address the issue of sufficient capacity, the issue of how the current tourism and hospitality education curriculum can meet the deeper needs of the Indigenous people of the region for sustainability, resilience, empowerment, and self-determination still needs to be considered.

In her discussion of tourism educational needs of the small island developing states (SIDS) of the Caribbean, Lewis (2004) argued that SIDS present a distinct context for tourism education, and as such, the challenge presented is how to develop a tourism curriculum that responds to the threats and challenges posed by globalization in SIDS, while at the same time placing the curriculum in the socio-economic and cultural context of the island destination. Conlin and Baum (2003) extend this challenge by suggesting that not only does the particular socio-cultural context need to be accounted for, in order to optimize opportunities for Indigenous people within the tourism industry, but a comprehensive, holistic approach to planning for tourism needs to be applied — one that integrates HRD and career planning as well as appropriate education. The following discussion will consider the need to position Indigenous tourism education in the Pacific within this broader planning context.

A Holistic, Integrative Approach to Indigenous Tourism Education

The case for the inclusion of Indigenous people in the tourism planning process is well documented in the tourism literature, as noted below. However, much of the emphasis of this inclusion has been to support a sympathetic understanding of the importance of tourism within the local community, often with the end objective to ensure tourism's success at the destination (Conlin & Baum, 2003). Conlin and Baum suggest that the positive impact of employment and career development for members of the local community often appears to be peripheral to the planning process, and when planning does incorporate considerations of employment, it is often restricted to the recommendation of the provision of education and training for the tourism industry. They go on to suggest that the emphasis on training for the development of lower-level skills and abilities for Indigenous people in tourism and hospitality fails to take into account cultural and traditional barriers which could mitigate against Indigenous people entering the tourism industry. Their argument warns that such an approach does not lead to the empowerment and self-determination of Indigenous people within tourism. In fact, somewhat paradoxically, it supports the status quo. They suggest that there is a need for a holistic, comprehensive process to plan for the development of Indigenous human resources in island tourism, one that considers and integrates five areas:

1   The tourism environment
2   Tourism and the labor market
3   Tourism and education
4   human resource practice in the industry
5   Tourism and the community

Liu and Wall (2006) echo this call for the integration of tourism human resources in the context of policy trends within planning paradigms, suggesting that future tourism plans in developing countries should give greater prominence to the development of human resources in a way that enables Indigenous people to participate in and benefit from tourism development in their region. They too suggest that HRD in tourism in developing countries has been constrained by attempts to meet international service standards with a disregard for cultural sensitivity and adequate adaptation to local societal and cultural contexts (p. 169). This limitation has been reflected in ongoing calls for greater involvement of Indigenous people in directing, participating in and benefiting from tourism development. They go as far as to suggest that '[i]f tourism is really to be a "passport to development" and a means to enhance the lives of destination residents, then greater attention must be given in tourism plans to their needs and capabilities' (p. 169). It is clear that Pacific Islanders need more than just technical and employability skills to realize their aspirations for the tourism industry. Cheer et al. (2018) and Movono and Scheyvens (2022) reference the need to align national (and we would suggest regional) tourism policy with Indigenous aspirations informed by Pacific voices. In this way, the opportunities for Pacific peoples to benefit from tourism will be enhanced in a way that allows them to maintain their culture and cultural values, reduce vulnerabilities, and build resilience.

Looking at it from a human resource perspective, one of the positive outcomes of adopting and applying an integrative planning process (both at the national and regional levels) is the development "'of' island communities and not just 'in' them" (Conlin & Baum, 2003, p. 117). Such an approach provides the platform from which tourism training and education can be prioritized within the national budget and the national educational and tourism plans. As a result, the opportunity to enhance Indigenous participation in tourism through broader educational opportunities is improved (TCSP, 1992).

Teaching and Learning the Pacific Way

Educators in the Pacific acknowledge large gaps between the culture of formal education, which is based mainly on colonial models, structures and institutions, and the culture of the majority of Indigenous students. This dissonance has resulted in teaching and learning difficulties, and in many cases, underachievement.

To address this gap, Thaman (2000, 2013) suggests that Pacific Island teachers and students must 're-claim their education by looking towards the sources of their identities and developing philosophies and teaching and learning strategies that are rooted in their cultural values and practices' (2000, p. 49). To achieve this, she suggests that higher education programs in the Pacific need to consider the culture from which students come - the ways in which they were raised and socialized - and value the knowledge that students bring to the educational setting (Thaman, 2000, 2004, 2013). This is starting to happen in Pacific education. For example, at the University of the South Pacific, many of the tourism and hospitality management (THM) courses e.g., TS209 Food & Beverage Management and TS309 Tourism Business Entrepreneurship require students to examine their cultural values and traditional practices as part of their learning and assessment and design authentic cultural experiences for visitors to share.

In relation to tourism and hospitality education, the starting point for this consideration needs to be the very definition and experience of tourism and hospitality. Although travel, in the forms of migration and mobilities, has been an important traditional feature of the Pacific Islands' culture, as discussed by Berno (1996, 1999), institutionalized tourism is not an Indigenous practice in the Pacific Islands (Trupp et al., 2022). This has resulted in non-Western travel and tourism mobilities in the Pacific being under-researched and therefore lacking theoretical development (Gibson et al., 2020; Trupp et al., 2022). The way in which many Indigenous people define or understand tourism, therefore, differs from what is commonly accepted in Western tourism and hospitality curricula. Many, if not most, tourism students in the Pacific have never been a 'tourist' themselves. Many have never flown on a plane, eaten in a restaurant or entered a hotel lobby. However, there are traditional concepts of leisure, ceremony, celebration, travel and hospitality, which can be utilized to bridge the gap (Blanton, 1981). Using Indigenous understandings of hospitality and personal experience as a platform, students can develop an understanding of tourist behavior, the institutionalization and commercialization of hospitality, the nature of tourist-host interactions, and cross-cultural differences (both tourist-host differences as well as national ones) (Berno, 1996, 1999; Blanton, 1981), as a means of understanding 'tourism' as practiced and experienced by visitors to the region. As suggested by an informant in Blanton's 1981 research,'… [in a developing country], one should start housekeeping training with the concept of a bed, not how to make it' (p. 111).

Botterill (1992) suggests that at the heart of the disappointment that many people have with tourism's relationship with education is the distinction that is made between education for tourism as opposed to education about tourism (p. 2). To make the distinction between these two approaches, the former, education for tourism, emphasizes courses in skill related areas of the tourism industry, typically with a management focus. The latter, education about tourism, considers the broader role and impact of tourism as a social phenomenon. This approach considers tourism as a tool for economic development, issues of cross-cultural understanding, and the impacts of human activity in tourism on the socio-cultural and natural environments.

In a region such as the Pacific, a strong argument can be made that there is a need for education both for and about tourism as part of the tourism curriculum. As King (1994) suggests, a blend of business studies and social sciences is helpful in developing an understanding of the impacts of tourism, as well as the techniques for the provision of high-quality service at the destination. More importantly, however, a balanced mix of educational platforms is needed to address the critical issue of empowerment and self-determination for Indigenous people in tourism. Mignolo (2009, p. 1) claims current perspectives are that "The First world has knowledge, the Third world has culture". This perspective reinforces colonial imperialism that has grown to dominate the colonized Pacific. It has embedded itself into education in the Pacific, evident with Western pedagogical approaches to tourism education and the focus on tourism education being on the subservient, subordinate and culturally entertaining roles which sees the employment of Indigenous people in the lower levels of tourism service rather than tourism management. Current tourism education reinforces this, given the lack of Indigenous entrepreneurship and Indigenous people in managerial roles (Shakeela et al., 2012). By providing business management education for tourism, Indigenous people are empowered by the skills and abilities to plan, develop and manage their own tourism product. It helps avoid what Burns (1992) criticizes as '… education in developing countries… [that aims] … at producing compliant employees for an industry dominated by expatriate managers and transnational corporations' (p. 8). For prospective Pacific entrepreneurs, hospitality is something they are born with, whilst tourism as a business must be learned.

By providing education about tourism that addresses critically the broader context of tourism as a social phenomenon, Indigenous people are further empowered through the skills to promote self-determination in the context of tourism. Specifically, as Burns (1992) suggests in his consideration of the development of a new tourism curriculum in Papua New Guinea,

> [i]f we recognize that tourism can sow the seeds of its own destruction, then we must also see that tourism education has the power to act as a regulating factor… [alerting] students to the problems of tourism while placing the problems in a specific cultural context…. [thus challenging] the tourism myths which masquerade as conventional? wisdom.
>
> (p. 8)

Burns goes on to suggest that this type of curriculum model, one which emphasizes culture and the fulfillment of Indigenous needs, could be used as a paradigm for the evaluation and reorganization of other programs of study.

Education for Empowerment

To further support the empowerment and self-determination of Indigenous people in tourism, tourism and hospitality education should address four separate, but not mutually exclusive, levels of tourism education: public awareness; vocational skills training; professional management education and entrepreneurial development (Charles, 1997; Echtner, 1995).

Public Awareness

In Pacific Island countries, like many other developed and developing nations, the tourism industry is not generally considered to be a mainstream career opportunity. It is often viewed as a short-term, transient type of work, with minimal financial and psychological benefits. This cynical view is exacerbated in many developing countries by the practices of multi-nationals, who give preference to expatriates for management positions, relying on Indigenous personnel to fill low-level vocational positions (Conlin & Baum, 2003). Additionally, there is often little appreciation by locals of the importance of tourism to their region, and career opportunities within the industry. In the Pacific (as in Western societies too), there is a tendency to regard employment in the public sector and the professions such as doctors, lawyers and accountants as the preferred option for advancement. Even though development opportunities in the Pacific Islands are often limited to tourism, tourism is not generally seen as a means to achieve the same ends as these other professions (TCSP, 1992). To address these issues, Kwmae (1997) suggests that tourism education in the school system can play an important role in creating public awareness of tourism in general, its importance to a region, and it can provide an opportunity for molding positive attitudes towards the industry and careers within it. New Zealand has begun the development of tourism studies as a subject in high schools. For the Pacific realm countries that implement the New Zealand school curriculum, tourism as a subject will also be an option in high schools. In tourist regions in Fiji such as: the Coral Coast and Yasawa Islands, tourism studies are also being taught in high school.

Vocational and Professional Training

With the rapid growth of tourism, in many developing countries, including those in the Pacific, there is a critical shortage of trained Indigenous people, both on the front lines as well as supervisory positions. Vocational training is essential to build a labor pool in support of these positions. As a complement to vocational training, there is a need for higher-level education (e.g., tertiary) to afford Indigenous people the opportunity to rise to management levels - positions that are currently often filled by expatriates.

Irrespective of whether it is vocational or professional tourism education, it is important that the content and delivery be contextualized and positioned within the culture and characteristics of the Pacific. For example, the content of courses needs to recognize the unique socio-political and cultural aspects of the region as they relate to tourism. This is not just a focus on nationality or regionality. It needs to be a comprehensive consideration and incorporation of the developmental issues relevant to the region (Thaman, 2004). As Craig-Smith (2005, p. 369) suggests, "Whilst in a global economy many skills and much knowledge should be, and is universal, an industry, and particularly the tourism industry, should be quick to adopt the "image" and "specialty" of the [Pacific] locale."

Professional Management and Entrepreneurial Education

The forms of tourism education discussed above, public awareness, vocational and professional, are all concerned for the most part with developing the skills and knowledge required to work for someone else (Echtner, 1995). The last component of tourism education, therefore, addresses the development of skills and knowledge to work for oneself. It has been suggested that one of the most critical needs for tourism in Pacific Islands countries is the fostering of entrepreneurship (Spann, 2021).

Echtner (1995) provides a range of salient benefits associated with indigenous entrepreneurship in developing countries. Along with the broader empowerment of Indigenous people, these benefits include: greater economic payback and local control; greater use of local supplies; less negative socio-cultural impacts; a more effective response to market changes, and cumulatively, the opportunity to enhance community stability.

That said, however, there are contextual issues in the Pacific in relation to entrepreneurship education that need to be considered. The characteristics of Pacific entrepreneurs differ from those of their Western counterparts (Saffu, 2003). Strong considerations should be given to indigenous connection to land and entrepreneurship (Scheyvens et al., 2017). Saffu, using Hofstede's (1980, 1991) cultural dimensions as a framework (including the dimensions of collectivism, power distance, masculinity/ femininity and uncertainty avoidance) suggests that the set of characteristics for success in the Pacific is sufficiently different from that of the West that a different approach to entrepreneurial education is warranted.

Studies by Saffu (2003) and Gibson (2012) argued that "the success of Indigenous entrepreneurs should be considered in the context of traditional community expectations, and not necessarily in accordance with western concepts of individualistic entrepreneurship" (Gibson, 2012, p. 118). Through research in Australia, Collins et al. (2017) reinforce that the focus of Indigenous culture and entrepreneurship is on communality and the benefit of many. They suggest that this is in contrast to Western models, where neoclassical free market economics dominates entrepreneurship and the focus is on individualistic wealth acquisition. Collins et al. further state that: "In this sense, Indigenous culture could be characterized as anti-entrepreneurial. However, … such characterization is a crude stereotype that offers little to understanding the contemporary relationship between Indigenous culture and entrepreneurship…" (p. 37). Saffu suggests that rather than teaching Pacific entrepreneurs how to run their businesses as individualistic, risk-taking profit maximizing businesspeople (the characteristics associated with successful entrepreneurship in the West), training should focus on providing skills that will help Pacific entrepreneurs become more adept at maneuvering the Pacific cultural and business environment. An appropriately designed training program should incorporate cultural nuances and norms and focus on how entrepreneurs can be adaptable and flexible, with the ability to blend the modern and the traditional. Saffu also suggests that training should equip aspiring entrepreneurs with the skills that will enable them to meet traditional social and familial demands, while at the same time, meet their business goals. Within the Pacific, conflicting value systems and concepts of 'success' demand that successful Indigenous entrepreneurs function at two different levels: *vanua* (ways of the land), which meets traditional commitments; and business, which commits to saving, budgeting, and future investment (Gibson, 2012).

Conclusion

Potentially, tourism in the Pacific provides opportunities for Indigenous people to engage in and benefit from the development process. A key issue, underpinning this involvement, is the education and training they currently receive to support this engagement.

Like many countries in the developing world, the small island states of the Pacific have been subject to educational systems adopted from previous colonial powers and/or external Western models. Tourism and hospitality education are no exception. This topic was originally discussed in 2006 (Berno). As we worked on the revision, we came to realize that despite the increase in tourism education providers in the Pacific, there has been little substantive change to the content and the way in which it is being taught. The tourism curricula and pedagogy remain essentially from a colonial, Western perspective. The continuing call for more Indigenous involvement in the tourism industry is testament to the failure of the traditional methods of HRD and education to address the needs of Indigenous people's aspirations.

Mobilities across the Pacific in the tourism and hospitality industries complicate tourism training and education. The Pacific is vast, consisting of many small island developing states and territories each with their own unique culture. This unique culture transpires to unique hospitality practices which have often been neglected in regional education. Unique hospitality practices may enhance the tourists' experience and ultimately contribute to destination competitiveness with return visitation and longer stays which contribute to a destination's economic development (Perovic et al., 2018). However, given the mobilities of Pacific people across the region for employment in the tourism industry, including hospitality, education needs to consider a general approach rather than be contextually specific. Although not culturally specific but acknowledging Pacific mobility, through a general approach to tourism education, knowledge of the unique hospitality practices of Pacific cultures should be provided for the possibility of the evolution of new dimensions of experiences that can enhance tourists' encounters.

Although this chapter has focused primarily on the SIDs of the Pacific, much of what we have discussed applies to Indigenous people elsewhere. We have argued in this chapter that providing tourism training and education alone is not enough to empower Indigenous people in tourism. To fully support the aspirations of Indigenous people, education and training must build a platform of cultural and contextual understanding that supports the aspirations of Indigenous people. In the Pacific, this challenge involves establishing a 'Pacific Way' for the future of tourism education - one that supports the development of international skills and understanding, while at the same time inculcates and incorporates the unique features and characteristics of the Pacific.

References

Apostolopoulos, Y., Leivadi, S., & Yiannakis, A. (2013). *The sociology of tourism: Theoretical and empirical investigations.* Oxon: Routledge.

Berno, T. (1996). Cross cultural research methods in the field - Content or context? A Cook Islands case study. In R. Butler & T. Hinch (Eds.), *Tourism and Indigenous peoples* (pp. 376–395). London: International Thomson Business Press.

Berno, T. (1999). When a guest is a guest - Cook islanders view tourism. *Annals of Tourism Research, 26*(3), 656–675.

Berno, T. (2001). Human resources development for the tourism sector in the South Pacific. Invited paper presentation at the *APETIT UN-ESCAP Conference*, Khajurho, India, 7–10 August.

Berno, T. (2006). Doing it the 'Pacific Way': Indigenous education and training in the South Pacific. In R. Butler & T. Hinch (Eds.), *Tourism and Indigenous Peoples* (pp. 28–39). London: Butterworth-Heinemman/ Elsevier.

Berno, T. (2007). Doing it the 'Pacific Way': Indigenous education and training in the South Pacific. In R. Butler & T. Hinch (Eds.), *Tourism and Indigenous Peoples* (pp. 29–39). London: Routledge.

Berno, T. (2007). Human resources development and capacity building for the tourism sector in the South Pacific: The challenges of being a regional institution. Invited presentation for *Workshop for Pacific Islands 2007 - Tourism Education and Cultural Heritage Management for a Sustainable Development in the Pacific Region, University of the Ryukyus and the Ministry of Foreign Affairs*, Okinawa, Japan, 30 January–1 February.

Bianchi, R. (2018). The political economy of tourism development: A critical review. *Annals of Tourism Research, 70*, 88–102.

Blanton, D. (1981). Tourism training in developing countries: The social and cultural dimension. *Annals of Tourism Research, 8*(1), 116–133.

Botterill, D. (1992). Tourism education: Changing attitudes – Facing the challenge. *In Focus, 3*, 2–3.

Britton, S. G. (1982). The political economy of tourism in the Third World. *Annals of Tourism Research, 9*(3), 331–358.

Burns, P. (1992). Going against conventional wisdom. *In Focus, 3*, 8.

Burns, P. M. (1999). Vaka pasifika: A regional approach to tourism training in the South Pacific Region. *Pacific Tourism Review, 3*(2), 101–118.

Chambers, D., & Buzinde, C. (2015). Tourism and decolonisation: Locating research and self. *Annals of Tourism Research, 51*, 1–16.

Charles, K. R. (1997). Tourism education and training in the Caribbean: Preparing for the 21st century. *Progress in Tourism and Hospitality Research, 3*(3), 189–197.

Cheer, J. M., Pratt, S., Tolkach, D., Bailey, A., Taumoepeau, S., & Movono, A. (2018). Tourism in Pacific Island countries: A status quo round-up. *Asia & the Pacific Policy Studies, 5*(3), 442–461.

Collins, J., Morrison, M., Basu, P. K., & Krivokapic-Skoko, B. (2017). Indigenous culture and entrepreneurship in small businesses in Australia. *Small Enterprise Research, 24*(1), 36–48.

Conlin, M. V., & Baum, T. (2003). Comprehensive human resource planning: An essential key to sustainable tourism in island settings. In C. Cooper (Ed.), *Classic reviews in tourism* (pp. 115–129). Clevedon: Channel View Publications.

Craig-Smith, S. J. (2005). Tourism education in Oceania. In C. Cooper and C. M. Hall (Eds.), *Oceania: A tourism handbook* (pp. 362–379). Clevedon: Channel View.

Dredge, D., Benckendorff, P., Day, M., Gross, M. J., Walo, M., Weeks, P., & Whitelaw, P. A. (2013). Drivers of change in tourism, hospitality, and event management education: An Australian perspective. *Journal of Hospitality & Tourism Education, 25*(2), 89–102.

Dowse, M. (1994). *Tourism studies at the USP. Discussion Paper by the Pacific ACP: EU Bureau for the Tourism Council of the South Pacific and the University of the South Pacific.* Unpublished manuscript. Suva, Fiji.

Echtner, C. M. (1995). Entrepreneurial training in developing countries. *Annals of Tourism Research, 22*(1), 119–134.

Gibson, D. (2012). The cultural challenges faced by Indigenous-owned small medium tourism enterprises (SMTEs) in Fiji. Case studies from the Yasawa Islands. *Journal of Pacific Studies, 32*, 106–132.

Gibson, D. (2019). More than smiles – Employee empowerment facilitating high quality, consistent services – The Wakaya Club, Fiji. *Journal of Pacific Studies, 39*(1), 6–29.

Gibson, D., Pratt, S., & Iaquinto, B. L. (2020). Samoan perceptions of travel and tourism mobilities – The concept of Malaga. *Tourism Geographies*, 1–22.

Higgins-Desbiolles, F. (2017). A pedagogy of tourism informed by Indigenous approaches. In P. Benckendorff & A. Sehrer (Eds.), *Handbook of teaching and learning in tourism* (pp. 439–454). Cheltenham: Edward Elgar Publishing.

Hofstede, G. (1980). *Culture's consequences.* Beverly Hills, CA: Sage.

Hofstede, G. (1991). Cultural constraints in management theories. *Academy of Management Executive, 7*(1), 81–94.

Hollinshead, K. (2013). Speaking of tourism Part II: The ongoing development of a conceptual glossary on fantasmatics. In *Conference presentation at the Welcoming Encounters: Tourism in a post-disciplinary era conference, Neutchatel, Switzerland, June* (pp. 19–22).

IDEEA Group, SPTO. (2021). *Pacific Tourism Statistics 2021–2030.* Paris: OECD Paris 21.

King, B. (1994). Tourism higher education in island microstates: The case of the South Pacific. *Tourism Management, 15*(4), 267–272.

King, B. (1996). A regional approach to tourism education and training in Oceania: progress and prospects. *Progress in Tourism and Hospitality 2*(1), 87–101.

Kwmae, R. D. (1997). Tourism education and training in the Caribbean"prearing foe the 21st century. *Progress in Tourism and Hospitality Education 3*, 189–197.

Lewis, A. (2004). Rationalising a tourism curriculum for sustainable tourism development in small island states: A stakeholder perspective. *Critical Issues in Tourism Education*, 9–15.

Liu, A., & Wall, G. (2006). Planning tourism employment: a developing country perspective. *Tourism Management, 27*(1), 159–170.

Medlik, S. (1989). *University studies in tourism.* Report by Prof S. Medlik within the Pacific Tourism Development Programme, financed by the European Economic Community, July. Suva: Tourism Council of the South Pacific.

Mignolo, W. (2009). Epistemic disobedience, independent thought and de-colonial freedom. *Theory, Culture and Society, 26*(7–8), 1–23.

Morris, P. (2015). *Review of the Pacific framework for technical and vocational education and training.* Noumea: South Pacific Commission.

Movono, A., Dahles, H., & Becken, S. (2018). Fijian culture and the environment: A focus on the ecological and social interconnectedness of tourism development. *Journal of Sustainable Tourism, 26*(3), 451–469.

Movono, A., Pratt, S., & Harrison, D. (2015). Adapting and reacting to tourism development: A tale of two villages on Fiji's Coral Coast. In D. Harrison & S. Pratt (Eds.), *Tourism in Pacific Islands: Current issues and future challenges* (pp. 118–133). London: Routledge.

Movono, A., & Scheyvens, R. (2022). Adapting and reacting to COVID-19: Tourism and resilience in the South Pacific. *Pacific Dynamics Journal of Interdisciplinary Research, 6*(1), 124–150.

Nabobo-Baba, N. (2012). Transformations from within: Rethinking pacific education initiative. The development of a movement for social justice and equity. *The International Education Journal: Comparative Perspectives, 11*(2), 82–97.

NUS (National University of Samoa). (2022*). Faculty of business and entrepreneurship. Department of Management Tourism and Hospitality.* https://nus.edu.ws/department-of-management-tourism-and-hospitality/

Perovic, D., Moric, I., Pekovic, S., Stanovcic, T., Roblek, V., & Bach, M. P. (2018). The antecedents of tourist repeat visit intention: Systemic approach. *Kybernetes, 47*(9), 1857–1871.

Saffu, K. (2003). The role and impact of culture on South Pacific island entrepreneurs. *International Journal of Entrepreneurial Behavior & Research, 9*(2), 55–73.

Scheyvens, R., Banks, G., Meo-Sewabu, L., & Decena, T. (2017). Indigenous entrepreneurship on customary land in the Pacific: Measuring sustainability. *Journal of Management & Organization, 23*(6), 774–785.

Shakeela, A., Ruhanen, L., & Breakey, N. (2012). Human resource policies: Striving for sustainable tourism outcomes in the Maldives. *Tourism Recreation Research, 37*(2), 113–122.

Spann, M. (2021). 'It's how you live' - Understanding culturally embedded entrepreneurship: An example from Solomon Islands. *Development in Practice*, 1–12. https://doi.org/10.1080/09614524.2021.1944988

Spivak, G. C. (1988). Can the Subaltern Speak? in Nelson, C. and Grossberg L. (eds) *Marxism and the Interpretation of Culture* (pp. 271–313). Urbana and Chicago: University of Illinois Press.

South Pacific Tourism Organisation (SPTO). (2002). *Regional tourism strategy for the South and Central Pacific.* Suva: SPTO.

South Pacific Tourism Organisation (SPTO). (2019). *2018 Annual review of visitor arrivals in Pacific Island Countries.* Suva: SPTO.

South Pacific Tourism Organisation (SPTO). (2020). *Pacific Tourism Organisation Strategic Plan 2020–2024.* https://southpacificislands.travel/wp-content/uploads/2021/07/SPTO-Strategic-Plan-2020-2024.pdf

South Pacific Tourism Organisation (SPTO). (2021). *2021 COVID-19 impact snapshot update.* Suva: SPTO.

Thaman, K. H. (2000). Towards a new pedagogy: Pacific cultures in higher education. In R. Teasdale & M. Rhea (Eds.), *Local knowledge and wisdom in higher education* (pp. 43–50). Oxford: Pergamon Press.

Thaman, K. (2004). *Le'o e peau: Towards cultural and cognitive democracy in sustainable development in Pacific Island communities.* Keynote address at Islands Conference, Kinmen Island, Taeain, 1–4 November.

Thaman, K. H. (2013). Quality teachers for Indigenous students: An imperative for the twenty-first century. *International Education Journal: Comparative Perspectives, 12*(1), 98–118.

Theuns, H.L. and Go, F. (1992). "Need led' priorities in hospitality education for the third world. *World Travel and Tourism Review* 2 2930302.

Tight, M. (2022). Internationalisation of higher education beyond the West: Challenges and opportunities – The research evidence. *Educational Research and Evaluation, 27*(3–4), 239–259.

Tourism Council of the South Pacific (TCSP). (1992). *Pre-feasibility study of tourism training facilities and institutional frameworks in the South Pacific.* Suva: Tourism Council of the South Pacific.

Tourism Council of the South Pacific (TCSP). (1995). *Human resources development and training needs assessment study.* Suva: Tourism Council of the South Pacific.

Tourism Council of the South Pacific (TCSP). (1999). *Proposals for the re-structuring of the human resource development division and review of industry training requirements.* Suva: Tourism Council of the South Pacific.

Trupp, A., Matatolu, I., & Movono, A. (2021). Gender and benefit-sharing in Indigenous tourism microentrepreneurship. In D. B. Morais (Ed.), *Tourism microentrepreneurship (Bridging Tourism*

Theory and Practice, Vol. 12) (pp. 51–63). Leeds: Emerald Publishing Limited. https://doi.org/10.1108/S2042-144320210000012005

Trupp, A., Pratt, S., Stephenson, M. L., Matatolu, I., & Gibson, D. (2022). Representing and evaluating the travel motivations of Pacific Islanders. *International Journal of Tourism Research*, 1–14. https://doi.org/10.1002/jtr.2528

Young, T., & Maguire, A. (2017). Indigenization of curricula: Trends and issues in tourism education. In P. Benckendorff & A. Sehrer (Eds.), *Handbook of teaching and learning in tourism* (pp. 455–464). Cheltenham: Edward Elgar Publishing.

14

STEWARDING MĀORI TAONGA FOR SUSTAINABLE INDIGENOUS TOURISM ENTERPRISE

Ashley Puriri and Alison McIntosh

Ki te kahore he whakakitenga, ka ngaro te iwi (*He Tongi o Kiingi Tawhiao Potatu Te Wherowhero*)
Without foresight or vision, the people will be lost (a proverbial saying of the Māori King Tawhiao)

Introduction

Māori are the Indigenous peoples of Aotearoa New Zealand and Māori businesses form an important part of the country's tourism offerings. Like most Indigenous cultures, the ethos of Māori culture is metaphysically interconnected by Whakapapa (genealogy) and history. The aim of this chapter is to illustrate how Māori tourism enterprise development embraces and is shaped by, the vision and stewardship placed upon the Māori entrepreneur by their Kuia and Kaumātua (Māori elders). The chapter draws on primary research of two Māori tourism cases, Taiamai Tours Heritage Journeys Limited and Maui Tourism Limited, to show how they embrace the stewardship placed upon them by their genealogy in their business decisions and actions to ensure cultural authenticity (Tika me Te Pono) and commitment to Māori taonga. In discussing these cases, the importance of Tino Rangatiratanga, or intergenerational responsibilities, from a Tōhua Te Ao position of caring for the world we live in, in a sustainable way, when developing an Indigenous sustainable tourism enterprise is highlighted.

Ko Wai a Tūturu Māori - Who Are Indigenous Māori

Māori are referred to as Tangata Whenua, which is the title Māori prefer to refer to themselves as. The term Māori however has a deeper meaning but is one of the names the colonists gave to Tangata Whenua. A loose translation for Tangata Whenua means, the people of the land. Māori traversed the great Pacific Ocean waters on many voyages to Aotearoa New Zealand from origins further north of the Triangle as part of the migrative movements southward. For the purpose of this chapter, all references to Indigenous tourism for Aotearoa New Zealand, will refer exclusively to Māori or Indigenous Tangata Whenua tourism.

DOI: 10.4324/9781003230335-17

Linda Smith (1999) discusses the very complex cultural epistemologies and ontologies of the Māori people and the Māori worldview. The complexities of a Māori worldview provide a distinctiveness to Māori tourism for Aotearoa New Zealand. A Māori worldview imbues everything with a Mauri, a life-force, which links things to the world we live in. This includes things that may not have a living life-force from a Westernised viewpoint, for example, a stone or building. Takiriranga Smith (2005) affirms that everything has a life-force (mauri) as part of Māori wisdom and that this Mauri is tapu (sacred) to Māori. As such, cultural perspectives and wisdom are fundamental to understanding Māori tourism.

An important enabler that grounds Māori cultural tourism experience and service is whakapapa (Puriri, 2017). In this context, whakapapa, means to provide a cultural connectedness of or for a Māori tourism business, enterprise or 'Pakihi' (Spiller, 2010) to something of greater cultural significance. As an example, cultural connectedness through Whakapapa can be metaphorically likened to a tree that is grounded to Papatūānuku[1] or Earth Mother through its roots, from whom the tree also relies on for nutrition, sustenance, stability, security and support. The importance of Whakapapa, in particular, for ensuring a Māori tourism enterprise is culturally authentic will be further discussed below.

Māori tourism is commonly discussed as being underpinned by significant cultural values, which include Wairua (spiritual affirmations), Whakapapa (genealogical connectedness), Aroha (love/passion), Manaakitanga (caring/hosting), Tōhua Te Ao (cultural sustainability), Tikanga Māori (Māori protocols, beliefs and customs), Tūturutanga (Authenticity), among other cultural values (see for example, Zygadlo et al., 2003). These values define how a tourism enterprise is uniquely Māori. Tikanga Māori, whakapapa and wairua according to Barlow (1994), are therefore essential to guide Māori tourism enterprises as to how to embrace these intrinsic values. Importantly, when imbued with a cultural lens, the enterprise can gain confidence in deciding if and how it shares cultural identity or authenticity with tourists. Historically, Māori tourism has been associated with inauthentic and stereotyped representations with little benefit to Māori themselves (Blair-Stahn, 2010). As such, consideration of case study Māori tourism enterprises, such as the two discussed here, is important for understanding the unique cultural underpinnings of an authentic Māori tourism enterprise.

The two Māori tourism enterprises that are drawn upon as case studies are briefly introduced below, followed by discussion of how the two case studies embrace the Māori worldview and values into their enterprises.

Case Study 1: Taiamai Tourism Heritage Journeys (Taiamai Tours)

Taiamai Tours Heritage Journeys is an authentic Māori waka (war canoe) tourism experience that operates from Paihia and Waitangi in the Far North of Aotearoa New Zealand (https://www.taiamaitours.co.nz/). Tourists undertake a guided cultural tour whilst paddling aboard a large traditional canoe and are introduced to ancient Māori customs, rituals and traditions. Taiamai Tours is a 100% Māori owned and operated enterprise and provides a cultural experience for tourists. They apply a Māori cultural philosophy at the forefront of their enterprise.

Case Study 2: Maui Tourism

Maui Tourism was a Māori tourism business concept that provided a cultural tourism experience based upon the Māori legend of Maui, a demi-god, who caught a giant fish, later to be referred to as Te Ika o Maui or Maui's Fish. The cultural experience included a lookout point on a 30-metre statue of Maui catching his fish, a Maui marae (traditional meeting complex) with a cultural performance depicting the story of Maui's catch, and a native bush scenery walk with a native bird aviary and Tuna (eels) based in the Wellington region in the North Island of New Zealand. Like Taiamai Tours, Maui Tourism was 100% Māori owned and was planned to operate a cultural experience for tourists based on the stories of Maui. It is a whānau (family) enterprise based on private Māori lands in the residential and farming area of Johnsonville, 11.5 kilometres from Wellington, the Capital City of New Zealand. The Māori tourism development was significantly guided by its elders, who ensured that Tikanga or customary practices, cultural underpinnings and processes were followed and that Kaupapa (primary aims) of the Whānau were adhered to and met. An extensive investment by the Whānau aimed at servicing the demands of the seasonal cruise industry in New Zealand.

The Māori Worldview and Tourism Enterprise: The Importance of Whakapapa

Whakapapa affords Māori the opportunity to ground their tourism development, product and or service and experience with a connectedness to cultural values that underpin the quality of the Māori Indigenous tourism. The practice of applying Whakapapa requires a culturally authentic adoption of these cultural values. Examples of these cultural values are harnessed by a sincere commitment to ensure that the cultural values are imbued within the organisation and delivery of the tourism experience. As such, they ensure the enterprise and tourism experience is culturally authentic.

The practice and application of Whakapapa enable a Māori tourism enterprise to be informed and guided by the principles of the cultural values that Māori choose to ground their tourism business by. McIntosh (2000) has provided further examples of how Māori tourism businesses have applied the cultural values of Manaakitanga (hosting/caring for others) and Aroha ki te Tangata (compassion to others) as significant cultural values that informed these Māori tourism businesses with how they delivered their tourism experience.

The founders of Taiamai Tours determined that they would have their employees tattooed with traditional Tā Moko tattoo patterns to demonstrate their commitment of presenting an authentic Māori warrior as the tour guide on their Māori canoe tours (see Figure 14.5). This painful commitment of having their face and bodies intensely tattooed with these traditional authentic Māori patterns, demonstrated the employee's commitment not only to the Taiamai Tours tourism business, and to Tikanga Māori, but also to their tourists who value the cultural enrichment of an authentic Māori tourism experience. To the founders of Taiamai Tours, the Tā Moko codify cultural meaning as well as an employment strategy, from a Māori perspective; an ancient strategy that pre-dates the arrival of Europeans in Aotearoa New Zealand. Even when confronted with the notion that tourists may find the tattoos 'aggressive looking', the founders are committed to being real, authentic and Pono (true) to their identity (Puriri & McIntosh, 2013).

Taiamai Tours, who are exempt from maritime legislation due to their Waka or canoes being authentic cultural canoes, choose to power their canoes with an outboard motor (rather than human padders) to assist the tourists with navigating on waters with strong currents. They also choose to provide their tourists with life jackets for safety reasons, albeit their employees remain in traditional costumes and choose not to wear life jackets as a commitment to providing their tourists with an authentic Māori waka experience.

Māori tourism businesses are often motivated to go beyond their normal duties of hospitality (Manaakitanga), that are fuelled by their commitment to sustain their cultural values. Maui Tourism as a whānau (family) business is committed to the inclusion and application of cultural values. This was evident at the enterprise planning stage, by ensuring that the needs of tourists were paramount in the decision-making applied by the whānau in their design and delivery. Manaakitanga or caring for others, requires putting others' needs before your own. The Maui Tourism whānau development ensured that a large shelter was incorporated in the build, to provide their tourists with protection from the harsh westerly winds, for this particular Wellington area which is known for its adverse weather that is typically cold with strong gales of up to 60-knots on a regular basis. The design of the complex was such that it met physical abilities of a mature aged demographic. This support and feature of manaakitanga were applied by incorporating an escalator that would run alongside a 25-metre-high staircase, to enable the elderly to experience a key feature of the Maui tourism complex by enjoying the vista overlooking the Wellington harbour. Four by four side cars were integrated into the planning to transport the tourists to various key points of the Maui tour which included access to nature bush walks, waterfalls and streams filled with large native eels plotted throughout the rough and hilly terrain of the Maui tourism experience.

A commitment that both Taiamai Tours Heritage Journeys and the Maui Tours enterprises are tasked to demonstrate is a sincere aim to incorporate a sustainable environmental strategy. Tōhua te Ao is a Māori theoretical philosophy that means, to care for the world we live in, in a sustainable way. This requires a commitment to ensuring that things of nature remain in pristine condition. This includes making certain that forests and bush walks, wildlife, waterfalls, and streams remain free from pollution and abuse from heavy volumes of foot traffic.

A Māori Perspective on Sustainability

Tōhua Te Ao

Māori exhibit a distinct approach to sustainability, guided by their Whakapapa. According to Puriri (2017), Tōhua te Ao is a unique cultural philosophy that ensures taonga important to Māori are cared for in a culturally appropriate way, sustained by Māori who believe they are guardians of their environment (Barlow, 1994). The concept of Tino Rangatiratanga (sovereignty) ensures Māori have the self-determination, self-reliance, rights and independence to protect and manage their resources.

For the Maui Tourism enterprise, a waste management system was incorporated that required that all waste was separated into different categories, and which provided different bins for plastic and food waste. Bush walks were maintained to ensure that the environment remained clean and rubbish free. Tōhua te Ao, also requires a Māori tourism enterprise to be conscious about the number of tourists that attend the experience. Managing the number of tourists that could attend and participate during the tour, helps with ensuring that natural assets like native trees, waterways, and bird life would not be affected by heavy foot traffic and impacts arising from overwhelming quantities of tourists.

A key principle of Tōhua Te Ao is to resist from changing the natural course of rivers and waterways, which would impact the wildlife living in those waters. Ensuring that natural assets, including the Whenua or land, are conserved allows the natural beauty to be included as a feature of the Māori tourism experience. For the Maui Tourism experience, they chose to build the giant 30-metre statue of Maui on the highest point of the cliffs and used the cliffs as a plinth for the Maui statue. This provided a magnificent view from the mid-point of the statue (see Figure 14.1).

For Māori Indigenous tourism, there is a cultural notion and responsibility to build the cultural tourism entity around our natural environment. Māori commitment to sustainability is

Figure 14.1   A black and white drawing of the design for a 30-metre-high statue of Maui to be situated on the top of the cliff (conceptualised by the first author).

Source: A. Puriri, author and artist.

demonstrated in the Tāpoi Poutama (Figure 14.2), which forms a staircase and derives from a history which Māori believe describes the origins of traditional knowledge. In this context, these Māori tourism enterprises positioned priorities according to cultural values and positioned them on the Tāpoi Poutama staircase. Notably, Māori place a significant cultural value on caring for the world we live in more than the value of making a business profit. Whilst economic motives are important, they are never at the expense of the higher cultural values, as shown in the Tāpoi Poutama. Moreover, this relates to a commitment these Māori tourism enterprises apply as a priority to ensure that the world they live in is sustainable for future generations. Taiamai Tours provides an example of this commitment by ensuring that their tourism activities do not negatively impact the ocean and waterways that their waka (canoe) tours traverse. These same waterways are where they harvest fish and seafood for their daily living needs. Therefore, they have a commitment to ensure that these waterways are pollution-free.

Māori Tourism Concepts of Strategy and Sustainability

A Māori perspective of sustainability is therefore distinct from a Western, non-Indigenous, perspective, as is the nature of their enterprise model. In fact, Taiamai Tours Heritage Journeys' strategy may be considered to be a controversial alternative approach to Western business strategies that prioritise an economic imperative. Taiamai, believe they do not need a strategy, rather the founders of Taiamai Tours chose to operate their tourism enterprise based on traditional values and their gut feeling, which Taiamai refer to as Puku Whakaaro, which was applied to provide an authentic experience that they felt their Tūpuna or ancestors would be proud of. The founders of the enterprise are deliberate in not letting their experience be marketed as a commodified cultural product that includes a Powhiri (traditional welcome), Karanga (traditional call), Hāka (traditional war dance) or other cultural components. This is a deliberate strategy to protect the taonga, integral to respecting Tōhua Te Ao. Instead, the cultural protocols that naturally flow during the waka tour

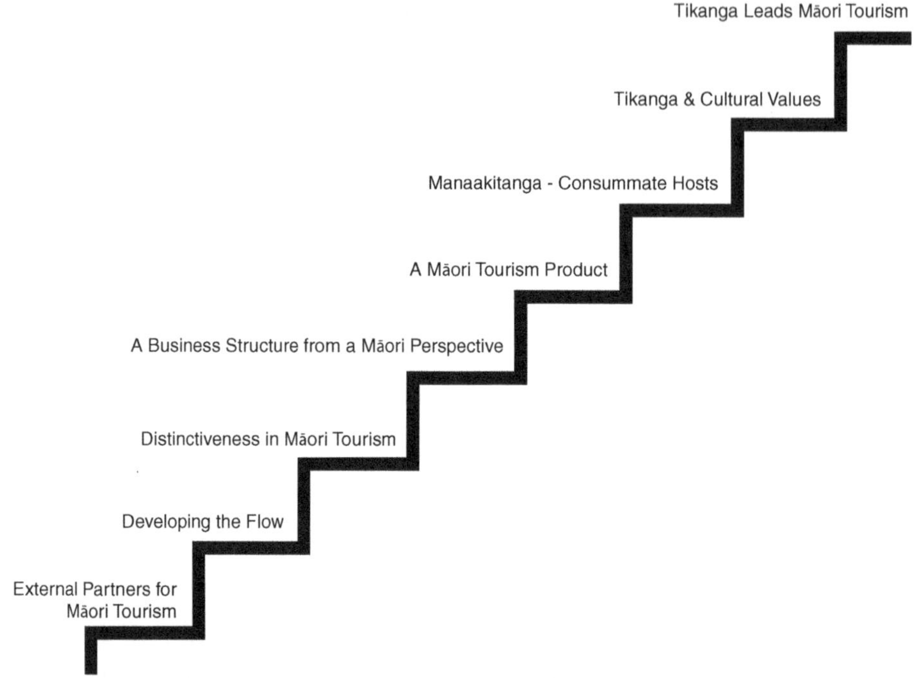

Figure 14.2   Tāpoi Poutama - Cultural lattice of significance.

are not scheduled or itemised as part of the tour experience. In this way, Māori aspects (taonga) of the tour are not 'for sale' and the founders of the enterprise remain in control of their Tikanga Māori (Māori philosophy) and their future.

Furthermore, Taiamai Tours operates their traditional double-hull waka tours in the open sea as well as in the famous Waitangi River, which runs from the mouth of the river inland to the Haruru Falls. These waterways are also a primary source for gathering food for whānau (family) of Taiamai, in addition to being used for the Māori tourism business. For Taiamai Tours, it is crucial that these waterways remain pristine and clean. Taiamai Tours share this integral attribute as part of their heritage story when they are talking to their tourists on their large double Waka (traditional canoes), to help tourists understand and gain an awareness of the importance of caring for their environment (Figure 14.3).

The founders of Taiamai Tours feel that they have a responsibility to ensure that their environment is clean and free from the impacts of their tourism activities. In earlier days of their operation, Taiamai Tours used to include diving for fresh green-lip mussels from within the Bay of Islands as an additional offering in their tourism experience but ceased doing this due to depleting numbers of these natural water purifying shellfish. Since they discontinued this over 15 years ago, the green-lip mussels have begun to return. The management team of Taiamai Tours has committed to never dive for green-lip mussels as part of their tourism experience but have chosen to reserve the gathering of the mussels for their personal consumption requirements.

Similarly, the Maui Tourism management group, under the direction of their Kaumātua, aimed to transform their tourism experience by grounding their Rangatahi (youth) with a commitment to environmental sustainability as a significant cultural value that adds intrinsic value to their tourism offering. The Whānau felt that implementing an environmentally sustainable strategy with their

Figure 14.3   Double hull canoes of Taiamai Tours.
Photographer, A. Puriri.

tourists would demonstrate a cultural responsiveness that tourists would appreciate. In a similar vein, in addition to providing their authentic Māori tourism experience for visitors, the founders of Taiamai Tours provide voluntary community training programmes for Māori youth from local gang affiliations to provide Wananga (Māori training programmes) about Waka (traditional Māori canoe), Whakapapa (genealogy) and Māoritanga (Māori customs and beliefs). Some of these youth have followed the leadership and mentoring of Taiamai Tours and have become employees or extended Whānau members of the enterprise. When the enterprise started out, many at Taiamai Tours did not know how to paddle or manage a Waka (canoe) or about the Moana (ocean). Now, they not only have this cultural knowledge but are also known nationally and internationally as an authentic Māori cultural tourism experience. For a Māori tourism enterprise therefore, sustainability requires them to consider a wider range of cultural, social and spiritual factors rather than running their business with an economic priority.

Kaumātua

As discussed by Puriri and McIntosh (2013), Kaumātua[2] are elders of an Iwi, Hapu and Whānau or Māori family unit. Kaumātua can guide the development of concepts for the cultural tourism design and make sure that tourism is experienced in a culturally authentic way and is grounded by the Tikanga or cultural principles of Māori. This is an important aspect of cultural sustainability.

According to Statistics New Zealand (2014, p. 11) Kaumātua are the matriarchal and patriarchal leaders of a Māori Whānau, respected by their Whānau as persons of experience who care for the interests of the whole Whānau (family). These interests cannot be brought and are not for sale, neither can they be compromised. Members of the Whānau trust their Kaumātua as being steeped in their cultural knowledge and always willing to share their years of experience and Mātauranga (intrinsic knowledge) with their Whānau. They provide significant leadership and mentoring for adherence to Tikanga Māori or traditional Māori protocols and for the passing on of genealogical knowledge.

In the case study of the Maui Tourism enterprise, the Kaumātua were also the parents and grandparents of four generations of Māori as illustrated in Figure 14.4. As elders of the Maui Tourism whānau enterprise, the Kaumātua guided the potential opportunities for the use of their lands and often would be an invisible hand by assigning roles to certain members of the Whānau tourism enterprise whom they believed had potential to lead in a particular capacity. This involved assigning the management of logistics, food preparation, sales, marketing, hosting, guiding tours and maintenance.

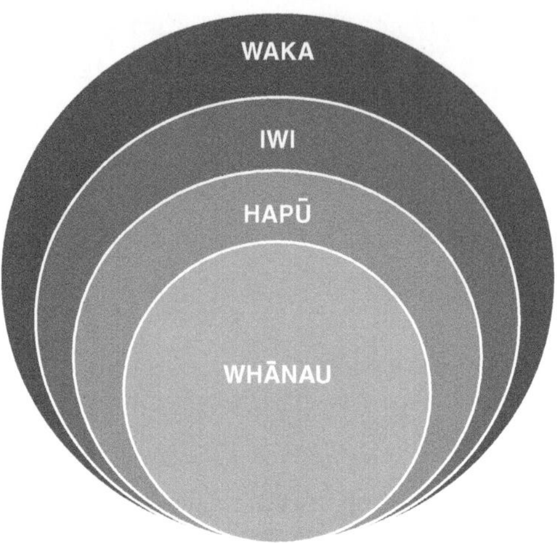

Figure 14.4   Māori social construct.

Figure 14.5   Tā Moko - cultural tattoos of Taiamai workers.

Source: A. Puriri, author.

For Taiamai Tours Heritage Journeys, the Kaumātua of the Hapu of Taiamai were approached by the tourism enterprise founder who requested permission to use the ancestral name of Taiamai as their company name. The Kaumātua gave their consent for the enterprise to use their tribal name. Hone Mihaka, Taiamai Tours Heritage Journeys founder, described this as a responsibility of high significance that he believed the enterprise needed to ensure that it carried the Mana (intrinsic cultural values) of Taiamai, thus it had the highest possible cultural esteem. This included respecting the Mana and the privilege of having Taiamai as the name of their business, which meant representing and putting their best foot forward at every opportunity as a conscious effort to uphold the name of their ancestor, Taiamai. This included wearing traditional costumes and authentic costumes. Male staff members elected to have Tā Moko or Māori tattoos on their skin with traditional Māori patterns as a commitment to honouring their ancestor Taiamai o te Whare Tapu o Ngāpuhi Nui Tonu (see Figure 14.5).

Kaumātua for these Māori tourism businesses demonstrated their wisdom, capacity and responsibility to ensure the businesses are culturally grounded with authenticity. Kaumātua (elders) provided essential cultural mentoring and prioritised the responsibility of Manaakitanga (caring for others in a consummate way). Kaumātua of the tourism businesses described their position as being one of responsibility to ensuring that their up-and-coming generations of their Whānau are educated and respect the cultural values that underpin their Māori tourism enterprise.

Upholding the wisdom of their Kaumātua, the founders of Taiamai Tours strive to ensure they follow the advice of their elders to 'Tiaki Te Taonga', or, to look after the rare and precious resources. This drives the enterprise's commitment to both cultural, social and environmental sustainability, honouring intergenerational responsibility so that the world they live in will be kept the same way they inherited it from their ancestors.

Conclusion

Māori strive to maintain their culture and language when presenting their distinctive Māori tourism experience. Māori according to Te Awekotuku (1981) have had a long history in Aotearoa New Zealand tourism and Māori enterprises form an important part of the country's contemporary tourism offerings. Māori tourism occurs within a complex cultural philosophy and Tikanga, guided by Māori values that follow ancestral lore. Drawing upon two case studies Māori tourism enterprises, Taiamai Tours Heritage Journeys and Maui Tourism, this chapter has highlighted the crucial nature of cultural underpinnings and processes, guided by Kaumātua (Māori elders), that shape and serve as the essence of an authentic Māori enterprise. These cultural underpinnings drive cultural authenticity and sustainability from a Māori perspective and demonstrate the Māori entrepreneurs' commitment to their Kaupapa and responsibility placed upon them by their genealogy. Embracing these cultural perspectives through a tourism enterprise provides an opportunity for tourists to gain authentic insights into Māori customs and protocols, and importantly, enable cultural empowerment and sustainability for Māori as the enterprise is managed according to the traditions and values of the Whānau. Above all, these perspectives enable the stewardship and guardianship of treasured Māori resources (Tikanga).

The two case study enterprises, both in development and in operation, are guided by the vision and stewardship placed upon the Māori entrepreneurs by their Kuia and Kaumātua (Māori elders) and Māori Tikanga. Understanding how authentic cultural perspectives shape Māori enterprises is not only important for understanding the unique cultural manifestations they may share with tourists as part of a cultural experience but is also significant for how the enterprises can instil Mana (self-empowerment) for current and future generations of Māori. In these two case studies, the importance of Whānau (family) and Kaumātua is integral to this objective. Māori Whakapapa (genealogical links), guided by Tikanga, and managed by Kaumātua, ensures cultural beliefs are followed to protect their treasured taonga and care for their visitors/tourists through tourism enterprises. These underpinnings also ensure a sustainable and intergenerational approach is at the forefront of Māori tourism enterprises and ensure the vision and perspectives of outsiders do not instead drive the enterprise opportunity.

Given the distinctive nature of authentic Māori tourism enterprises, and Māori approaches to sustainability in a tourism context, there is a need for future Māori tourism researchers to further examine Māori tourism entrepreneurship to enhance understandinging of Māori perspectives of tourism enterprise and commitment to sustainability. As tourism in Aotearoa New Zealand increasingly seeks to embrace alternative approaches to achieving sustainable tourism, such as regenerative and wellbeing models, an understanding of authentic Māori approaches are vital to achieving sustainable tourism that benefits the communities on which it relies, helping shape meaningful partnerships, and inspiring future Māori entrepreneurs to continue their important cultural stewardship for future generations to come.

Notes

1 See Puriri (2017) p. 80.
2 https://www.tandfonline.com/doi/full/10.1080/03036758.2019.1656260

References

Barlow, C. (1994). *Tikanga whakaaro: Key concepts in Maori culture* (Reprint, with corrections. ed.). Auckland, New Zealand: Oxford University Press.

Blair-Stahn, C. (2010). Ha: Breath of life [Theater review]. *The Contemporary Pacific, 22*(2), 492–494.

McIntosh, A. J. (2000). *Tourist experiences of Maori culture in Aotearoa, New Zealand.* Dunedin, New Zealand: Centre for Tourism University of Otago.

Puriri, A. R. (2017). *Maori Indigenous tourism development* (Doctoral). University of Waikato.

Puriri, A. R., & McIntosh, A. J. (2013). Indigenous tourism and heritage: A Māori case study. In B. Garrod & A. Fyall (Eds.), *Contemporary cases in heritage* (pp. 79–102). Oxford, England: Goodfellow.

Smith, L. T. (1999). *Decolonizing methodologies: Research and Indigenous peoples.* London, England: Zed Books.

Smith, T. (2005). *Whakapapa korero, tangata whenua and turangawaewae: A case study of the colonisation of Indigenous knowledge* (Unpublished doctoral thesis). University of Auckland, New Zealand.

Spiller, C. (2010). *How Māori cultural tourism buisnesses create authenticity and sustainable well-being* (Unpublished doctoral thesis). University of Auckland, New Zealand.

Statistics New Zealand. (2014). *He Arotahi Tatauranga.* Retrieved from www.stats.govt.nz

Te Awekotuku, N. (1981). *The sociocultural impact of tourism on the Te Arawa people of Rotorua, New Zealand* (Doctoral thesis). Univeristy of Waikato, New Zealand. Retrieved from http://hdl.handle.net/10289/7389

Zygadlo F, Simmons DR, McIntosh AJ, Matunga H, Fairweather JR. 2003. *Māori tourism: concepts, characteristics and definition* (no. 36). Christchurch: Tourism Recreation Research and Education Centre, Lincoln University.

15

INDIGENISING TOURISM BY INDIGENISING THE LANDSCAPE

A Digital Marketing Case Study of Wagiman Ethnobiology for Tourism and Conservation

Gabrielle McGinnis

Introduction

According to Turner et al. (2022, p. 1), the study of ethnobiology has traditionally focused on "the dynamic relationships among peoples, biota, and environments" and is a growing field of interest among Indigenous peoples worldwide. The field of ethnobiology encompasses Indigenous peoples' traditional ecological knowledge (TEK) and leadership in researching and conserving biocultural diversity while also sustaining local and global ecosystems. In Australia, TEK can also be known as Australian Indigenous biocultural knowledge (AIBK), which focuses on the cultural use and significance of local and national biota and landscapes (McGinnis, 2018; Pert et al., 2015). McGinnis (2018, p. 71) introduced the concept of DAIBK, or Digital AIBK, "whereby Aboriginal biocultural knowledge is shared and interpreted via digital platforms such as websites, apps and digital mapping systems i.e. Google Maps" as a way of conserving biocultural knowledge for a long term while also diversifying locally led Indigenous tourism marketing and products.

Before McGinnis (2018), there was little literature concerning the overlap of ethnobiology with the field of tourism. Ethnobiology, TEK, and AIBK have traditionally been studied in fields such as anthropology, natural sciences, and conservation science. In recent years, Indigenous-led ethnobiological research has become increasingly popular among scientists, policymakers, NGOs, and other organisations as a way to effectively conserve local biota for a long term while also empowering local, Indigenous communities. For example, World Wildlife Fund (WWF) Australia works with local Aboriginal ranger programmes across Australia to research, monitor, develop, and manage conservation projects, including WWF's "Regenerate Australia" campaign, as well as to empower Indigenous Australians in leading such projects. After the devastating 2019–2020 bushfires in Australia, WWF Australia sought the expertise of local Aboriginal rangers to look after and heal the country. In a web article on the WWF Australia website (2021, np), they quote Sonya Takau, a Jirrbal Traditional Owner: "You need us. You need our knowledge. We've been here for so long, and that knowledge has been passed down from generation to generation". In WWF Australia's *Regenerate Australia: A Roadmap to Recovery and Regeneration* (2020), they promote the leadership and integration of Indigenous methods and knowledge into conservational science research and practice while also providing resources and services to help empower local Indigenous rangers and communities, including

DOI: 10.4324/9781003230335-18

women and women rangers. WWF Australia acknowledges the importance of AIBK in conservation and employs the use of innovative technologies to help Aboriginal rangers look after the country, including drones, Geographic Information System (GIS) mapping, and other conservation tools (WWF Australia, 2020, 2021). The use of such technologies is becoming increasingly useful for Indigenous communities in Australia and around the world in both conservation and tourism research and development (McGinnis et al., 2020).

According to an article by Emma Ruben (2022), researchers from Charles Darwin University (CDU) in the Northern Territory (NT) of Australia are partnering with First Nations rangers in the NT to help monitor the health and population of freshwater turtles throughout the territory over the long-term. The CDU researchers are holding various workshops with Indigenous rangers throughout the NT to learn Indigenous methods and knowledge regarding identifying, trapping, and testing freshwater turtle populations. The workshops also include training exercises in using virtual reality technology as well as drones to help monitor and find turtles. Workshops and programmes like these are helping to Indigenise conservation practices and policies while also accelerating the effective, long-term conservation of various species (Ruben, 2022; Turner et al., 2022). However, the use of ethnobiology in tourism research has yet to be fully explored.

The Indigenisation of tourism research, policy, and practice is becoming prevalent in the tourism literature. The typological research of Nielsen and Wilson (2012) regarding "Indigenous tourism" or "Indigenous-driven tourism" can be conceptualised into four separate positions to help identify the level of Indigenous "presence, role and voice" in tourism research and development (p. 67). The four positions of Indigenous involvement are classified as follows:

1 **Invisible:** The first position of research is labelled as "invisible", whereby there is an outside focus on Indigenous people in research but a lack of Indigenous voice or participation in the said research (Nielsen & Wilson, 2012, p. 69).
2 **Identified:** The second position of research is classified as "identified" whereby there is a promotion of more Indigenous benefits from tourism, but there is still a lack of Indigenous involvement or leadership in the research process (Nielsen & Wilson, 2012, p. 69).
3 **Stakeholder:** The third position is classified as "stakeholder" whereby Indigenous people are still the focus and may have some increased involvement or participation, but their presence is still limited and objectified (Nielsen & Wilson, 2012, p. 69).
4 **Indigenous-Driven:** Finally, the fourth position is conceptualised as "Indigenous-driven" tourism research whereby Indigenous people are in control of their involvement in tourism research and development and are in charge of the benefits that may be derived from involvement in tourism (Nielsen & Wilson, 2012, p. 69), especially regarding the social and economic benefits such as capacity building.

Following Nielsen and Wilson (2012), Chambers and Buzinde's (2015) critical review of the tourism literature regarding the decolonisation of tourism research suggests that even though current research has done well to unveil the issues of postcolonial tourism research and development, there is a lack of Indigenous-driven research and development in current tourism discourse which they claim is still "predominantly colonial" (p. 1). However, more recent research by Holmes et al. (2016) in Canada follows the suggestions of Chambers and Buzinde (2015) and breaks the dominant "post-colonial" narrative in Indigenous tourism research and development by integrating and focusing on the voices and stories of the Denesoline people.

The Denesoline act as leaders in tourism research and development in their boreal forest and the protected areas within. Holmes et al. (2016) utilise community-based, participatory, narrative-based methods with the Denesoline people to help foster Denesoline empowerment in matters of Indigenous tourism practice and policy in the area. This Indigenous methodology-based study is revisited in the works of Grimwood et al. (2017) in an effort to further decolonise Indigenous tourism research planning and development in the lands of the Denesoline people of Canada. This study also follows the works of Chambers and Buzinde (2015) by attempting to put decolonisation methodology theory into practice. They find that decolonisation of research is possible through Indigenous-based research methodologies, i.e. narratives and participatory, community-based tourism research, planning, and development.

Tribe and Liburd (2016) also follow the works of Chambers and Buzinde (2015) by incorporating more Indigenous-knowledge-based systems and methodologies into tourism research and discourse for more sustainable tourism research and practices. Tribe and Liburd (2016) argue that the Chambers and Buzinde (2015) study regarding the dominant post-colonial voice in tourism research and development can be decolonised through Indigenous-knowledge-based systems in a sociological approach to tourism research and development.

Tribe and Liburd (2016, p. 51) take a participatory, interdisciplinary, multidisciplinary, and community-based systems approach in their discourse on "web 2.0" as do Holmes et al. (2016) and Grimwood et al. (2017) in their anthropological research and development projects in Canada. They argue for the incorporation and evolution of studies in anthropology and sociology, as do Nielsen and Wilson (2012), and into tourism discourse and practice in order to decolonise research and empower Indigenous peoples in tourism development. Nielsen and Wilson (2012), Chambers and Buzinde (2015), Holmes et al. (2016), Grimwood et al. (2017), and Tribe and Liburd (2016) all argue for Indigenous methodologies and knowledge-based systems for the decolonisation of research and the emancipation of Indigenous peoples in research and development. Tribe and Liburd (2016) also call for the incorporation of innovative, reflexive, participatory technologies in tourism discourse and development, which, in my study, are coupled with anthropological, sociological, and Indigenous methodologies and narratives of the Wagiman people of Pine Creek for community empowerment in tourism development. The concept of reflexivity in anthropological, Indigenous-driven tourism research and development is explored further below.

The autoethnographic research by Tomaselli et al. (2008) aligns well with Indigenous methodologies of research in that they use personal narratives, i.e. storytelling, to disseminate the knowledge learned from Indigenous colleagues. According to Tomaselli et al. (2008),

> Ethnography is not simply a collection of the exotic "other"; it is reflective of our own lives and cultural practices even when discussing another culture. Auto ethnography involves the use of cultural richness for self- reflection and understanding the nature of the encounter... Instead of only questioning why people react as they do to the presence of researchers, we must also question social assumptions about the nature of research.
>
> (p. 348)

This notion of questioning the "self" as well as the "other" is also discussed by the Maori researcher Tuhiwai-Smith (2012) in "decolonizing methodologies". Notably, she also addresses the social importance and significance of Indigenous research following the previously discussed literature.

Tuhiwai-Smith (2012) discusses how tourism research and development has been focused on "trading the other" as an exchange of goods and knowledge. She notes the tendency for tourism to "sell Indigenous knowledge and goods to other, outside cultures and peoples" (Tuhiwai-Smith, 2012, pp. 92–93), but states that it is important to give Indigenous knowledge back to communities, especially to the younger generations of Indigenous peoples (Tuhiwai-Smith, 2012). She also argues that taking a reflexive, subjective approach to research and recognising the "spiritual" aspect of Indigenous methodologies, especially when researching "country", can help decolonise research and Indigenise methodologies in tourism and other sectors and fields of study.

Following Tomaselli et al. (2008), taking on an Aboriginal lens Indigenises the researcher, as well as offering opportunities to change how they see the world, the people, the land, and even the project and research, thereby driving the researcher to bond with locals and give back to the community (in contrast to the extractive, disengaged research of the past criticised by Foley, 2014; Tobias et al., 2013; and Tuhiwai-Smith, 2012).

Tuhiwai-Smith's (2012) Indigenous research agenda focuses on the need to decolonise methodologies in order to politically and socially empower Indigenous peoples in tourism research as well as policy and development. She emphasises the importance of Indigenous methodologies in research and suggests that Indigenous and non-Indigenous researchers and tourism stakeholders need to work together to create a beneficial tourism product for all (Tuhiwai-Smith, 2012). She states that

> Sharing is a good thing to do…it is a human quality…. To create something new through that process of sharing is to recreate the old, to reconnect relationships and to recreate our humanness.
>
> (Tuhiwai-Smith, 2012, p. 110)

Maintaining the consistent skillsets and relationships required to support tourism ventures often presents major challenges for smaller Indigenous communities over the longer term (Botterill & Platenkamp, 2012; Foley, 2014; Jacobsen, 2017; McGinnis et al., 2020). Additionally, most knowledge keepers within Aboriginal communities are Elders, they are ageing and increasingly immobile, and therefore, many Aboriginal communities are at risk of losing knowledge, language, and heritage due to limited on-site visitation opportunities and various issues with accessibility to the country (McGinnis et al., 2020; McGinnis, 2018). Issues in mobility necessitate more mixed-methods approaches in on-site visitations of the country (Tobias et al., 2013). McGinnis (2018) and McGinnis et al. (2020) argue that creating off-site digital knowledge systems for the community and by the community can help address such issues through virtual interaction, engagement, and visitation of the country.

As discussed by Tobias et al. (2013), with regard to research, direct engagement with tourists may also be an issue as this option may not be welcome or possible within certain Indigenous communities, or among certain individuals of that community. Engagement with outsiders, in general, has become increasingly problematic due to the recent COVID-19 pandemic; therefore, the use of digital technologies during such times has become not only more relevant but more necessary (Carr, 2020). As previously discussed, digital options for cultural interpretation have recently begun to take shape through digital mapping of knowledge and skill-sharing to transmit Indigenous wisdom authentically and transparently, to monitor the impacts of the destination on the environment and people, and to empower the community and improve biocultural conservation (Brown & Weber, 2013; Hunter, 2014; McGinnis, 2018; McGinnis et al., 2020; Turner et al., 2022).

Case Study Background

This study took place in Pine Creek, NT, Australia. The township of Pine Creek is located in what is commonly known as the "Top End" of the NT of Australia. Pine Creek lies on the western side of the Stuart Highway, which runs from Darwin in the Top End down to Alice Springs in the "Red Centre" of Australia. Pine Creek is located at the junction of the Stuart and Kakadu Highways, roughly 230 km from the city of Darwin and 90 km from the town of Katherine. Geographically, Pine Creek is located close to some of the most famous and visited tourist destinations in Australia (61 km from Kakadu National Park, 68 km from Nitmiluk National Park, and 163 km from Litchfield National Park). Figure 15.1 illustrates the geographic location of Pine Creek in relation to these Top End tourist attractions and major cities of NT.

Pine Creek is a small town with a current population of roughly 330 people (Australian Bureau of Statistics [ABS], 2016). The town's population dropped significantly (by approximately 50%) in the late 1990s when the gold mines located in the hill country around the town closed down. Since then, reflecting global trends in the transition from traditional industry to service industries, Pine Creek has diversified its economy by focusing increasingly on tourism and urban developments. These developments include the regular maintenance of museums, historical buildings, interpretive signs, accommodation and caravan parks, and other attractions, such as the Water Gardens and a small reptile house located in the Lazy Lizard Tourist Information Centre. In 2017, the town's recreational centre and tennis courts were renovated and are open to the community as well as tourists visiting the town. Pine Creek also houses two petrol stations, which are important stopping points for tourists driving long distances between NT towns and cities (McGinnis, 2018; McGinnis et al., 2020).

The majority of tourism-based products and promotions in Pine Creek focus on the railway and mining heritage of the town. There is evidence of Aboriginal quarries around Pine Creek, as many of the local rocks and stones have been used for making tools, such as axes, for centuries by the Aboriginal groups of the area (Pine Creek Railway Resort, 2018a). According to the ABS, in 2016, 149 people in Pine Creek identified as Aboriginal, which accounts for approximately 45% of the Pine Creek population. Of these 149 Aboriginal people, most tend to identify as part of either the Jawoyn or Wagiman language groups (ABS, 2016; Pine Creek Railway Resort, 2018a). The Wagiman Elders consider Jawoyn land to be located on the eastern side of the Stuart Highway and old railway line while Wagiman country is considered to be located on the western side of the highway and railway. These areas are significant to the Wagiman people of Pine Creek and were mapped in my research. Despite the extensive history and population of Aboriginal people in the town and surrounding area, there is a significant gap in the tourism product of Pine Creek as Aboriginal histories are not central to the town's tourism heritage interpretation and the current tourism product.

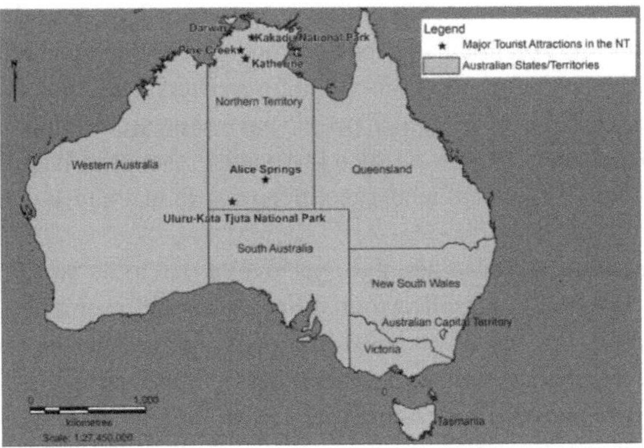

Figure 15.1   Map of Australia, locating Pine Creek and Top End tourist attractions.

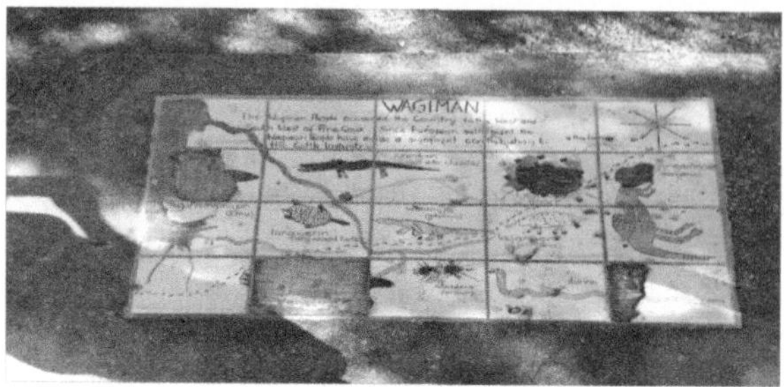

Figure 15.2 Wagiman mosaic map from the Walk Through Time project.
Source: G. McGinnis.

There is one Aboriginal tourism feature in the town, which was constructed as a community-based school project in the early 2000s, called the *"Walk Through Time"*, which focuses on the names of local plants and animals significant to the Aboriginal groups of the area. This collection of mosaics depicts the cultural and historical heritage of Pine Creek over the years and includes the non-Aboriginal, colonial heritage of the town, featuring prominent Pine Creek miners and pastoralists of the 1960s (Pine Creek Railway Resort, 2018b). The *Walk Through Time* feature is a static display that, unfortunately, also illustrates the lack of attention given to this single Wagiman interpretation of the country, as this feature is disintegrating and obviously poorly maintained (see Figure 15.2).

In 2016, the local council replaced the signs on the historical buildings with new versions; however, throughout the extended period of my research, nothing has yet been done to repair or renovate the *Walk Through Time* mosaics, leaving the one Aboriginal-focused tourism product needing revitalisation. Many of the Aboriginal members of the Pine Creek community have sought alternative means for engaging and developing tourism products that showcase their heritage, in their way, and in ways that can be sustainable for the long term.

The Aboriginal people mostly reside within two Aboriginal communities located in the Pine Creek area. One is situated on the north-eastern border of the main town, while another, Kybrook Farm, is located 8 km south-west of Pine Creek towards Umbrawarra Nature Park. Both communities have a fluctuating population of 25–60 occupants, who tend to move between Darwin, Katherine, and Pine Creek throughout the year.

The research described in this study emerges from ongoing research and engagement with the Wagiman community. Since the late 1980s, my principal supervisor, Mark Harvey, has been working with the Wagiman Elders on linguistic and anthropological research to help them pass on their language and knowledge to future generations and researchers. Based on the Australian Research Council (ARC) Linkage Grant, the Wagiman people have expressed an interest in developing tourism products as an opportunity to pass on such knowledge about the Wagiman culture, including language and knowledge of the local landscape, flora and fauna, and Indigenous histories of the railway and droving.

This study concerns the Wagiman heritage specifically due to its unique representation of the area and the lack of Wagiman-controlled tourism products and promotion opportunities, which are more available to the Jawoyn community who own and operate the successful tourist parks of Nitmiluk. A research proposal entailed conversing with the Wagiman people of Pine Creek to develop digital materials aimed at generating tourist visitation to Pine Creek while educating the public on the Wagiman peoples' unique and vibrant knowledge, values, landscape, and bioculture through virtual storytelling.

Regarding drivers in Indigenous tourism, my research seeks to provide options for community development and empowerment as, currently, the main means for economic income for the Indigenous communities of Pine Creek are typically provided through welfare or the Wagiman (Guwardagun) Ranger Program.

This study aims to investigate the Indigenisation of tourism research through ethnobiology as a way to not only further promote conservation of local biota, but also promote conservation of the local Wagiman culture, language, heritage, and history while employing digital technologies. The findings suggest that Indigenous-led tourism research coupled with Indigenous DAIBK may help empower local communities not only through education about local bioculture and conservation management by the local Wagiman rangers, but also through digital marketing, promotion, and diversification of the local tourism economy of Pine Creek.

Methods

Interviews with the Wagiman and non-Aboriginal residents of Pine Creek took two forms: semi-structured, recorded interviews and informal, *in situ* conversational interviews, which were documented in the form of note-taking with the permission of the interviewees. Most of the informal, *in situ* interviews, as well as audio-recorded, formal interviews, were used for the purpose of providing data for the study. All formal, video-recorded interviews were used to provide content for the websites and maps, as well as data for the study. Fieldwork for this study occurred during the dry seasons (May–September) of 2015–2017. Participant observation was also used to understand the lived experiences of the residents and tourists at Pine Creek, which were recorded through photos and note-taking following the ethnographic research literature.

From 2015 to 2017, 11 Wagiman people engaged in semi-structured interviews, informal interviews, and participant observations. Of these 11 interviewees, 6 were identified as Wagiman Elders. In Aboriginal culture, an Elder is a leader and well-established, older member of the community or language group who safeguards the country, knowledge, and heritage of the group. Interviews were recorded using audio recording and video camera devices.

After working with the Wagiman community during the three fieldwork periods in 2015, 2016, and 2017, a total of 165 video/audio recordings and 150 sites were mapped. The mapped sites were uploaded to the tourism and Women's and Children's Healthcare Australasia (WCHA) websites created as part of the research and project outcomes. These recordings and maps were uploaded to Vimeo, Google Maps, and websites for tourists and the Wagiman community to view for many years to come.

Additionally, following the works of Hunter (2014) in Canada as well as previous work carried out by Harvey with the Wagiman peoples (Harvey, 2016; Liddy et al., 2006), the Northern Land Council (NLC), and the Australian Aboriginal Protection Authority (AAPA), GIS data were also recorded with the Wagiman community. The "GPS tour" app was used to collect GIS data, while Google Maps/Google Earth Professional was used to analyse the GIS data and present it on the websites. These were created using the website development platform wix.com. Reasons why these particular platforms were selected stem from the results related to local use and knowledge of digital media, as presented by McGinnis et al. (2020).

During the fieldwork in Pine Creek in 2015, 2016, and 2017, a total of 265 tourist surveys were also collected and analysed. The paper-based surveys of tourists in Pine Creek consisted of 20 questions regarding tourist demographics; where they were coming from in the NT; who with; their reason for travelling; their transportation; their accommodation; the (technological) devices they used and social media platforms they engaged with to locate information when travelling in general as well as to Pine Creek; what sites they were visiting in the town and the surrounding area; and where they are travelling within the NT. The results from the surveys were analysed using Survey Monkey and Microsoft Excel.

While visiting Darwin and Katherine, paper-based, electronic and photographic GIS, ethnobiology and ethnographic data were collected from AAPA, Glenn Wightman, the Northern Territory Archives, the Darwin National Library, the Northern Territory Wildlife Park, and the Katherine Tourism Board to support this study. These field visits and data materials were provided with the express permission of the communities, who are the owners and knowledge keepers of the information, heritage, and significance within.

Findings

Findings from interviews with the Wagiman Elders indicate that many of the Wagiman people who live in rural areas would rather tell stories about the land and heritage to their kin than interact directly with tourists. This may be owing to interactions with tourists in the past not being as efficient or fruitful as they may have hoped. Some of the Elders have since developed signs for tourism interpretation and for managing their sacred sites – some of which are in very close proximity to tourist parks. Signage has not been very efficient as many tourists still tend to venture into the sacred sites, mainly by accident. An example can be seen in the discussion with Tammy[1], below, taken at the Douglas-Daly Hot Springs in July 2015:

Tammy: Yeah. That's why we all move them (tourists) over there *points to designated Tourist Park*. They make a big round billabong. You can see that water boiling up. It's really hot. And them tourists still go there. Tell them, don't go there, bring them here *points to Tourist Park*…there have cold water and warm water. But at night, they still go around. But sometimes you can't see them. You know people will go around the other way and they won't see the sign. They have that, they've put in that thing, mobile? With the walk, that thing go like that *makes beeping/stopping sign with hand*.

Many of the Wagiman Elders and younger Wagiman people that I spoke to expressed concern over tourists visiting sacred sites and wanted to make sure that tourists knew where they could and could not go, through the maps.

Me: do you have any views on tourists?
Penny: I'm not really into it, but um, I guess any type of tourism ventures would be a bonus to the people so, it'll showcase a lot more of the landscape and stuff like that, so I think it is great!
Me: Is there anything you wouldn't want tourists to see or do out here?
Penny: I guess as long as they know where they can and can't go. Um…I'm not sure if there are any sacred sites or anything like that…make sure they're not going where they shouldn't be. That would be my only thing to address.
Me: And what about putting them on the website? Showing them where there are sacred sites, would that be OK?
Penny: I probably wouldn't see a problem with it, but you probably need to speak to the Elders about it. Like Jimmy and Tammy and them. See how they feel about that.
Jimmy: We can show them birds, trees, country. Them cattle stations we've been doing, but sacred sites, we keep those just Wagiman, just us.

Tammy: Yeah, I wouldn't mind telling them, but we bit frightened. To lose my going, and to lose something. You know, tourists might say, 'Oh, it's not really', you know? Keep it closed. Can't tell anybody about it. Tourists, you know? They will make a big mess, you know.

The Elders and younger Wagiman people expressed a fear of people physically visiting sacred sites as well as hearing the stories, which mainly depended on the gender of the tourist, but also on the behaviour or reaction tourists may have to the sacredness of an area. This shows that the Wagiman people were more interested in safeguarding their social and cultural values rather than undertaking a possibly economically beneficial endeavour. This was especially true with regard to tourism at sacred sites. This notion of social benefits in tourism development adds to the discourse on the importance of the social-value-based approach in Indigenous tourism. We can also see that the Wagiman Elders were most enthusiastic and comfortable to share knowledge about local biota, e.g. birds, trees, and other plants and animals, with tourists in the area.

Importantly, discussions about plants and animals between me and the Wagiman Elders also allowed for stronger and better conversational flow in our interviews, thereby creating stronger bonds of trust and friendship between us. This is mainly due to the Elders teaching me "language" by teaching me about the local plants and animals, something they wish to also engage in with tourists. That type of engagement with outsiders could help not only educate tourists and safeguard local heritage, but also provide avenues for stronger bonds of trust and friendship between outsiders and the community.

In terms of interviews concerning flora and fauna, the following questions were initially asked:

Me: What do you call a "kangaroo" in Wagiman? Do you use that animal for anything?
Tammy: Kangaroo? Him go * hand gesture for hopping* hop hop hop.

When more local jargon and language are used when referring to animals and plants, greater conversational flow could be fostered. Importantly, during this initial meeting with the Wagiman Elders, they spoke the Wagiman language while discussing local flora and fauna, which encouraged me to learn Wagiman words and phrases. The use of Wagiman words and phrases led me to Indigenise my research questions and discussions from that point forward. For example,

Me: What you call 'em tree in language? Them bolomin? What you do with them tree?
Tammy: Him called wakkala. Use em for make em mattress. Wrap em up and keep em warm at night. Em good for sleeping.

Tammy and Jimmy reiterated the importance of teaching the young ones "language" so that they would know the stories and the country better as well. The Wagiman Elders, especially Tammy and Jimmy, wanted to teach me "language" as well and began to do so by teaching the names of the plants and animals that we would encounter in the bush. While the majority of the interviews were recorded in Kriol and Aboriginal English so that tourists and young Wagiman could understand the stories of the area, the Elders also thought it is important to record some stories, especially dreaming stories, in Wagiman as well, and to record ethnobiology studies of place names and plants and animals in Wagiman, which led to the digitisation of the *Wagiman Plant and Animal Book* online as part of the Wagiman Community Heritage Archive website: https://pcaaa7.wixsite. com/wagimanpinecreek/wagiman-plants-and-animals.

The bird page of the book is accessible to the public and is included on the tourism website as many tourists to Pine Creek are avid birdwatchers. Here are included not only the names for the biota in Wagiman, English, and their scientific names, but also photographs and interactive maps that link to citizen science data maps of various species in Australia provided by the Living Atlas of Australia. Providing a more interactive and innovative platform to educate young Wagiman residents as well as tourists can aid in conservation operations in NT among ranger programmes as well as Indigenise the citizen science data available online via the Living Atlas of Australia.

Currently, only the bird page of the Wagiman Plant and Animal Book is publicly accessible and linked to the tourism website, as there are many avid birdwatchers that come to visit the town and local area. The WCHA does state "Welcome to Wagiman country" and is designed almost identically to that of the tourism website. The Jawoyn people do not seem to mind this as the Wagiman Archive is specifically marketed towards Wagiman people without much explicit reference to Pine Creek as a destination.

More contextualised questions with the Wagiman about flora and fauna provided greater insights into the biodiversity and ecological and conservational health of the region. For example, when discussing goannas with Tammy, findings show that their numbers are dwindling due to the invasive cane toad species, which is now common in the area:

Tammy: Tell people that tree. That culture you know?

Me: So, you like to tell other people too? About the honey? And the fish?

Tammy: Yeah. I like to tell people about honey, sugar bag grass. Goanna too. They would say 'what that name?' for goanna and we say 'walanjya'

Me: *Walanga?*

Tammy: wal-an-ja

Me: Walanjya.

Tammy: yeah. And them blue tongue too. They'd say, 'call him now' I reckon, and I say 'gungarak' Blue tongue. But we don't see them blue tongue. Or that goanna. Because of that frog. That wortngong frog now…Yeah all been killed. No goanna. Goanna left now with blue tongue.

Me: Oh no! They left because of the frogs?

Tammy: Yeah. They have that big mob here. In the ground underneath. When rain come, they all come out. Them big one too.

By applying local bioculture and language in the research process and questions, greater conversational flow was inspired with the Wagiman participants, who led the Indigenisation of the interviews by incorporating more language and teaching me about the Wagiman names and significance of plants and animals in the area.

When asked about cattle stations and working on the homesteads, the Wagiman people also led this process and inspired more conversational flow as this work was strongly tied to family history and personal narratives about the cattle stations and activities there. For example,

Me: Can you tell me a little bit about this place here Jimmy?

Jimmy: Well, his father used to manage this old house, this homestead here…And we used to have camp there…used to camp every year. Knock all them tree down now. That road wasn't here. The road, a bit further up, a long way from the house. We used to work here. That old yard there, look, when I was a kid. Far in, you see that there?

Me: Yeah, way down there.

Jimmy:	It used to be an old yard there. We used to work…Bring all the cattle from the yard… Where pig yard there, over there. And goat yard up here…We had a big place, clean place. Not all this weed and growth now. We had with…Big shed in that - Big turtle shed. We used to have several to 11 of them. We used to work here…We used to pump water from the billabong…To the tank. And we had a garden that would've been right here. Right here… Cabbage, melon, pumpkin…And sweet potato…Used to grow 'em right here.

Notably, the above conversation with Jimmy occurred on-site on Claravale Road where Jimmy walked the old homestead as he relayed information about what the place used to look like, illustrating his picture-perfect memory of the area. The conversation with Debbie occurred off-site at her house in Pine Creek. For example,

Me:	Was that at Claravale Station or?
Debbie:	Well when mom and dad was together, yeah, we used to stay at Claravale. Go down Jinbat and Jabiru area. That's where my dad … where… when he was a kid.

While she was willing to talk about this familiar subject and its relevance to her family history, she did not provide as much detail as Jimmy had on-site at Claravale. This may be due to various factors:

1 that Jimmy had more knowledge of the country and was more comfortable being on video;
2 that Debbie was off-site during the conversation above and was usually shy during videotaped conversations; and
3 that the conversation with Jimmy was videotaped for the website, while the conversation with Debbie was audio-recorded for the written research.

If she had been on-site, perhaps she would have provided a more descriptive narrative as she did when visiting and videotaping at Umbrawarra Gorge about the local flora, which she did mostly in the Wagiman language.

However, even in this case, she was shy about being videotaped and turned the conversation over to Jimmy shortly after filming. For example,

Debbie:	*laughs from shyness*… the mango flower come. So yeah, especially during the wet season.
Me:	This is the green plum yeah?
Debbie:	Yeah, green plum. Man, this is pretty good plum here. Is good Vitamin C. And it's good for eating and it's healthy. You can do it (to Jimmy)
Jimmy:	this one here? It is real strong you know? You can make firewood too, break them leg and…

Jimmy continued to lead the conversation above at Umbrawarra Gorge while Debbie listened nearby. Debbie was more active in the conversations on-site than off-site but was still shy in front of the video camera and tended to show more confidence during audio-recorded interviews off-site and on-site.

This demonstrates the context-specific nature of fieldwork and research within Aboriginal communities and the importance of not using the findings as a one-size-fits-all solution for all individuals or communities, as argued by Tobias et al. (2013). This conversation also demonstrates the need for further education of the Wagiman youth in local ethnobiology and why the Elders are enthusiastic to create an archive for the young people to access as they get older and traditional knowledge becomes more endangered.

However, when audio-recorded discussions or participant observations were made on-site at places such as the Peanut Farm along Claravale Road, Debbie was more active in the conversation than she was off-site at her house. For example,

Debbie: What they been growing here? Them fruit?
Jimmy: Yeah, they been growing everything. Peanut, chili, pumpkin, all that.
Debbie: Oh, that's real good eh.

Notably, Debbie led this discussion about the Peanut Farm by asking the Elders questions as I had been doing throughout the road trip in the area. This shows that, even with Debbie in the conversations above, the on-site interviews were shown to stimulate conversation more than off-site interviews, especially when talking about a certain place of significance.

When interviewing the younger Wagiman people of Pine Creek, I found that many expressed interest in becoming a Wagiman ranger. A couple of younger members of the Wagiman community, Jessica (age 19) and Penny (age 34), also expressed interest in website design, content, and management. They especially expressed interest in the content and management of the "Wagiman Website", which we designed with the Wagiman Elders in tandem with the "Tourism Website". Discussions with the Wagiman rangers also showed that they were keen on developing and using an online archive of local plants and animals, with interactive maps so they could look after the country and train new rangers more efficiently.

Vanessa: A map and website of our plants and animals would be good. Could teach them young ones more about country and how to be a ranger, like us. Might even get them more interested in becoming a ranger if they see this.

Concerning the tourist surveys, interestingly, respondents in the 35–44 age group indicated more interest in Aboriginal culture, as well as natural attractions including birdwatching and the water gardens. This is an interesting finding because, according to Ryan and Huyton (2000), this group aligns with the "active information seekers" in their study who frequent natural attractions in the Top End and indicate interest in authentic Aboriginal tourism experiences. Further, cultural tourism researchers such as Timothy (2011) find that older, more educated tourists in this age category, as well as in the Grey Nomad category, are interested in cultural tourism including Indigenous tourism experiences that are considered authentic. These findings align with Ryan and Huyton (2000) in that few Grey Nomads were categorised as "active information seekers" as they did not indicate much interest in Aboriginal culture nor natural tourist experiences such as birdwatching or the water gardens. Many of the Grey Nomads may not have indicated an interest in Aboriginal culture in Pine Creek owing to the lack of Aboriginal cultural interpretation, tourism infrastructure, or promotion in Pine Creek, aside from the *"Walk Through Time"* feature.

Thus, if Aboriginal cultural experiences were digitally interpreted in Pine Creek's current tourism infrastructure, i.e. museums, then both markets could be targeted and diversified simultaneously – the Grey Nomads may become more interested in technology and Aboriginal culture owing to their interest in frequenting the museums of Pine Creek, and active information seekers could gain an authentic Aboriginal experience by visiting the established museums in the town, as well as the tourism website, which includes a page for birdwatchers that also highlights the Aboriginal names and knowledge of the birds in the area (Liddy et al., 2006; Wagiman Communities of Pine Creek and Kybrook, 2017).

Tourists noted the main tourism offerings, ventures, and attractions in the town, as well as the town's condition and how it is viewed by tourists, such as "Lazy Lizard"; "gold"; "Pine Creek"; "Pussycat Flats"; "town"; "spot"; "hooded parrots"; "place"; "caravan park"; "signs"; "history"; "little"; "accommodation"; and "railway museum", and also the surrounding attractions that brought them to Pine Creek: "stop"; "travelling"; "Litchfield"; "Kakadu"; "Darwin"; and "touring NT". Importantly, tourists hardly mentioned anything about "Aboriginal"-, "Indigenous"-, or "Wagiman"-related content in their comments, even when discussing travelling to surrounding areas, such as Kakadu, suggesting that their lack of mention of Aboriginal-related content was owing to the lack of Aboriginal interpretation in the town.

While the tourist survey yielded results about what tourists are looking for when visiting Pine Creek, more information was needed from the Wagiman and non-Aboriginal residents about how they would like to market their heritage of the area to tourists. Interviews with the members of the Wagiman community in Pine Creek indicate that, overwhelmingly, the development of AIBK and the integration of this knowledge digitally (DAIBK) into the existing tourism infrastructure is a viable option for engaging with tourists. This alternative would allow greater opportunities to benefit the Wagiman community as well as the non-Aboriginal community who manage the tourism industry in Pine Creek. They provided feedback on how they would like to market their heritage, teachings, and websites to tourists in Pine Creek and beyond, enhancing their existing stand-alone entrepreneurial ventures by integrating them into the existing tourism infrastructures of the town, following Whitford and Ruhanen (2014) and Botterill and Platenkamp (2012).

The interview participants of the Wagiman community were enthusiastic about more DAIBK integration as it could allow them to be involved in tourism in a way that benefits them, on their terms, through the development and management of their AIBK tourism product and promotion. By integrating Wagiman knowledge into the tourism developments of the town virtually, their knowledge is not only conserved and passed on for generations to come, but also makes the Wagiman people more active members of the tourism community and can provide further options and developments for them to become more active in tourism in the town in the future. For example, the online Wagiman plant and animal book could also be sold as hard copies in the town for the birdwatchers who visit. The Elders can talk with the council and the Pine Creek Aboriginal Advancement Association (PCAAA) about doing bushwalks through the town's Water Gardens and get paid as bush guides there, as Tammy proposed in 2017. They could also possibly do walks and talks at Umbrawarra where Jimmy expressed interest in 2017 in giving tours. Tammy also expressed interest in 2017 in giving talks and tours at Douglas Daly Hot Springs, if she had a car to get there. Debbie also expressed interest in building a cultural centre in the town in 2017 for tourists to buy bracelets, soaps, and other souvenirs while learning more about the country from Wagiman Elders.

Conclusion

As a result of conversations with the Wagiman Elders and their families, a tourism website and a Wagiman Community Archive website were developed to showcase the underrepresented

Wagiman culture and knowledge of the town and surrounding area while also conserving Wagiman knowledge online. While the community employed me as an "outside expert" in technological, Indigenous tourism developments, the control of the study and developments remains in the charge of the community. The community works together on maintaining the websites and is making further developments and changes to the websites.

By integrating more DAIBK into the Wagiman community archive, the Indigenous and non-Indigenous residents of Pine Creek were able to diversify their tourism products and promotion in a socio-culturally inclusive manner, while the Wagiman residents were able to fully share and conserve their knowledge and heritage digitally for many generations to come. They also helped create a tool that could aid local rangers in monitoring, identifying, and managing local species and their conservation for the long term.

This research found that Wagiman and non-Aboriginal co-management can be developed through the sustainable application, promotion, and conservation of AIBK and vital Indigenous values and heritage. Benefits of DAIBK include the creation of opportunities for longer-term heritage conservation; ongoing knowledge-sharing; and horizontally and/or vertically integrating that knowledge into local, established tourism infrastructures and management systems. These benefits could help create more options for Indigenous employment, diversified non-Indigenous and Indigenous products, as well as creating authentic Indigenous experiences for "active" and even "inactive" information seekers and other tourists.

With consistent monitoring and adjustment of digital products and their promotion, further diversification of the products can take place and can make sure that the tourism content remains innovative while also conserving the knowledge, values, and natural heritage of the Indigenous community. This could be possible through continuous Indigenous control, engagement, and involvement in the development, promotion, and management processes of tourism, which could help foster community empowerment and social, as well as economic, vitality for all in the years to come.

Acknowledgements

This research was led by the Wagiman Elders and community as well as the wider Pine Creek community under the supervision of Dr. Mark Harvey from the University of Newcastle, Dr. Tamara Young from the University of Newcastle, and Dr. Ian D. Clark from Federation University. This research was supported by the Pine Creek Aboriginal Advancement Association (PCAAA), in Kybrook Farm, the Wagiman (Guwardagun) Ranger Program of the Northern Land Council (NLC), as well as the Australian Aboriginal Protection Authority (AAPA) and Dr. Glenn Wightman of the Herbarium in Darwin.

Note

1 Pseudonyms have been used to protect the identity and anonymity of the participants.

References

Australian Bureau of Statistics (ABS). (2016). *2016 Census QuickStats: Pine Creek.* Retrieved 21 April 2022 from http://www.censusdata.abs.gov.au/census_services/getproduct/census/2016/quickstat/ILOC70 400106?opendocument

Botterill, D., & Platenkamp, V. (2012). *Key concepts in tourism research.* London: SAGE Publications.

Brown, G., & Weber, D. (2013). Using public participation GIS (PPGIS) on the Geoweb to monitor tourism development preferences. *Journal of Sustainable Tourism, 21*(2), 192–211.

Carr, A. (2020). COVID-19, indigenous peoples and tourism: A view from New Zealand. *Tourism Geographies, 22*(3), 491–502.

Chambers, D., & Buzinde, C. (2015). Tourism and decolonisation: Locating research and self. *Annals of Tourism Research, 51*(1), 1–16.

Foley, D. (2014). Too many didgeridoos. *Journal of Australian Indigenous Issues, 17*(2), 56–71.

Grimwood, B. S. R., King, L., Holmes, A., & the Lutsel K'e Dene First Nation. (2017). Decolonizing tourism mobilities? Planning research within a First Nations community in Northern Canada. In J. Rickly, K. Hannam, & M. Mostafanezhad (Eds.), *Tourism and leisure mobilities: Politics, work, and play* (232–247). New York, NY: Routledge.

Harvey, M. (2016). Stones and grinding: Wagiman ethnogeology. *Australian Aboriginal Studies, 1*, 12–23.

Holmes, A., Grimwood, B. S. R., King, L., & the Lutsel K'e Dene First Nation. (2016). Creating an indigenized visitor code of ethics: The development of Denesoline self-determination for sustainable tourism. *Journal of Sustainable Tourism, 24*(8–9), 1177–1193.

Hunter, J. (2014). Oral history goes digital as Google helps map ancestral lands. *The Globe and Mail.* Retrieved 21 April 2022 from https://www.theglobeandmail.com/news/british-columbia/oral-history-goes-digital-as-google-helps-map-ancestral-lands/article19566302/

Jacobsen, D. (2017). Tourism enterprises beyond the margins: The relational practices of Aboriginal and Torres Strait Islander SMEs in remote Australia. *Tourism Planning & Development, 14*(1), 31–49. https://doi.org/10.1080/21568316.2016.1152290

Liddy, L. G., Martin, L. D., Huddlestone, J. G., Liddy, L. J., Liddy, H. I., McMah, C. G., & Wightman, G. (2006). *Wagiman plants and animals: Aboriginal flora and fauna knowledge from the mid Daly River basin, Northern Australia.* In Department of Infrastructure (Ed.), *Northern Territory Botanical Bulletin* (30). Darwin and Diwurruwurru Jaru, Katherine.

McGinnis, G. (2018). We speak for country: Indigenous tourism development options for community engagement in Australia. PhD thesis, Department of Anthropology and Sociology, *University of Newcastle Australia: NOVA Theses.* 1–348.

McGinnis, G., Harvey, M., & Young, T. (2020). Indigenous knowledge sharing in Northern Australia: Engaging digital technology for cultural interpretation. *Tourism Planning & Development, 17*(1), 96–125, https://doi.org/10.1080/21568316.2019.1704855

Nielsen, N., & Wilson, E. (2012). From invisible to Indigenous-driven: A critical typology of research in Indigenous-tourism. *Journal of Hospitality and Tourism Management, 19*(1–9), e5.

Pert, P. L., Ens, E. J., Locke, J., Clarke, P. A., Packer, J. M., & Turpin, G. (2015). An online spatial database of Australian Indigenous biocultural knowledge for contemporary natural and cultural resource management. *Science of Total Environment, 534*, 110–121.

Pine Creek Railway Resort. (2018a). Our history. Retrieved 21 April 2022 from http://www.pinecreekrailwayresort.com.au/about-us/our-history

Pine Creek Railway Resort. (2018b). Things to see and do. Retrieved 21 April 2022 from http://www.pinecreekrailwayresort.com.au/what-to-see-do/events

Ruben, E. (2022). First Nations knowledge to aid turtle conservation in NT. *National Indigenous Times.* 15 April 2022. Retrieved 21 April 2022 from https://nit-com-au.cdn.ampproject.org/c/s/nit.com.au/first-nations-knowledge-to-aid-turtle-conversation-in-nt/?amp

Ryan, C., & Huyton, J. (2000). Who is interested in Aboriginal tourism in the Northern Territory, Australia? A cluster analysis. *Journal of Sustainable Tourism, 8*(1), 53–88.

Timothy, D. J. (2011). *Cultural heritage and tourism.* Bristol: Channel View Publications.

Tobias, J. K., Richmond, C. A. M., & Luginaah, I. (2013). Community-based participatory research (CBPR) with Indigenous communities: Producing respectful and reciprocal research. *Journal of Empirical Research on Human Research Ethics, 8*(2), 129–140. https://doi.org/10.1525/jer.2013.8.2.129

Tomaselli, K. G., Dyll, L., & Francis, M. (2008). "Self" and "other". In N. K. Denzin, Y. S. Lincoln, & L. T. Smith (Eds.), *Handbook of critical and indigenous methodologies* (347–372). Los Angeles, CA: SAGE Publications.

Tribe, J., & Liburd, J. J. (2016). The tourism knowledge system. *Annals of Tourism Research, 57*, 44–61.

Tuhiwai-Smith, L. (2012). *Decolonizing methodologies: Research and Indigenous peoples* (2nd ed.). London: Zed Books Ltd.

Turner, N. J., Cuerrier, A., & Joseph, L. (2022). Well grounded: Indigenous Peoples' knowledge, ethnobiology and sustainability. *People and Nature, 4*(3), 1–25. https://doi.org/10.1002/pan3.10321

Wagiman Communities of Pine Creek and Kybrook. (2017). Wagiman community heritage archive. Retrieved 21 April 2022 from www.pcaaa7.wixsite.com/wagimanpinecreek.

Whitford, M., & Ruhanen, L. (2014). Indigenous tourism businesses: An exploratory study of business owners' perceptions of drivers and inhibitors. *Tourism Recreation Research, 39*(2), 149–168. https://doi.org/10.1080/02508281.2014.11081764

World Wildlife Fund (WWF) Australia. (2020). *Regenerate Australia: A Roadmap to Recovery & Regeneration*. WWF Australia. October 2020. 1–13. Retrieved 21 April 2022 from https://apo.org.au/sites/default/files/resource-files/2020-10/apo-nid308991.pdfWorld Wildlife Fund (WWF) Australia. (2021). Why Australian nature needs 'Our Knowledge.' *WWF Australia Blogs*. 26 September 2021. Retrieved 21 April 2022 from https://www.wwf.org.au/news/blogs/australian-nature-needs-indigenous-knowledge

16

CONVERSATIONS ABOUT CULTURE

The Need to Integrate Indigenous Voices into the Development of Sustainable Cultural Heritage Tourism Opportunities in the Pacific

Anne Ford, Anna Carr, Nyssa Mildwaters, Dionne Fonoti,
Gregory Jackmond and Glenn Summerhayes

Introduction

Tourism has been identified as an important industry for Pacific nations (Harrison & Prasad 2013), with entities such as the UN World Tourism Organisation (UNWTO 2017), World Bank (2017) and the Asian Development Bank (Everett et al. 2018) focusing upon the need for these smaller countries to develop sustainable tourism that draws upon their rich natural and cultural heritage (Scheyvens 1999; Rayel 2012). In the Pacific, tourism is often positively portrayed as having multiple benefits for developing nations, including the stimulation of economic growth; the creation of employment opportunities and local enterprises; the promotion of infrastructure development; the provision of tax revenue to governments; the import of foreign currency and potentially foreign investment; and finally, the potential spread of economic opportunities into rural areas and local communities (Cole 2006; Scheyvens 2015; Telfer & Sharpley 2016).

However, Miller and Twining-Ward (2005) note that, in the Pacific, there are also risks to developing tourism, including high economic leakage due to the need to import goods and services; the increased need for foreign labour; environmental issues when developing areas of limited resources such as water and land; and the commercialisation of culture. Other issues include high rates of foreign ownership of resorts (Harrison & Prasad 2013). Because of this, a key area of interest for Indigenous peoples of Pacific nations is increasing local community involvement and emphasising the need for the development of sustainable tourism practices (Gabriel et al. 2017; Harrison 2003; Kau 2014; N'Drower 2014; Scheyvens 2004; Tauaa 2010; Taumoepeau & Addison 2016; Zeppel 2014).

Community Based Tourism (CBT) can be defined as "alternative forms of tourism development which maximize local benefits and advocate capacity building and empowerment as means of achieving community development objectives" (Tolkach & King 2015: 388–389). There are various forms of CBT already in practice within the Pacific, which usually involve staying in local communities in traditional forms of accommodation. For example, Samoa has a long history of

205   DOI: 10.4324/9781003230335-19

DOI: 10.4324/9781003230335-19

CBT through the development of beach *fale* tourism (Scheyvens 2004). *Fale* operations are usually owned by local Indigenous families, and they are considered to be economically attractive as they bring in money, diversify livelihoods and create employment. Tourists staying in *fale* are paying accommodation fees directly to the local community, and there is flow on effects as guides and workers are also usually hired locally. There is also little economic leakage as tourists tend to have no expectations of foreign or imported goods, so *fale* owners buy/hire goods and services locally, such as fresh fish and vegetables from other members of the community.

While generally considered as positive, in practice CBT can be complex to implement and the comparatively few studies that critique CBT have noted that it has met with varying degrees of success in achieving sustainable tourism objectives for communities (Cole 2006; Muhanna 2007; Sakata & Prideaux 2014; Salazar 2012; Telfer & Sharpley 2016). For example, many CBT projects have been ecotourism based, where a large conservation organisation has implemented a tourism project as part of an effort to provide an economic value to a conservation area, thus encouraging communities to protect these areas (Manyara & Jones 2007). The difficulty with CBT projects that are developed by external organisations using this top-down approach is that they can face issues of community buy-in and sustainability once the initial funding for the establishment of the project has been exhausted (Salazar 2012). This is a common problem with any developmental project that is externally rather than internally controlled (Benson 2012). It is essential that local Indigenous communities are included in all stages of planning and development (Reggers et al. 2016) and that this planning is organised in line with the values and needs of the local community (Taumoepeau & Addison 2016). Other issues identified have included creating a dependence on tourism, shifting communities away from traditional livelihoods such as subsistence agricultural economies, and making them vulnerable to wider economic and social forces, while at the same time creating social and cultural change as communities are now reliant on case economies rather than traditional exchange systems (Reggers et al. 2016). This demonstrates that CBT projects need to be developed in a way that explores sustainability from different angles, including cultural, environmental, and financial/ economic contexts.

One of the potential growth areas for CBT in the Pacific is cultural heritage tourism that is led by or with Indigenous communities. Although globally cultural heritage tourism is worth 40% of the total tourism market, this is an undeveloped area in the Pacific, which is generally marketed as a tropical beach destination (SPTO 2014). Although cultural heritage tourism is becoming advocated for in national policies (PNGTPA 2006; STA 2014), there appears to be resourcing to grow local community awareness of the potential for this type of tourism to be developed at a CBT level. The research in this chapter explores this issue by asking the following questions: (1) What level of awareness do local communities have in the Pacific regarding cultural heritage tourism; (2) would cultural heritage tourism be considered a possible benefit for sustainable local community development; and (3) if this was the case, what would be the opportunities and constraints that local communities perceive in the development of locally-based cultural heritage tourism?

This chapter will explore these questions through two case studies drawn from the South Pacific: Madang in Papua New Guinea (PNG) and Samoa. This will include describing the case studies in terms of their archaeological and cultural backgrounds to understand the range of Indigenous or cultural heritage tourism opportunities available, before outlining the methods used for the current research and finally presenting the results of the fieldwork. However, prior to introducing the case studies, it is essential to explain what cultural heritage tourism is, as well as the benefits/difficulties that it may present local Indigenous communities in terms of sustainable development.

Cultural Heritage Tourism in PNG and Samoa

Cultural heritage is a legacy of the past to the people of today. It is the material and intangible attributes that a group inherits from previous generations that are maintained in the present and for the future (Timothy 2011). Material or tangible cultural heritage includes archaeological sites, monuments, museums, settlements, or landscapes/places that have historical and cultural significance (ICOMOS Charter 2008). These are places that can be physically visited and experienced. In contrast, intangible heritage includes the knowledge and practice of culture, including oral traditions and histories, performing arts, rituals, festivals, and knowledge about traditional craftsmanship (UNESCO 2003).

Cultural heritage tourism can incorporate both tangible and intangible aspects. The South Pacific Tourist Organisation (SPTO 2014: 7) defines cultural heritage tourism as

> all tourist trips that include cultural activities, such as visiting monuments, sites, and museums, as well as experiences and interaction with local communities, such as attending festivals, local cultural events and visiting markets. It involves travelling to experience the places and activities that authentically represent the stories and people of the past and present.

Although globally cultural heritage tourism is worth 40% of the total tourism market (UNWTO 2017), tourism operators tend to market the Pacific as a tropical beach destination, with a lack of understanding of the diversity and richness of Pacific cultural heritage (SPTO 2014). Papua New Guinea is an exception in this regard; tourist operators recognise the diversity and authenticity of cultures present (SPTO 2014; World Bank 2017), however, a lack of developed cultural heritage tours or destinations, plus concerns regarding security and tourist infrastructure, hinder current tourist participation (Imbal 2010).

The PNG Tourism Master Plan 2007–2017 (PNGTPA 2006) acknowledges the importance of niche market tourism to PNG, in particular natural, Indigenous and cultural heritage tourism. The PNG Tourism Promotion Authority (PNGTPA) considers cultural heritage tourism to include culture and village-based tourism, cultural shows/ singsings and WWII history. The Samoa Tourism Sector Plan 2014–2019 (STA 2014) also recognises the importance of cultural heritage tourism, with a goal to increase community management and participation in commercial enterprises focused on attraction sites, including cultural sites. The Plan notes "that well-managed attractions can provide sustainable income for local communities and other accommodation products can benefit from the development of linkages to cultural and environmental tour products and community-based tourism opportunities" (STA 2014: 35).

One of the reasons that cultural heritage tourism is recognised as being significant for Pacific nations is because Indigenous cultures tend to be context and place-specific, with visitors having to travel to the place where that culture originates in order to experience it (Butler & Hinch 2007; Pfister 2000; Timothy 2011). This is important for rural and local communities in particular because these are the types of communities that cultural heritage tourists are looking to interact with. The ability to draw tourists into rural or remote areas in search of 'authentic' cultural experiences is considered one of the advantages for local and Indigenous communities focusing on cultural heritage tourism (Butler & Hinch 2007; Du Cros & McKercher 2015; Timothy 2011).

The advantage of attracting cultural heritage tourists as a niche in themselves is that not only are they usually high-value tourists (meaning they are usually high spenders) they also tend to be discerning about their money, wishing to contribute to the communities where they stay and pay towards maintaining and invigorating cultural heritage (SPTO 2014; Timothy 2011). However, there are also several other potential markets for local communities to assess when developing cultural heritage products. Incidental cultural heritage tourists are tourists who visit primarily for other reasons, such as surfing or beach tourism, but also engage in cultural heritage activities or visit attraction sites as part of their overall stay (Du Cros & McKercher 2015). As most visitors to the Pacific visit for the tropical holiday experience, such as the beaches and (hopefully) the weather, this market is potentially one to consider when developing cultural heritage experiences. Other important potential cultural heritage customers are the local domestic and related Visiting Friends and Relative (VFR) markets (Hall & Duval 2004), as diaspora communities are often looking to reengage with their Indigenous cultural heritage or introduce/educate their children about their ancestry and heritage.

Apart from economic benefits, cultural heritage tourism is also perceived as providing social/cultural advantages over other types of tourism. The PNGTPA (2006) acknowledges that culturally-based tourism products can assist in the preservation of Papua New Guinean cultural heritage by providing an economic value to those products. Allowing communities to receive economic benefits from staying connected to their traditional cultural values is important and provides incentives for younger generations to learn about and value their culture (Butler & Hinch 2007; Sakata & Prideaux 2013). At the same time, cultural heritage tourism, if managed appropriately, can allow tourists to engage more directly with the Indigenous local communities they visit, thus creating real interactions between locals and tourists (Du Cros & McKercher 2015; Pfister 2000). This engagement can result in knowledge sharing, community empowerment and increased empathy between tourists and communities (Carr et al. 2016; Whitford & Dunn 2014).

Other advantages of cultural heritage tourism include encouraging more people to participate in community level tourism, as cultural heritage products tend to be community-based activities such as handicrafts, singsings, and sharing of traditional knowledge. There are also lower start-up costs required as the communities already have the traditional knowledge, tools and skill sets needed to produce products and therefore do not need to spend initial capital on infrastructure development, although this will vary depending on the type of tourism being developed. At the same time, cultural heritage tourism tends to have low economic leakage as the products are produced within the village by the villagers and often do not require additional external spending, meaning that income distribution can more readily stay within the community.

Cultural heritage tourism is not without its risks, however. For materially-based cultural heritage assets, poorly managed tourism can lead to the deterioration or destruction of the asset itself. Local communities may not have sufficient funds or resources to be able to develop tourism markets themselves and are therefore at risk of losing control and access to local assets through the involvement of external providers (Butler & Hinch 2007; Prideaux & Timothy 2008). For cultural heritage assets that have an intangible base, concerns have focused upon the problem with turning cultural heritage into a tourism 'product' and potential issues relating to inauthenticity, trivialisation, commodification, and exploitation (Carr et al. 2016; Gabriel et al. 2017; Prideaux & Timothy 2008; Tolkach & Pratt 2021; Whitford & Dunn 2014). These factors mean that cultural heritage tourism needs to be carefully designed in ways that are suitable and appropriate for the host Indigenous community and the cultural heritage itself, not just for the tourists.

Case Studies

This section will introduce the two case studies for this research: Madang, the capital of Madang Province, which is located on the north coast of Papua New Guinea, and Samoa. A brief archaeological and historical background will be included to provide a context for the range of cultural heritage opportunities complementing the Indigenous communities inhabiting these areas. This nuanced background is essential as it outlines both the shared but also diverse histories and narratives within Pacific nations, with relevance for how cultural heritage tourism can be developed in these areas in a way that can benefit Indigenous communities.

Archaeological and Historical Background

The Pacific has a long history of human occupation, with PNG first settled some 45,000 years ago with the first movement of people into the Pacific region, reaching as far east as Buka in the Solomons Islands (Summerhayes et al. 2010). By 3,300 years ago, a new wave of people had arrived on the shores of PNG: the Austronesian speaking Lapita peoples. These seafaring horticultural people came from South East Asia, bringing with them pigs, pots, and new horticultural crops, with some groups settling in PNG, while others continued out into the Pacific, reaching as far east as Samoa by 2,850–2,700 years ago. Samoa is significant as this is the last place that we see the Lapita culture, with voyaging stopping here for another 1,800 years before venturing further into the eastern Pacific (Wilmshurst et al. 2011).

This gap in voyaging places the Fiji-Samoa-Tonga triangle as the area where the ancestral Polynesian cultures developed before the peoples headed further east to inhabit Rapanui, Hawai'i and eventually Aotearoa New Zealand. This migration event is important as not only is it the greatest seafaring migration ever undertaken, it also links the Indigenous Pacific nations today through a shared cultural narrative of migration and voyaging. As further voyaging began, there was an explosion in the building of extensive monumental architecture in Samoa. Built from stone, these structures include hundreds of star mounds, house platforms, walkways, terraces, ovens, and fortifications, as well as large earthen mounds and ditches, recorded across Upolu, Savai'i and Manono. The largest ancient Polynesian structure dating to this time period (1100–1800 AD) is the famous Pulemelei mound on Savai'i (alternatively called the Tia Seu Ancient Mound). Archaeological surveys by Gregory Jackmond since the mid-1970s have recorded 3,000 features including stone platforms, stone fences, pathways and earth ovens (Jackmond et al. 2018) (Figure 16.1).

In the 1800s, both PNG and Samoa were colonised by Europeans. Samoans were converted to and quickly adopted Christianity, a major feature of *fa'a Samoa* the Samoan way of life, today. Along with the missionaries, colonial powers soon became interested in the economic potential of Samoa, particularly as a shipping port, with the British, Americans and Germans all wishing to establish control. The Tripartite Treaty was signed in 1899, which gave control of Western Samoa (what is Samoa today) to the Germans.

In 1908, an independence movement, the Mau movement, emerged and rebelled against colonial rule in Samoa. In 1914, following the outbreak of the First World War, New Zealand sent the Samoa Expeditionary Force, to German Samoa to seize the local wireless station. The administration of Samoa was then taken over by New Zealand until the end of the First World War. At the end of the War, New Zealand was officially handed the mandate for Samoa by the League of Nations. Ill-conceived policies by the New Zealand administration (which was largely run by administrators from military backgrounds), such as the ability to banish chiefs and remove titles, and the exclusion of Samoans from any real representation in governance, led to a revitalisation of the Mau movement in 1927. This movement paved the way for Western Samoa to become the first Pacific Island nation to gain political independence in 1962.

Today Samoa is renowned for its strong adherence to *fa'a Samoa*, which is based on a chiefly system of rule, with service to *matai* (chief), *aiga* (extended family) and the church being

Figure 16.1 Pulemelei Mound, Savai'i.

Source: Anne Ford Photographer.

paramount principles. However, less well known is Samoa's rich archaeological history, particularly its monumental archaeological landscape and the role that the islands played in being the last voyaging place of the Lapita peoples and subsequently, during the 1,800-year interval before voyaging began again, part of the homeland where the ancestral Polynesian cultures developed. While Disney has recently recognised the significance of these stories, using them as a basis for the *Moana* film, there is little evidence for the stories or archaeological heritage being drawn upon to complement existing Indigenous and cultural heritage tourism offerings in Samoa.

West of Samoa, Papua New Guinea, at the time of European colonisation, was notably home to people who undertook some of the longest-distance trading voyages in the world, where large canoes traded goods across hundreds of kilometres. These include the *kula* of the Trobriand Islands, written about by anthropologist Bronislaw Malinowski and instrumental in Marcel Mauss's book *The Gift* which became the seminal work on understanding how gift exchange works as a political and social process. In Madang, the Bel people who inhabit many of the local islands and coastal areas are also famous for their long-distance trading voyages, undertaken in large balangut or lalong canoes, trading pots along the northeast coast, from Karkar Island in the north to the Vitiaz Strait in the east. These pots were exchanged for many goods, including food, stone tools, personal ornaments and jewellery, and wooden items such as bowls and drums, making the Bel affluent traders.

The first European to develop connections with this area was Nicholai Miklouho-Maclay, a Russian naturalist and ethnographer who arrived aboard the Russian ship *Vitiaz* in 1871 (Mennis 2018). Miklouho-Maclay lived in this area on and off until 1883, recording the customs and language of the local Bel people, and making natural history and anthropological collections for the area. In 1881 and 1884, Otto Finsch, a German ornithologist visited the Madang area. Although Finsch was making natural history collections, he was also advocating for the establishment of a

German colony. In November 1884, Germany proclaimed the northeastern part of PNG as German New Guinea. From 1884 to 1899 this was run by the German Neu Guinea Kompagnie, who established a settlement at Madang called Friedrich-Wilhemshafen, named after the German Crown Prince. The main purpose of the German Neu Guinea Kompagnie was business, and during this time many plantations across German New Guinea were set up for copra production. These were largely unsuccessful, however, and from 1899 to 1914 the colony was run by the Imperial German Administration with its headquarters in Rabaul, in New Britain.

German occupation ceased at the end of the First World War when the Australians were handed German New Guinea under a mandate from the League of Nations. The Australians administered German New Guinea until 1942, when the Japanese bombed Madang during the Second World War, destroying much of the old town. The Japanese occupied Madang itself from April 1943 until April 1944, when Madang was liberated by the Australian forces. Following the end of the Second World War, Papua (British New Guinea) and New Guinea (German New Guinea) were combined into a single administrative territory, the Territory of Papua and New Guinea, to be run by an Australian trusteeship.

In 1975, after a period of self-government, PNG achieved independence and Madang became a province in 1978, with its own provincial government. Today, PNG is considered one of the most culturally diverse countries in the world, with more than 800 Indigenous languages, spread through a range of environmental zones, from islander to coast, lowlander to Highlands. Much of the country still practices traditional agricultural subsistence strategies and maintains strong connections to the past as evidenced through craft production, oral histories/stories, and singsing performances. The legacy of generations of anthropologists working in Papua New Guinea, from Bronislaw Malinowski to Margaret Mead writing on the diverse cultures and languages within Papua New Guinea, and having their work made into documentaries and published in educational textbooks and popular media, such as National Geographic, has meant that Papua New Guinea has retained an internationally renowned reputation for Indigenous cultural heritage.

Indigenous Cultural Heritage Potential

Since European colonization in the 19th century, Madang has been a drawcard for tourism in PNG being referred to as the *Pearl of the Pacific* or the *Prettiest Town in the Pacific*. The Lonely Planet describes the province as "PNG in miniature" (Brown et al. 2016: 92). It is very diverse in cultures, with people representing island, coastal, river and mountain Indigenous communities, and this diversity is expressed in languages, with close to 200 languages present in this province alone.

Tourism is an established industry for Madang with visitors being attracted to the region for ecotourism, diving, surfing, island cruising, cultural attractions, and military/colonial heritage (N'Drower 2014). Cultural attractions vary between generic tourism sites and cultural immersive experiences. Heritage sites include a local museum (the Madang Visitors and Cultural Bureau); local markets where tourists can buy handicrafts and local produce; and colonial/military heritage sites, places such as Alexishafen Catholic Mission which preserves evidence of the German colonisation and military history of the region.

Another attraction of significance is Bilbil Village, one of the last villages in PNG where Lapita pottery making has survived as an important cultural tradition and where tourists can witness pottery demonstrations (Figures 16.2 and 16.3).

Culturally immersive opportunities include village visits and homestays, where tourists can try local foods, learn about customs and lifeways, or learn how to make traditional handicrafts,

Figure 16.2   Bilbil Village pottery.

Source: Anne Ford Photographer.

Figure 16.3   Pottery demonstration Bilbil Village, Madang, PNG.

Source: Anne Ford Photographer.

including staying overnight and receiving traditional hospitality. Cultural shows/festivals/singsings can also be immersive Indigenous experiences. Like elsewhere in PNG, these shows are performances and dances representative of the different cultures within the local region and further afield (Whitford & Dunn 2014). The Madang Festival (now renamed the Mabarosa Festival) is the most well-known, and recently a smaller festival, the Karkar Island Bilum Festival has begun, focusing on the production and display of woven bags (bilums).

Cultural heritage opportunities are also abundant within Samoa, including tangible and intangible Indigenous heritage. Existing cultural heritage opportunities include two museums: the Museum of Samoa (Falemata'aga), the Robert Louis Stevenson Museum and the Samoan Cultural Village, where visitors can see different Samoan cultural traditions, including tattooing, *umu* demonstrations, and *siapo* making, before finishing with music/dance and an *umu* lunch. Indigenous Samoan and Fa'a Samoan heritage are also present as visitor experiences within local villages. Accommodation providers offer the opportunity to experience cooking demonstrations or traditional *umu* food, *fiafias* or cultural performances, as well as visits to villages or opportunities to learn about traditional handicrafts.

A major difference between Samoa and Madang is the rich monumental architectural landscape spread across the islands (including Upolu, Savai'i and Manono) that consists of star mounds, house platforms, walls, walkways and fortifications (see for example Samoa Observer 2017; Jackmond et al. 2018). These are highly evocative sites that speak to the pre-European history of Samoa. Some star mounds are visited as part of tourist operations, largely at Falealupo; however, there is a relatively low visitation rate and little interpretation is provided to enable visitors to understand the Indigenous and cultural significance of these sites in Samoan and Polynesian history.

Talanoa and TokTok: Conversations about Cultural Heritage Tourism

To understand local Madang and Samoan communities' perspectives towards the development of cultural heritage tourism as sustainable IT or CBT offerings, five research trips were undertaken as a collaboration between New Zealand and local researchers. In August 2017, an initial visit to Samoa strengthened a research partnership with the Centre for Samoan Studies (CSS) at the National University of Samoa (NUS). This was followed by a second visit in December 2017 to interview local village communities in the Palauli district of Savai'i with CSS, asking questions about their perceptions and awareness of cultural heritage tourism. These questions included how they perceive to be the positives and negatives of developing these types of activities. The interviews were completed in Samoan and translated by CSS scholars and staff at NUS.

In March 2018, a third trip to Samoa was completed that involved formal interviews with both cultural heritage and tourism stakeholders, including the Samoa Tourism Authority, Savai'i Samoa Tourism Association, the Cultural Division of the Ministry of Sports, Education and Culture, Robert Louis Stevenson Museum, Museum of Samoa and local tourism providers.

In February 2018, a two-day workshop on cultural heritage tourism was held in Madang that included national and local stakeholders from the National Museum and Art Gallery of Papua New Guinea (NMAG), Papua New Guinea Tourism Promotion Authority, Tourism and Anthropology staff from Divine Word University, National Research Institute, Madang Visitors and Cultural Bureau, and Melanesian Tourist Services.

In June 2018, a second visit to Madang was conducted to interview members from local communities currently participating in tourism in the Madang area. These visits were undertaken with the assistance of staff from NMAG and Divine Word University. Similar questions to those included in the Samoan community interviews were asked. This research also involved follow-up formal interviews with cultural heritage and Indigenous tourism stakeholders in Madang and Port Moresby that had previously attended the Madang workshop. The local community interviews were completed in Tok Pisin and translated by staff from NMAG.

A key part of this research was the returning of findings to community members and stakeholders. Follow-up visits were undertaken to Madang in November 2018 to present transcripts and findings back to the local communities who had participated in the initial interviews. Research team members returned to Samoa in August 2018 to present findings at the public Samoa Conference IV, hosted by NUS. A final report for both Madang and Samoa was also circulated to stakeholders for comment prior to completion and being made publicly available.

Findings

The study findings have been complex due to the very different nature of the two case studies. The following section will focus on the questions that originally drove this research: (1) What level of awareness do local communities have in the Pacific regarding cultural heritage and Indigenous tourism?; (2) Would cultural heritage tourism be considered a possible benefit for sustainable local Indigenous community development?; and (3) If this was the case, what are the opportunities and constraints that local Indigenous communities perceive in the development of locally-based cultural heritage tourism?

Community Perceptions and Awareness of Cultural Heritage Tourism

There were similarities in how the local Indigenous communities perceived tourism as a whole, and cultural heritage tourism more specifically. Local communities in both case studies supported community-based tourism ventures, as tourism provides important economic gains for community groups through accommodation fees, access/entry fees, employment, or handicraft sales, which could enable the community to support schools, medical needs, and other community activities. In particular, tourism was seen by younger members of the community as a way to access employment while also being able to remain in the village. This was particularly noted in Samoa, where many of the employment opportunities available for young people are located in the capital Apia, on the island of Upolu, which meant needing to leave home villages for such employment.

However, local communities were less aware of the concept of cultural heritage tourism. While many tourism operations incorporated Indigenous cultural heritage to varied extents within their tourism products, it was important to note that the term 'cultural heritage' was unfamiliar to participants (reflecting the Eurocentric/western understanding of heritage). In Madang, it was noted that intangible cultural heritage which included daily community activities was too familiar to them to be considered something that might be valued or sought after by tourists. Once community participants were engaged in discussions about what cultural heritage was, there was raised awareness of diverse cultural activities around cuisine, crafts and natural resource usage that could be an important opportunity for future tourism growth. In Samoa, some respondents noted that tourists interested in culture and heritage were their preferred types of tourists as they were more likely to be more respectful of their traditional cultural values and lifeways, particularly during village visits.

Communities perceived that the benefits of cultural heritage tourism activities compared to other types of tourism included that it requires low levels of investment in capital and infrastructure and builds upon an already understood asset – their own cultural heritage. Cultural heritage tourism was therefore seen as potentially being a sustainable industry which could be readily incorporated into the Indigenous communities' current livelihoods, and which they were already empowered in many ways to participate in as they already had control over the cultural knowledge that was sought after by potential tourists. This is particularly important in areas such as Madang where tourism has fluctuated, particularly in recent times.

In both Madang and Samoa connections to place were also mentioned as a positive. As cultural heritage is often associated with a particular place, it encourages tourists to visit rural and remote communities, thus allowing these communities to earn economic opportunities within their home location. Connections to culture were also discussed, where sharing culture with international and domestic tourists can result in knowledge sharing and increased empathy and understanding between Indigenous or local communities and tourists. Samoa traditionally prides itself on the sharing of its culture (see NZTRI 2018). For Madang this was also considered important as creating links between tourists and communities has value for promoting PNG and can challenge stereotypes such as those around security issues.

Finally, cultural heritage tourism was potentially advantageous as a trigger to enhance pride in the cultural heritage of the Indigenous communities themselves. In both Madang and Samoa, there was an emphasis on valuing traditional or Indigenous knowledge that would potentially further protect and preserve culture. This was considered important for the archaeological sites in Samoa, which currently have no formal protection. By placing an economic and heritage value on these sites, it may ensure their long-term protection from destruction and deterioration. The same opportunities were considered for intangible heritage. The women demonstrating pottery making at Bilbil Village are one of the remaining villages in PNG to still practice a craft originating from Lapita culture. Having economic outlets for their pottery and the potential for demonstrations or village visits/tours encourages younger women to learn the craft. The motivation was not just to continue the pottery tradition but to also benefit economically so that the villagers could access education and health services.

Not all perceptions of tourism were positive. In Samoa, in particular, it was noted by local communities that they wanted to ensure that tourists were respectful of village customs (i.e. by wearing respectful or modest clothing) and did not bring negative impacts and influences to the community. They also wanted to prevent being taken advantage of by tourists, for example, by ensuring that tourists paid appropriate access fees to visit sites.

Opportunities and Constraints in the Development of Cultural Heritage Tourism

Despite similar attitudes towards tourism, opportunities and constraints vastly differed between the two case studies and support the essential need to complete in-depth studies on every local situation before developing community-based cultural heritage tourism opportunities. For example, Madang (and PNG in general) already has an international reputation for developing cultural heritage tourism, therefore tourists visiting the area seek out these opportunities at cultural festivals and villages. Constraints to tourism in Madang (and wider PNG) are more likely due to concerns around security and crime, perceived health risks, lack of tourism infrastructure and high cost of travel. In contrast, Samoa is seen by tourists as an affordable and relaxed beach holiday destination with lower awareness of its cultural heritage, therefore it needs to target marketing to broaden visitor perceptions, improve cultural heritage or Indigenous tourism offerings and continue to market these as a point of distinction compared to other South Pacific destinations.

Currently, in both Madang and Samoa, the main focus on cultural heritage tourism is related to tangible heritage, with groups in Samoa incorporating traditional elements such as handicrafts (*siapo*, fine mat weaving), *fiafia* nights or *umu* demonstrations, and groups in Madang also incorporating handicrafts (pottery, bilum making), festivals/singsings or *umu* feasts. There is less awareness in both places of intangible heritage although in Madang, WW2 history is an important part of tourism products, but these are usually operated by external tour guides rather than local Indigenous communities. In Samoa, the archaeological sites involving monumental architecture were noted as being an underutilised cultural heritage resource in terms of tourism. Few of these sites are available for visitation by tourists currently, yet their addition to the tourism offerings for Samoa would increase the diversity and uniqueness of experience for tourists visiting Samoa.

Numerous constraints were noted in both case studies, including a need for further investment by the government at all levels to provide finances, resources and training to the communities that need assistance with infrastructure (including basic community infrastructure around transport, health and education), capacity building, product development, marketing and guide training around how to design engaging, authentic and, from a long-term perspective, regenerative sustainable tourism experiences.

Other local considerations when developing cultural heritage tourism opportunities include traditional community structures and how this will affect governance issues and the overall implementation of projects. For example, projects in Samoa will need to consider *fa'a Samoa* principles, while projects in Madang need to consider clan relationships. Other factors that will come into play include the level of support from the government, how tourism and cultural heritage authorities and bodies are structured in-country, the specific legislation and regulations involved in protecting cultural heritage and implementing tourism, as well as the wants and needs of the specific communities involved.

Conclusion

Cultural heritage tourism has been described as a potential niche industry for Pacific nations developing sustainable Indigenous tourism products, particularly at a local community level. However, there is a need for advocacy and awareness training amongst communities to become familiar with the notably Western concept of cultural heritage and how this may work for Indigenous community development. While many advantages were noted by the communities, there are also constraints that they face, requiring support in product development, education/training, capacity building and marketing. These types of support, as well as the opportunities available for communities, however, need to be identified on a case-by-case, place-specific basis. This research has demonstrated that for sustainable development to occur in Samoa and PNG, it is essential to dedicate time and resources to talanoa (Samoa) and toktok (PNG) to ensure that the voices of the communities are adequately heard and listened to.

Acknowledgements

The authors wish to acknowledge the villagers of Bilbil and Madang (PNG) and Palauli district of Savai'i (Samoa) for their generous participation in this research.

References

Benson, C. (2012). Conservation NGOs in Madang, Papua New Guinea: understanding community and donor expectations, *Society and Natural Resources* 25(1), 71–86.

Brown, L., Carillet, J-B., & Kaminski, A. (2016). *Papua New Guinea and Solomon Islands*, 10th edition, New York: Lonely Planet Global Limited.

Butler, R.W., & Hinch, T.D. (Eds.) (2007). *Tourism and Indigenous Peoples*, London: Routledge.

Carr, A., Ruhanen, L., & Whitford, M. (2016). Indigenous peoples and tourism: the challenges and opportunities for sustainable tourism, *Journal of Sustainable Tourism* 24(8&9), 1047–1069. https://doi.org/10.1080/09669582.2016.1206112

Cole, S. (2006). Information and empowerment: the keys to achieving sustainable tourism, *Journal of Sustainable Tourism* 14(6), 629–644.

Du Cros, H., & McKercher, B. (2015). *Cultural Tourism*, Oxon: Routledge.

Everett, H., Simpson, D., & Wayne, S. (2018). *Tourism as a Driver of Grown in the Pacific: Issues in Pacific Development No. 2*, Manila: Asian Development Bank.

Gabriel, J., Filer, C., Wood, M., & Foale, S. (2017). Tourist initiatives and extreme wilderness in the Nakanai Mountains of New Britain, *Shima* 11(1), 122–143.

Hall, C.M., & Duval, D.T. (2004). Linking diasporas and tourism: transnational mobilities of Pacific Islanders resident in New Zealand, pp 78–94 in T. Coles & D.J. Timothy (Eds.) *Tourism, Diasporas and Space*, London: Routledge.

Harrison, D. (2003). Themes in Pacific Island tourism, pp 1–23 in D. Harrison (Ed.) *Pacific Island Tourism*, London: Cognizant Communication Corporation.

Harrison, D., & Prasad, B. (2013). The contribution of tourism to the development of Fiji and other Pacific Island countries, pp 741–761 in C. Tisdell (Ed.) *Handbook of Tourism Economics*, Singapore: World Scientific Publishing Co.

Imbal, J. (2010). Contemporary challenges facing the development and management of culture tourism in Papua New Guinea, *Contemporary PNG Studies, DWU Research Journal* 12, 12–28.

International Council on Monuments and Sites (ICOMOS) (2008). *The ICOMOS Charter for the Interpretation and Presentation of Cultural Heritage Sites*. Available online at https://www.icomos.org/en/resources/charters-and-texts (Accessed 23 September 2023).

Jackmond, G., Fonoti, D., & Matavai Tautunu, M. (2018). Sāmoa's hidden past: LiDAR confirms inland settlement and suggests larger populations in pre-contact Sāmoa, *Journal of the Polynesian Society* 127(1), 73–90.

Kau, T.W. (2014). *The role of tourism promoting community participation in the development of Jiwaka Province in Papua New Guinea*, Thesis, Master of Management Studies, University of Waikato, Hamilton, New Zealand.

Manyara, G., & Jones, E. (2007). Community-based Tourism enterprises development in Kenya: an exploration of their potential as avenues of poverty reduction, *Journal of Sustainable Tourism* 15(6), 628–644.

Mennis, M. (2018). *The Flagged History of Madang 1871–2018*, Madang: The Melanesian Foundation.

Miller, G., & Twining-Ward, L. (2005). Samoa Sustainable Tourism Indicator Project, pp 233–259 in G. Miller & L. Twining-Ward (Eds.) *Monitoring for a Sustainable Tourism Transition: The Challenge of Developing and Using Indicators*, Wallingford: CABI Publishing.

Muhanna, E. (2007). The contribution of sustainable tourism development in poverty alleviation of local communities in South Africa, *Journal of Human Resources in Hospitality and Tourism* 6(1), 37–67.

N'Drower, F. (2014). Sustainable tourism practices in Madang, *Contemporary PNG Studies: DWU Research Journal* 20, 90–102.

New Zealand Tourism Research Institute (NZTRI) (2018). *Samoa International Visitor Survey January – June 2018*. Auckland: New Zealand Tourism Research Institute (AUT).

Papua New Guinea Tourism Promotion Authority (PNGTPA) (2006). *Papua New Guinea Tourism Sector Review and Master Plan (2007–2017), "Growing PNG Tourism as a Sustainable Industry"*. Madang: Papua New Guinea Tourism Promotion Authority.

Pfister, R.E. (2000). Mountain culture as a tourism resource: aboriginal views on the privileges of storytelling, pp 115–136 in P.M. Godde, M.F. Price & F.M. Zimmermann (Eds.) *Tourism and Development in Mountain Regions*, New York: CABI Publishing.

Prideaux, B., & Timothy, D.J. (2008). Themes in cultural and heritage tourism in the Asia Pacific region, pp 1–14 in B. Prideaux, D.J. Timothy & K. Chon (Eds.) *Cultural and Heritage Tourism in Asia and the Pacific*, London: Routledge.

Rayel, J.J. (2012). Local community involvement in trekking and residents' perceptions of implications of tourism and the proposed mining operations: a case of Kokoda Track, Papua New Guinea, *TEAM Journal of Hospitality and Tourism* 9(1), 27–38.

Reggers, A., Grabowski, S., Wearing, S.L., Chatterton, P., & Schweinsberg, S. (2016). Exploring outcomes of community-based tourism on the Kokoda Track, Papua New Guinea: a longitudinal study of Participatory Rural Appraisal techniques, *Journal of Sustainable Tourism* 24(8–9), 1139–1155.

Sakata, H., & Prideaux, B. (2013). An alternative approach to community-based ecotourism: a bottom-up locally initiated non-monetised project in Papua New Guinea, *Journal of Sustainable Tourism* 21(6), 880–899.

Sakata, H., & Prideaux, B. (2014). Community-based ecotourism: opportunities and difficulties for local communities and link to conservation, pp 199–212 in B. Prideaux (Ed.) *Rainforest Tourism: Conservation and Management*, London: Routledge.

Salazar, N.B. (2012). Community-based cultural tourism: issues, threats and opportunities, *Journal of Sustainable Tourism* 20(1), 9–22.

Samoa Observer (7 May 2017). Unearthing prehistoric Samoan history. Available online at http://www.sobserver.ws/en/07_05_2017/local/19707/Unearthing-prehistoric-Samoan-history.htm (Accessed 24 September 2023).

Samoa Tourism Authority (STA) (2014). *Samoa Tourism Sector Plan 2014–2019*. Suva: STA.

Scheyvens, R. (1999). Case study: ecotourism and the empowerment of local communities, *Tourism Management* 20, 245–249.

Scheyvens, R. (2004). Growth of Beach Fale Tourism in Samoa: the high value of low-cost tourism, pp 188–202 in C.M. Hall & S.W. Boyd (Eds.) *Nature-based Tourism in Peripheral Areas: Development or Disaster?* Bristol: Channel View Publications.

Scheyvens, R. (2015). Tourism and poverty reduction, pp 118–140 in R. Sharpley & D.J. Telfer (Eds.) *Tourism and Development: Concepts and Issues*, Bristol: Channel View Publications.

South Pacific Tourism Organisation (SPTO) (2014). *South Pacific Niche Market Report – Cultural Heritage Tourism*. Suva: SPTO.

Summerhayes, G., Leavesley, M., Fairbairn, A., Mandui, H., Field, J., Ford, A., & Fullagar, R. (2010). Human adaptation and plant use in highland New Guinea 49,000 to 44,000 years ago, *Science* 330, 78–81.

Tauaa, S. (2010). Tourism issues in the Pacific, *eJournal of the Australian Association for the Advancement of Pacific Studies* Issues 1.2 and 2.1.

Taumoepeau, S., & Addison, A. (2016). Cultural and heritage subjects in a national tourism curriculum framework: a Samoan case study *Open Journal of International Education* 1(2), 29–49.

Telfer, D.J., & Sharpley, R. (2016). *Tourism and development in the developing world*, 2nd edition, Abingdon, Oxon and New York: Routledge.

Timothy, D.J. (2011). *Cultural Heritage and Tourism: An Introduction*, Bristol: Channel View Publications.

Tolkach, D., & King, B. (2015). Strengthening community-based tourism in a new resource-based island nation: why and how? *Tourism Management* 48, 386–398.

Tolkach, D., & Pratt, S. (2021). Globalisation and cultural change in Pacific Island countries: the role of tourism, *Tourism Geographies* 23(3), 371–396.

UNESCO (2003). Convention for the Safeguarding of the Intangible Cultural Heritage. Available online at https://ich.unesco.org/en/convention (Accessed 18 September 2023).

United Nations World Tourism Organisation (UNWTO) (2017). *Compendium of Tourism Statistics Dataset* [Electronic], UNWTO, Madrid.

Whitford, M., & Dunn, A. (2014). Papua New Guinea's Indigenous cultural festivals: cultural tragedy or triumph? *Event Management* 18, 265–283.

Wilmshurst, J.M., Hunt, T.L., Lipo, C.P., & Anderson, A.J. (2011). High-precision radiocarbon dating show recent and rapid initial human colonisation of East Polynesia, *PNAS* 108(5), 1815–1820.

World Bank (2017). *Tourism: Pacific Possible Series*, Washington D.C.: World Bank Group.

Zeppel, H. (2014). Indigenous peoples and rainforest tourism: canopy walkways as tourist attractions, pp 226–238 in B. Prideaux (Ed.) *Rainforest Tourism: Conservation and Management*, London: Routledge.

17

TOURISM ENTERPRISES IN THE SOUTH PACIFIC

Culturally Centered Adaptation in the Face of Covid-19

Jason Mika, Apisalome Movono, Sophie Auckram, Suzanne Hepi and Regina Scheyvens

Introduction

As with most travel-based enterprises, Indigenous-run tourism enterprises in Aotearoa New Zealand and elsewhere in the Pacific were severely affected by Covid-19. The collapse of global travel forced companies to respond in different ways to cope with the pandemic. However, few studies have examined how Indigenous enterprises reacted to the volatility of tourism and the measures Indigenous people adopted both to protect their people and to stay afloat. Based on empirical research conducted in Aotearoa New Zealand and the Cook Islands, this chapter sheds light on the difficulties the pandemic-enforced border closures brought for Indigenous enterprises, but it also focuses on how Indigenous owners and managers of tourism enterprises have re-prioritised their employees and communities and adapted in the face of pressure. Furthermore, this chapter exposes the values underpinning Indigenous enterprises' decision-making, which influence how they adapt. For some, their purpose has been culturally embedded, determining what is best for their *whānau* (family), their people, and the *whenua* (land). When a shock such as a global pandemic strikes, however, it is apparent that over-reliance on this relatively fickle and highly seasonal industry can be a risky business. Importantly, this chapter highlights that where Indigenous people retain ownership of customary land, this resource can provide alternative livelihoods, sustaining people through challenging times.

We start by reviewing Indigenous peoples' entrepreneurship in tourism. We then focus on Indigenous involvement in the tourism sector in Aotearoa New Zealand and the Cook Islands, two unique countries whose Indigenous people have strong ancestral and cultural ties. Following this, the methodology is discussed. The central part of the chapter follows: here, we seek to demonstrate how specific Indigenous tourism enterprises in Aotearoa and the Cook Islands have adapted to the global pandemic. Indigenous enterprises have adopted a range of approaches to exercising *tino rangatiratanga* (self-determination). For many, their purpose has been culturally-centred,

DOI: 10.4324/9781003230335-20

determining what is best for their *whānau* (family), their people in general, and the *whenua* (land). This chapter asserts that in response to shocks such as the global coronavirus pandemic, Indigenous enterprises have exercised their *tino rangatiratanga*.

Background

There is perhaps no industry more emblematic of the impact of Covid-19 on economic development than tourism. From February 2020 onward, a domino effect occurred whereby one country after another chose to close its international borders to halt the inflow of visitors who could potentially harbour the Covid-19 virus. Restrictions on significant events and gatherings, ordering enterprises to close unless they could offer contactless service, and "lockdowns" confining millions to their homes were among the health-related measures imposed. These protective mechanisms were economically and socially devastating for an industry reliant on human mobility and contact.

Tourism has, for decades, offered a lifeline to many Indigenous peoples. These groups have often struggled in the face of historical injustices associated with confiscating their assets, particularly land, and the establishment of laws and policies which undermine their rights and wellbeing (Bunten & Graburn, 2018). Certainly, Indigenous people are often the subject of intense interest by tourists (de-Lima & Weiler, 2015). Yet, in recent decades, they have also played significant roles as tourism enterprises' entrepreneurs and owners. Where their rights are enshrined in law and once alienated assets have been returned to them and compensation paid, Indigenous groups have had the opportunity to invest in their self-determined development, using tourism as a mechanism to secure more comprehensive benefits (Amoamo et al., 2018; Mika & Scheyvens, 2021).

Indigenous Entrepreneurship

Indigenous peoples have been engaged in entrepreneurial activity for millennia (Cachon, 2012; Dana & Anderson, 2007) but have done so according to a worldview and tradition grounded in an ontology of the interrelatedness of all things (Harris & Wasilewski, 2004). When entrepreneurial activity is approached on this basis, it occurs within the philosophy of kinship-based enterprise in the broadest possible sense because all things – earth, sky, land, water, flora, fauna, and people – are related (Kawharu & Tapsell, 2019; Rout et al., 2021). The dead are remembered as unseen faces whose legacy is a knowledge system of self-sufficiency in balance with nature (Mead, 2003), while the unborn are rationalised as the true beneficiaries of existing endeavours (Royal, 2003). Enterprise established and conducted on these terms aspires to a system of kinship rights and obligations that have at their heart belongingness, where self-identity and wellbeing are tied to the identity and wellbeing of the group (Colbourne & Anderson, 2020; Mika, 2021; Reid, 2021). This means that entrepreneurship, typically understood as an individual pursuit with a focus on material gain as an end in and of itself (Deakins & Scott, 2020), in Indigenous terms, more appropriately understood as a collective pursuit (Henry et al., 2020). In this view, materiality (money) is more of a means to an end, with those ends primarily concerned with the wellbeing of family, environment, and community (Mika et al., 2022).

In Aotearoa New Zealand, the notion of belongingness is enunciated in *whakapapa*, which refers to the layering of genealogical lines of descent of humanity from earth and sky (Te Rito, 2007). Kinship rights and obligations constitute ethical coordinates for business articulated in Indigenous values (Bargh, 2012; Hēnare, 2015; Stewart et al., 2017). These values give Indigenous peoples the authority to act, govern, manage, and use nature's bounty to sustain the group in a responsible, reciprocal, and regenerative way (Mika et al., 2020; Rout et al., 2021; Trosper,

2009). When a relational worldview is applied to the management of the environment and business conducted in this context, three implications for entrepreneurship arise. First, entrepreneurship is conducted within a conceptualisation of economy that is fundamentally environmental, in the sense that "all economic practices… are fundamentally rooted in the wider spiritual-socioecology" (Rout et al., 2021, p. 3). Second, entrepreneurship and economic development are accepted and valued forms of human activity, but not at the expense of human and ecological wellbeing (Mika, Dell, Newth, et al., 2022). Third, entrepreneurial success is defined in terms of what one does to enhance the well-being of others, known locally as a *mana*-enhancing business practice, oriented to the pursuit of economies of well-being (Wolfgramm et al., 2019).

Mana (authority), *utu* (balance), *kaitiakitanga* (guardianship), and *whakapapa* (genealogy) are values that typify an Indigenous ontology of entrepreneurial endeavour in Aotearoa and encourage enterprises to be sustainable (Bargh, 2012) in the myriad industries in which Indigenous enterprises operate (Nana et al., 2021). Despite the promise of equilibrium between nature and humanity in this Indigenous ideology, Indigenous enterprise is not immune to the risks of business failure, whether causes are attributed to intrinsic (entrepreneurial capabilities) or extrinsic (environmental constraints and enablers) factors (Joseph et al., 2016; Mika, 2014; Reid et al., 2019). The coronavirus pandemic is just the latest extrinsic factor to challenge Indigenous businesses. Business failure in Indigenous enterprises can come at great personal cost, which can discourage Indigenous entrepreneurs from owning and operating Indigenous enterprises (Mill & Millin, 2021).

Ingram (1990) found that successful Indigenous entrepreneurship and tourism development in small island states, such as many Pacific islands, required harmonious integration of economic, social, cultural, and political factors. Indigenous entrepreneurship in the Pacific is often focused on people creating opportunities for their families and wider communities to thrive. This means many such enterprises are integrally concerned with treading lightly on mother earth, treating their customers as guests, and treating their employees with respect (Mika et al., 2022).

An essential challenge confronting Indigenous entrepreneurship is emancipation from the ongoing effects of colonisation (Hill, 2021; Reid et al., 2017). The apparatus of colonial political economies has historically excluded Indigenous peoples from participating in, benefitting from, and directing their economic activity using Indigenous lands and other resources according to their ideas of what business is and should be (Galperin Bella et al., 2021; Henry, 2017; Miller et al., 2019). Due to assimilation, urbanisation, and globalisation, Indigenous knowledge about traditional ways of living and providing for people and the environment has been lost but is now undergoing a process of recovery, reclamation, and reanimation in modern Indigenous enterprises. Thus, the ethical coordinates of Indigenous values are being re-learned as modalities for responsible, sustainable, and self-determining Indigenous enterprise. The heterogeneity of Indigenous peoples means that entrepreneurship's modality, capabilities, and outcomes will look different in Aotearoa, the Cook Islands, and elsewhere in the Indigenous world (Colbourne, 2021; Ingram, 1990; Vao'iva Tofilau, 2018; Vunibola & Scheyvens, 2019).

Indigenous Tourism Entrepreneurship in Aotearoa and the Cook Islands

Homogeneity in Indigenous entrepreneurship is not to be expected. Differences in approaches to entrepreneurship on Indigenous lands between Aotearoa and the Pacific are illustrative. In Aotearoa, Māori land – land owned by Māori people under the jurisdiction of the Māori Land Court – comprises about 5.6% of the country's total landmass (Kingi, 2007). Its capacity as a culturally significant economic resource to sustain the Māori population was stripped away during the New Zealand wars between 1863 and 1873, which involved the confiscation of large tracts of

Māori land (Mika, Dell, Elers, et al., 2022). While Māori enterprise is still conducted on Māori land by entities collectively known as Māori authorities (Kingi, 2008), most Māori do not derive their primary source of income from their ancestral lands (NZIER, 2003). While Māori land represents a *taonga tuku iho* (valued cultural legacy), its productive capacity has been limited relative to the growing size and needs of the Māori people and the relatively small proportion of this land that is suitable for pastoral uses (Phillips et al., 2014). Māori authorities have become highly adept at farming, forestry, and fishing. They do so to create sufficient income to retain the land, provide dividends and other benefits to owners near and far, and be competitive in global markets (Awatere et al., 2017; Reid & Rout, 2020; Stats NZ, 2021, 2022).

Māori have been integral to the tourism industry since its inception in Aotearoa and have been central to the growth and success of the wider industry (Ransfield & Reichenberger, 2021). Māori guides, cultural attractions, and geothermal activity were the basis of the earliest tourism enterprises around Rotorua, dating back to the Pink and White Terraces (Puriri & McIntosh, 2019). In recent decades, importantly, Māori have taken on vital governance roles in the sector and ownership of major tourism enterprises. For example, Ngai Tahu owns premium tourism brands and products such as Shotover Jet, The All Blacks Experience, and The Dark Sky Project (Tahu, n.d.). Aotearoa had a thriving tourism industry pre-pandemic, bringing in over USD 11 billion in revenue in 2019 and providing 8.4% of jobs (Table 17.1). Note that this is only referring to direct revenue and employment; indirectly, tourism pours more money into economies through related industries and jobs, e.g., construction firms and taxi businesses. Tourism growth had been so strong in Aotearoa that concerns had started to grow about 'over tourism' associated with environmental degradation and social disturbance caused by too many tourists in particular places during peak periods (Insch, 2020).

In the Pacific islands such as the Cook Islands, Fiji, Samoa, and Vanuatu, customarily owned land comprises over 80% of all land. This land still represents a necessary means for daily livelihoods and food security through individual and collective social enterprise (Movono & Dahles, 2017; Scheyvens et al., 2017; Vunibola, 2021; Vunibola & Scobie, 2022). While these countries have also tried to develop their agriculture and fishing industries, tourism is their primary source of employment and foreign exchange earnings. Concerning this, the Cook Islands have a heavy reliance on tourism. Before the Covid-19 pandemic, there were over ten visitors per year for every resident on the island of Rarotonga, which only has a population of 14,000 (see Table 17.1). There is also heavy ownership of enterprises by Indigenous Cook Islands people, thanks partly to laws that require a majority shareholding in all businesses to be held by Cook Islanders.

Methodology

Researcher Positionality and Ethical Issues

The ideas and findings presented in this chapter derive from a research project that commenced in 2020. The team comprised five scholars: Movono is Fijian, Mika and Hepi are Māori, and Scheyvens and Auckram are Pākehā/Palagi. Collectively, we have over 50 years of experience learning from, about and with Indigenous tourism enterprises. Each of us played distinct roles in the research, from conceptualising the project to planning the methodology, connecting us to possible research communities, and collecting data. We locate ourselves as researchers concerned about social and environmental justice. We were concerned about how Indigenous peoples in the South Pacific, including Aotearoa New Zealand and the islands of the Pacific, were coping with a drastic decline in tourism income in the early stages of the global coronavirus pandemic due

Table 17.1 Comparison of tourism in the Cook Islands and Aotearoa NZ pre-pandemic (2019)

	Population	*Visitor arrivals*	*Tourism employment*	*Tourism receipts*	*Tourism revenue as a % of GDP*
Cook Islands	18,000	168,760	2,386 (34% of total employment)	USD $253 million	66–87%
Aotearoa New Zealand	4.979 million	3.8 million	229,566 (8.4% of total employment)	USD $11.2 billion	5.8%

Source: Tourism Satellite Account: 2019 | Stats NZ (n.d.).

Note: While tourism's contribution to GDP is a universally used indicator of the value that tourism adds to an economy, there are limitations to this indicator in terms of inconsistencies in the way related data is collected and reported. Thus, there is a significant difference in some of the estimates we accessed and a GDP range is provided here for the Cook Islands.

to border closures. The overarching research question was how had Covid-19 impacted people involved in tourism, and in what ways had they responded?

This work is inspired by Smith's (1999) decolonised research approach that places Indigenous people at the research centre. This encouraged the research team to focus on the well-being of tourism workers, Indigenous tourism business owners, and the wellbeing of people living in destination areas.

This research is committed to Massey University's ethics procedures and Pacific Research Principles (Massey University, 2017). Respect for knowledge holders in terms of upholding the *mana* (dignity, prestige) of participants was demonstrated by taking a strengths-based approach aligned with appreciative critical inquiry (Ridley-Duff & Duncan, 2015). Rather than being positioned as "victims" of the global pandemic, participants were approached as "actors" who would have a range of capabilities, knowledge, networks, and resources they could draw on when responding to the economic challenge.

With significant border restrictions preventing the key researchers from travelling for fieldwork and localities experiencing lockdowns, we had to take an adaptive approach to data collection. Three main methods were employed, incorporating an online survey, face-to-face interviews and Zoom interviews that took place in the communities (usually conducted in a *talanoa* style involving fluid discussions and the sharing of ideas and stories) (Nabobo-Baba, 2008).

Finally, we reviewed secondary data about the impacts of the pandemic on Indigenous-owned tourism enterprises in the chosen countries, including mainstream and social media articles and reports by governments or business associations.

Data Analysis and Limitations

Initial data analysis took place during the Zoom meetings with Pacific-based researchers in August and September 2020. The same process was repeated after the wellbeing survey in late 2021/early 2022. Each researcher was asked to reflect on the data they had collected and whether any common ideas were being expressed, as well as highlight anything that stood out to them as significant or surprising. We engaged in an organic, circular process of thematic analysis involving coding data, searching for meaning, and interpreting the data (O'Leary, 2021).

Findings

An initial summary report of findings was provided to participants, and all articles, news items and other research outputs were shared in a timely fashion on a website: www. reimaginingsouthpacifictourism.com. Because of the cultural links between Aotearoa and the Cook Islands, only data from findings in these two countries is used in this chapter. The following findings are based on case studies of an Indigenous-owned tourism enterprise in the Cook Islands, and a Māori Trust in Aotearoa New Zealand, which focused on culturally centred means of adaptation in the face of various pandemic-related challenges. We present, firstly, the voices of key people from these enterprises who comment on how they adapted over time. Secondly, we consider their reflections on what they learned when dealing with both the shock of border (and business) closures and having more time on their hands during the "pandemic pause", and on what they want from future tourism.

Responses to the Pandemic by Māori Traditional Landowners at Waitangi

Two of the authors of this chapter (Suzanne Hepi and Sophie Auckram) conducted interviews among members of the Te Tii (Waitangi) B3 Trust in November 2020.

The Te Tii B3 Trust is an Ahu Whenua Trust (one which administers and promotes the use of the land in the interest of its owners). The trust administers 70 freehold blocks located in Waitangi, 85% of the land is zoned commercially, with the remainder classed as residential. The trust exists to administer the assets and revenue for the benefit of those descendants of the original 251 tupuna/ ancestors as listed in the scheduled owners list. Due to Covid-19 and the associated lockdowns, the Trust had to think differently about how they protected their people as well as how they operated their enterprises. Their initial reaction was to ensure the safety of their people, then over time they adopted techniques which diversified their operations demonstrating resilience.

As Covid-19 hit the shores of Aotearoa in early 2020, so did the restrictions. *Hapū* (extended family) members wanted to keep Waitangi coronavirus-free. Thus, when New Zealand went into lockdown on March 26, 2020, the Trust immediately responded to protect the community. The quotes below show the response of the Tai Tokerau border control which stepped in:

> … when Covid turned up, we were able to be an essential service in terms of the holiday park [providing accommodation], but we chose not to do that, because we did not want to put our community at risk, because we didn't know who was out there, who had Covid. And so, we chose to shut Waitangi down. We put gates and fences and everything up, just to protect our families and the people living here, and our enterprises.
>
> (Administrator at Te Tii B3 Trust, 2020)

> [There was] the coming together of the local hapū, in terms of setting up roadblocks, and taking a more serious approach to it. We blocked off the whole block…We thought protect our papakāinga, particularly our kaumātua, kuia.
>
> (Operations Manager at Te Tii B3 Trust, 2020)

This did not mean that they turned away *hapū* members who were in need, however:

> We have used a couple of our units as sort of interim accommodation [for beneficiaries].
>
> (Director of Risk and Audit Committee at Te Tii B3 Trust)

Because we are [a] Trust, everything that we do is for the people and for the community.
(Marketing and Sales Manager of Te Tiriti Motel Waitangi, 2020)

Protecting vulnerable members among the residents at Waitangi while also being open to *hapū* members returning home – sometimes because they had lost jobs in distant towns or were facing other hardships due to the pandemic – was especially significant for some Māori who had not connected strongly with their culture for some time:

[I returned home because of Covid]… so being Māori, being home, and on our tūrangawaewae and being heavily involved in the marae, heavily involved in the community development work and doing lots of stuff here, that's me. That is my identity.
(Community member Waitangi, 2020)

The Trust took measures to enhance their business and keep it resilient during challenging times. One strategy was to recruit community members who were recently unemployed so that even if hospitality enterprises had to close, other work could continue:

I guess the positive was that we went out looking for opportunities, and snapped up, luckily, essential workers, working on various projects…and we based them out here. So we managed to keep some income coming in.
(Operations Manager at Te Tii B3 Trust, 2020)

Some major upgrades to operational systems were also put into place during lockdown to ensure the enterprises would be in a stronger position and be able to operate efficiently when business resumed:

The first thing that we did [after the initial lockdown] we signed the contract with Secom Company and we updated our PMS [property management system], which is a management system for the hospitality enterprises, and it was really big change for us. It gives us a lot of opportunity, and we updated our website as well to improve our direct sales on the website.
(Marketing and Sales Manager at Te Tii B3 Trust, 2020)

Other changes to business practice evolved, allowing ongoing communication among those involved in the governance or management of the Trust, without risking people's health and also reducing costs:

Before, we would never have virtual meetings. We would fly people all over the country to come in for a three or four hour meeting, face-to-face kanohi-ki-te-kanohi; but now we've accepted the technology…
(Administrator at Te Tii B3 Trust, 2020)

People around Aotearoa were eager to travel once local lockdowns ceased, yet they were unable to leave the country. The Trust thus embraced the opportunity to diversify into the domestic market, resulting in large influxes of New Zealand tourists:

This is the busiest winter we've ever had. In terms of Waitangi, we're up about 30 percent, in terms of this time last year. We're beating all last year…and its mainly because we've got a domestic market.
(Administrator at Te Tii B3 Trust, 2020)

There was, meanwhile, a clear priority to ensure employees were well catered for and that their wellbeing was protected:

> Aspirations in terms of our work - I think we're working on a worker wellbeing scheme, making sure everyone's mental and spiritual wellbeing is taken care of. Going into this holiday period [ensuring] that they feel secure in their job, secure in their finances…and just knowing that they're looked after. That sort of culture.
>
> (Office Manager at Te Tii B3 Trust, 2020)

Reponses to the Pandemic by Pacific Resorts in Rarotonga: A Focus on Staff Wellbeing

Pacific Resorts in the Cook Islands provides an excellent example of a business which took the personal wellbeing of its staff seriously during the pandemic period. There are four boutique beachfront properties across the islands of Rarotonga and Aitutaki which come under the Pacific Resorts umbrella. The land on which the resort properties are based is leased from the traditional landowners, who have also invested in the business so that they are the main directors. Some of these people are extremely "hands-on" and have been known to help out at every level of the business, including doing the gardening. The current CEO changed their slogan a few years ago from "absolute guest satisfaction" to "absolute guest and staff satisfaction". That ethos runs through the business.

We interviewed a manager at Pacific Resorts whose, role has a particular focus on staff wellbeing. As she posited, "…everyone needs to feel good in order to do their jobs". When asked why she cares so much about the resort staff, she noted:

> A lot of it is maternal practice…a maternal point of the view is the safest way to go, because the most genuine way to go is reaching out emotionally.
>
> (Manager, Pacific Resorts, 2021)

Ani was concerned about the mental health of her staff, especially foreign workers who were in the Cook Islands without their families and couldn't travel home for long periods of time due to the pandemic:

> With COVID…I was very curious about the mental health of my staff. Um, clearly because some of them were showing signs of depression. And - just illnesses that had popped up that doctors couldn't really diagnose, other than say that it must just be something that they're mentally inflicting on themselves…I had two staff that were very ill… the way they felt about COVID was impacting them physically. So, you know, they were losing a lot of weight… some of them have got children back in their islands whether it's Fiji or the Philippines - I think the whole weight of that made them feel even more [anxious].
>
> (Manager, Pacific Resorts, 2021)

So this manager decided to send out a quick "How are you feeling?" survey to staff. The result surprised her: "most of them had some level of anxiety or stress you know, [both] Cook islanders and expats…". Thus, during the long months of border closures in 2020 and 2021, the resort sought to support the physical, mental, and spiritual wellbeing of staff in a variety of ways.

In a later staff survey, our respondent suggested starting face-to-face staff briefing sessions with a prayer. She was a little nervous when doing so, as she explains: "I wanted to see if they would be offended, you know, because everyone's got different religions...". In fact, those who filled in this survey were all supportive of the idea. Thus, a system was established whereby a staff member could voluntarily open a briefing meeting with a prayer, but there was a rotational approach, so no denomination dominated in this space. Staff responded well to this:

> I've gotten so much feedback from everyone... [they liked that] they could pray in their own language. They could, they could just send wishes, if they didn't feel like they would need to acknowledge their God in front of the rest of us. It was a moment just to share something... you know, and I feel that small thing just made a massive difference and gave them a bit of their own time.
>
> (Manager, Pacific Resorts, 2021)

This manager believed it also helped staff to have a little more empathy regarding the challenging circumstances others were facing: "It made us feel like we need each other, and we need to be more patient and forgiving..." (Manager, Pacific Resorts, 2021).

Another way in which resort management supported staff was by encouraging them to get involved in a range of training opportunities that were on offer. Many of them completed first aid training with the Red Cross, for example, or undertook electrical or woodwork courses offered for free by the Cook Islands government during the closed border period, proudly displaying their certificates on the resort staff's private Facebook page. Our respondent explained how these expanded skillsets could stand staff in good stead in the future:

> ...so it's great cross training, because I think moving forward...it's gonna rely heavily on the current staff that we have been cross trained in all areas of the business....
>
> (Manager, Pacific Resorts, 2021)

The number of team-building and fitness activities for staff was increased during the Covid period. Mangers organised various fitness exercises to take place on a field adjacent to the resort: these were open to employees, their family members, and community members alike. They were popular and served the dual purpose of supporting the physical and mental wellbeing of staff, while providing a fun vibe in the community: old aunties and nannies would sit and watch, teasing or calling out support for those doing the fitness classes. This manager also established fun activities and games to keep morale up, for example, a trivia night centred around one of their training modules. Similar activities with a slightly competitive element, encouraged healthy banter among the staff.

Another activity that resort managers continued during the lockdowns was providing assistance to the broader community. Pacific Resorts has provided various forms of support to the Elderly Association, the Autism Association and schools in the area, providing opportunities for their staff to volunteer in these altruistic experiences:

> [We] didn't want too much focus on winning all the time ... [we wanted to] grow an interest and a heart in the island. We need social connections to keep this island going.
>
> (Manager, Pacific Resorts, 2021)

When asked why their business was so committed to the wider community, the manager expressed the desire for all on Rarotonga to be connected: "It's a lot of community involvement – it's exhausting but it's rewarding at the same time … it's all about enhancing those community relationships" (Manager, Pacific Resorts, 2021).

A Time for Reflection and Resetting Priorities

While undoubtedly many Indigenous business owners across Aotearoa and the Cook Islands have struggled with financial stress during the pandemic, some felt that the net effect of the "pandemic pause" in international tourism was positive because it made people a) explore other options to meet their needs, b) be more careful with their money management, and c) reflect on what was really the most important thing in life, rather than always chasing the dollars:

> I know this sounds a bit mean but I think this was a good thing…that opened people's eyes to what really, you know, what's really important… we just started to realise that money wasn't everything.
>
> (Cook Islands Male, 2022)

> People are just trying to make use what we have now on the island. A lot of people are going back to getting their seafood, making up their own food, you know cooking and the planting, the crops now being sold.
>
> (Cook Islands Elder, 2020)

While reflecting on their priorities in life, they have realised that over-reliance on the tourism industry is not good for the economy, and nor for their people:

> *And then at the beginning of this, just seeing Muri Beach just completely dead, you're like:* "What are we without these people?"…And then over time you see the locals taking their families to the beach, you know, people are spending time with their families and stuff and then everyone's back in the taro patches and cleaning their yards and…it's become a moment of self-reflection. It's given people the time to self-reflect and made them realise how reliant we have become on the tourism industry….
>
> (Cook Islands Male, 2020)

Having time for reflection – a rare thing for many business owners – combined with economic necessity definitely encouraged people to explore new ideas for their enterprises:

> In terms of myself; I'm definitely smashed, exhausted, but it's also given me a different outlook on work life balances, and it's changed our structure, our Trust, the governance levels, how we look at things.
>
> (Administrator at Te Tii B3 Trust, 2020)

> Families have been brought closer – before, [when we had] occasional gatherings, everyone was occupied with work, always busy, busy, busy. Such closeness has brought about ideas of how to survive – sharing ideas of how to set up enterprises….
>
> (Cook Islands Male, 2022)

> We also looked at other ways to make money from garage sales, selling produce, making crafts and turning latent skills or voluntary positions into job income earning potential.
>
> (Cook Islands Female, 2022)

Considering the challenges and lessons from the Covid-19 pandemic, interviewees in both Aotearoa and the Cook Islands were facing the future with a new outlook on life, one based on utilising the land and resources sustainably and in a way that benefits their people.

Many interviewees in Aotearoa and the Cook Islands spoke about diversifying the tourism product and target markets. For several participants at Waitangi, the opportunity to tell more of their own stories and to weave their culture more into the tourism offering was important:

We need to develop a better tourism industry. More culture [-based tourism]...or maybe [tours] for the schools as well and the camps.

(Marketing and Sales Manager at Te Tii B3 Trust)

I think Māori tourism is very lacking up here in the north...[we can reinvigorate it with] our marae, our groups and working together, setting out trails, Māori tourism trails.

(Administrator of Te Tii B3 Trust)

The community would like to centre "Māori" back in the Aotearoa tourism sector: "What's here? What's in the Bay of Islands? It's us telling our stories" (Chairman of Te Tii B3 Trust).

Looking beyond tourism was seen as important to Te Tii Trust members to make themselves more resilient to future shocks:

We are looking at diversifying our portfolios, tourism... at the moment we are heavily reliant on our holiday sector. So, looking at how we can [have]... maybe a joint venture with other Māori organisations locally.

(Administrator at Te Tii B3 Trust)

We've got papakāinga [community housing] prospects in mind. All of those things can now be explored...the next thing is developing our supermarket.

(Chairman of Te Tii B3 Trust)

In the less economically diverse the Cook Islands, the heavy reliance on tourism for a large part of their GDP, and the devastating impacts on their economy after almost two years of border closures, meant that there was also a lot of interest in diversification beyond tourism:

This Covid-19 has educated me the dangers of us relying solely on tourism.... Every time I talk to somebody, [we discuss] how are we going to diversify?

(Cook Islands Elder, 2020)

I think we just need to be very wise about our tourism development.

(Cook Islands Elder, 2020)

There appears to have been an increase in pro-environment sentiments among the Cook Islands people during the pandemic period and this will likely influence their future aspirations and the extent to which mass tourism fits into their plans: "This time to me is about restoring and renewing things [and] relationships and giving our environment time to restore and breathe again before it gets busy" (Cook Islands Elder, 2020).

People expressed that they want to bring back tourism in a more sustainable way. In particular, their treasured lagoon was a central point of concern for many people worried about how it had been impacted by tourism:

> Our village carries a lot of mana, especially because it has the lagoon. Muri is the backbone of the economy for tourism.
>
> (Interview – Cook Islands Elder, 2020)

> Tourism is filling our environment, that's killing our lagoon, and the other thing is waste....
>
> (Cook Islands Elder, 2020)

> There has to be a better game plan for when the tourists come through, maybe just get all the big guns, even the resorts just to butt heads and [explore] how we can better maintain our lagoon.
>
> (Cook Islands Male, 2020)

Conclusion

Very few studies have examined how indigenous enterprises have reacted to the volatility of tourism in relation to the coronavirus pandemic, and the various measures they have adopted to stay afloat. This chapter has discussed various ways in which the pandemic has challenged two Indigenous tourism enterprises in Aotearoa New Zealand and the Cook Islands, as well as discussing how they have adapted, often in culturally centred ways, and embodied resilience. We have also reflected on what people said they learned from having more time on their hands, during the "pandemic pause", and on what they want from future tourism.

In the face of shocks such as the global pandemic, Indigenous enterprises have taken different approaches, exercising their *tino rangatiratanga*. For many, the purpose has been to determine what's best for the people – both their family and staff and the land. With border closures and limitations on movement literally preventing external inputs and solutions to the challenges faced, it was clear that Indigenous entrepreneurs looked inwardly for ideas and solutions: they leaned on traditional knowledge and skills, drew strength from their culture and ancestors, and utilised their customary resources such as access to the land and sea to support their livelihoods. These are signs of an endogenous development approach that demonstrates the strengths and power of Indigenous people (Brohman, 1996).

Strong social and cultural bonds and access to customary land seemed to provide outstanding support to many entrepreneurs. More importantly, this chapter amplifies the need to grow recognition for business practices that encourage reciprocity and care (Harris & Wasilewski, 2004). In essence, Indigenous enterprises are culturally-oriented and focused on contributing to the family, local community, and other essential relationships within the operating environment. As tourism seeks practical lessons from the global pandemic, Indigenous businesses in Aotearoa New Zealand and the Cook Islands offer pathways to operationalising tourism in ways that are more considerate, sustainable, and equitable in how they operate (Puriri & McIntosh, 2019). They are not purely fixated on profit but recognise the value of protecting and investing in their relationships with the community and their collective social and ecological wellbeing as a means to function as a going concern.

It is time that government and business leaders listen to the voices of Indigenous entrepreneurs who have demonstrated great adaptive capacity and resilience during the pandemic and whose values often shine through in their business approach. There is much we could learn from them in terms of planning for more inclusive and sustainable tourism in the future. There is also a need to diversify beyond a heavy reliance on the tourism industry, where this exists. We thus advocate that, in line with international declarations, Indigenous entrepreneurs should be central to discussions on ways of reimagining tourism post-pandemic:

> …the reality is that Indigenous voices are not always given the opportunity to be heard [in planning for and managing tourism], despite this right being enshrined in the United Nations Declaration on the Rights of Indigenous Peoples 2007 (UNDRIP) and expressed specifically in respect of Indigenous tourism through the Larrakia Declaration (2012).
>
> (Hutchison et al., 2021)

References

Amoamo, M., Ruckstuhl, K., & Ruwhiu, D. (2018). Balancing indigenous values through diverse economies: A case study of Māori ecotourism. *Tourism Planning & Development, 15*(5), 478–495. https://doi.org/10.1080/21568316.2018.1481452

Awatere, S., Mika, J. P., Hudson, M., Pauling, C., Lambert, S., & Reid, J. (2017). Whakatipu rawa mā ngā uri whakatipu: Optimising the 'Māori' in Māori economic development. *AlterNative, 13*(2), 1–9. https://doi.org/10.1177/1177180117700816

Bargh, M. (2012). Rethinking and re-shaping indigenous economies: Māori geothermal energy enterprises. *Journal of Enterprising Communities, 6*(3), 271–283. https://doi.org/10.1108/17506201211258423

Brohman, J. (1996). New directions in tourism for third world development. *Annals of Tourism Research, 23*(1), 48–70. https://doi.org/10.1016/0160-7383(95)00043-7.

Bunten, A. C., & Graburn, N. H. H. (Eds.). (2018). *Indigenous tourism movements.* University of Toronto Press.

Cachon, J.-C. (2012). *Paleo aboriginal entrepreneurship: Evidence from Turtle Island.* International Conference on Small Busienss (ICSB), 11–13 June 2012, Michael Fowler Centre, Wellington, New Zealand.

Colbourne, R. (2021). Indigenous entrepreneurship. In T. M. Cooney (Ed.), *The Palgrave handbook of minority entrepreneurship* (pp. 319–348). Palgrave.

Colbourne, R., & Anderson, R. B. (Eds.). (2020). *Indigenous wellbeing and enterprise: Self-determination and sustainable economic development.* Routledge.

Dana, L.-P., & Anderson, R. B. (Eds.). (2007). *International handbook of research on indigenous entrepreneurship.* Edward Elgar.

Deakins, D., & Scott, J. M. (Eds.). (2020). *Entrepreneurship: A contemporary & global approach* (1st ed.). Sage.

de-Lima, I., & Weiler, B. (2015). Indigenous protagonism in tourism operations and management in Australia, Brazil, and New Zealand. *ASR CMU Journal of Social Sciences and Humanities, 2*(1), 7–37.

Galperin Bella, L., Chavan, M., & Muhidin, S. (2021). Indigenous entrepreneurs in Australia: Past, present, and future. In C. S. James, M. Adela, & M. Mark (Eds.), *Clan and tribal perspectives on social, economic and environmental sustainability* (pp. 35–47). Emerald Publishing Limited. https://doi.org/10.1108/978-1-78973-365-520211006

Harris, L. D., & Wasilewski, J. (2004). Indigeneity, an alternative worldview: Four R's (relationship, responsibility, reciprocity, redistribution) vs. Two P's (power and profit). Sharing the journey towards conscious evolution. *Systems Research and Behavioral Science, 21*(5), 489–503. https://doi.org/10.1002/sres.631

Hēnare, M. (2015). Tapu, mana, mauri, hau, wairua: A Māori philosophy of vitalism and cosmos. In C. Spiller & R. Wolfgramm (Eds.), *Indigenous spiritualities at work: Transforming the spirit of enterprise* (pp. 77–98). Information Age Publishing.

Henry, E. (2017). The creative spirit: Emancipatory Māori entrepreneurship in screen production in New Zealand [Article]. *Small Enterprise Research, 24*(1), 23–35. https://doi.org/10.1080/13215906.2017.1289853

Henry, E., Mika, J. P., & Wolfgramm, T. (2020). Indigenous networks: Broadening insight into the role they play, and contribution to the academy. *Academy of Management Proceedings, 2020*(1), 18715. https://doi.org/10.5465/AMBPP.2020.18715abstract

Hill, C. (Ed.). (2021). *Kia whakanuia te whenua: People, place, landscape.* Mary Egan Publishing.

Hutchison, B., Movono, A., & Scheyvens, R. (2021). Resetting tourism post-COVID-19: Why Indigenous peoples must be central to the conversation. *Tourism Recreation Research, 46,* 261–275. https://doi.org/1 0.1080/02508281.2021.1905343

Ingram, P. T. (1990). *Indigenous entrepreneurship and tourism development in the Cook Islands and Fiji* [Massey University]. Palmerston North, New Zealand.

Insch, A. (2020). The challenges of over-tourism facing New Zealand: Risks and responses *Journal of Destination Marketing & Management, 15,* 10037. https://doi.org/10.1016/j.jdmm.2019.100378

Joseph, R., Tahana, A., Kilgour, J., Mika, J. P., Rakena, M., & Jefferies, T. P. (2016). *Te pae tawhiti: Exploring the horizons of economic performance.* https://www.waikato.ac.nz/__data/assets/pdf_file/0006/322188/MAIN-FINAL-Te-Pai-Tawhiti-Report-Nov-11-2016.pdf

Kawharu, M., & Tapsell, P. (2019). *Whāriki: The growth of Māori community entrepreneurship.* Oratia.

Kingi, T. (2007). *Māori land ownership and management in New Zealand: Pacific Land Program Case Study 4.4.* Institute of Natural Resources, Massey University.

Kingi, T. (2008). Maori landownership and land management in New Zealand. In Australian Agency for International Development (Ed.), *Making land work volume two: Case studies on customary land and development in the Pacific* (Vol. 2, pp. 129–152). Australian Agency for International Development (AusAID), Canberra.

Massey University. (2017). *Pacific research principles.* Pacific Research and Policy Centre. https://communication.massey.ac.nz/massey/fms/Human%20Ethics/Do

Mead, H. M. (2003). *Tikanga Māori: Living by Māori values.* Huia.

Mika, J. P. (2014). Manaakitanga: Is generosity killing Māori enterprises? In P. Davidsson (Ed.), *Proceedings of the Australian Centre for Entrepreneurship Research Exchange Conference 4–7 February 2014, UNSW, Sydney, Australia* (pp. 815–829). Queensland University of Technology.

Mika, J. P. (2021). *Māori perspectives on the environment and wellbeing.* https://www.pce.parliament.nz/media/197165/mika-m%C4%81ori-perspectives-on-the-environment-and-wellbeing-pdf-21mb.pdf

Mika, J. P., Colbourne, R., & Almeida, S. (2020). Responsible management: An Indigenous perspective. In O. Laasch, R. Suddaby, E. Freeman, & D. Jamali (Eds.), *Research handbook of responsible management* (pp. 260–276). Edward Elgar.

Mika, J. P., Dell, K., Newth, J., & Houkamau, C. (2022). Manahau: Toward an Indigenous Māori theory of value. *Philosophy of Management,* 1–23. https://doi.org/10.1007/s40926-022-00195-3

Mika, J. P., Dell, K. M., Elers, C., Dutta, M. J., & Tong, Q. (2022). Indigenous environmental defenders in Aotearoa New Zealand: Ihumātao and Ōroua River. *AlterNative: An International Journal of Indigenous Peoples,* 1–13. https://doi.org/https://doi.org/10.1177/11771801221083164

Mika, J. P., Felzensztein, C., Tretiakov, A., & Macpherson, W. G. (2022). Indigenous entrepreneurial ecosystems: A comparison of Mapuche entrepreneurship in Chile and Māori entrepreneurship in Aotearoa New Zealand. *Journal of Management & Organization,* 1–19. https://doi.org/10.1017/jmo.2022.15

Mika, J. P., & Scheyvens, R. (2021). Te Awa Tupua: Peace, justice and sustainability through Indigenous tourism. *Journal of Sustainable Tourism, 30*(2–3), 637–657. https://doi.org/10.1080/09669582.2021.1912056

Mill, A., & Millin, D. (2021). *He manukura: Insights from Māori frontier firms.* https://www.productivity.govt.nz/assets/Documents/he-makukura/He-Manukura-Insights-from-Maori-frontier-firms.pdf

Miller, R. J., Jorgensen, M., & Stewart, D. (Eds.). (2019). *Creating private sector economies in Native America: Sustainable development through entrepreneurship* (1st ed.). Cambridge University Press.

Movono, A., & Dahles, H. (2017). Female empowerment and tourism: A focus on businesses in a Fijian village. *Asia Pacific Journal of Tourism Research, 22*(6), 681–692.

Nabobo-Baba, U. (2008). Decolonising framings in Pacific Research: Indigenous Fijian Vanua research framework as an organic response. *AlterNative: An International Journal of Indigenous Peoples, 4*(2), 140–154. https://doi.org/10. 1177/117718010800400210

Nana, G., Reid, A., Schulze, H., Dixon, H., Green, S., & Riley, H. (2021). *Te ōhanga Māori 2018: The Māori economy 2018.* https://berl.co.nz/sites/default/files/2021-01/Te%20%C5%8Changa%20M%C4%81ori%202018.pdf

NZIER. (2003). *Māori economic development: Te ōhanga whanaketanga Māori.* Te Puni Kōkiri.

O'Leary, Z. (2021). *The essential guide to doing your research project* (4th ed.). Sage Publications Ltd.

Phillips, T., Woods, C., & Lythberg, B. (2014). *Māori farming trusts: A preliminary scoping investigation into the governance and management of large dairy farm businesses.* DairyNZ.

Puriri, A., & McIntosh, A. (2019). A cultural framework for Māori tourism: Values and processes of a Whānau tourism business development. *Journal of the Royal Society of New Zealand, 49*(suppl), 89–103. https://doi.org/10.1080/03036758.2019.1656260

Ransfield, A. K., & Reichenberger, I. (2021). Māori Indigenous values and tourism business sustainability. *AlterNative: An International Journal of Indigenous Peoples, 17*(1), 49–60. https://doi.org/10.1177/1177180121994680

Reid, J. (2021). *Adopting Māori wellbeing ethics to improve Treasury budgeting processes.* https://www.pce.parliament.nz/media/197163/reid-adopting-m%C4%81ori-wellbeing-ethics-to-improve-treasury-budgeting-processes-pdf-12mb.pdf

Reid, J., & Rout, M. (2020). The implementation of ecosystem-based management in New Zealand – A Māori perspective. *Marine Policy, 117*(2020), 1–6. https://doi.org/https://doi.org/10.1016/j.marpol.2020.103889

Reid, J., Rout, M., Mika, J. P., Gillies, A., Ruwhiu, D., & Awatere, S. (2019). *Whenua, life, values programme: Report four—Testing a values-centred decision-support tool for Māori agribusiness.* University of Canterbury Ngāi Tahu Research Centre.

Reid, J., Rout, M., Tau, T. M., & Smith, C. (2017). *The colonising environment: An aetiology of the trauma of settler colonisation and land alienation on Ngāi Tahu whānau.* Ngāi Tahu Research Centre.

Ridley-Duff, R. J., & Duncan, G. (2015). What is critical appreciation? Insights from studying the critical turn in an appreciative inquiry. *Human Relations, 68*(10), 1579–1599. Https://doi.org/10.1177/0018726714561698

Rout, M., Awatere, S., Mika, J. P., Reid, J., & Roskruge, M. (2021). A Māori approach to environmental economics: Te ao tūroa, te ao hurihuri, te ao mārama—The old world, a changing world, a world of light. *Oxford Research Encyclopedia of Environmental Science.* https://doi.org/10.1093/acrefore/9780199389414.013.715

Royal, T. A. C. (Ed.). (2003). *The woven universe: Selected writings of Rev Māori Marsden.* Estate of Rev Māori Marsden.

Russell-Mundine, G. (2012). Reflexivity in Indigenous research: Reframing and decolonising research? *Journal of Hospitality and Tourism Management, 19*(1), 85–90. https://doi.org/10.1017/jht.2012.8

Scheyvens, R., Banks, G., Meo-Sewabu, L., & Decena, T. (2017). Indigenous entrepreneurship on customary land in the Pacific: Measuring sustainability. *Journal of Management & Organization, 23*(6), 774–785. https://doi.org/https://doi.org/10.1017/jmo.2017.67

Smith, L. T. (1999). *Decolonising methodologies: Research and indigenous peoples* (1st ed.). Zed Books.

Stats NZ (2019). *Tourism Satellite Account: 2019.* www.stats.govt.nz.

Stats NZ. (2021). *Tatauranga umanga Māori – Statistics on Māori businesses: 2020 (English).* https://www.stats.govt.nz/information-releases/tatauranga-umanga-maori-statistics-on-maori-businesses-2020-english

Stats NZ. (2022). *Māori authorities show resilience despite COVID-19 impacts.* https://www.stats.govt.nz/news/maori-authorities-show-resilience-despite-covid-19-impacts

Stewart, D., Verbos, A. K., Black, S. L., Birmingham, C., & Gladstone, J. S. (2017). Being Native American in business: Culture, identity, and authentic leadership in modern American Indian enterprises. *Leadership, 13*(5), 549–570.

Tahu, T. R. o N. (n.d.). Ngāi Tahu Annual Reports. *Te Rūnanga o Ngāi Tahu.* Retrieved July 3, 2022, from https://ngaitahu.iwi.nz/investment/ngai-tahu-annual-reports/

Te Rito, J. S. (2007). Whakapapa: A framework for understanding identity. *MAI Review LW, 1*(3), 10.

Trosper, R. (2009). *Resilience, reciprocity and ecological economics: Northwest Coast sustainability* (1st ed.). Routledge.

Vao'iva Tofilau, M. a. S. (2018). *Lupe fa'alele - releasing the doves: Factors affecting the successful operation of Samoan businesses in New Zealand* [Doctor of Philosophy in Sociology, Massey University]. Auckland, New Zealand. https://mro.massey.ac.nz/bitstream/handle/10179/15183/02_whole.pdf?sequence=2&isAllowed=y

Vunibola, S. (2021). *Eda dravudravua e na dela ni noda vutuni-i-yau: Customary land and economic development: Case studies from Fiji* [Doctor of Philosophy, Massey University]. Palmerston North, New Zealand. https://mro.massey.ac.nz/bitstream/handle/10179/16302/VunibolaPhDThesis.pdf?sequence=1&isAllowed=y

Vunibola, S., & Scheyvens, R. (2019). Revitalising rural development in the Pacific: An itaukei (indigenous Fijian) approach. *Development Bulletin, 81*(November), 62–66.

Vunibola, S., & Scobie, M. (2022). Islands of Indigenous innovation: Reclaiming and reconceptualising innovation within, against and beyond colonial-capitalism. *Journal of the Royal Society of New Zealand, 1*–14. https://doi.org/10.1080/03036758.2022.2056618

Warren, L., Mika, J. P., & Palmer, F. R. (2018). How does enterprise assistance support Maori entrepreneurs? An identity approach. *Journal of Management & Organization, 23*(6), 873–885. https://doi.org/10.1017/jmo.2017.73

Wolfgramm, R., Spiller, C., Henry, E., & Pouwhare, R. (2019). A culturally derived framework of values-driven transformation in Māori economies of wellbeing (Ngā hono ōhanga oranga). *AlterNative: An International Journal of Indigenous Peoples*, 1–11. https://doi.org/10.1177/1177180119885663

SECTION 4

Indigenous Knowledge and Rights

Ensuring the incorporation of Indigenous knowledge into the management of tourism development is essential for achieving the desired relationships between Indigenous communities, tourists, and the tourism industry. The changing Ainu situation in Japan is an attempt to increase Indigenous participation in creating and offering Indigenous forms of tourism but there are arguments (Hashimoto) that the new law does not go far enough. Unsuitable development can cause conflicts between groups, result in unacceptable impacts (cultural, social, environmental and economic) on Indigenous communities and often prove unsuccessful in providing lasting and desired tourism enterprises.

Incompatible activities such as extractive industries (mining, forestry, fishing) as Calfucura notes, can cause major conflicts and problems both with resource management and development, and with government policies that fail to respect Indigenous rights and fail to incorporate Indigenous knowledge. Other land and resource uses are often allowed which impact negatively on traditional activities, as in the case of the Sami, where both land resources and cultural traditions have been assigned to non-Indigenous parties, creating protests over loss of access and control of land and artefacts as Viken illustrates. Even where attempts have been made to improve situations, Hashimoto points out there often still remain elements of appropriation and dominance that prevent the full acceptance of Indigenous knowledge and the primacy of Indigenous rights.

The establishment of national parks, creations which, perhaps more than any other type of protected areas, one would expect to rely heavily on local and Indigenous knowledge and to be protective of Indigenous rights have sometimes proved just the opposite. While intending to provide supplementary and replacement sources of income for Indigenous groups, national parks can often have little positive impacts on such groups, and as Degarege points out may aggravate rather than ease existing problems such as inadequate food production. The lack of Indigenous involvement in the establishment of such bodies as national parks is well illustrated by Mason et al. with their example of the disregard for traditional Indigenous activities at the time of the creation of the Rocky Mountain National Parks in Canada, and the difficulty in gaining recognition of Indigenous needs and legitimate claims for access for traditional gatherings and other activities in the parks.

235

DOI: 10.4324/9781003230335-21

Only in recent years have things improved, along with the creation of alternative forms of protected areas where Indigenous groups are involved in the creation and management of such areas from before their establishment. Paskova et al. note the ability of Indigenous groups and their knowledge and traditions to support the sustainability component involved in the establishment and operation of Geoparks in South America. The creation of geoparks reflects the desire to maintain and preserve specific examples of landscape and landscape forms and processes, features which are inescapably linked to Indigenous beliefs and activities and dependent on Indigenous knowledge for their conservation and survival as viable entities.

18

THE INTEGRATION OF INDIGENOUS KNOWLEDGE IN CANADIAN PROTECTED AREAS TO FOSTER CONSERVATION, RECONCILIATION, AND TOURISM DEVELOPMENT

Courtney W. Mason, Bill Snow and Jason W. Johnston

Introduction

For many Indigenous communities around the globe, the histories of tourism development around parks are filled with experiences of displacement. In Canada, the nation's earliest protected areas, the Canadian Rocky Mountains National Parks, play a central role in the Canadian imaginary of what parks should represent: beautiful landscapes, wilderness, and conservation. This romantic view of the Rocky Mountains National Parks erases the traumatic legacies around the formation of the parks system and early tourism development in Western Canada (Snow 2005). Collaborative research with Nakoda First Nations has demonstrated that the creation of the Rocky Mountains National Parks (especially Banff and Jasper) had serious impacts on their communities (Mason 2014; Johnston & Mason 2020a). The histories of Indigenous experiences in many Canadian parks and tourism industries have been fraught with exploitation and cultural loss (Cruikshank 2005; Sandlos 2007, 2014). The global experiences of Indigenous peoples around the formation of parks are equally as problematic (Rangarajan 1996; Neumann 1998; Keller & Turek 1999; Ruru 2010, 2012; Spence 1999).

Currently in Canada, many Indigenous communities are building tourism and recreation infrastructure in or near parks as a mechanism to enhance sustainable economic development. Despite the problematic histories of park formation, current co-management (Indigenous nations and levels of government) practices have proven to be beneficial alternatives for a number of Indigenous communities in Canada (Zurba et al. 2019). Colonial practices of park management in Canada are being replaced by consultation processes that favor Indigenous management frameworks (Mason 2018). Led by ground-breaking legislation supporting Indigenous land rights, shifting dynamics of colonial power are fostering both new designations of parks to emerge and management practices in established parks to evolve, that provide cultural and socio-economic opportunities (Youdelis et al. 2021). In this chapter, we profile both of these developments.

DOI: 10.4324/9781003230335-22

Despite the number of histories of the Canadian Rockies, there are few scholarly works that consider Indigenous histories, particularly those that collaboratively consult Indigenous communities for their perspectives (Snow 2005). Little is known about the impacts in Nakoda communities of being displaced from their lands which include consequences for Indigenous-centered approaches to education, health and cultural continuities, and also their abilities to continue their subsistence practices (Mason 2014). More research is required at the grassroots level to understand how these conservation and tourism histories influence the contemporary lives of Indigenous peoples and the current decision-making processes in communities.

Guided by Indigenous methodologies (IM), this research examined the following key questions: How have colonial policies impacted Indigenous communities through tourism development in early Canadian parks? What is the future for Indigenous conservation models and tourism facilitated by new park designations and shifting relations in established parks?; How can tourism in parks be a key facilitator of reconciliation and decolonization processes in Canada? While this chapter centers on the foundational Rocky Mountains National Parks in Western Canada, where possible we try and extend the analysis to broader Canadian Indigenous contexts. We argue that Indigenous-led conservation practices in parks have the potential to enhance healthy ecosystems, regional economies, and the preservation of cultural values.

Methods and methodological approach to foster a research process that is collaborative in orientation and that holds Indigenous perspectives at its core throughout the entirety of the research project, IM guided the collaborative community-based research. IM focus on Indigenous research paradigm ideas of trust, respect, reciprocity, and inclusion (Kovach 2009). IM highlights inequitable power relationships (Tuhiwai Smith 2012) and helps ensure that communities' interests are recognized and access to sensitive material is appropriately guarded (Battiste & Henderson 2000). The authors have built strong researcher-community relationships by spending extended time in Nakoda communities, listening to and learning first-hand about local perspectives over a number of years. This paper is based on a secondary review of literature, analysis of government policy and legal documents, and the collective research experience of the co-authors who have worked at the grassroots community level on tourism development and Indigenous/park relations with Nakoda communities in the Canadian Rocky Mountains for decades. Most of this work relies on oral histories with Nakoda Elders and knowledgeable land users. It is important to note that one of the co-authors (Bill Snow) is Nakoda and has been working within his own community and cultural context.

The Histories of Colonial Power and Tourism Development in the Canadian Rocky Mountain Parks

As Indigenous peoples throughout Canada and internationally consider tourism development through parks as tools for economic growth, self-empowerment, and the preservation of their cultural practices, it is imperative to assess the history of these experiences of participation in tourism industries. The Banff-Bow Valley was a key gathering place for diverse groups of Indigenous peoples for centuries prior to the development of a tourism economy or even a European presence in the region. The Nakoda, and other groups of Indigenous communities, lived throughout the Banff–Bow Valley in what would become the western Canadian province of Alberta. For example, the Banff-Bow Valley and the Bow River, called *Mînî Thnî Wapta* (Cold Water River), have been the traditional spiritual center of the Nakoda peoples since time immemorial (Mason 2014). As the giver of life, the valley provides traditional foods, medicinal plants, shelter, animals to hunt, as well as sacred areas and vision quest sites for Nakoda peoples. The Bow River forms the center of

Nakoda culture, economies, family, and ways of living off the land (Snow 2005). Beginning with the arrival of Europeans to the valley in the late 18th century, Indigenous peoples were forced to undergo a series of significant changes that would alter aspects of a well-established way of life that had persisted for millennia.

Increasing significantly with the emergence of the package travel business in the 1860s and the creation of the British travel firm Thomas Cook (Walton 2010), Western elites toured mountain ranges like the European Alps and later further afield to the Canadian Rockies. Elite tourists were attracted to these regions as recreational experiences began to be on offer. Urban-elite conceptions of these environments also arrived with the affluent tourists and the flow of economic capital that facilitated the development of the tourism industry (Mason 2008). The Canadian Pacific Railway (CPR), became one of the world's largest travel companies by the turn of the 20th century (Hart 1983).

From the 1880s until the middle of the 20th century, the development of a tourism economy in the Banff–Bow Valley initiated a dynamic period in the region's history. Local Indigenous peoples participated in the tourism industry which was central to the marketing of the Banff townsite. By selling certain images of the region while actively concealing others, tourism entrepreneurs promoted the Banff–Bow Valley as an international tourist destination. Nakoda community members influenced these tourism promotional campaigns through their involvement in the tourism industry as guides, cultural performers, and sporting participants (Mason 2020). The 1887 formation of Rocky Mountains Park (which became Banff National Park in 1930) was the beginning of tourism infrastructure development in the Canadian Rocky Mountains. As a joint venture between the Canadian federal government and the CPR, the national park was the first of its kind in Canada and was originally established as a means of generating railway tourism visitation with few conservation objectives considered (Hart 1983). The park became a convenient way to establish a monopoly on transportation access to the region that effectively controlled development. Although this is not a glamorous account of the formation of the Canadian park system, these park decisions were designed to centralize control of the lands and restrict access to the region (McNamee 1993).

Indigenous communities were displaced from their lands through the formation of Banff and later Jasper (1907) National Parks. Research (Mason 2008) has demonstrated that throughout the first few decades of the 20th century, Nakoda peoples were continually denied access to the region because their subsistence practices (hunting, fishing, gathering) were in direct conflict with Euro-Canadian perspectives of "conservation" and the objectives of an emerging tourism industry For Nakoda peoples, whose traditional territories include the areas of Banff and Jasper National Parks, the struggles of first being displaced from, and then later being denied access to, the newly formed parks were instrumental to numerous forms of cultural loss. A pass system was introduced whereby Nakoda peoples had to apply to local government officials to leave the reserve for any purpose, this included opportunities to visit family and friends on neighboring reserves in the region or access sacred sites (Snow 2005). Entering the foothills and mountains of the Banff-Bow Valley to hunt, fish, and gather was not viewed positively by officials. This process of displacement and denial was facilitated by park management with the support of a variety of actors that included the police, missionaries, government officials, and tourism entrepreneurs. These agents of the colonial bureaucracy were motivated by two objectives. First, it was imperative to ensure that Nakoda subsistence practices did not interfere with growing tourism economies. While mountaineering and skiing were important activities in Banff and Jasper later in the 20th century, sport hunting and fishing were the primary recreational practices that initially attracted urban elite tourists to the region. This was the case until the National Parks Act (1930) which established Banff National Park (BNP) boundaries at 6,642 km^2 and made hunting, fishing, and gathering illegal or highly regulated (Mason 2020). Nakoda subsistence practices put local communities in direct competition with local tourism businesses and conflict with park rangers. Second, it was also critical that Nakoda community members remained on reserves to be exposed to assimilatory institutions like the church and residential schools.[1] Similar to many Indigenous communities across the nation, Nakoda peoples are still healing from the separation from their sacred territories and the cultural repression they endured (Snow 2005). While justifications included protecting wildlife, it was these two reasons that explain why agents of the colonial bureaucracy collaborated to deny local Indigenous communities from accessing the newly formed national parks (Mason 2014; Johnston & Mason 2020b).

To parallel what occurred internationally for Indigenous peoples during this period, colonial governments, particularly in Africa, Asia, and the Pacific, were also pursuing similar policies concerning Indigenous peoples. Under the guise of "conservation" and "wildlife management," Indigenous communities were displaced, access to lands was denied, and their subsistence practices were either severely restricted or in some cases rebranded as illegal. There are a number of excellent international studies on the exclusion of Indigenous peoples from parks and protected areas, some of which also clearly involved tourism development objectives (Chatty & Colchester 2002; Neumann 2008; Rangarajan 1996; Ruru 2012; Carr 2017).

The Banff Indian Days Festival and early Indigenous tourism economies Ironically, while tourism was used as a justification to exclude Indigenous peoples from the parks, it was also the main way that communities regained access to sacred sites and traditional territories inside the Rocky Mountain National Park boundaries (Banff and Jasper). Throughout the first few decades of the 20th century, Nakoda peoples refused to accept some colonial policies and continued to access the Rocky Mountain Parks for cultural purposes, as well as to hunt, fish, and gather. Nakoda leaders also sought out opportunities for their people to engage in Banff's developing tourism industries and some community members found work as guides (Mason 2020). The Banff Indian Days tourism festivals are an example of how tourism economies were significant spaces for Indigenous peoples during a period in Canadian society when they faced incredible levels of marginalization and discrimination (Mason 2014). The Indian Days tourism festivals began in 1894, but occurred annually from 1910 to 1972 and were one of the most important tourist events in the entire region. At their height in the 1920s, the festivals attracted over 70,000 tourists which facilitated substantial growth of the local tourism economy (Meijer-Drees 1993). The Indian Days consisted of numerous activities that profiled the sporting and cultural practices of Indigenous peoples. The festivals originally involved only Nakoda peoples, but throughout the decades several other Indigenous groups from the region also participated, including Cree, *Ktunaxa* (Kootenay), *Tsuu T'ina* (Sarcee), *Pikunni* (Peigan), *Siksika* (Blackfoot), and *Kainai* (Blood) and *Secwépemc* (Shuswap) (Meijer-Drees 1993). At the turn of the 20th century, very few Indigenous peoples lived in Banff townsite as they were effectively displaced from the surrounding national park. There were also few spaces during this period that afforded extensive interaction between local Indigenous peoples, non-Indigenous Canadian visitors to BNP, who mostly arrived from urban centers, and international tourists. It is for these reasons that the Indian Days became an essential location for the exchange of intercultural knowledge (Mason 2015). Through their engagement in the festival, The Nakoda peoples challenged prevailing stereotypes of their cultures and returned to important locations for ceremonial purposes. The Indian Days offered unique socio-economic, political, and cultural opportunities, including the space to celebrate and reinforce Indigenous identities. It was through participation in the burgeoning tourism industries within the Rocky Mountains that some Indigenous community members refused colonial structures and defied limiting definitions of their cultural practices (Mason 2008).

Jasper National Park: Disposession and Indigenous Stewardship

Similar to the cultural contexts of Banff, the mountains and valleys of what is now Jasper National Park (JNP) were home to diverse nations of Indigenous peoples. There are 26 Indigenous communities that have traditional ties to the area redefined as JNP. These include communities from the Beaver (Dane-zaa), Cree, Ojibway, Secwépemc (Shuswap), Sioux, Aseniwuche Winewak, the Métis Nations and of course the Nakoda (Johnston & Mason 2020a). Established in 1907, Jasper was first created as a forest park. It officially became a national park in 1930 (MacLaren 2011), covering over 10,878 km². The park was formed without consideration for, or consultation with, the Indigenous peoples who called these lands home. As occurred in the formation of Banff, Indigenous peoples were forcibly removed from Jasper under the guise of conservation concerns by park wardens, deputized non-Indigenous locals, and the police. Park management claimed that Indigenous hunters were the cause of dwindling wildlife populations in the park, while the roles of forestry, mining, railway construction, and tourism on wildlife and key habitats were ignored (Binnema & Niemi 2006). Since the Supreme Court decisions that recognized Aboriginal title in the 1970s (Langdon et al. 2010), Indigenous peoples have been asserting some measure of control over their traditional lands in Jasper and working to incorporate their cultures back into the landscapes from which they were displaced (Johnston & Mason 2020c).

Consultation Processes and the Integration of Indigenous Knowledge in Banff and Jasper

Despite these problematic colonial histories, progress is being made in what many would argue are the oldest and most conservative parks in Canada. In BNP, the Nakoda in particular have consistently been increasing their presence in the park as they were invited by Parks Canada to return to sacred locations in 2001 and gained plant and medicinal harvesting rights in 2004. Nakoda Elders became involved with elk culls in 2007 and the monitoring of grizzly bears in 2016. These experiences encouraged further collaboration on BNP conservation initiatives and Nakoda peoples led the Buffalo Treaty in 2015 which enabled new conservation partnerships around the successful reintroduction of plains bison to the park in 2017 and the cultural monitoring of bison by Nakoda researchers in 2022. The 2013 redesign of the Cave and Basin National Historic Site, which included extensive consultation and collaboration with Nakoda artists and Elders on Indigenous interpretation of the Banff Hot Springs, is another example of how Parks Canada is working on reconciliation issues with local Indigenous communities (Mason 2018).

In Jasper, similar types of progress on consultation and the assertion of Indigenous rights inside park spaces have been occurring. The formation of the Jasper Indigenous Forum, as an unofficial Indigenous advisory group, occurred in 2006. The Indigenous Forum was created to work with Jasper National Park management on various projects and consultations processes (Youdelis 2016). Despite some notable challenges, the Jasper Indigenous Forum helped to forge new relationships between Indigenous nations who have traditional connections to Jasper and the management team. These consultations included tourism development projects. Following the establishment of the forum, the Indigenous Cultural Interpretation Program was developed in 2008. It soon became a priority of the forum to identify a Cultural Use Area within Jasper that would be suitable for small gatherings and ceremonies. In 2011, the Cultural Use Area was established in Jasper and was considered a milestone achievement (Johnston 2018). The creation of this site was a reconciliation initiative for Indigenous groups who were displaced from the park earlier in the 20th century. The Cultural Use Area was a great accomplishment for the Indigenous Forum who had long been at

Table 18.1 The integration of Indigenous knowledge in Banff and Jasper National Parks

Banff National Park	Jasper National Park
1887: Formation of Rocky Mountain National Park	1907: Formation of the Jasper Forest Park
1930: Establishment of Banff National Park under the *National Parks Act*	1930: Establishment of Jasper National Park under the *National Parks Act*
2001: Nakoda communities invited by Parks Canada to return to sacred sites which they were previously restricted from accessing	2006: Formation of the Jasper Indigenous Forum to work with park management to consult on various tourism projects, traditional use areas, and conservation initiatives
2004: Nakoda gained plant and medicinal harvesting rights within park boundaries. In 2012, a Park Pass Ceremony (lifetime passes) was conducted that allowed Nakoda peoples to collect medicines in Banff National Park	2008: Indigenous Cultural Interpretation Program developed
2007: Nakoda peoples involved in elk culls in collaboration with Parks Canada staff	2011: Establishment of the Indigenous Cultural Use Area for gatherings and ceremonies within park boundaries
2013: Redesign of Cave and Basin National Historic Site in consultation with Nakoda artists and Elders	2015: Indigenous Cultural Awareness Training for Parks Canada staff
2015: The Buffalo Treaty signed, led by Nakoda peoples and supported by other US-Canada Indigenous Nations	2017: Simpcw First Nation harvesting of mammals in cooperation with Parks Canada
2016: Nakoda peoples involved in cultural monitoring of grizzly bears in the Banff Bow Valley (in collaboration with Environment Canada)	2018: Consultation between the Jasper Indigenous Forum and park management to create and plan for an Indigenous Cultural Exhibit
2017: Reintroduction of plains bison into Banff National Park (a collaboration with Parks Canada staff in Banff and Nakoda peoples). The Nakoda peoples complete a cultural monitoring of bison study (2022) that informs that current reintroduction program.	2021: Construction begins on the new Indigenous Cultural Exhibit (anticipated completion in 2022)

odds with the park over the inability of local Indigenous groups to practice their ceremonies as well as collect medical plants on ancestral lands (Johnston & Mason 2020a). Further developments include the introduction of an Indigenous Cultural Awareness Training for Parks Canada staff in 2015 and the planning for the creation of an Indigenous Cultural Exhibit in Jasper in 2018. Another notable achievement was the Simpcw First Nation Traditional Hunt, which first took place inside the park in October 2017. For the hunt, members of the Simpcw, including Elders, youth, men, and women, reconnected with their traditional lands during the sustainable harvest period. The Simpcw considered the hunt as an exercise of their constitutionally protected rights to hunt on their traditional territory. Park management worked to ensure there was public awareness of the hunt, proper consultation with Simpcw First Nation, and the sustainability of the wildlife in the park (Johnston 2018).

There have also been some changes to Indigenous representation in the park and some progress is being made to repair past relationships in reconciliation efforts, but more work is needed in this realm (Johnston & Mason 2020b). Members of the Jasper Indigenous Forum have expressed the need for traditional place names to be incorporated into the park and the inclusion of public signage that recognizes that the park is on their traditional lands (Johnston 2018). The incorporation of respectful Indigenous content into national parks' interpretive programming and signage, including materials used in tourism industries, will foster reconciliation efforts and further improve relationships between levels of government and local Indigenous Nations. These initiatives will help Indigenous communities to generate a greater sense of ownership over, and expression of, their cultural heritage. It is also important to note that some forum members want their cultures and histories incorporated into the park, not for visitors, but for themselves (Johnston 2018). These changes could increase self-determination, reclaim histories, languages, and cultures, and improve community health (Shultis & Heffner 2016).

More Indigenous cultural content will also deconstruct negative stereotypes created through misrepresentations and facilitate broader education on Indigenous histories of the region for Euro-Canadians and international tourists alike. Incorporating Indigenous representations is one way that the tourism economy can support bridge-building and larger reconciliation initiatives. However, much more significant changes are afoot that are impacting the roles and responsibilities of Indigenous peoples in key decision-making processes in new park designations. The shifting dynamics of colonial power are fostering new opportunities that simultaneously support Indigenous land rights, conservation targets, and socio-economic development. As demonstrated below, tourism, and especially Indigenous tourism, features prominently in many of these new park designations that are designed to achieve multiple community and conservation objectives.

Contemporary Indigenous Park Management Frameworks and Indigenous Tourism

Indigenous involvement and leadership in protected area management varies in structure and context across Canada. As part of this shift in protected area establishment and management, a range of joint decision-making processes have been employed under the broad term co-management (Clark & Joe-Strack 2017). Since the 1970s, Indigenous peoples have increasingly asserted their constitutionally protected rights and title through a variety of legal and political means in the face of mounting tensions between natural resource extraction industries and Indigenous communities (Mabee et al. 2013). One aspect of this has been the involvement in co-management agreements ranging from advisory roles, to shared governance and consensus-based decision-making (Artelle et al. 2019; Hawkes 1996). Alongside many of these agreements, Indigenous nations have continued to educate visitors, steward areas, and protect sacred sites through Watchmen or Guardian programs (Artelle et al. 2019). Most notably, the Haida Watchmen in the cooperatively managed Gwaii Haanas National Park Reserve, have been active for over 20 years, as caretakers of sacred areas and throughout the busy summer tourist season. In these positions, they educate visitors about Haida culture, the heritage of village sites, and land and sea safety (Thomlinson & Crouch 2012). These collaborative initiatives have been important avenues for Indigenous communities to assert their rights and communicate their knowledge to visitors through partnerships with tourism operators and park management (Armitage et al. 2011). However, not all co-management arrangements share nation-to-nation relationships, and some state governments involved in these agreements are criticized for not supporting Indigenous-led decision-making processes (Finegan 2018).

There are also other forms of Indigenous protected area leadership that exist independently from state-governance structures. One of the main examples in Canada that is especially prevalent in British Columbia, is Tribal Parks, which are projections of Indigenous sovereignty and nations' assertion of rights and responsibilities as caretakers to their traditional territories. This designation of park is not currently recognized by provincial or federal protected area structures (Murray &

King 2012). Instead, existing Tribal Parks have looked to local allies, tourism operators, and non-government organizations to work with and support Indigenous governance over these areas (ICE 2018; Murray & Burrows 2017). For example, the *Tla-o-qui-aht* Tribal Parks have created a Tribal Parks Ally Program, whereby participating organizations in the Tofino area of Vancouver Island undergo a certification process. This certification includes acknowledgement of the *Tla-o-qui-aht* First Nation's continued management of their traditional territories, the need for an effective Guardians Program, and an agreement to contribute a 1% Ecosystems Service Fee to the Nation for the ongoing environmental management of their territories. In return, the *Tla-o-qui-aht* First Nation provides ongoing advice, educational resources, and consultation throughout the partnership (*Tla-o-qui-aht* First Nations 2022).

In 2017, following a landmark Supreme Court of Canada ruling that found the *Tsilhqot'in* Nation have title to traditional territories, the *Dasiqox* Tribal Park was announced. Located in central British Columbia, the park covers 3,120 km² and permits locally-driven tourism development. It protects the cultural and ecological values of the *Tsilhqot'in* as it asserts land rights on the basis of Indigenous law. This provides an alternative model of governance, land management, and local economy that notably includes Indigenous-led tourism. Tribal Parks support sustainable livelihoods of local peoples that allow communities to facilitate tourism experiences and other economic initiatives in ways that align with Indigenous goals for their ancestral territories (Bhattacharyya & Dasiqox Tribal Park 2018). The application of this diversity of protected areas set the scene for the creation of a broad classification of park designation referred to as Indigenous Protected and Conserved Areas (IPCAs).

The term IPCAs emerged from the Indigenous Circle of Experts, who are an Indigenous group formed to advise federal, provincial, territorial, and Indigenous levels of governments on how to support biodiversity targets while recognizing Indigenous knowledge systems. The Indigenous Circle of Experts' report, *We Rise Together* (2018), was developed in response to Canada's commitment to international biodiversity protection targets. Similar to many nations across the globe, Canada has been experiencing biodiversity declines, ineffective protected area conservation, and environmental degradation from unsustainable resource extractive industries. Concurrently, Indigenous nations continue to assert their rights and responsibilities to ancestral territories through a variety of means, both within and outside of state-recognized structures. In response to these challenges, international agencies formally called for a change in conservation paradigms through the Durban Accord (2005) which explicitly highlighted the need to involve Indigenous peoples in creating and managing protected areas (IUCN 2005). Additionally, in 2010, the Convention on Biological Diversity created a revised *Strategic Plan for Biodiversity 2011–2020*, which clearly emphasized the importance of including Indigenous communities in biodiversity protection, specifically in Targets 14 and 18 (CBD 2010). The creation of the 2007 United Nations Declaration of Indigenous Rights (UNDRIP) has further outlined the need for nations to respect and support Indigenous peoples' self-determination (UNGA 2007).

Although late to develop a national strategy to meet biodiversity objectives, in 2016 the Canadian Federal government adopted "*The 2020 Biodiversity Goals and Targets for Canada*," which addresses international calls for working with Indigenous nations. Canada's first target reflects the Aichi Target 11 to protect 17% of terrestrial areas and inland water, and 10% of coastal and marine areas by 2020 (ECCC 2016). This area-based target was met by Indigenous nations already involved in stewardship and conservation of their traditional territories and opened the door for other communities to formally begin this process. The Indigenous Circle of Experts and Canadian parks and protected area bodies propose IPCAs as one way to support both conservation and reconciliation goals (Parks Canada 2018).

In Canada, IPCAs represent a variety of different land protection initiatives in governance structures and management systems. These protected area designations have been proposed as mechanisms to formally recognize existing Indigenous management areas and guide the creation of new protected areas. At their core, IPCAs are Indigenous-led, represent a long-term commitment to conservation, and support Indigenous livelihoods, rights, and responsibilities. Specifically in Canada, they are opportunities for reconciliation and restoring nation-to-nation relationships, building sustainable local economies with an emphasis on tourism, and an acknowledgement of Indigenous rights and title, which includes constitutional and treaty rights (ICE 2018). In the 2018 Federal Budget, there was support for a $1 billion (CDN) nature fund to resource the creation of new IPCAs.

On October 11th, 2018 the Dehcho First Nations Assembly designated *Edéhzhíe* (*eh-day-shae*), the first official IPCA in Canada, in their traditional lands of the southwestern part of the Northwest Territories. At 14,218 km², it covers an area more than twice the size of BNP. *Edéhzhíe* is ecologically important to the Dehcho Dene culture, language, and ways of life. By forming *Edéhzhíe* as an IPCA, the management board will make its decisions by consensus while encouraging an Indigenous presence on the land. Elders have often referred to the significance of the area as a critical food harvesting location that has sustained communities for many generations. As local communities encounter even more barriers to food security, such as climate change and mounting food production and shipping costs, it is essential to protect these lands from further industrial development by establishing an IPCA (Mason 2018).

Since *Edéhzhíe* was established in 2018, there have been numerous announcements of new IPCAs. Many of these are located in Canada's north where vast, unfragmented forest and tundra, land-claim agreements, and prevailing Indigenous land-use planning have provided favorable circumstances (Coristine et al. 2018). In addition to the rural north, the interior of the province of British Columbia is also being targeted for IPCA development due to the presence of unceded Indigenous territories that exist outside of formal Treaty agreements.

Despite increased resource allocation for IPCAs and new biodiversity targets, Canada failed to protect 17% of lands and inland waters and 10% of marine and coastal areas by 2020. In response, the Federal government recommitted to these goals and proposed to protect 30% of Canada's lands, fresh water, and oceans by 2030. This means that if biodiversity is the focus, it is crucial to preserve ecosystem services that target biodiversity hotspots (Mitchell et al. 2021). IPCAs are a key strategy to accomplish this, and they will play a critical role if Canada is to reach those targets because Indigenous territories overlap significantly with regions of high biodiversity and carbon sequestration potential (Schuster et al. 2019).

Outside of the biodiversity goals, the importance of IPCAs for the sustainability of Indigenous communities, including tourism development as a central strategy to support socio-economic growth, cannot be overlooked and must be recognized alongside conservation objectives (Tran, Ban, & Bhattacharyya 2020). Importantly, tight timelines to secure area-based targets through the creation of IPCAs could reify colonial processes if they lack the consultation required to build strong relationships that enhance conservation and rural livelihoods, while also meeting biodiversity and Indigenous self-governance objectives (Zurba et al. 2019).

Conclusion

For well over a century, Indigenous communities globally have engaged in forms of tourism in and around parks as a strategy to meet diverse objectives. In this chapter, we unravel some of the complexities of past and current Indigenous participation in tourism industries and assess how they are linked to contemporary understandings of park development, land-use, and Indigenous rights in Canada. Such research can provide valuable knowledge to assess the viabilities of parks and related tourism industries in Indigenous communities. Comprehensive understandings are needed of how Indigenous experiences of being displaced and denied access to parklands inform contemporary decisions on the viabilities of tourism in parks as productive strategies. These understandings can help Indigenous peoples throughout Canada, and internationally, to develop their lands in ways that minimize risks to local ecosystems and increase economic opportunities, while supporting cultural continuities. More research in this area could inform policy decisions concerning the development of Indigenous lands and resources, which will have national and global relevance as well as significance at grassroots community levels.

In Canada, there are 77 different designations of parks and protected areas. This diversity leads to serious complications over land use management decisions and the stakeholders invested in them. Co-management structures, where Indigenous and Crown governments partner to jointly share decision-making, currently involve only 3% of our protected areas across the country (Mitchell et al. 2021). It is imperative to learn from both Indigenous approaches to conservation and those repressive colonial policies that have shaped current relations between colonial governments and Indigenous nations. Histories of displacement in parks are directly linked to the contemporary issues faced by many Indigenous communities, including gross health inequalities, disproportionate levels of food insecurity, and barriers to pursue tourism development. Colonial histories have impacted land use management decisions and, in some cases, have significantly constrained consultation processes and tourism development. Park designations that situate Indigenous communities as key decision-makers can foster alliances, navigate political corridors, and collaborate on tourism development for more effective conservation practices and stronger economies to support local peoples.

Acknowledgements

This research has been generously supported by the Canadian Mountain Network. We would like to recognize all the Elders and knowledgeable land users that have made this community-based research possible. We would also like to acknowledge the fine research and editing skills of Emalee Vandermale.

Note

1 Residential schools have a long history in Canada. Missionaries began some schools for Indigenous children in the late 17th century. The school networks were expanded significantly under the 1876 Indian Act. Funding and facilities were mainly provided by the federal government, but teachers were from several religious denominations. Children were forcibly removed from their homes from ages 6–15 for up to 10 months a year. The health conditions in schools were deplorable as there were high rates of disease, mortality, and at times horrendous levels of physical and sexual abuse. In 1948, mandatory attendance was abolished (Miller 1996).

References

Armitage, D., Berkes, F., Dale, A., Kocho-Schellenberg, E., & Patton, E. (2011) "Co-management and the co-production of knowledge: Learning to adapt in Canada's Arctic," *Global Environmental Change* 21 (3), 995–1004. Available from: https://doi.org/10.1016/j.gloenvcha.2011.04.006

Artelle, K.A., Zurba, M., Bhattacharrya, J., Chan, D.E., Brown, K., Housty, J., & Moola, F. (2019) "Supporting resurgent Indigenous-led governance: A nascent mechanism for just and effective conservation," *Biological Conservation* 240 (November), 108–284. Available from: https://doi.org/10.1016/j.biocon.2019.108284

Battiste, M., & Henderson, J.S.Y. (2000) *Protecting Indigenous Knowledge and Heritage: A Global Challenge*, Saskatoon, SK: Purich Pub.

Bhattacharyya, J., & Dasiqox Tribal Park (2018) *Nexwagweẑʔan: Community Vision and Management Goals for Dasiqox Tribal Park (Summary)*. Available from: http://dasiqox.org/wp-content/uploads/2018/04/DTP_VisionSummary-April-2018-web.pdf [Accessed 11th April 2022].

Binnema, T.T., & Niemi, M. (2006) "'Let the line be drawn now': Wilderness, conservation, and the exclusion of Aboriginal people from Banff National Park in Canada," *Environmental History* 11 (4), 724–750.

Carr, A. (2017) "Māori tourism in New Zealand," in M. Whitford, L. Ruhanen, & A. Carr (eds) *Indigenous tourism: Cases from Australia and New Zealand*, Oxford: Goodfellow Publishers Ltd., 145–161.

CBD (Convention on Biological Diversity) (2010) *Strategic Plan for Biodiversity 2011–2020 and the Aichi Biodiversity Targets*. Available from: https://www.cbd.int/doc/strategic-plan/2011-2020/Aichi-Targets-EN.pdf [Accessed 11th April 2022].

Chatty, D., & Colchester, M. (2002) *Conservation and Mobile Indigenous Peoples: Displacement, Forced Settlement, and Sustainable Development*, Oxford: Berghahn Books.

Clark, D., & Joe-Strack, J. (2017) "Keeping the 'co' in the co-management of Northern resources," *Northern Public Affairs* 5 (1), 71–74. Available from: https://research-groups.usask.ca/human-wildlife-interaction/documents/clark--joe-strack-2017-keeping-the-co-in-the-co-management-of-northern-resources.pdf

Coristine, L.E., Jacob, A.L., Schuster, R., Otto, S.P., Baron, N.E., Bennett, N.J., Bittick, S.J., Dey, C., Favaro, B., & Ford, A. (2018) "Informing Canada's commitment to biodiversity conservation: A science-based framework to help guide protected areas designation through Target 1 and beyond," *FACETS* 3 (1), 531–562. Available from: https://doi.org/10.1139/facets-2017-0102

Cruikshank, J. (2005) *Do Glaciers Listen?: Local Knowledge, Colonial Encounters, and Social Imagination*, Vancouver, BC: University of British Columbia Press.

ECCC (Environment and Climate Change Canada) (2016) *2020 Biodiversity Goals and Targets for Canada*. Availbale from: https://biodivcanada.chm-cbd.net/2020-biodiversity-goals-and-targets-canada [Accessed 10 April 2022].

Finegan, C. (2018) "Reflection, acknowledgement, and justice: A framework for Indigenous protected area reconciliation," *International Indigenous Policy Journal* 9 (3). Available from: https://doi:10.18584/iipj.2018.9.3.3

Hart, E.J. (1983) *The Selling of Canada: The CPR and the Beginning of Canadian Tourism*, Banff, AB: Altitude Publishing Ltd.

Hawkes, S. (1996) "The Gwaii Haanas agreement: From conflict to cooperation," *Environments* 23 (2), 87–100.

ICE (Indigenous Circle of Experts) (2018) *We Rise Together: The Indigenous Circle of Experts' Report and Recommendations*. Available from: https://www.iccaconsortium.org/wpcontent/uploads/2018/03/PA234-ICE_Report_2018_Mar_22_web.pdf

IUCN (International Union for Conservation of Nature) (2005) *Benefits Beyond Boundaries: Proceedings of the Vth IUCN World Parks Congress*. Available from: https://portals.iucn.org/library/sites/library/files/documents/2005-007.pdf [Accessed 10 April 2022].

Johnston, J.W. (2018) *Incorporating Indigenous Voices: The Struggle for Increased Representation in Jasper National Park*. [Masters thesis]. Kamloops, BC: Thompson Rivers University. Available from: https://www.tru.ca/__shared/assets/Jason_Johnston_thesis44603.pdf

Johnston, J.W., & Mason, C.W. (2020a) "The paths to realizing reconciliation: Indigenous consultation in Jasper National Park," *International Indigenous Policy Journal* 11 (4), 1–27. Available from: https://doi.org/10.18584/iipj.2020.11.4.9348

Johnston, J.W., & Mason, C.W. (2020b) "Rethinking representation: Shifting from a Eurocentric lens to Indigenous methods of sharing knowledge in Jasper National Park, Canada," *Journal of Parks and Protected Areas Administration* 38 (3): 1–19. Available from: https://doi:10.18666/JPRA-2020-10251

Johnston, J.W., & Mason, C.W. (2020c) "The struggle for Indigenous representation in Canadian National Parks: The case of the Haida Totem Poles in Jasper," *Journal of Indigenous Research* 8 (1), 1–14. Available from: https://doi.org/10.26077/7t6x-ds86

Keller, R.H., & Turek, M.F. (1999) *American Indians and National Parks*, Tuscon: University of Arizona Press.

Kovach, M. (2009) *Indigenous Methodologies: Characteristics, Conversations and Contexts*, Toronto, ON: University of Toronto Press.

Langdon, S., Prosper, R., & Gagnon, N. (2010) "Two paths one direction: Parks Canada and Aboriginal peoples working together," *The George Wright Forum* 27 (2), 222–233.

Mabee, H.S., Tindall., D.B., Hoberg, G., & Gladu, J.P. (2013) "Co-management of forest lands: The cases of clayoquot sound and Gwaii Haanas," in D.B. Tindall, P. Perreault & R. Trosper (eds) *Aboriginal Peoples and Forest Lands in Canada*, Vancouver, BC: UBC Press, 242–259.

MacLaren, I.S. (2011) "Rejuvenating wilderness: The challenge of reintegrating Aboriginal peoples into the 'playground' of Jasper National Park," in C.E. Campbell (ed) *A Century of Parks Canada, 1911–1921*, Calgary, AB: University of Calgary Press, 333–370.

Mason, C.W. (2008) "The construction of Banff as a "natural" environment: Sporting festivals, tourism and representations of Aboriginal peoples," *Journal of Sport History* 35 (2), 221–239.

Mason, C.W. (2014) *Spirits of the Rockies: Reasserting an Indigenous Presence in Banff National Park*, Toronto, ON: University of Toronto Press.

Mason, C.W. (2015) "The Banff Indian days tourism festivals," *Annals of Tourism Research* 53, 77–95. Available from: https://doi.org/10.1016/j.annals.2015.04.008

Mason, C.W. (2018) "Indigenous protected areas are the next generation of conservation," *The Conversation*. 30 November 2018. Available from: https://theconversation.com/indigenous-protected-areasare-the-next-generation-ofconservation-105787 [Accessed 10 April 2022].

Mason, C.W. (2020) "Colonial encounters, conservation, and sport hunting in Banff National Park," in C. Adams (ed) *Sport and Recreation in Canadian History*, Champaign, IL: Human Kinetics, 77–100.

McNamee, K. (1993). "From wild places to endangered spaces: A history of Canada's National Parks," in P. Dearden & R. Rollins (eds) *Parks and Protected Areas in Canada*, Oxford: University of Oxford Press, 15–44.

Meijer-Drees, L. (1993) "'Indians' bygone past: The Banff Indian Days, 1902–1945," *Past Imperfect* 2, 7–28.

Miller, J.R. (1996) *Shingwauk's Vision: A History of Native Residential Schools*, Toronto, ON: University of Toronto Press.

Mitchell, M., Schuster, R., Jacob, A., Hannah, D., Dallaire, C., Raudsepp-Hearne, C., Bennett, E., Lehner, B., & Chan, K. (2021) "Identifying key ecosystem service providing areas to inform national-scale conservation planning," *Environmental Research Letters* 16, 014038. Available from: https://doi.org/10.1088/1748-9326/abc121

Murray, G., & Burrows, D. (2017). "Understanding power in indigenous protected areas: The case of the Tla-o-qui-aht Tribal Parks," *Human Ecology* 45 (6), 763–772. Available from: https://doi.org/10.1007/s10745-017-9948-8

Murray, G., & King, L. (2012) "First nations values in protected area governance: Tla-o-qui-aht Tribal Parks and Pacific Rim National Park Reserve," *Human Ecology* 40 (3), 385–395. Available from: https://doi:10.1007/s10745-012-9495-2

Neumann, R.P. (1998) *Imposing Wilderness: Struggles over Livelihood and Nature Preservation in Africa*, Berkeley: University of California Press.

Parks Canada. (2018) *Canada's Conservation Vision: A Report of the National Advisory Panel*. Available from: http://publications.gc.ca/collections/collection_2018/pc/R62-549-2018-eng.pdf

Rangarajan, M. (1996) *Fencing the Forest: Conservation and Ecological Change in India's Central Provinces, 1860–1914*, Delhi: Oxford University Press.

Ruru, J. (2010) "A cloaked landscape: Legal devices in Mount Aspiring National Park," in J. Stephenson, M. Abbott, & J. Ruru (eds) *Beyond the Scene: Landscape and Identity in Aotearoa*, Dunedin: Otago University Press, 137–150.

Ruru, J. (2012) *Settling Indigenous Place: Reconciling Legal Fictions in Governing Canada and Aotearoa New Zealand's National Parks* [PhD dissertation]. Victoria, BC: University of Victoria.

Sandlos, J. (2007) *Hunters at the Margin: Native People and Wildlife Conservation in the Northwest Territories*, Vancouver, BC: UBC Press.

Sandlos, J. (2014) "National parks in the Canadian north: Comanagement or colonialism revisited?" in S. Stevens (ed) *Indigenous Peoples, National Parks, and Protected Areas: A New Paradigm Linking Conservation, Culture, and Rights*, Tucson: University of Arizona, 133–149.

Schuster, R., Germain, R.R., Bennett, J.R., Reo, N.J., & Arcese, P. (2019) "Vertebrate biodiversity on indigenous-managed lands in Australia, Brazil, and Canada equals that in protected areas," *Environmental Science and Policy* 101 (January), 1–6. Accessed from: https://doi.org/10.1016/j.envsci.2019.07.002

Shultis, J., & Heffner, S. (2016) "Hegemonic and emerging concepts of conservation: A critical examination of barriers to incorporating Indigenous perspectives in protected area conservation policies and practices," *Journal of Sustainable Tourism* 24 (8–9), 1227–1242. Accessed from: https://doi.org/10.1080/09669582.2016.1158827

Snow, J. (2005) *These Mountains Are Our Sacred Places: The Story of the Stoney People*, Calgary, AL: Fifth House Ltd.

Spence, M.D. (1999) *Dispossessing the Wilderness: Indian Removal and the Making of National Parks*, New York: Oxford University Press.

Thomlinson, E., & Crouch, G. (2012) "Aboriginal peoples, Parks Canada, and protected spaces: A case study in co-management at Gwaii Haanas National Park Reserve," *Annals of Leisure Research* 15 (1), 69–86. Accessed from: https://doi.org/10.1080/11745398.2012.670965

Tla-o-qui-aht First Nation (2022) *ƛuⱡst'aⱡšiⱡ- To Make an Alliance: Tribal Parks Allies' Protocol Agreement 2022*. Available from: https://tribalparks.com/wp-content/uploads/2022/01/OFFICIAL-2022-PROTOCOL-AGREEMENT.pdf

Tran T.C., Ban, N.C., & Bhattacharyya, J. (2020) "A review of successes, challenges, and lessons from Indigenous protected and conserved areas," *Biological Conservation* 241 (November 2019), 108–271. Accessed from: https://doi.org/10.1016/j.biocon.2019.108271.

Tuhiwai Smith, L. (2012) *Decolonizing Methodologies: Research and Indigenous Peoples*, 2nd edition, London: Zed Books Ldt.

UNGA (United Nations General Assembly) (2007) *United Nations Declaration on the Rights of Inidenous Peoples: Resolution Adopted by the General Assembly*. Accessed from: https://www.refworld.org/docid/471355a82.html

Walton, J. (2010) "Thomas Cook: Image and reality," in R. Butler & R. Russell (eds) *Giants of Tourism*, Wallingford: CABI, 81–92.

Youdelis, M. (2016) "'They could take you out for coffee and call it consultation!': The colonial anti politics of Indigenous consultation in Jasper National Park," *Environment and Planning A: Economy and Space* 48 (7), 1374–1392. Accessed from: https://doi.org/10.1177/0308518X16640530

Youdelis, M., Townsend, J., Bhattacharyya, J., Moola, F., & Fobister, J.B. (2021) "Decolonial conservation: Establishing Indigenous Protected Areas for future generations in the face of extractive capitalism," *Journal of Political Ecology* 28 (1). Accessed from: https://doi.org/10.2458/jpe.4716

Zurba, M., Beazley, K.F., English, E., & Buchmann-Duck, J. (2019) "Indigenous Protected and Conserved Areas (IPCAs), Aichi target 11 and Canada's pathway to target 1: Focusing conservation on reconciliation," *Land* 8 (1), 1–20. Accessed from: https://doi.org/10.3390/land8010010

19

TOURISM APPROPRIATION

"Taking" Land and Culture in Sámi Areas

Arvid Viken

Introduction

Cultural appropriation has been addressed frequently during the past two decades, in many fields: literature (Morton 2020), arts and music (Young 2005, 2021; Matthes 2016, 2019), fashion (Chatterjee 2020; Kaiser and Green 2021), and change of cultural practices and objects in general (Jackson 2021). Young (2005, 136) defines cultural appropriation as "the taking of something produced by members of one culture by members of an other," making some sort of performance or business thereof in another context. There are many similar definitions, among these. Some adding that taking becomes appropriation if it hurts or offends people from the culture of which it is taken (Young 2005, 2020; Matthes 2016, 2019), or when the taking takes place without permission (Scafidi 2005). Morton, discussing cultural appropriation in literature, claims that the rule is that cultures take from each other without asking or receiving of permission. As he says, those being appropriated most often are groups that "… don't have representatives authorized to grant it [permission]." (Morton 2020, 86). He also claims that:

> There is no product of culture that isn't the result of mixing – that isn't the result of taking things without permission – from the meals we make to the music we enjoy to the language that I'm using to write this article.
>
> (Morton 2020, 93–94)

Being an American scholar, he raises the issue of power structures related to the history of America, which is very much a history of cultural and other forms of appropriation, with the land taken from the native peoples, the labor of the enslaved, and corporations rejecting labor rights and decent salaries. These sorts of appropriation matter more than copying or making an artwork inspired by a culture the artist is not a member of, Morton claims (2020, 90).

Since tourism is a way of making money by displaying culture, appropriation should be examined in this field, concerning the exploitation of both land and culture. Tourism is a business involved with travel and often includes sojourns in foreign territories. Most of what goes on in the sector is mere business, but part of it is presentations and experiences related to culture. It is in this context that cultural appropriation appears. The topic has been discussed particularly in

DOI: 10.4324/9781003230335-23

relation to minority or indigenous groups (Young 2008, 2020; Matthes 2019; Chatterjees 2020). Cultural appropriation tends to occur when members of a majority culture take a cultural expression from a minority culture, exposing it, and making business from the taking. This is also a pattern observed in Sápmi (Viken 2022), and the focus of this chapter. In a study that addressed the tourism and identity nexus, locals expressed concern about the fact that the only hotel in the village (Karasjohka), owned by a Scandinavian hotel chain, was generating income by exposing Sami culture, but did not reinvest the profit locally (Viken 2006). When interviews were conducted in the same community in 2019, with a more explicit focus on cultural practices and objects appearing in tourism contexts, the data provided many examples of cultural appropriation within tourism.

This chapter goes into the discourse of cultural appropriation, illustrated by data from the Norwegian part of Sápmi, based on a study undertaken within the ARCTISEN research project (Olsen et al. 2019). The project focused on the sensitivity of and towards indigenous cultures in tourism, among them the Sámi, the indigenous group of Norway, Finland, Sweden, and Russia. In the Norwegian part of this project, we interviewed 23 people situated in the urban Tromsø area, the district of Northern Troms, and in Karasjohka which is the location of the Sámi Parliament and its administration in Norway. The interviewees represented 18 small and medium-sized tourism companies, and 5 other tourism-related organizations. The fieldwork took place during March and April 2019. In the interviews the issue of cultural appropriation frequently came up, giving us examples and perspectives on the topic that will be presented later.[1]

The chapter starts with a short description of the area studied, followed by a discussion of cultural appropriation as an analytical term. Then the results from the study of Sámi culture and tourism are summarized in two sections, one focusing on the appropriation of land and space, and another addressing cultural appropriation. In the discussion section, appropriation is discussed as different levels of belonging to Sámi culture, and in the conclusion, three levels of cultural appropriation are suggested.

Sámi and Sàpmi

The Sámi are recognized as indigenous groups located in Norway, Sweden, Finland, and Russia. This area, without firm borders, is called Sápmi. Although Sámi is a Finno-Ugrarian language, there are different languages and distinct cultural differences within the culture. In Norway the Sámi live all over, but particularly in Northern areas and as migrants in towns like Oslo, Tromsø, and Alta. Reindeer herding is reckoned to represent the core of their culture, but this industry involves less than 10% of the Sámi population. Most Sámi live ordinary modern lives, working in various industries. However, the history of the Sami is strongly related to reindeer herding which in Norway is a privilege for the Sámi and regulated by law. However, their belonging and rights to the land are contested, recently in disputes over windmills and mining projects (cf. Hovelsrud et al. 2021).

Sámi rights in Norway are established in the constitution and in laws, and also by the ratification of the International Labour Organization Convention no 169 (ILO 169). ILO 169 is a convention that states the indigenous people's rights "to participate in governance in relation both to bio-based and mineral resources," according to Ravna (2014, 302). Articles 6 and 15 in ILO 169 say that indigenous people shall be consulted in cases of importance to them. Ravna also highlights the International Covenant on Civil and Political Rights (ICCPR), which Norway has adopted in a Human Rights Act (from May 21, 1999, no. 30). This covenant secures indigenous peoples their rights to practice their language and culture, including traditional livelihoods. According to Ravna (2014, 305), this "means that pastures, waters, and other natural resources of importance for traditional livelihoods all enjoy legal protection." Reindeer herding is also protected by Norwegian laws (in the Constitution and the Reindeer Herding Act). According to Lawrence and Åhren (2017) the Sámi have a right to the land as a landowner, as international law claims that indigenous groups are the legal owner of land they have used for centuries. Thus, the Sámi herders are not merely a stakeholder in the land. This is not, however, the way the Norwegian and the Swedish governments have treated these territories. Lawrence and Åhren (2017) discuss how the Swedish authorities, through defining the north as *Terra Nullius* – no-man's land, over time have colonized these areas. A similar process has been going on in Norway, and most of the north has status as state land. In 2006 the state land of Finnmark, covering 95% of this Northern most region of Norway (previously a county), was handed over to a foundation, Finnmark Real Estate, with equal representation from Sami Parliament and the county council, on its management board. This was not a transfer of ownership to the Sámi people, although it is often presented as such. Thus, different sorts of state appropriation have taken place over a period of several hundred years.

Sami Tourism

Tourism is a significant industry in Sápmi. "Sámi tourism" is a term used for the industry accommodating and providing experiences for tourists. This is not following the stricter definitions suggested in discourses of indigenous tourism (Butler and Hinch 2007). For instance, the Sámi are not in control of tourism development, the tourism industry, nor tourism management in the Sámi area (Viken 2016). North Cape is the major destination and attraction, located in the very middle of the Sámi territory and in a grazing area, and received 284,000 tourists in 2019. However, Since 1927, these operations has been run by a Norwegian company with a monopoly position, keeping the Sámi reindeer herders apart from the area (Viken 2022). In general, the Sámi work in national tourism companies, or in small family-owned companies providing Sámi experiences, most often as part of day-excursion programs. In this, the providers play on Sámi traditions such as food, listening to songs (*yoik*), and narratives about the old ways of living, and also experiencing contemporary Sámi life (see Kramvig and Førde 2020). Some visitors take part in feeding the reindeer, and in sledding with reindeer, and snowmobiling. In the winter the Northern Lights are often the main attraction, the Sámi culture being a supplement or the context. Some providers bring the tourists to the reindeer herd, and a few arrange and offer participation in the spring migration. However, it is difficult to combine reindeer herding and tourism, due to strict regulations on the use of motorized vehicles outside the trails, which are only allowed for reindeer herding (defined as meat production). In addition, there are several museums and a small theme park focusing on Sámi history and culture.

Tourism has long traditions in the Sámi areas, and the Sámi were involved in the first modern tourism operations in Northern Norway – cruise tourism, towards the end of the 19th Century. Visits to Sámi camps were among the experiences offered (Petterson and Viken 2007). Currently, tourism in the north is primarily related to Arctic nature and particular iconic places (e.g., Lofoten, Lyngen and North Cape), and Sami culture. Some of the landscapes (Lofoten, Lyngen and Finnmarksvidda) are also settings for outdoor recreation and extreme sports.

Appropriation as Cultural Change

Cultural appropriation takes place when people display, present, or trade cultural expressions originating from cultures to which they are outsiders, in ways that offend people from those cultures (Matthes 2019; Young 2008, 2020). Jackson (2021), discusses cultural appropriation compared to other forms of cultural change. As a departure, he writes about innovation, where cultural appropriation tends to occur. Imitation, copying, and inspiration are elements in innovation, that all tend to bring about some form of cultural change. Cultures also change through wars and genocides, and colonization has been one of the most important gamechangers in many cultures. Also within contemporary societies cultures influence each other and change. The problem arises when the parties involved have unequal positions and power.

Jackson (2021) discusses some of the ways in which such cultural change takes place. First, cultures change through *diffusion*, which takes place when a culture picks up elements from another, such as terms, tools, and practices, without awareness of how and when it happens. In most languages there are many words imported from other languages. The Sámi language has many loan-words from Swedish or Norwegian. Diffusion takes place widely, making the cultures more alike. The homogenization of culture has been an issue for a long time (cf. Tomlinson 1995). Second is *acculturation*, the fact that cultures in contact, pick up elements from each other. The term is broader than diffusion, referring to how long-lasting contact between cultures results in similarities. It is more like "an aggregation of diffusion events...," Jackson (2021, 85) and tends to take place in relations characterized by relative power imbalances (Jackson 2021, 22). Due to processes of acculturation (and diffusion) the differences between Norwegian and the coastal Sámi culture are not distinct. *Assimilation*, Jackson's third category of cultural exchange, tends to be a result of intentional policies or decisions. Throughout history, pressure has been put on groups of people, most often minority groups, to behave according to the patterns of a majority culture. The Sámi have experienced hard assimilation. From about 1850 to the 1970s, the Norwegian government had a goal to assimilate the Sámi (similar in many respects to the situation in Canada over roughly the same period, editors' note), applying different strategies, like forbidding Sámi names for land owners, and banning Sámi language and song (*yoik*) in schools and in public places (Minde 2003). The Sámi should become Norwegian. This is also a background for looking at contemporary cultural appropriation, for instance within tourism. What happens today can often be tracked back to previous policies and world views (Viken 2022). Lastly, Jackson (2021, 88), mentions *cultural appropriation*, a "structural inversion of assimilation." "In a framework of appropriation, in contrast, the powerful group takes aspects of the culture of the subordinate group, making them its own." As Jackson notes, such maneuvers often are proclaimed as diffusion or multiculturalism, and as cultural appreciation, not appropriation. However, cultural appropriation has many troublesome aspects and some of these can also be found in tourism.

There are both positive and negative perceptions of cultural exchange, and different opinions about when and where it takes place, but not all cultural changes are due to cultural appropriation. Lenard and Balint (2020) point out two other ways foreign cultures impact negatively on Indigenous culture. One is cultural offense, which hurts or upsets those being hit. In Norway, an example is *samehets*, prerogative utterances about Sámi people, which is still quite common, with examples occurring in media from time to time. The other form is misrepresentation or talking about a group that gives incorrect connotations, for instance, stereotypes or insulting portrayals, often legacies from the colonial epoque. A common conception is that Sámi are reindeer herders, although more than 90% are not. Traits like "dirty", "drunk", and "trying to cheat you", are other prejudices expressed about the Sámi and such perceptions may hurt (Lingaas 2021).

Cultural appropriation is something else, as Lenard and Balint (2020) see it. To distinguish the term from other terms, they suggest four elements that relate to cultural appropriation: "[I]n order for an act to be cultural appropriation, it must meet four conditions: (1) a taking condition, (2) a value condition, (3) a knowledge, or culpable ignorance condition, and (4) a contested context condition" (337). All these elements can be identified within Sámi tourism. The first one, taking, happens when tourists are shown something related to another culture than the one the presenter originates from. It can be discussed whether "taking" is the right word – as tourists we mostly leave behind what we consume. If we "take" something along, we pay for it, even for visual consumption. However, those guiding us through landscapes of culture and nature, may somehow "take" what they present, narrating from or about a culture they are not part of. This frequently happens in Sápmi – for instance, there is no requirement for local knowledge for guiding in the area. Therefore, the taking nature of tourism cannot be denied.

The second condition is that what is in focus should be valued by the members of its originating culture. The narratives, the singing, clothes, and jewelry, are all of value to the Sámi culture, and therefore, many Sámi are not indifferent to how these elements are represented. For instance, there are serious debates about the use of traditional clothing across cultures within fashion, theatre, and fairs (see Chatterjee 2020), and also concerning the Sámi garment, the *gáhkti* (Kramvig and Flemmen 2019a,b), and tourism in general (Viken 2022). However, tourism is also an exception; as the souvenir trade is accepted as an area of copying and artificial appropriation. It is a separate genre, which only slightly pretends to be "culture." Trading of arts and handicrafts is different, due to the emotions, values, traditions, and history inscribed. This is also why accusations of cultural appropriation come up, as it concerns identities and emotions.

The third condition, relating to knowledge, can be problematic in a tourism context. Often, the excitement of what we observe is based on the novelty of the experience for us, and thus our lack of knowledge. Acquiring knowledge is currently recognized as an important part of a tourist experience. Since the 1960s tourism has been criticized for its ignorance (Boorstin 1964). There is currently also a normative stance related to tourism, asking for a higher level of sensitivity, including knowledge about the sins of the past, from all participants involved in tourism encounters (Viken et al. 2021).

The fourth condition, that there is somebody contesting the external use of a cultural element, is often how the issue of cultural appropriation emerges. For instance, in 2019, a can of reindeer meatballs decorated with a "Sámi" boy, which had been on the market for years, became a controversial issue in the media in Tromsø (Nordlys 2019). Other debates have been about the selling of a Sámi hat, and the use of Sámi garments by non-Sami, and their use as decorations (Viken 2022). Similar issues are recorded from other places, for instance concerning Inuit representations (Ren et al. 2021). However, there may also be a desire for elements from a culture to be exposed – as a way of promoting the culture and the land it represents. People tend to be flattered by the fact that tourists come and admire the Sámi culture (Viken 2006), but still there are valid reasons for raising the question of cultural appropriation as a tourism feature in the Sámi area.

The above discussion has noted different forms of cultural appropriation without addressing what the appropriated elements are. Young (2005) makes a distinction between subject, object, and content appropriation, that can be fruitful in tourism contexts. Subject appropriation is the use of characters or stories from another culture. This is happening frequently, when non-Sámi tourism providers tell stories about the Sámi, or present shows with characters pretending to be Sámi. This has been a tradition in amateur theaters in the region, ridiculing the Sámi, their way of speaking Norwegian, their costumes, and the reindeer relations. Elements from such shows were common as entertainment for tourists.

Object appropriation, the taking of objects, also occurs, but is not frequently reported concerning Sámi tourism. In Alta, one of the towns in Sápmi, rock carvings and ancient paintings – all relating to ancient (and probably Sámi) culture – are kept unexposed, due to risks of damage and theft. Once, in the 1980s, a tourist in Alta was caught trying to load a stone with rock carvings into his car – the stone was situated outside the local museum, a location that itself represented a removal from its original context.

The most common cultural appropriation relates to content. Lenard and Balint (2020, 338) observed "Cultural content includes symbols and practices, as well as forms of especially traditional dress and physical presentation." They add that cultural appropriation as a negative occurrence, implies "… a presumptive degree of moral blameworthiness: someone who, despite the protestations of those appropriated from, knowingly or culpably takes and uses something valuable from their culture has … done something wrong." However, they also claim that this "does not mean that all cultural appropriation is seriously wrong, or even something that should be avoided or stopped." Cultural appropriation, as Jackson (2021) maintains, is one of the ways through which cultures interact and evolve.

Concerning the Sami culture it may often be difficult to differentiate between object and content, as it is the object that has the symbolic power. The most significant symbolism is related to the landscape, reindeer, the tent, the *siedi* (offering site), the drum, the *yoik*, and the costume. This is also in accordance with a Sámi epistemology, that does not make such distinctions. The Sami identity lives these elements, as is elegantly described by Joks, Østmo, and Law (2020) concerning the *meahcci* – the Sami land. This is why the outsiders' use of Sámi garments, the gáhkti, tends to offend – they are parts of Sámi bodies and identity, as a former Sámi president once said (Viken 2006).

Appropriation of Land and Space in Sápmi

In Sápmi, the most profound appropriations have been the annexation (taking) of land and space. From the 1600s on, reindeer herding has been an industry, where the herders follow the animals' natural migration patterns and routes, moving from inland to the coast in springtime, and back in the autumn. The area where reindeer herding has taken place covers areas from south of Trøndelag (and Trondheim) and the entire north of Norway. In the 1700s the number of inhabitants in Norway increased significantly, there was a need for more land for food production, and the Norwegian authorities supported the clearing and cultivating of land, and land not already owned was primarily located in the north (Andresen et al. 2021), in areas where reindeer herding had been dominant. This colonialization did not take place without opposition from the Sámi, and some conflicts and struggles occurred (Niemi 1994). With the establishment of the national borders in 1751, through a treaty, signed by the governments of Denmark (of which Norway was a colony), and Sweden. The treaty had an addendum, called *Lappecodisillen,* that regulated the border in the Sami areas which acknowledged that the reindeer herding industry could proceed with its traditional migration patterns. However, with changing borders, and wars in Europe, the migration of the reindeer and their owners has been, and still is, an issue in border negotiations.

One central issue, throughout history, has been ownership of the land in the Sámi areas. Traditionally, the Sami reindeer herders did not own the land, but used it for grazing. But since the 1600s the state authorities created several legal constructs treating the Sami areas as state-land. From time to time, and currently more than ever, these state annexations of the land have been contested, but the state still behaves as an owner. The state, or the public bodies in charge of state properties, make money from commercial activities, such as selling and leasing lots for cabins, selling hunting and fishing rights, and from taxes on mining companies. As representatives of the owners, these organizations also are central in processes of the planning of industrial development – for instance windmill plants, and decisions about how areas should be regulated, for instance concerning outdoor recreation. Secondly, the state is in charge of the management of the law regulating outdoor recreation, *Friluftslova* (The Outdoor Recreation Act; LOV-1957-06-28-16). This is an important law, stating the freedom to roam (*everyman's right*) as a legal principle allowing everybody to walk around, put up a tent, and pick berries and mushrooms, in areas that are not inhabited or cultivated as farm or grazing land. Here, one can hike, cycle, go dog sledding, and snowmobile along designated trails, but not with motorized vehicles outside the trails. The reindeer grazing areas are not conceived as cultivated and are regulated according to the freedom to roam law. These rights tend to trump other principles and practices, including those related to reindeer herding (Skavhaug 2020).

Thirdly, the *Reindriftslova* (LOV-2007-06-15-40, The Reindeer Herding Act, together with *Motorferdselslova*, which strictly regulates the use of motorized vehicles in natural areas, make it difficult to combine herding with tourism, despite the fact that it has been part of it for more than a century. Fourthly, the state is also in charge of the protected areas, with rather strict regulations, often claimed to be hindering reindeer herding in areas in which herders have operated for centuries, places they think of as their homes (Skavhaug 2020). Tourism seems to have a higher priority than reindeer herding in the management of many of these areas (Skavhaug 2020). Fifthly, modern public infrastructure such as railways, roads and electricity cables cross grazing lands, and areas are provided for tourism resorts and real estate development, and individual tourism takes place widely. One aspect of this relates to media expositions of the Sámi areas. In recent years, there have been many reality shows from the Sámi areas, specializing on survival in the outdoors, presented by non-Sámi on national TV-channels. They may be good tourism promotion, but perpetuate the image of the Sámi areas as playgrounds for others, thus adding to the transformation of Sámi land into a playground for tourists.

When we collected our data in 2019–2020, we also came across other problems concerning Sámi tourism activities involving land use. One example was a tour guide who tried to transgress an area with grazing animals. A dispute took place, but the reindeer herder phoned the company manager and the case was solved by changing the route of the tourist group. Another controversy was over a farm that a tourism provider had bought, the purpose being to set up a reindeer sledding experience company there. The farm was located in a reindeer district of which the new proprietor was not a part. A third example was a national park board that banned dogsledding, which often involves tourists, in their park – as it was seen as disturbing wildlife in the park. The most significant concern, however, is about the cumulative impacts of all modern activities. These are summed up in a report and included: infrastructure, mines, hydropower plants, windmills, second-home resorts, and outdoor and nature-based tourism activities (Riseth and Johansen 2018).

Tourism as a growing activity also leaves traces in towns. Many cities have been transformed into tourism hubs, like Rome, London, and Paris (where, due to their size, tourism is not dominant), but in smaller towns like Reykjavik and Tromsø, the transformation into tourism sites is more obvious. Here, one may talk about transformation of space. The main streets are filled with tourist shops and tourists, and the local services that used to be there before, have been moved to shopping centers outside the city centers. In Tromsø, as part of this process, Sámi symbols are strikingly present, dominating window displays: reindeer images in all sizes, garments, tents, and Sámi handicrafts, are sold together with Norwegian trolls and other cheap souvenirs. There are obviously, fewer options and less appeal for the locals. However, this is intentional development – the shops are leased by those able to pay the highest prices. The center of the town has been appropriated by tourism.

One issue of the governing of Sámi land concerns holy sites, *sieidis*, places people used or use for sacrifice. According to Olsen (2017, 227) "archeologists broadly divide Sámi sacred sites into three class." These are *structures* as "carved stubs, erected stones, wooden pole and stone circles"; *terrain formations* that "consist of mountains, fell tops, rock formations, islands, lakes and headlands; and *natural objects* as "cleaved boulders, stones, wooden poles and small caves or clefts". Such sites can be found all over Sápmi (Äikäs and Salmi 2013), and some have become tourism attractions. Alieriesto (2020) shows how Ukko, a sieidi on a small island outside Inari in Finland, became an example of contamination, deprived and secularized, and seemingly treated without respect by tourism actors. When this issue arose, the manager of the outdoors in Finland, Metsähalitus, stripped the island of facilities, and the tourism industry stopped excursions to the site. In Norway, holy Sami sites are protected by law, but it is difficult to find out whether a specific stone is part of this system or not – there are usually no on-site markings. Such sites are still visited today, and sacrificing still takes place. Olsen (2017), who over the years observed a sieidi in Alta, found coins and small stones placed on the sieidi, and gradually more and more items, including bottles, a boll, flowers, and jewelry. The siedi, Áhkku, "is close to being overshadowed by its own success,…," Olsen writes (2017, 235) and he also notes that it is difficult to say who the sacrificers are, but foreign coins and bottles indicate that tourists are involved. The respect for a sieidi also made a committee promoting Tromsø as a candidate town for the winter Olympics in 2018, drop the idea of using the town's majestic mountain, Tromsdalstind, as an arena for the downhill ski competitions. The holiness of the mountain and protests from Sámi stakeholders were respected (Kraft 2010). Tour guides operating in Sámi landscapes quite often face the challenge of how to handle sacred sites, whether to tell about them, show them, "sacrificing", or not? This was not a topic in our 2019–2020 fieldwork, but was stressed in an earlier study (Viken 2006). A tour guide confronted with this, answered that he thought of what his grandmother would have said, if she saw him, or if he had asked her. In general, moral compasses tend to be rooted in identity and culture as such – moralities that in general are challenged by business thinking and market economics.

Tourism and Cultural Appropriation in Sápmi

The research behind this chapter represents ongoing discussions about the encounters between cultures and Indigenous cultures, and in particular how the Sámi culture is cared for in tourism contexts. One of the issues revealed was related to cultural appropriation as the Sámi involved in tourism felt its relevance. During our fieldwork whilst talking to them, we came across some incidents of cultural appropriation – cases where people were offended or where debates about cultural appropriation had emerged. There are several examples of debates addressing the misuse of Sámi symbols. One is about the sun sign, central to Sámi culture. A hotel chain in Finnmark, owned by external investors, used the Sámi sun sign in their logo in the 1990s. It did not raise any public debate. More recently, a Sámi silversmith tried to patent the sun sign as their brand symbol, giving the company an exclusive right to the sign. This was dismissed in court. In a tourism park in Karasjoka, the Sámi *Noaidi* (shaman) drum is used as a model for information signposts (the drums are often decorated with figures of humans, animals, and the sun). This was viewed as *trampling* by some of the locals (Viken 2006). Nowadays these figures are frequently used as models for jewelry or souvenirs. As Mathisen (2020, 12) ironically says, "… the symbolic figures of the Sámi Noaidi's drum have left the drum skin and started appearing on nearly all kinds of other souvenir products". The purchasing of drum copies is relatively new in Norway – it was long avoided out of respect for the culture, but this has changed. One of the informants in our study had the drum and drumming as a central point in her narrative to tourists. She was aware of its sensitivity and wondered whether her Sámi background justified her use – as she said, she did not reckon herself a shaman. This might be an example of what others have called self-appropriation (Ruiz-Ballesteros and Hérmández-Ramírez 2010).

There are some elements in Sami culture that are closer to the Sámi identity than others. Yoik, the traditional Sámi singing, is one example. This is a singing where the song is addressed to someone or something. People are typically yoiking (not yoiking to) a relative, a friend or a reindeer (see Aubinet 2020). It may also be a present. A yoik is filled with emotions and meaning. However, in these days it is a music genre, that has also been incorporated into pop music, jazz, and recently in film music (the opening theme of the Disney movie *Frozen*, composed by Frode Fjellheim). Most of those performing it are Sámi, but it is also a genre in national song contests (*Stjernekamp, NRK*), where there are Sámi contestants only occasionally. It also has become part of almost all tourism presentations where the Sámi culture is involved. Not all of those performing have learned to yoik in the traditional way, while being raised in a Sámi context. However, this sort of appropriation has not been widely discussed (cf. Ren and Thisted (2021), as it does not seem to offend. Neither has the fact that most shops selling "Sámi" souvenirs are run by non-Sámi. Only one of about ten souvenir shops in the center of Tromsø in 2019 was run by a Sámi. Not surprisingly, the operator had had several thoughts about some of the objects he was selling. Particularly, his selling of Sámi star hats (*stjernelua*) had raised a debate in a local newspaper. He had chosen to still sell the hats and tourists buy and wear them in the streets of Tromsø. Most people ignore this, but some think it is offensive. Another incident that caused offence was a group of sales representatives who met in Tromsø dressed up in Sámi-like "garments" (Kramvig and Flemmen 2019a,b). The event was criticized, but based on these reactions, a local newspaper had on several occasions blamed the Sámi for being too easy to take offence (Nordlys 2020). The newspaper was then accused of not being sensitive to the ways Sámi have been and still are patronized by the majority of society (Bratland and Viken 2020). This illustrates the variety of opinions that exists and indicates it may be a complicated matter to maneuver in the terrain of Sámi culture.

Based on conversations with people representing Sámi tourism, cultural appropriation obviously is part of the picture. Examples of provocative and debatable taking of Sámi cultural expressions came up. Whether all is cultural appropriation is not obvious. Thomas (2021) discusses belonging in a way that is relevant to this study, according to him, to perceive something as appropriation, the practice or objects at stake should robustly belong to the culture it is taken from, and meaningfully and exclusively relate to that culture (Thomas 2021, 283). However, there are grades or levels of belonging, some practices or objects are more strongly related to a culture than others. Most places, land and nature as such, robustly belong. In Sápmi, this includes the reindeer, and this is also why the reindeer is so central to Sámi culture despite relatively few Sámi now being reindeer herders or owners. Obviously, the *gáhkti* (garment) and its star hat, the *yoik* (song genre), and items related to former religious practices, like the *runebom* (drum) and *sieidis* (offer stones), clearly belong to the Sámi culture and identity. There are also cases of repatriation that illustrate the robustness of Sámi belonging. A Sámi drum that was taken from Karasjoka by Danish priests in the 1600s, and later became part of a collection in a Danish museum, was recently repatriated after requests from the Norwegian Sámi Parliament, supported by the national government. Thus, as Thomas (2021) also stresses, there is a difference between belonging and formal ownership. The conversations with Sámi tourism operators also revealed concern for other items and practices, and in particular those stemming from a handicraft tradition or handicrafters. These items, for example, the *guksi* (cup), the knife, the coffee bag made of skin, often also appear as copies made of other materials in factories and tend to be characterized as inauthentic or fake. Some also think of this as cultural appropriation, but most people are pragmatic about it.

Conclusion

In the discussion above, examples of cultural appropriation within Sámi tourism have been given and most, but not all, of them were seen as offensive. People are engaged in their culture in varying ways and to varying degrees, and as a consequence, their reactions to cultural appropriation also vary. Most cultures are not "clean", but have evolved through different forms of influence from other cultures (Jackson 2021). Assimilation has been the most troublesome factor for the Sámi in Norway, as it was state policy to eliminate their culture until the last part of the previous century. Legacies from this period still appear as prejudices and stereotypes related to the Sámi. In addition, there are cultural elements that have been taken for the benefit of the takers. Some of these losses offend many Sámi, and represent cultural appropriation.

However, appropriation depends on "what it signifies, what is its story" (Thomas 2021, 279). A few such stories have been discussed here, stories that relate to a history of land annexation, exploitation, and patronage. They represent a sad history. Some of the contemporary cultural markers, such as the Sámi gáhkti (the garment), the Sámi star hats, the yoik, and the runebom (shaman drum) are reminders of the times of oppression, but are also markers of cultural recovery and survival. Therefore, many Sámi feel offended when outsiders treat these items or practices disrespectfully as they are filled with meaning. Thomas (2021, 180) categorizes the different meanings of meaning as origin-meaning, impact-meaning, and purpose-meaning. The gáhkti for example, although most of its materials are imported, is sewn by locals in ways and with techniques that reveal their origin. The purposes and impacts of the garment are to keep the wearer warm in the winter, protecting them from the sun in the summer – making the Arctic livable. They have strong symbolic meaning – exposing identity and belonging, which also are parts of their origin-meaning. Thus, there are several layers of meaning, but these do not include decoration on walls in souvenir shops, or on foreign tourists' heads.

As stated in the beginning of this chapter, contemporary culture is a mix – wherever one is – a product of intercultural encounters over hundreds of years. Most modern people think that this is the way it should be. Cultures will continue to develop and evolve through diffusion, acculturation, assimilation, and appropriation (Jackson 2021). The result is, and has for a long time, been hybrid cultures (Bhahba 1996). One argument could be that cultural appropriation does not matter. But it does. However, with varying opinions and practices related to culture, there obviously different levels of cultural robustness, and different levels of cultural appropriation. There are elements in most cultures that go to the heart of their members – their very identities. Such cultural elements are defining structuring and stabilizing our lives. Therefore, cultural expressions related to these processes should not be treated carelessly. People should be sensitive towards the cultures they visit. People should as a norm, reciprocally recognize and respect each other's cultures (Viken et al. 2021). This is also a way to limit or prevent cultural antagonisms from growing. Disrespect is cultural appropriation.

There are also elements in most cultures that are valued differently to the most robust cultural belongings, for instance tools and practices, that represent traditional skills and knowledge. The tent, the river boat, the lasso (for catching reindeer), the wooden coffee cups and the knives, and much more, are sold as the tools or agents they are. Some are also sold as souvenirs at affordable prices. Much of this is treated as handicraft (see Schilar and Keskitalo 2016; Viken 2022), *duodji*, which is a strongly respected Sámi livelihood and tradition, and essential to the culture as such. Being handmade, such products are valued differently from mass-manufactured items and are often imbued with care and emotions (see Kupapi and Hoeckert 2020). This is a layer of cultural belonging where the discourse of cultural appropriation represents a reminder and quest for sensitivity. There is also a further category in the tourist market, items that often are called kitch (or junk), mostly souvenirs signaling Sámi connections through labels like "Sápmi" or "Sámi", the Sámi flag, or drawings of reindeer (see Keskitalo et al. 2021). The appropriation element in this is not the items themselves, but the image of Sámi as a touristic, commercial, and commoditized culture that the items representand that this trade creates, which is wrong, because it is "taken" to provide an impression of Sámi culture.

Thus, there is a hierarchy of practices and objects that are taken within tourism, for the matter of money-making by exposing Sámi culture. At the top, there are objects and practices that robustly belong to the culture, important for people's identity, including the land, the reindeer, garment items, religious remnants, and some traditional practices. At a level beneath these are the taking of tools and remedies related to Sámi life and traditions, including handicrafts. At the bottom, are the taking of materials, models and words, using them to mark souvenirs from Sápmi, giving the Sámi culture a commercial flavour. However, the challenge, often disputable, is where the borders between these levels lie, and into which category a cultural expression belongs. There is a need for a high level of sensitivity to maneuver in these waters. Now and then the waves are quite high, which, in fact is a sign of a vital culture.

Acknowledgement

Thanks to the project, culturally sensitive tourism in the Arctic – ARCTISEN, funded by Northern Periphery and Arctic program (EU), and its participants, and Sámi tourism providers, for inspiring discussions.

Note

1 The methods and methodological implication of the study is discussed more in detail in Viken (2022).

References

Andersen, A., Evjen, B. & Ryymin, T. (2021) *Samenes historie. Fra 1751-2010.* Oslo: Cappelsen Damm Akademisk.

Aubinet, S. (2020) "Enchantment and non-human voices in the Sámi *yoik*." Enchantement et voix non humaines dans le yoik Sámi. *Chier de Littérature*, 195–215. https://doi.org/10.4000/clo.8663

Bhabha, H. (1996) "Culture's in-between." In S. Hall & P. De Guy (eds.) *Questions of cultural identity.* London: Sage, Ch. 4, 53–60.

Boorstin, D. J. (1964) *The image. A guide to pseudo-events in America.* New York: Atheneum.

Brattland, C. & Viken, A. (2020) "Joikakaker og turistanlegg som kulturell og materiell appropriasjon." *Nordlys*, April 29, 2020.

Butler, R. & Hinch, T. (2007) *Tourism and Indigenous Peoples. Issues and Implications.* Oxford: Butterworth-Heineman.

Chatterjee, D. (2020) "Cultural appropriation: Yours, mine, theirs or a new intercultural?" *Studies in Costume & Performance* 5(1), 52–71.

Friluftslova (1957) LOV-1957-06-28-16. *Lov om motorferdsel i utmark og vassdrag (motorferdselloven)* - Lovdata. Oslo: Klima- og miljødepartementet.

Hansen, L. I. & Olsen, B. (2004) *Samenes historie fram til 1750.* Oslo: Cappelen.

Hovelsrud, G. K., Risvoll, C. Riseth, J. Å. Tømmervik, H., Omazic, A. & Albihn, A. (2021) "Reindeer herding and coastal pastures: Adaptation to multiple stressors and cumulative effects." In D. C. Nord (ed.) *Nordic perspectives on the responsible development of the Arctic: Pathways to action.* Cham: Springer Polar Sciences. https://doi.org/10.1007/978-3-030-52324-4_6

Jackson, J. B. (2021) "On cultural appropriation." *Journal of Folklore Research* 58, 77–122.

Joks, S. Østmo, L. & Law, J. (2020) "Verbing *Meahcci*: Living Sámi lands." *The Sociological Review Monographs* 68, 305–321. https://doi.org/10.1177/0038026120905473

Kaiser, S. B. & Green, D. N. (2021) *Fashion and cultural studies.* London: Bloomsbury Publishing.

Keskitalo, E. C. H., Schilar, H., Cassel, S. H. & Pashkevich, A. (2019) "Deconstructing the Indigenous in tourism. The production of indigeneity in tourism-oriented labelling and handicraft/souvenir development in Northern Europe." *Current Issues in Tourism* 24, 16–32. https://doi.org/10.1080/13683500.2019. 1696285

Kraft, S.-E. (2010) "The making of a sacred mountain. Meanings of nature and sacredness in Sápmi and northern Norway." *Religion* 40(1), 53–61.

Kramvig, B. & Flemmen, A. B. (2019a) "Turbulent indigenous objects: Controversies around cultural appropriation and recognition of difference." *Journal of Material Culture* 24, 64–82. https://doi.org/10.1177/1359183518782719

Kramvig, B. & Flemmen, A. B. (2019b) "What alters when the Sámi traditional costume travels? Affective investments in the Sámi." In J. Frykman & M. Povrazanovic Frykman (eds.) *Sensitive objects: Affects and material culture.* Lund: Nordic Academic Press, 179–196.

Kramvig, B. & Førde, A. (2020) "Stories of reconciliation enacted in the everyday lives of Sámi tourism entrepreneurs." *Acta Borealia* 37(1–2), 27–42. https://doi.org/10.1080/08003831.2020.1752463

Kugapi, O. & Höckert, E. (2019) "Affective entanglements with travelling mitten." *Tourism Geographies.* https://doi.org/10.1080/14616688.2020.1801824

Lawrence, R. & Åhrén, M. (2016) "Mining as colonisation: the need for restorative justice and restitution of traditional Sami Lands." In L. Head, H. L Saltzman, G. Setten & M. Stenseke (eds.) *Nature, temporality and environmental management: Scandinavian and Australian perspectives on landscapes and peoples.* London: Taylor and Francis, 149–166.

Lenard, P. T. & Balint, P. (2020) "What is (the wrong of) cultural appropriations?" *Ethnicities* 20(2), 331–352.

Lingaas, C. (2021) "Hate speech and racialised discrimination of the Norwegian Sámi: Legal responses and responsibility." *Oslo Law Review* 8, 88–107. https://doi.org/10.18261/issn.2387-3299-2021-02-02

Lyngnes, S. & Viken, A. (1998) *Samisk turisme.* Report BI. Oslo: BI.

Mathisen, S. R. (2020) "Souvenirs and the commodification of Sámi spirituality in tourism." *Religions* 11(429), 1–13. https://doi.org/10.3390/rel11090429

Matthes, E. H. (2016) "Cultural appropriation without cultural essentialism?" *Social Theory and Practice* 42(2), 343–366. https://doi.org/10.5840/ soctheorpract20164221

Matthes, E. H. (2019) "Cultural appropriation and oppression." *Philosophical Studies* 176, 1003–1013. https://doi.org/10.1007/s11098-018-1224-2

Minde, H. (2003) "Assimilation of the Sámi – Implementation and consequences." *Acta Borealia* 2, 121–146. https://doi.org/10.1080/08003830310002877

Morton, B. (2020) "All shook up: The politics of cultural appropriation dissent." *Muse* 67(4), 84–94. https://doi.org/10.1353/dss.2020.0075

Motorferdselslova (1977) LOV-1977-06-10-82. Lov om motorferdsel i utmark og vassdrag (motorferdselloven) - Lovdata. Oslo: Klima- og miljødepartementet.

Nordlys (2020) Ta det rolig [Calm down]. *Nordlys.* Unsigned newspaper article. April 20. Tromsø: Nordlys.

Olsen, K (2017) "What does the *sieidi* do? Tourism as part of a continues tradition." In A. Viken & D. Müller (eds.) *Tourism and indigneity in the Arctic.* Bristol: Channel View Publications, 225–245.

Olsen, K. O., Avildgaard, M. S., Brattland, C. & Müller, D. K. (2019) *Looking at Arctic tourism through the lens of cultural sensitivity: ARCTISEN – A transnational baseline report.* Multidimensional Tourism Institute. Rovaniemi: Lapland University.

Petterson, R. & Viken, A. (2007) "Sami perspectives on indigenous tourism in northern Europe: commerce or cultural development?" In R. Butler & T. Hinch (eds.) *Tourism and Indigenous Peoples. Issues and Implications.* London: Butterworth-Heinemann, 176–187.

Ravna, Ø. (2014) "The fulfilment of Norway's international legal obligations to the Sami – Assessed by the protection of rights to lands, waters and natural resources." *International Journal on Minority and Group Rights* 21, 297–329.

Reindriftslova (2007) LOV-2007-06-15-40. *Lov om reindrift (reindriftsloven)* - Lovdata. Oslo: Landbruks- og matdepartementet.

Ren, C., Jóhannesson, G. T., Kramvig, B., Pashkevich, A. & Emily Höckert, E. (2021) "20 years of research on Arctic and Indigenous cultures in Nordic tourism: A review and future research agenda". *Scandinavian Journal of Hospitality and Tourism* 21, 111–121. https://doi.org/10.1080/15022250.2020.1830433

Ren, C. & Thisted, K. (2021) "Branding Nordic indigeneities." *Journal of Place Management and Development* 14, 301–314. https://doi.org/10.1108/JPMD-01-2020-0007

Riseth, J.-Å. & Johansen, A. (2018) *Inngrepskartlegging for reindrifta i Troms fylke.* Rapport no. 23. Narvik: Norce.

Scafidi, S. (2005) *Who owns culture?: Appropriation and Authenticity in American Law.* London: Rutgers University Press.

Schilar, H. & Keskitalo, E. C. H. (2018) "Elephants in Norway: Meaning and authenticity of souvenirs from a seller/crafter perspective." *Tourism, Culture and Communication* 18, 85–99. https://doi.org/10.3727/109830418X15230353469483

Skavhaug, I. (2020) *Mangfoldige verneområder. Å utforske forståelser av vernet natur og landskap i Nord-Troms.* Ph.D. thesis. Institutt for reiseliv og nordlige studier. Tromsø: UiT Norges arktiske universitet.

Thomas, J. L. (2021) "When does something 'belong' to a culture?" *British Journal of Aesthetics* 61, 275–290.

Tomlinson, J. (1995) *Homogenisation and globalisation. History of European ideas.* London: Taylor & Francis.

Viken, A. (2006) "Sámi tourism and Sámi identity – An analysis of the tourism-identity nexus in a Sámi community." *Scandinavian Journal of Hospitality and Tourism* 6(1), 7–24. https://doi.org/10.1080/1502225060056154

Viken, A. (2016) "Marginalisering av sámisk turisme." In E. Angell, S. Eikeland & P. Selle (eds.) *Nordområdene i endring urfolkspolitikk og utvikling* Oslo: Gyldendal Akademisk, 235–257.

Viken, A. (2022) "Tourism appropriation of Sámi land and culture." *Acta Borealia*, forthcoming.

Viken, A., Höckert, E. & Grimwood, B. (2021) "Cultural sensitivity: Engaging difference in tourism." *Annals of Tourism Research*, 103223. https://doi.org/10.1016/j.annals.2021.103223

Young, J. O. (2005) "Profound offense and cultural appropriation." *Journal of Aesthetics and Art Criticism* 63, 135–146.

Young, J. O. (2021) "New objections to cultural appropriation in the arts." *British Journal of Aesthetics* 61, 307–316. https://doi.org/10.1093/aesth/ayeb009

Ziff, B. H. & Rao, V. (1997) *Borrowed power: Essays on cultural appropriation.* New Brunswick, NJ: Rutgers University Press.

20
AN EVALUATION OF THE NEW AINU LAW

Tourism Promotion Policy and Indigenous Rights of the Ainu People in Japan

Atsuko Hashimoto

Introduction

The Ainu people are one of the two Indigenous peoples in today's Japan. They have lived in the *Ainu Mosir* [Ainu Homeland] for over a millennium until Japan colonised and assimilated them into the *Wajin* [ethnic Yamato people – mainstream Japanese] society. Those of Ainu lineage outside Hokkaido tend to hold strong affinity to their hometowns (or parents' hometowns) in Hokkaido (the original Ainu homeland) and form Ainu societies to embrace their heritage (Watson 2014). While this chapter will not examine all people of the Ainu lineage separately, urban Ainu people outside Hokkaido have been treated differently, are more disadvantaged and as such are subjected to increased assimilation into Wajin society. With this history in mind, the evolution of the New Ainu Law, which came into effect in May 2019, will be examined along with its impacts on the Ainu and the promotion of their culture through tourism.

In May 2019, the Japanese government passed the *Promotion of Measures to Realize a Society That Will Respect the Pride of the Ainu Act* (hereafter the New Ainu Law), officially recognising the Ainu as Indigenous Peoples. This replaced the *Hokkaido Former Aborigines Protection Act* of 1899 and came 22 years after the *Ainu Culture Promotion Act* (ACPA). There were high expectations for the New Ainu Law, formulated after consultations with the Ainu and non-Ainu advocates in line with the UN Declaration of Indigenous Rights (UNDRIP). However, the New Ainu Law has been criticized as the Indigenous right to autonomy and self-governance was not explicitly stated. A key component of this law for tourism is that it grants opportunities for communities to develop tourism products featuring Ainu culture and tradition, for the purposes of educating non-Ainu Japanese and international tourists. A loophole in the law is that there is no stipulation that the tourism applicants must be persons of Ainu heritage.

While the focus here is on the New Ainu Law, many lenses can be applied to similar Indigenous legislation and practices elsewhere. A political ecology lens, for example, has been utilized in Indigenous studies and related issues of colonisation, assimilation, neo-colonialism, environmental degradation, and commodification, all referred to in this chapter. The Indigenous peoples of the world have been oppressed and assimilated through colonisation. Separation of Indigenous peoples from their land and resources, which is part of their spirituality and culture, is articulated

DOI: 10.4324/9781003230335-24

as "primitive accumulation" by Marx. People who live and produce in the natural environment are dispossessed by (1) commodification and privatisation of land, (2) conversion of communal property to private property and (3) suppression of rights to the commons. As they lose their land and the commons, they lose their place of production, suddenly becoming cheap labour for someone else and their culture is commodified. This "accumulation by dispossession" (Harvey 2003) is evident globally within tourism development as national forests, protected parks, nature reserves and marine parks are created, displacing Indigenous peoples and dispossessed locals from their resources and livelihoods. These parks are then used for the conservation of wildlife, ecotourism, or community-based tourism to generate profits (Pellis et al. 2018, see also Chapter 18, editors' note). Governments or park authorities can evict local inhabitants, hiring a few as rangers or tour guides. The legacy of this system is still at work today, and instead of colonists, multinational enterprises and developed nations are operating within the structure of neo-colonialism (Simmons 2017).

In writing this chapter, the author is aware of the challenges a non-Indigenous researcher like herself faces in interpreting Ainu representation of culture, even though I grew up as part of the *Wajin* and now live as a racialised minority in the country I choose to live in. With this understanding, this chapter will employ a political ecological discourse from the perspective of a genderised minority person. As the political economy of Indigenous rights is a product of social construct as well as environmental politics, the chapter will begin with a brief review of Ainu assimilation strategies, how these strategies impacted the evolution of Ainu tourism, the emergence of legislation and policies regarding Ainu peoples, and how this background led to the stipulation of the 2019 New Ainu Law. Lastly, reviewing criticisms of the 2019 New Ainu Law, the chapter will discuss this law, regarding tourism development.

History of Ainu Assimilation – "Vanishing People"

The Japanese myth of Japan as a homogenic ethnic nation has significantly influenced its government's stance on assimilation of Indigenous peoples into mainstream *Wajin* ('ethnic' Japanese) society. The *Kaikoku* (opening the nation) to the outside world in 1854 was a reluctant beginning, nonetheless, the decision to invent a modernized, 'homogenous,' centrally governed state was a driving force to assimilate the Ainu, Okinawan, and other minority nationals in Japan (Maruyama 2014). Japanese National identity was firmly established around 1895 during the Sino-Japanese War period (Choi 2011), and assimilation of non-*Wajin* in Japan was of national importance as much as the creation of the 'precarious Other' (Gittings 2018) as an opposing entity highlighted through *the Hokkaido Former Aborigines Protection Act*. This section briefly reviews Ainu assimilation through colonisation as this has had a significant impact on the *Wajin* and the government understanding of the Ainu today.

The Ainu lived in the north-eastern lands of Honshu and Hokkaido, southern Sakhalin, on the Kuril Islands, part of the Lower Amur River Basin, and in the southern part of Kamchatka – the Northern territories disputed between Japan and Russia (Ural Federal University 2019; Uzawa 2019). Colonisation and assimilation of the Ainu started during the Edo period (1603–1867) and became more hostile during the Meiji period (1868–1912). Since the 15th century, the Ainu have resisted, engaging in numerous unsuccessful battles against Wajin settlers in Hokkaido (Umeda 2019). During the Edo period, the Wajin were prohibited from entering Ainu communities, and as such, the Ainu culture was protected from Wajin cultural influences (Choi 2012). As Japan opened its door to the outside world after 200 years of *Sakoku* (closed nation policy), Japan tried to invent a unified, monolingual nation (Maruyama 2014). In 1869, the Meiji government officially annexed

Hokkaido and founded the Hokkaido Development Commission (Umeda 2019). Ainu place names were replaced by Japanese names. In 1871, the prohibition of hunting/fishing and forced engagement with agriculture and the prohibition of women's tattoos and men's earrings were imposed on the Ainu. The Meiji government's total Ainu assimilation policy led to the annihilation of Ainu culture and economy, the recognition of which later became the basis of developing Ainu tourism (Choi 2012).

With the 1875 Treaty of St. Petersburg between Russia and Japan regarding the Kuril Islands, the Ainu people on these islands were forced to move to Shikotan Island in 1884 (Saito 1994) becoming physically, economically, and culturally displaced. Rather than following the treaty's provision allowing the Ainu to select Russian or Japanese nationality, the Japanese government relocated the Russian Ainu to Hokkaido with the objective of supplementing its local labour force (Zaman 2020). This forced relocation and the imposition of Japanese nationality were part of the Ainu assimilation strategy. Forced labour on infertile lands, epidemics, and inhumane living conditions decimated half the population within a decade, driving more Ainu to depression, alcoholism, and suicide (Zaman 2020).

The education system under *the 1899 Hokkaido Former Aborigines Protection Act* instilled in the Ainu that they were a conquered race by the Wajin and thus doomed to extinction. This teaching brainwashed many Ainu to abandon their heritage and assimilate with the Wajin (Arai 2012). Assimilation in this context meant the Ainu's identity could be confirmed only when they pledged eternal subjugation to Imperial Japan (Kido 2019). As part of the assimilation policy during the Meiji period, abandonment of the Ainu language and culture, and marriages to *Wajin* thereby diminished blood ties amongst the Ainu population (Kido 2019). To counter this, young Ainu volunteered to be pioneers in Manchuria, a newly established Japanese territory in China (Choi 2012) in an effort to build a nation without discrimination (Arai 2012).

Due to discrimination, generations of Ainu hid their heritage, did not teach children the Ainu language and culture, and tried to assimilate into mainstream Wajin society. It is believed the actual Ainu population size is much higher than that listed in the national census as there is no clear official definition or guidelines as to who can identify as Ainu (Jozuka 2019), (unlike for example the Métis people in Canada who are of mixed heritage and recognised as Aboriginal peoples – c.f. Hashimoto 2021). Fuelled by systemic racism amongst the *Wajin*, the assimilation of the Ainu under the *1899 Former Aborigines Protection Law* was assumed complete by the end of the 1920s for almost no one spoke the Ainu language (Stevens 2001; Hashimoto 2021). During the 1930s, many Ainu publications appeared marking a cry for freedom from subjugation to Imperial Japan and an anti-commodification stance on Ainu culture through tourism (Kido 2019; Choi 2012). Studying the narratives in a documentary film on the Ainu in 1959, Choi (2011) revealed the Ainu were considered by the *Wajin* narrator to be completely assimilated into mainstream Japanese society, and those who staff Ainu tourist attractions are 'fake' Ainu – dressing up for the tourists, demonstrating craft-making, posing for photographs – while their daily lives are no different from the rest of the Japanese people.

The Ainu poet and writer, Masao Sasaki (Winchester 2009, 1) described the Ainu as a "situation" rather than a race or people. Sasaki's commentary published in 1973 raised concern that Japanese attitudes towards Ainu were a "forever repressed minority… in need of the governmental protection, aid and respect" (Winchester 2009, 2). In a 1980 report to the United Nations (*Consideration of reports submitted by States parties under article 40 of the Covenant: International Covenant on Civil and Political Rights*), the Japanese government acknowledged every citizen had rights to enjoy culture, religion, and language (ICCPR Article 27), however, there no longer existed an ethnic minority group in Japan (Stevens 2014; United Nations 1980; Hashimoto 2021).

It was not until 1991 the Japanese government admitted the existence of the Ainu in accordance with the *International Covenant on Civil and Political Rights* (ICCPR) (Maruyama 2014). In 2008, the Ainu were recognised as an Indigenous People of Japan, and there was a new set of policies reflecting the UNDRIP in 2009 (Winchester 2009).

Historical View of Ainu Tourism

With another chapter in this Handbook (Okado and Edelheim) focusing on contemporary trends in Ainu tourism, this section will briefly touch on historical perspectives under the New Ainu Law. Since the annihilation of Ainu culture in 1869, the Ainu people have been used as a tourism attraction. The sacred ritual of *Iomante* was used as an important 'event' to invite *Wajin* guests in 1881, which then toured mainland Japan as an exhibit (Choi 2012). Royal visits to Hokkaido in the Meiji period (1896-1912) by the Emperor and the Crown Prince also led to numerous visits by researchers and inspectors (Uchida 2015). This could be considered as an incubation period for the current Ainu tourism format of inviting visitors to the house, displaying hospitality with traditional entertainment, and explaining Ainu culture (Saito 1994).

The Taisho period (1912–1926) introduced a new way to display Ainu people and their lifestyle at exhibitions. At the 1913 *Meiji Memorial Colonisation Exhibition* in Osaka, an Ainu house was set up in the pavilion where Ainu people lived and demonstrated craft making for sale during the exhibit (Saito 1994). The Hokkaido Tourism Organisation was established in 1946, and Hokkaido Tourism began promotion in earnest. The Ainu were one of the main attractions contributing to the success of the 'Hokkaido Tourism Boom' (Uchida 2015; Choi 2012). This 'tourist's gaze' (Urry and Larsen 2011) and consumption of Indigenous peoples as an attraction is a part of the discourse on imperialism raised by tourism anthropologists and sociologists. As a racialized contrivance, the spectacularization of culture (Stanley 1998) describes the creation of startling differences between the audience and the attraction, fuelling cultural politics. While early Ainu exhibitions were used as a symbol of commodified culture, their good will to promote their culture was not part of the attraction. It was clearly a 'show,' easily recognisable, yet the audience was unaware of the staged authenticity or showmanship of the performers. These exhibits would later become tourism of 'Symbolised Spaces of Ethnic Harmony' discussed later in this chapter.

The 1958 Hokkaido Exhibition took place in Osaka but unlike Taisho period exhibitions, Ainu individuals were not based in the Ainu house alongside Ainu culture (e.g. clothing, jewellery, weapons, Ainu hut, etc.) (Choi 2012). Yet this still represented a view of "fake" Ainu culture owing to the misconception of Ainu assimilation by the *Wajin* (Choi 2012) and "fake" Ainu art-and-crafts due to the commodification of cultural objects for tourism (Uchida 2015). Even so, the exhibition received 1.95 million visitors (Choi 2012) and the many organised tours to Hokkaido which soon followed were a great success (Uchida 2015). It was only after the promulgation of UNDRIP in 2008 that Indigenous tourism style and stance changed from being 'a subject to be gazed' to one of 'showing' or 'sharing' lifestyle and culture with outsiders.

Historical Evolution of Laws Regarding the Ainu

The formal relationship between Ainu and *Wajin* goes back to the early 17th century. The period between the Matsumae clan of the Tokugawa Shogunate being granted land in Hokkaido in 1604, and the Meiji government's Land Ordinance Regulation in 1872 which recognised Hokkaido as *terra nullius* and appropriated Ainu land (Stevens 2001), the two groups (as separate identities) enjoyed trade, especially when *Wajin* emissaries and businesspeople began to migrate to Hokkaido

(Saito 1994). Saito (1994) however, identified in a document written in 1783 that it was illegal for the *Wajin* to give or trade weapons to the Ainu. The Japanese government incorporated the Ainu peoples and their land without any formal treaties or agreements (Sonohara 1997), and this annexation led to rapid colonisation (today's Hokkaido). The Ainu were forced to adopt Japanese nationality in 1871 (Umeda 2019).

The 1875 Treaty of St. Petersburg gave the Japanese government the right to relocate the Ainu in Russian territories to the southern Sakhalin Island and Hokkaido. In 1878, the Ainu people were included in the Japanese family registration system and identified as "Former Aborigines" (Choi 2011). Such stigmatisation in the official register forced many Ainu to undeclare their heritage (Choi 2011). Under the *Hokkaido Former Aborigines Protection Act* promulgated in 1899, the Ainu suffered colonial injustices including the prohibition of traditional Ainu practices and lifestyles, limited education in the Japanese language (four years instead of six for Japanese children) and adopting *Wajin* names. Some argued that at least the 1899 *Act* recognised Ainu as non-Japanese (Aborigines), nevertheless, this law was designed to extract the indigeneity out of the Ainu people. In the 1899 *Act*, the Ainu were to receive some land for farming, nonetheless, the most fertile lands were appropriated for *Wajin* immigrants, leaving only inhospitable lands for the Ainu (Umeda 2019). It took nearly a century before the *1899 Act* was replaced.

After the defeat of World War II, the Imperial Constitution was replaced by the Constitution of Japan, in which every individual's rights and sovereignty were stipulated and the myth of the homogeneity of the nation persisted (Maruyama 2014). With the 1983 Cobo study on the *Problem of Discrimination Against Indigenous Populations*, the United Nations was moving towards finalising a declaration on Indigenous People's Rights, in consultation with various Indigenous Peoples' groups. In 1988, the Hokkaido Prefectural Government, driven by Ainu lobbying, requested the National Government enact a new law to protect the rights of the Ainu people and promote Ainu culture. The result of this movement was the 1997 *Act on the Promotion of Ainu Culture and Dissemination and Enlightenment of Knowledge About Ainu Tradition, etc.* (Act No. 52 of 1997 - or (ACPA)) which replaced the 1899 *Act*. Under ACPA, the promotion of Ainu culture was directed by promotion policies led by the Minister of Land, Infrastructure, Transportation and Tourism, and the Minister of Education, Culture, Sports, Science and Technology, rather than Ainu representatives or organisations (Umeda 2019).

There were two important incidents around this time demanding rights for the Ainu: The Ainu Association of Hokkaido adopted a 1984 draft of the New Ainu Law; and Ainu plaintiffs filed a 1997 court case over the Nibutani Dam. The court decision on the Nibutani Dam had a significant impact. The Court acknowledged the Ainu's unique culture intertwined with their spirituality/religion and that construction of the dam without adequate impact assessment was illegal; however, the Court also found it "inappropriate to declare the expropriation decision null and void" because the dam was completed, and the land was submerged (Sonohara 1997). The Court decision drew on international legal text, and the relevance of the Constitution of Japan. In the end, the Court dismissed the plaintiffs' claim, yet this was a significant step for the rights of the Ainu, and a warning for the Japanese government for failing to protect individuals rights according to the Constitution. The *1997 ACPA* is believed to be the Japanese government's response to the Nibutani Dam decision as well as to the draft bill by the Ainu Association (Sonohara 1997).

In 2008, Japan voted to support the *United Nations Declaration on the Rights of Indigenous Peoples*, or UNDRIP (A/RES/61/295) (Sept. 13, 2007) and acknowledged the Ainu as an Indigenous Group of Japan. This acknowledgement led the Cabinet Office to further enhance Ainu policies and the promotion of Ainu culture by forming the Advisory Council for Future Ainu Policy. This Advisory Council consisted of "experts in constitutional law, international law, history and cultural anthropology; a former Minister of Education, Culture, Sports, Science and Technology; the Governor of Hokkaido; and the Executive Director of the Ainu Association of Hokkaido" (Tsunemoto 2019, 1).

"The Advisory Council also stated

> Just as the histories and current situations of the world's 370 million indigenous people are diverse, so are the countries in which they live. These individual conditions cannot be ignored as far as the UN Declaration is concerned. In this respect Japan should establish its Ainu policy in line with the current conditions of the country as well as of Ainu people themselves.

(Tsunemoto 2019, 2)

This stance by the Advisory Council implicitly suggests the rights of Ainu may still be secondary while not openly opposing UNDRIP's provisions. Although Tsunemoto's commentary noted the Japanese government would not hinder those Ainu who want to live as Japanese nor those who want to live as an Ainu, the New Ainu Law did not define the Ainu or recognise self-determination of the Ainu as Indigenous Peoples. All promotional projects for the Ainu, including those for tourism, had to be approved by the Japanese government.

The New Ainu Law 2019 (Act No. 16 of 2019)

In April 2019 the Japanese Diet(a bicameral legislature, composed of House of Representative and the House of Councillors) [1] passed the '*Act on Promoting Measures to Realize a Society in Which the Pride of the Ainu People is Respected*' becoming effective on 24 May 2019. There are 45 Articles in this law with clauses detailing procedures, limitations and delimitations of Ainu promotion projects and possible adjustments in associated regulations. This New Ainu Law has already been the subject of much controversy. As the title suggests, this law is on 'promotion measures' and is not meant to identify the Rights of Indigenous Peoples (the Ainu), nor to guarantee their rights (Ichikawa 2019a). Although Japan supported UNDRIP in 2008, the New Ainu Law does not reflect Indigenous Peoples' collective rights and individual rights. Nakamura (2019) posits that the New Ainu Law "fails to directly address welfare policies, including life protection insurance, employment support schemes, and existing gaps in education levels." This law is a little different from the *1997 ACPA* (c.f. Sonohara 1997), does not legally identify the Ainu people, does not provide cultural rights recognizable by law, nor recognize Japan's injustice towards the Ainu people; the Law focuses only on the promotion and dissemination of Ainu culture to non-Ainu populations, which has implications for Ainu tourism.

The New Ainu Law encourages Ainu culture promotion through the development of 'Symbolic Spaces for Ethnic Harmony'. These 'Symbolic Spaces' (known elsewhere as Cultural Centres) are spaces for *Wajin* and international tourists to encounter Ainu culture in various formats. These 'Spaces' will be under the control of the Ministry of Land, Infrastructure, Transport and Tourism. Proposals for 'Symbolic Spaces' must be submitted through relevant municipalities or prefectural governments for permission from the national government and access to grants. There is no restriction in the New Ainu Law on who can submit an original proposal – it could be submitted by Ainu individuals, or non-Ainu individuals. The promotion of Ainu culture through a 'Symbolic Space' is very much a tourist attraction with some educational components. These 'Symbolic Spaces' can be viewed through political ecology as the destructive force, i.e. Indigenous fishing and hunting is displaced in favour of a 'more benign' form of tourism 'Space' employing Ainu. In this sense, examining the foci outlined in the New Ainu Law below will provide insights on the potential influence this law brings to Ainu tourism in Japan.

Ainu as a Homogenous Collective Term

Although the New Ainu Law does not define who the Ainu are, Article 4 clearly states that the Ainu must not be discriminated against for being Ainu. The definition of "being Ainu" must come from within Ainu communities. Due to assimilation strategies, there are Ainu descendants of mixed heritage, those who hid their heritage for generations, or those who grew up in a *Wajin* environment not knowing their heritage until adulthood. With the Ainu assimilation strategy and globalisation, many Ainu descendants also live outside Hokkaido (Watson 2014; Sunazawa 2014; Uzawa 2014). Furthermore, there are personal accounts of people of the Ainu lineage wanting to return to the Ainu community but who are not welcomed by some Ainu (Jozuka 2019).

There are two main premises for promulgating the New Ainu Law that complicate the effectiveness of the Act. First, the Ainu are addressed as a homogenous group of people. Yet historically the Ainu spread throughout Hokkaido and the Kuril Islands, and by forced relocation of the 1875 Treaty of St. Petersburg, Russian Ainu were moved to Hokkaido and the southern Sakhalin Island. The Ainu's identity is by the unit of *Kotan* (village) which controls exclusive hunting and fishing rights and ergo, wars between *Kotans* were not unusual, and each *Kotan* has its own legal system (Ichikawa 2019a, b). In addition to this disregard for the *Kotan* of the Ainu, the New Ainu Law is based on a second assumption, that the Ainu are near extinction and such village units no longer exist.

Ainu as a Vanishing People

The second premise of this Act and the various Ainu promotion strategies is that the Ainu are a vanishing people, assimilated into mainstream Japanese society unlike other countries where Indigenous peoples reside communally, Ainu people are not able to maintain their culture including their language (Cabinet Office 2011). Ichikawa (2019a) advocates that this stance of the Japanese Government led to no acknowledgement of Indigenous self-determination in the Act for a non-existent Indigenous population. As noted above, the Japanese government informed the United Nations in its 1980 report that no ethnic minority group existed in Japan. As in other countries, acknowledgement of collective rights of Indigenous people requires official apologies to Indigenous peoples and admittance of a state's historical injustices, and redress for the grievances of Indigenous peoples, i.e., return or recompense of appropriated land and associated hunting and fishing rights, and self-determinacy. Hideki Uemura, an activist and advocate for the Ainu in Japan pointed out that the process of establishing UNDRIP since 1980 was meant to force states to reflect on and reconsider their colonial actions (cited in Winchester 2009, 3). Nevertheless, the Japanese government actions and their 'Advisory Council for Future Ainu Policy' do not seem to fully embrace UNDRIP.

Procedures for 'Symbolic Spaces for Ethnic Harmony' and Implications for Tourism

Article 2, Clause 3 outlines the creation of "Symbolic Spaces for Ethnic Harmony." Such a space is the Upopoy National Ainu Museum and Park opened in 2020 at Shiraoi township, Lake Potoro, Hokkaidō, under the directive of the Ministry of Land, Infrastructure, Transport and Tourism (MLIT 2021). Ichikawa (2019a) comments such spaces are "a type of theme park" including museums, recreational parks, and memorial sites for Ainu remain which had been exhumed and taken away by universities and museums for forensic anthropology research. According to a Cabinet Office report, the concept of the 'Spaces' does not restrict the benefits to the Ainu but extends also to domestic and international tourists; the planning and creation of the 'Symbolic Spaces' is to be approved by the Japanese government; education, research and exhibition facilities are under Japanese government initiatives; and the design of the main spaces will "attempt" to retain the designs and concepts based on Ainu's spirituality and natural worldview (Cabinet Office 2011). Upopoy opened prior to the 2020 Tokyo Summer Olympic Games, "in pursuit of realizing a target of 1 million visitors" (MLIT 2019, 69). It is expected that many more 'Symbolic Spaces for Ethnic Harmony' will be approved by the Japanese government under the New Ainu law as Articles 7, 8, and 10 encourage prefectural governments and municipalities to propose Ainu promotion projects. Grants for Ainu promotion projects will be given to municipalities from the national government (Article 15) however, grants will not be offered directly to the Ainu population (Ichikawa 2019a). So, while these 'Spaces' provide opportunities for tourism, they are not controlled by the Ainu.

New Ainu Law and Ainu Tourism

Many Ainu activists and experts argue the New Ainu Law is merely a strategy guideline for Ainu cultural promotion. The Cabinet Office announced in September 2019 that

> 470 million yen [US$ 4.2 million] was to be delivered for tourism and other industry promotion projects; 150 million yen [US$ 1.36 million] for regional, inter-regional, and international exchange promotion projects; 20 million yen for Ainu culture preservation projects; and 20 million yen for projects for raising awareness of Ainu culture.
>
> (Osakada 2021, 1059)

The main argument against this New Ainu Law is that it is not a true reflection of the Ainu. Ainu activists demanded the withdrawal of the New Ainu Law during a March 2019 rally in Sapporo, Hokkaido (Kyodo 2019) and at a press conference at the Foreign Correspondents' Club of Japan (Higashiyama 2019). Nonetheless some, including Ainu and Japan-based Ainu researchers, admit it does represent a small step forward in recognising the Ainu people as an Indigenous People of Japan (Murakami 2019; Jozuka 2019).

Lack of 'Self-Determination' of Ainu People

Winchester (2009) claims the Japanese government agreed to UNDRIP because of the wording of Article 46 ensuring "the territorial integrity of political unity of sovereign and independent States" (UN 2007). The Japanese government's resistance against legally extending rights to the Ainu according to UNDRIP concerns compatibility with the Constitution and what would follow with official redress (Winchester 2009). It is a common concern of states where Indigenous populations live and fight for their rights, that acknowledging and legalising Indigenous Peoples could compromise state sovereignty. However, the Nibutani Dam court case has proven that the Ainu as Japanese nationals are entitled to the protection of rights according to the Constitution of Japan (Sonohara 1997). The resultant *1997 ACPA* failed to acknowledge the Ainu as an Indigenous People of Japan – it only acknowledged the Ainu as a people (Sonohara 1997). It was not until the New Ainu Law in 2019, that the Ainu were finally acknowledged as Indigenous people. Prior to the stipulation of the law, an Ainu survey with dubious results "reported that the majority of the Ainu participants declined a national apology for past injustices" (Osakada 2021, 1058). Most importantly what the New Law did not provide was unreserved self-determination and livelihood support for Ainu elders.

Nature of Ainu Culture

Although the New Ainu Law encourages Ainu culture promotion projects and the provision of grants to approved projects, there are a few points that contradict such promotion. The first point is about the nature of Ainu culture. The persisting assumption here is that the Ainu as an ethnic group does not exist as a collective due to the near complete assimilation of the Ainu Peoples into the Wajin culture. The Ainu culture is portrayed as extinct and needing to be revived under the careful supervision of the government. The 'culture' to be preserved is a culture of the past, yet culture is dynamic and evolves over time. Uzawa (2014), an Ainu artist reveals that new interpretations of traditional Ainu dance is her way to revere ancestors; however, tourists' anticipation and expectations of authenticity have typecast her to perform established 'traditional dance,' undermining her new interpretation, and the fact that culture is dynamic. Preservation of traditions is undoubtedly important, however, there must be acknowledgement and protection of 'living culture' which continues to evolve in this era of globalisation. Rather than careful supervision and approval from the government on which parts of Ainu culture to preserve, there must be dialogue amongst the Ainu peoples and between Ainu and *Wajin* experts. As Uzawa (2014, 90) advocates, the Ainu "are and can only be the authors of our own culture and society." Ultimately it must be the Ainu people who have the final say in these decisions.

Division of Spirituality/Religion and Culture

The second point concerns the division of spirituality (and religion) and culture. The Ainu culture is intertwined with spirituality and from a legal perspective, Ichimura posits that the New Ainu Law promises public grants for Ainu culture promotion projects, which clearly violates the Japanese Constitution on the separation of politics and religion (Article 20). He argues it is impossible to remove spirituality from the Ainu culture and one cannot use a memorial site in a 'Symbolic Space' as a non-religious facility. Every ritual, reverence, meaning, and use of 'cultural items' exhibited in 'Museums and Symbolic Spaces' without the appropriate spiritual context is only a shell of culture. This applies not only to ceremonial events, but also to reverence of sacred places, or fishing rituals for example, which cannot be carried out as a spiritual performance for visitors and guests. How much of their spirituality the Ainu people want to share with outsiders must be decided by the Ainu and not by Wajin experts or the government. The Ainu must be given the power of self-determination to make these decisions.

Japanese Government Control over Ainu Culture

The third point is the power imbalance among the involved actors in Ainu culture promotion. According to the New Ainu Law, final decisions on Ainu promotion projects remain in the hands of the Japanese government. Ainu elder Shimizu and Ainu musician Kano state that under the New Ainu Law, the Ainu must seek "approval for state-sponsored cultural projects" (Jozuka 2019). In addition, once the government deems the project to have deviated from its original concept, it has the power to rescind approval (Article 14). Ainu promotion projects are entrusted to prefectural governments and municipalities and the New Ainu Law does not allow Ainu individuals or groups to apply directly for project funding. The management of approved projects must be controlled by registered public organisations. If an Ainu community does not set up a registered corporation (profit or non-profit), they cannot manage the promotion of Ainu culture. Akan Consulun Inc. was registered by incorporation in August 2019 for the purpose of protecting Ainu intellectual rights (Hirono and Okada 2021), but not all Ainu organisations have managed to register as a corporations. In addition, there is overreliance on *Wajin* experts on Ainu culture and Ainu affairs in decision-making. Rather than focusing on profit-oriented Ainu culture promotion or emphasising educational spaces for *Wajin* visitors to understand the precariousness of a minority people, the New Ainu Law could have included provisions for the education of Ainu people to nurture Indigenous experts on law and politics, education, and other professions. Only 33% of the Ainu youth advance to university education, which is much lower than the overall Hokkaido average (Kyodo 2019; Jozuka 2019). Kano advocates without "Ainu lawyers, film directors and professors," the Ainu culture will always be under the control of the Japanese (Jozuka 2019). This is not meant to dismiss collaboration with non-Indigenous researchers, nevertheless, the Ainu must determine their own Indigenous methodologies to prioritise their own interpretation and understanding of the world (Uzawa 2014).

'Double-Edged Sword' Provisions

The ambiguity of the language in the New Ainu Law is a 'double-edged sword' – simultaneously creating more flexibility and opportunity in protecting Ainu culture, but also creating loopholes as to who can be involved. By not clearly defining who the Ainu are, or clearly limiting the qualifications for applicants for Ainu culture projects, the law may be interpreted as being widely inclusive in this endeavour and detrimental to the Ainu. The underlying premise of using minority culture as the main tourist attraction is clear in the New Ainu Law, rather than focusing on legalising and protecting Ainu rights. On the other hand, it may be more advantageous for non-Ainu persons and organisations who have more resources and connections to compete with and outdo Ainu applicants. For instance, the town of Biratori, that includes Nibutani with a significant number of Ainu, received a national government grant for the temporary employment of Ainu people for a cultural project (Nakamura 2019). Allowing non-Ainu to propose and receive funding for Ainu culture promotion projects raises the question as to whom this law was designed for. This is not just a matter of trade and intellectual copyright (Article 18) but rather determining who are the main beneficiaries of Ainu culture promotion.

Geographic Interpretation of the New Ainu Law

The New Ainu Law does not consider the different geographic realities of the Ainu that have emerged over time and how these have impacted traditional rights. Sunazawa (2014) reflects on the reality that the Ainu must get permission from the Japanese government to organize ceremonies and conduct fishing or hunting for rituals. While the New Ainu Law makes allowances for simplifying procedures to obtain permission for fishing salmon and gathering timber for ceremonies (Articles 16 and 17), Ainu leaders regret that this is not the same as the return of fishing rights to the Ainu people (Kyodo 2019). This also raises the question of whether this simplification of procedures also applies to Ainu communities outside Hokkaido. Urban areas like Tokyo host several different Ainu organisations that work closely with their corresponding Ainu organisations in

Hokkaido (Watson 2014). Can Ainu communities in Tokyo for example, receive the same treatment from the Tokyo government and the National Government to fish in a local river if it is for an Ainu ceremony? Noting past discriminations against the Ainu in Tokyo (Watson 2014), it is not hard to imagine these being repeated without a clear statement on the geographical interpretation of the Ainu in the New Ainu Law. It is notable that in December 2018 the Russian Government agreed with the proposal to recognize the Ainu as an Indigenous people of Russia, with one possible motive being special fishing rights from the Japanese government for the Ainu in Russia (Ural Federal University 2019; Uzawa 2019). Such recognition of Ainu people as Indigenous people outside Japan may have impacts on the geographical interpretation of the Ainu people in the New Ainu Law.

A Society Where Ainu Cannot Freely Fight

The last point focuses on the oppressive society in which the Ainu find themselves. Even though the New Ainu Law claims to realise a society where Ainu dignity and pride are respected, the fact that this law does not guarantee self-determination and autonomy for the Ainu means that it is merely a revised version of the *1997 ACPA*. Sunazawa (2014), who now resides in Malaysia and contributes to Ainu campaigns from afar, voices concern over current power structures. She understands that Ainu people feel they have nothing to fight for because the Ainu have lost land and resources to the *Wajin* and the Japanese government and faced assimilation. What is more, the Ainu organisations and their leadership style have been "influenced by the Japanese government and society in general" (Sunazawa 2014, 97) which is top-down. The adopted top-down hieratical structure of the Ainu organisation, according to Sunazawa, is a part of the history of Ainu campaigns against the government that have not resulted in substantial improvements. Higashiyama (2019) further explains that the policy development group that produced the Ainu policy framework had a very limited number of Ainu participants. Since this procedure was carried out confidentially, the Ainu who were not part of the process knew few details. Osakada (2021) also commented on the lack of transparency about Ainu inclusion in the process of policymaking. "Authorities have used the same method since first drawing up the culture promotion law of slipping their intended policies into a report commissioned from the advisory panel, a private consultation body" (Higashiyama 2019 n.p.). The Government's imposiing 'Japanese-style' policy on Indigenous peoples instead of abiding by the UNDRIP, according to Higashiyama, reflects the Japanese government's disinterest in granting collective rights (self-determination, land rights, fishing rights, etc.) and making a show of crafting an inclusive society.

Conclusions

The Ainu in Japan can be examined through a political ecology lens, having faced the processes of colonisation, assimilation, neocolonialism, environmental degradation, and commodification. The New Ainu Law came into effect in 2019 to address past tribulations, and yet many are cautious that it is still too early to judge its outcomes. On the other hand, those Ainu people and experts who campaigned for a new law to protect their human rights as Indigenous people are bitterly disappointed. They view the New Law as a revised version of the *1997 ACPA* and are disheartened that the bill which Ainu organisations drafted for Indigenous rights, land rights, and self-determination was not incorporated. Even among Ainu people there are divisions over claims and expectations, i.e., an official apology for past injustices and redress for the appropriation of lands (Osakada 2021). Unlike older Ainu generations who suffered discrimination, younger generations with wider experiences may call for a movement to fight for their rights in a different, possibly more effective, way with international collaboration (c.f. Sunazawa 2014; Uzawa 2014).

The New Ainu Law, despite its name, is not a law guaranteeing Indigenous rights, but rather a procedure for cultural tourism development using Ainu culture as the main attraction over which the Ainu have little control. It is understandable to some extent that the government's intention to create cultural tourism spaces for edutaining non-Ainu populations is to compensate for past injustices and disseminate more accurate information about the Indigenous peoples of Japan. It is the duty of governments and the citizens of mainstream society to admit past injustices and work towards understanding minority Indigenous peoples. Nevertheless, such a cultural tourism space, 'Symbolic Spaces for Ethnic Harmony' must be developed on Ainu peoples' terms. While some Ainu may be content with the prospect of more permanent employment through the creation of these 'Symbolic Tourism Spaces', others may oppose further commodification of Ainu culture. As long as the Japanese government is at the top of the hierarchy supervising Ainu culture, the Ainu people do not have ownership of their culture. The basic strategy under the New Ainu Law has not changed since 1899. The New Ainu Law acquires Ainu culture as an asset for the development of cultural tourism under a façade of inclusion and innovation under a central government, while still obfuscating Ainu identity in favour of promoting the image of a *Wajin* 'homogenous' nation. Whether Ainu tourism under the New Ainu Law is a success remains to be seen.

References

Arai, K. (2012) "Ainu Shuuraku ga Mizukara no Rekishi o Katarihajimeru Koto: Kaizawa Tadashi ga Henshuu suru Chiikishi 'Nibutani' no Totatsu," *Ouyou Shakaigaku Kenkyu* 54, 219–236.

Cabinet Office (2011) "'Minzoku Kyosei no Shocho to naru Kukan' Sagyoubukai Hokokusho," accessed 31 July 2021. https://www.kantei.go.jp/jp/singi/ainusuishin/shuchou-kukan/houkokusho.pdf

Choi, E. (2011) "Tourism and identity: Ethnic representations of the Ainu and otherness of television documentaries in 1950's," *Shakaigakubu Ronshu* 53, 1–18.

Choi, E. (2012) "What is 'Kankou Ainu'?: A prospect on the historical changes of the gazing," *The Society of Socio-Informatics* 1(2), 93–108.

Gittings, C.E. (2018) "Indigenous Canadian cinemas: Negotiating the precarious," in C. Brucúa and C. Sitnisky (eds.) *The precarious in the cinemas of the Americas*. Global Cinema. Cham: Palgrave Macmillan. 221–244. https://doi.org/10.1007/978-3-319-76807-6_12

Harvey, D. (2003) "Accumulation by dispossession," in *The new imperialism*. Oxford: Oxford University Press. 173–182.

Hashimoto, A. (2021) "Indigenous people's rights and tourism," in A. Hashimoto, E. Härkönen, and E. Nkyi (eds.) *Human rights issues in tourism*. London: Routledge. 296–325.

Higashiyama, T. (26 April 2019) "No rights, no regret: New Ainu legislation short on substance," *Nippon. com*, accessed 30 July 2021. https://www.nippon.com/en/in-depth/d00479/no-rights-no-regret-new-ainu-legislation-short-on-substance.html

Hirono, H. and Okada, M. (2021) "Kanko o Toshite Ainu Bunka no Keisho o Mezasu – Akanko Ainu Kotan no Chosen," *CITS Sosho* 14, 107–139. http://hdl.handle.net/2115/81460

Hokkaido Former Aborigines Protection Act (1899)

Ichikawa, M. (2019a) "Senjuken Naki 'Ainu Shinpo' dewa naku……" *Jumin to Jichi (30 August 2019)*, accessed 30 July 2021. https://www.jichiken.jp/article/0125/

Ichikawa, M. (2019b) *Ainu no Hou-teki Chii to Kuni no Fuseigi*, Sapporo: Ju-Rousha.

Jozuka, E. (22 April 2019) "Japan's 'vanishing' Ainu will finally be recognized as indigenous people," *CNN*, accessed 30 July 2021. https://www.cnn.com/2019/04/20/asia/japan-ainu-indigenous-peoples-bill-intl/index.html

Kido, S. (2019) "Senzenki ni okeru Ainu Minzoku no Doka o Meguru Senryaku – 'Ezo no Hikari' to 'Ekashi to Fuchi' ni Chumoku shite," *Gendai Shakaigaku Kenkyu* 32, 51–68.

Kyodo (19 April 2019) "Japan enacts law recognizing Ainu as indigenous, but activists say it falls short of U.N. declaration," *Japan Times*, accessed 30 July 2021. https://www.japantimes.co.jp/news/2019/04/19/national/japan-enacts-law-recognize-ainu-indigenous-despite-criticism-ethnic-group/

Maruyama, H. (2014) "Japan's policies towards the Ainu language and culture with special reference to north fennoscandian Sami policies," *Acta Borealia* 32(1), 152–175. http://dx.doi.org/10.1080/08003831.2014.967980

MLIT (2019) *White paper on land, infrastructure, infrastructure, transport and tourism in Japan, 2019*, accessed 27 July 2021. https://www.mlit.go.jp/common/001325161.pdf

MLIT (2021) "Chapter 4 Section 5 promotion of the comprehensive development of Hokkaido," *Kokudo Kotsu Hakusho* [MLIT White Paper], accessed 02 February 2022. https://www.mlit.go.jp/hakusyo/mlit/r01/hakusho/r02/pdf/np204000.pdf

Murakami, S. (25 February 2019) "Japan's Ainu recognition bill: What does it mean for Hokkaido's indigenous people?" *Japan Times*, accessed 30 July 2021. https://www.japantimes.co.jp/news/2019/02/25/reference/japans-ainu-recognition-bill-mean-hokkaidos-indigenous-people/

Nakamura, N. (2019) "Does the new Ainu bill truly symbolise progress? Another attempt at improving Indigenous rights in Japan," *Asia and the Pacific Policy Society*, accessed 30 July 2021. https://www.policyforum.net/japans-northern-territories-v-russias-kuril-islands/

Osakada, Y. (2021) "An examination of arguments over the Ainu Policy Promotion Act of Japan based on the UN declaration on the rights of Indigenous peoples," *The International Journal of Human Rights* 25(6), 1053–1069. https://doi.org/10.1080/13642987.2020.1811692

Pellis, A., Pas, A. and Duineveld, M. (2018) "The persistence of tightly coupled conflicts. The case of Loisaba, Kenya," *Conservation and Society* 16(4), 387–396. https://www.jstor.org/stable/26500653

Promotion of Measures to Realize a Society That Will Respect the Pride of the Ainu Act (2019)

Saito, R. (1994) "Hoppo Mizonku Bunka Kenkyu ni okeru Kanko Jinruigaku-teki Shiten (1) Edo-Taisho-ki ni okeru Ainu no Baai," *Hokkaido-ritsu Hoppo Minzoku Hakubutsukan Kenkyu Kiyou* 3, 139–160.

Simmons, D. (2017) "Neoliberal politics and the fate of tourism," in R. Butler and W. Suntkul (eds.) *Tourism and political change*. Oxford: Goodfellow. 9–24.

Sonohara, T. (1997) "Toward a genuine redress for an unjust past: The Nibutani Dam Case," *Murdoch University Electronic Journal of Law* 4(2). http://classic.austlii.edu.au/au/journals/MurUEJL/1997/16.html

Stanley, N. (1998) *Being ourselves for you: The global display of cultures*, London: Middlesex University Press.

Stevens, G. (2001) "The Ainu and human rights: Domestic and international legal protections," *Japanese Studies* 21(2), 181–198.

Stevens, G. (2014) "The Ainu, law, and legal mobilization, 1984–2009," in M.J. Hudson, A-E. Lewallen and M.K. Watson (eds) *Beyond Ainu Studies: Changing Academic and Public Perspectives*. Honolulu: University of Hawaii Press, 200–222.

Sunazawa, K. (2014) "As a child of Ainu," in M.J. Hudson, et al. (eds.) *Beyond Ainu studies: Changing academic and public perspectives*, Honolulu: University of Hawai'i Press. 92–98.

Tsunemoto, T. (2019) "Overview of the Ainu Policy Promotion Act of 2019," *Hokkaido University*, accessed 30 July 2021. https://fpcj.jp/wp/wp-content/uploads/2019/11/b8102b519c7b7c4a4e129763f23ed690.pdf

Uchida, J. (2015) "How to transmit Ainu culture in the context of tourism: Case studies of Shiraoi and Nibutani Districts," *Kokuritsu Rekishi Minzoku Hakubutsukan Kenkyu Houkoku*, 193. https://doi.org/10.15024/00002196

Umeda, S. (2019) "Japan: New Ainu law becomes effective," *Library of Congress*, accessed 30 July 2021. https://www.loc.gov/item/global-legal-monitor/2019-08-05/japan-new-ainu-law-becomes-effective/#:~:text=Article%20Japan%3A%20New%20Ainu%20Law,effective%20on%20May%2024%2C%202019.&text=The%20Ainu%20are%20indigenous%20people,part%20of%20Japan%2C%20especially%20Hokkaido

United Nations, The (1980) "Initial reports of States parties due in 1980: Japan," *UN Doc CCPR/C/10/Add 1. 14 November 1980*, accessed 08 May 2017. https://digitallibrary.un.org/record/20463?ln=en

United Nations, The (2007) *Universal declaration on the rights of Indigenous peoples*. https://www.un.org/esa/socdev/unpfii/documents/DRIPS_en.pdf

Ural Federal University, The (26 March 2019) "Ainu people claims Japan Ingenious people rights," accessed 23 February 2022. https://urfu.ru/en/news/26574/

Urry, J. and Larsen, J. (2011) *The tourist gaze 3.0*. London: Sage Publications.

Uzawa, K. (2014) "Charanke," in M.J. Hudson, et al. (eds.) *Beyond Ainu studies: Changing academic and public perspectives*. Honolulu: University of Hawai'i Press. 86–91.

Uzawa, K. (24 April 2019) "Indigenous World 2019: Japan," *IWGIA*, accessed 07 December 2021. https://www.iwgia.org/en/japan/3427-iw2019-japan.html

Watson, M.K. (2014) "Tokyo Ainu and the urban Indigenous experience," in M.J. Hudson, et al. (eds.) *Beyond Ainu studies: Changing academic and public perspectives*. Honolulu: University of Hawai'i Press. 69–85.

Winchester, M. 2009. "On the Dawn of a New National Ainu Policy: The 'Ainu as a Situation' Today," *The Asia-Pacific Journal* 7(41), 1–20.

Zaman, M. (2020) "The Ainu and Japan's colonial legacy," *Tokyo Review*, accessed 30 July 2021. https://www.tokyoreview.net/2020/03/ainu-japan-colonial-legacy/

21

INDIGENOUS HOSTS AND FOOD SECURITY

A Case Study from Simien Mountains National Park

Gebeyaw Ambelu Degarege

Introduction

Ensuring Indigenous peoples' food and nutrition security sustainably and equitably is a compelling human rights issue (Degarege, 2019) and is a well-recognised global development challenge. Food insecurity is a situation in which people are unable to access adequate healthy food, experience hunger, and consume less nutritious foods due to limited ability and options (Degarege, Lovelock, & Tucker, 2018; Gallegos, Ramsey, & Ong, 2014; Gundersen & Kreider, 2009; Gundersen & Ziliak, 2015). In other words, food insecurity is a limited or uncertain availability of nutritionally adequate and safe foods or a limited ability to acquire personally acceptable foods that meet cultural needs in a socially acceptable way (Ministry of Health, 2019; Coleman-Jensen, 2010). However, there is still limited research and understanding of Indigenous tourism communities, food security in general, and Indigenous people's food security in developing countries. This chapter examines the interplay between Indigenous community tourism and food security in the case of Indigenous communities in Ethiopia using in-depth interviews as a data collection tool. It explores perceptions of Indigenous communities towards food insecurity, problems with their food security undertakings, and potential policy options for improving food security outcomes through sustainable Indigenous tourism. It will shed light on food security innovations in policy and practice for addressing the food security concerns of tourism-dependent Indigenous peoples.

Food insecurity is most prevalent in low-income areas of developing countries; however, it is also a concern for developed countries. Food security or 'zero hunger' is one of the 17 Sustainable Development Goals (SDGs) established by the United Nations in 2015, with eight targets to eradicate world hunger (United Nations, 2015; see Table 21.1). It aims to end hunger and ensure universal access to safe, nutritious, and sufficient food all year round for all people (United Nations, 2015). Despite the global commitment to achieve 'zero hunger' in 2030 (United Nations, 2015), insights thus far suggest that the world has not generally been progressing towards ensuring access to safe, nutritious, and sufficient food for all people all year round or eradicating all forms of malnutrition (FAO, 2021).

In 2020, between 720 and 811 million people were exposed to severe levels of hunger – as many as 161 million more than in 2019 (FAO, 2021). Compared with 2019, more than 46 million

277

DOI: 10.4324/9781003230335-25

Table 21.1 Zero hunger targets

Food security target	Description
Target 2.1 Universal access to safe and nutritious food	End hunger and ensure access by all people, in particular the poor and people in vulnerable situations, including infants, to safe, nutritious and sufficient food all year round.
Target 2.2 End all forms of malnutrition	End all forms of malnutrition, including achieving, by 2025, the internationally agreed targets on stunting and wasting in children under 5 years of age, and address the nutritional needs of adolescent girls, pregnant and lactating women and older persons.
Target 2.3 Double the productivity and incomes of small-scale food producers	Double the agricultural productivity and incomes of small-scale food producers, in particular women, Indigenous peoples, family farmers, pastoralists and fishers, including through secure and equal access to land, other productive resources and inputs, knowledge, financial services, markets and opportunities for value addition and non-farm employment.
Target 2.4 Sustainable food production and resilient agricultural practices	Ensure sustainable food production systems and implement resilient agricultural practices that increase productivity and production, that help maintain ecosystems, that strengthen capacity for adaptation to climate change, extreme weather, drought, flooding and other disasters and that progressively improve land and soil quality.
Target 2.5 Maintain the genetic diversity in food production	Maintain the genetic diversity of seeds, cultivated plants and farmed and domesticated animals and their related wild species, including through soundly managed and diversified seed and plant banks at the national, regional and international levels, and promote access to and fair and equitable sharing of benefits arising from the utilisation of genetic resources and associated traditional knowledge, as internationally agreed.
Target 2.6 Invest in rural infrastructure, agricultural research, technology and gene banks	Increase investment, including through enhanced international cooperation, in rural infrastructure, agricultural research and extension services, technology development and plant and livestock gene banks in order to enhance agricultural productive capacity in developing countries, in particular least developed countries.
Target 2.7 Prevent agricultural trade restrictions, market distortions and export subsidies	Correct and prevent trade restrictions and distortions in world agricultural markets, including through the parallel elimination of all forms of agricultural export subsidies and all export measures with equivalent effect, in accordance with the mandate of the Doha Development Round.
Target 2.8 Ensure stable food commodity markets and timely access to information	Adopt measures to ensure the proper functioning of food commodity markets and their derivatives and facilitate timely access to market information, including on food reserves, in order to help limit extreme food price volatility.

Source: United Nations (2015).

people in Africa, almost 57 million more in Asia, and about 14 million more in Latin America and the Caribbean were affected by hunger in 2020 (FAO, 2021). The debate over the root causes of food insecurity has fuelled highly contested viewpoints. A multitude of environmental, social, economic, and political disruptions and interactions lead to food insecurity (Alexandratos, 1999; Burchi & De Muro, 2012, 2016; Coad & Pedley, 2020; Devereux & Maxwell, 2001; Ericksen, 2008; McDonald, 2010; Misselhorn, 2005). Above all, the COVID-19 pandemic has drastically altered food supply chains and worsened food insecurity globally (Deaton & Deaton, 2020;

Kim, Kim, & Park, 2020; Laborde, Martin, Swinnen, & Vos, 2020). It has exacerbated existing socio-economic inequalities and disproportionately impacted those already experiencing food insecurity (Fitzpatrick, Harris, Drawve, & Willis, 2021; Mishra & Rampal, 2020). The food security situation of tourism-dependent communities has already been a subject of discussion, as tourism development is expected to bring additional challenges to the food security of Indigenous people in tourism destination sites (Santafe-Troncoso & Loring, 2021). Understanding the positive and negative food security impacts derived from tourism on Indigenous peoples is crucial to designing appropriate bottom-up development policies and plans. This chapter uses Simien Mountains National Park (SMNP) in Ethiopia as a case study. This site was chosen because it is a drought-prone and food-insecure area with the recent addition of tourism as an economic driver. Traditional mixed agriculture that consists of crop production and livestock rearing is the major economic activity of local communities in SMNP. People raise livestock traditionally for trade and cash reasons. Accessibility is difficult, most places can only be reached on foot or horseback, and basic infrastructure facilities such as access to safe water and health services are still unavailable. Ensuring livelihoods and food security for Indigenous people through improving natural resource management practices and protecting the environment are crucial development issues. Tourism is widely recognised as a livelihood diversification tool, but there has not been a strong emphasis on its implications for food security.

This chapter explores the options for improving food security outcomes through sustainable Indigenous tourism in the context of a tourism destination with considerable tourism potential but with a very acute food insecurity situation.

Food Security, Communities, and Indigenous Tourism

The concept of food security and related approaches to addressing food insecurity have evolved in accordance with the evolution of thinking about development and understanding the nature of the food problem (Devereux & Maxwell, 2001; D. Maxwell, 1996; Maxwell & Slater, 2003). Since the 1970s, three fundamental shifts in food security thinking have been identified (Devereux & Maxwell, 2001; D. Maxwell, 1996). These shifts are (a) from the global and the national levels of food supply to the household and the individual levels of food access; (b) from a food-first perspective to a livelihood perspective; and (c) from objective indicators to subjective perceptions of food security. These shifts in thinking significantly add to the existing understandings (Devereux & Maxwell, 2001; D. Maxwell, 1996). The initial focus of food security discussions was primarily on food supply problems at the international and national levels (Burchi & De Muro, 2012; FAO, 2003; Jones, Ngure, Pelto, & Young, 2013). This exclusive focus on the explanation of food insecurity stemming from insufficient food supply is labelled as food availability decline (FAD) theory (Devereux, 1988). Despite adding other dimensions of food security (access, utilisation, and stability), the food availability approach is still the most influential one in the international food security discourse and practice (Burchi & De Muro, 2012, 2016). FAD theory attributes food insecurity mainly to agricultural sector production failure and explains that people starve because of insufficient food supplies (Devereux, 1988). Stamoulis and Zezza (2003) highlight that improved food availability is important but does not necessarily result in food security unless the underlying causes of extreme food insecurity such as the lack of access to resources, insecurity of tenure, low returns to labour, and education are understood and addressed. That is to say, adequate food availability may not automatically translate into food security unless the underlying causes of food insecurity are resolved (Frankenberger & McCaston, 1998; Sen, 1981b).

Ensuring food security and requirements of vulnerable and Indigenous peoples sustainably and equitably is a recognised human right that requires urgent policy actions (Degarege & Lovelock, 2019; S. Maxwell, 1996; Pérez-Escamilla, 2017). The UN Committee on Economic, Social and Cultural Rights, in its general comment 12, noted that the right to adequate food would be realised when every man, woman, and child, alone or in a community with others, had physical and economic access at all times to adequate food or means for its procurement (1999). Indigenous communities represent most of the world's cultural diversity and conserve about 80% of the global biodiversity (Krystyna & Philippa, 2020). In contrast, several studies indicated that Indigenous peoples generally experience social disadvantage and are recognised as having higher food insecurity than non-Indigenous people (Cidro, Adekunle, Peters, & Martens, 2015; Santafe-Troncoso & Loring, 2021). Due to long historical processes of colonisation and destruction of Indigenous food systems (Bagelman, 2018; Cidro et al., 2015; Kuhnlein & Chotiboriboon, 2022), income inequality (Coad & Pedley, 2020; Kuhnlein & Chotiboriboon, 2022), reliance on colonial policies (Bagelman, 2018; Kuhnlein & Chotiboriboon, 2022), climate change (Barnett, 2011; Myers et al., 2017; Tapsell et al., 2011), and political and sociocultural exclusions (Carter, Lanumata, Kruse, & Gorton, 2010; Ministry of Health, 2019), Indigenous peoples are among the most vulnerable to hunger, malnutrition, and food insecurity.

Even in the case of developed countries like Aotearoa (New Zealand), it is reported that food insecurity is particularly evident among Māori and Pasifika households, in large and low-income families and families in high-deprivation areas (Carter et al., 2010; Cidro et al., 2015; Ministry of Health, 2019; Parnell, 2005; Utter, Izumi, Denny, Fleming, & Clark, 2018). This is reflected in statistics for *tamariki Māori* where over one in four (28.6%) lived in food-insecure households in 2015/16 (Ministry of Health, 2019). Moreover, the prevalence of food insecurity was much higher in females than in males (Carter et al., 2010). Increased local community control over food systems is identified as a valuable means of addressing food insecurity (Cidro et al., 2015; Kuhnlein & Chotiboriboon, 2022). These include respecting Indigenous peoples' traditional ways of living, strengthening traditional food systems, and protecting subsistence activities such as agriculture, hunting, and fishing (Kuhnlein & Chotiboriboon, 2022). Equally important is the adjustment and restructuring of economic and political systems and structures vital to building Indigenous food security (Cidro et al., 2015; Kuhnlein & Chotiboriboon, 2022). Some studies have revealed that food security is a complex phenomenon that requires integrated actions of everyone involved in food production, processing, and consumption (Candel, 2014; Charlton, 2016; Degarege & Lovelock, 2019).

Indigenous Tourism, Sustainability, and Food Supply

Indigenous tourism development has emerged as an alternative development approach in response to neocolonialism, neoliberalism, and conventional mass tourism (Pearce, 1992; Scheyvens, 2011; Telfer, 2015). The importance of community-based tourism for an equitable flow of benefits to local community members and other stakeholders through a system of local development control and consensus-based decision-making is recognised (Afenyo & Amuquandoh, 2014; Pearce, 1992). Supporting this, Lapeyre (2010) suggests three primary advantages of community-based tourism: (1) minimising leakages occurring at the local level; (2) generating significant linkages for the local economy; and (3) providing an opportunity for community members to gain institutional and managerial capacity (see also Johnson, 2010; Tolkach & King, 2015; Zapata, Hall, Lindo, & Vanderschaeghe, 2011). In light of this, community-based tourism is acknowledged for its contribution to rural poverty alleviation and environmental protection (Lapeyre, 2010; Zapata et al., 2011). Zapata et al. (2011, p. 742) noted that community-based tourism can be an "excellent catalyst to support the poor rural community". Cognizant of these facts, many multilateral organisations, government, and non-governmental bodies have advocated for tourism's contribution to poverty reduction. Beyond looking at the commodity values of Indigenous food, the provision of Indigenous food experiences by Indigenous people can form an integral part of Indigenous tourism experience.

However, the relationship between Indigenous tourism and sustainable development is not linear nor free of debate (Carr, 2020; Santafe-Troncoso & Loring, 2021; Scheyvens et al., 2021). Scheyvens, in 2002, asserted that "Elites often dominate community-based development efforts and monopolise the benefits of tourism" (p. 9). In such a context, ensuring that the benefits reach the poor becomes one of the more difficult undertakings (Ashley & Roe, 2002; Choi & Sirakaya, 2006; Scheyvens, 2002). The extent of Indigenous tourism development and its potential role in food security may be determined by the type of tourism policy and planning framework utilised in the context of food requirements (Degarege & Lovelock, 2019) and the way tourism is synchronised with the Indigenous traditional food system (Degarege & Lovelock, 2021; Degarege et al., 2018). Degarege and Lovelock (2021) in their studies of the effects of tourism on food security in two tourism destinations in Ethiopia found that tourism benefits in general appear to be small and limited to a very few community members.

Although local community members consider tourism as a means of income to support their food security, they seldom rely on tourism alone. Owing to the recognised importance of tourism for the sustainable development of destination communities, it is important to find a means to harness the development potential of Indigenous people and their food systems perspective. Without appropriate Indigenous development perspectives, tourism may result in the commodification of food and Indigenous food systems. Nevertheless, there is a dearth of studies about the nature of food insecurity in the context of tourism-dependent Indigenous communities. Scheyvens et al. (2021), in their study of Indigenous tourism and the SDGs, highlighted the critical relevance of research on how the implementation of the SDGs can respond to Indigenous peoples' priorities and aspirations.

This chapter, in discussing the experiences and beliefs associated with food security goals in the case of Indigenous communities, is guided by Amartya Sen's (1981b) entitlement approach to food security in order to explore tourism-dependent communities' food security contexts. Sen's entitlement approach challenged the food availability thinking and argued that people could starve in the midst of plentiful food owing to a collapse in their means of command over food (Sen, 1981b). Unlike the long-lasting view of food security as a problem of food availability, this approach re-addresses the problem of hunger and famine by giving more relevance to the socio-economic conditions of people (Burchi & De Muro, 2012, 2016). Sen's entitlement approach to food security shifted the focus to people's inability to acquire food (Burchi & De Muro, 2012; Devereux, 2001; Sen, 1981b), and it concentrates on "each person's entitlements to commodity bundles including food and views food insecurity as resulting from failure to be entitled to any bundle with enough food" (Sen, 1981a, p. 434). Food entitlement depends on personal endowments that include the resources that a person legally owns, such as a house, livestock, land, and intangibles such as knowledge and skills, and the set of commodities that individuals have access to through trade and production (Osmani, 1993; Sen, 1981b). Sen (1981b) identified four general sources of food entitlement: (1) production-based entitlement, which refers to those who produce food for themselves; (2) trade-based entitlement, which encompasses individuals who may sell or barter physical assets to secure their food requirements; (3) labour-based entitlement, where landless labourers and urban employees all need to buy or barter food in the market; and (4) transfer entitlement, which includes individuals who may receive informal gifts from individuals and formal transfers such as aid from governments. Accordingly, people become food insecure because of the failure in entitlement, be it "personal endowment" or "exchange" entitlement failure (Osmani, 1993; Sen, 1981b). The entitlement approach addresses the multidimensional nature of food security (Burchi & De Muro, 2012, 2016).

A discussion of the dynamics of Indigenous food security in the tourism development context is relevant for several reasons. First, understanding how Indigenous people construct their traditional food system will help to systematically connect the tourism sector's food requirements with the local food system and vice versa (Degarege et al., 2018; Santafe-Troncoso & Loring, 2021). Second, understanding mechanisms through which tourism could support or detracts from the traditional food system's performance and efficiency is crucial for sustaining Indigenous food systems (Degarege et al., 2018; Santafe-Troncoso & Loring, 2021). Third, it may provide better insight into how Indigenous communities' food security and sustainable development concerns might be assisted through tourism as a means of livelihood at a grassroots level (Carr, 2020; Degarege & Lovelock, 2021; Degarege et al., 2018; Scheyvens et al., 2021).

Methods

This study was conducted in SMNP (Figure 21.1), located 885 km north-west of Addis Ababa, the capital city of Ethiopia. It is a UNESCO-registered nature-based tourism destination with a community-based tourism initiative. The park was formally established in 1966 and gazetted in 1969 (Order No 59), after being recommended by UNESCO in 1965 (EWCA, 2013). The park was inscribed on the list of World Natural Heritage Sites in 1978 for its unique landscapes and rich biodiversity resources (EWCA, 2016). The highest peak in the country (4,533 Metres), *Ras Dashen* Mountain, is found in this park (Amhara Regional State Tourism and Parks Development Bureau, 2009). Visitors to SMNP have increased significantly since 2000, and in 2016, there were 25,999 arrivals (Table 21.2). Correspondingly, revenue generated from the park and income generated by the local community have increased proportionally. Most visitors come during the drier months, October to April.

The area is the most drought-prone in the region, and agriculture is almost entirely rain-fed and highly vulnerable to water shortage. The farming practices employed in the study area are characterised by traditional oxen plough and hoe culture. Mixed agriculture systems (crop–livestock) characterise the study site. Major crops grown in the area are barley, wheat, teff (*Eragrostis tef*), sorghum, maize, pulses, pea, bean, and lentil. At present, there is no clear initiative to harness the agritourism potential of the area. The main tourism development initiative is focused on hiking and trekking in and around the National Park.

Qualitative semi-structured interviews were conducted with eight purposefully selected (Creswell, 2013; Teddlie & Tashakkori, 2009) Indigenous community members and leaders from SMNP to generate information for a deeper understanding of the Indigenous community's food security. Tourism forms a recent form of livelihood opportunity in the study site and is currently occurring in the form of community tourism associations/cooperatives, as well as individual business entrepreneurship. Tourism stands out as the most important activity in terms of its potential to increase the local communities' incomes through employment in multiple tourism-related activities including scouting, luggage portering, muleteer provision, renting pack animals, local tour guiding, and cooking. On average, a 40-minute in-depth interview with each participant was audio-recorded, after which they were transcribed into Amharic and translated into English. A six-step process of thematic data analysis that involved becoming familiar with the data (which

Table 21.2 Tourist arrivals and receipts in *Debarq*

Year	Number of tourists	Revenue earned (in ETB)		Total
		Government	Community	
2000	1,289	210,303	230,000	440,303
2001	1,825	205,678	241,031	446,709
2002	2,652	307,083	355,997	663,080
2003	3,495	402,036	395,907	797,943
2004	3,767	454,215	517,848	972,063
2005	5,074	516,147	736,567	1,252,714
2006	6,019	593,678	745,049	1,338,727
2007	6,991	828,724	930,398	1,759,122
2008	8,460	956,071	1,161,992	2,118,063
2009	11,648	1,439,790	2,602,000	4,041,790
2010	14,016	3,281,275	2,523,484	5,804,759
2011	17,566	4,314,591	4,127,423	8,442,014
2012	15,883	4,260,510	4,498,875	8,759,385
2013	15,948	4,441,518	3,910,435	8,351,953
2014	22,020	4,544,020	7,210,810	11,754,830
2015	22,457	4,152,450	9,510,120	13,662,570
2016	25,999	5,960,697	19,773,060	25,733,757

Source: *Debarq* District Culture and Tourism Office (2017).

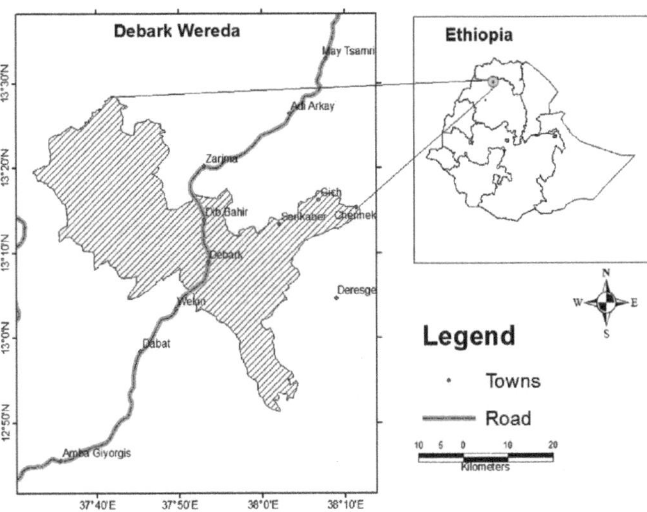

Figure 21.1   Debarq District study area map

involved transcribing, reading and rereading, and noting down initial ideas), generating initial codes, searching for themes, reviewing themes, defining and naming themes, and finally producing the reports was used (Braun & Clarke, 2006). The themes of analysis were pre-established based on the existing theory and an emerging theory that evolved gradually out of the data itself.

Findings

In general, tourism development initiatives lack a focus on food security targets and broader economic, social, and environmental dimensions. Diverse perspectives regarding food security implications of Indigenous tourism are discussed below.

Indigenous Voice Matters: Essential, Not Optional

Commitment to Indigenous communities' participation in tourism development issues that influence their food security and livelihood conditions is fundamental. The foundation of tourism development in the study area is always externally driven by the national government, private investors, and non-governmental organisations (NGOs). NGOs are involved in several development initiatives related to tourism-supported livelihood development and park resource conservation, but most of the initiatives are not Indigenous in terms of their approach and practice. Participants indicated that resources such as land, forest, and wildlife, which used to be part of the food system, are no longer freely available because of tourism development and conservation-related constraints. The findings indicate that the food insecurity status of local communities has worsened because of both human and natural causes including land, agricultural, socio-economic, and hydrological factors. Local communities are aware that food insecurity in their locality has become rampant because of the dynamic processes explained by these interrelated factors.

This is exemplified in the following interview excerpts that outline the factors that place considerable pressure on farmers:

> *Efforts to address food insecurity and other problems in our neighbourhood are always there but much of the interventions are not practical to our situation. You can take fertiliser, we used to have productive land in the past...ever since we started using fertiliser everything got complicated... only very few people afford fertiliser (INT1)*
> *...a lot of things are going on in our society... and yet life is getting more complicated and challenging. We don't have much say on what is going on around (INT3)*

The above excerpts indicate that 'Indigenous voice matters' is not an integral element of local development in the study area. It is understandable that the adoption of modern farming tools and inputs without much awareness and community acceptance is identified as another cause of food insecurity. These excerpts also show that participants had a feeling that their living in general was better in the past. This indicates the need for the corroboration of context and system-specific support and development interventions to improve the living standards of local communities. Currently, efforts to harness the agritourism potential of the area are limited to only fruit selling. Tourism development that recognises the existing Indigenous agriculture civilisation and harnesses its agrotourism potential as a resource or experience may bring better awareness of environmental and resource management and thereby better food security.

Solutions that Do Not Fit Local Food Production

The land and other natural resources remain the main drivers of food insecurity. Some food-producing farmers become food insecure because of failure in their food production. For most local communities, crop and livestock production are fundamental and the culturally established means of their household food security. Recognising the dependence on farm-based livelihood activities for food security, the interviews indicated that food insecurity in the study areas is mainly a result of inherent shortcomings in a household's crop and livestock production, which was attributed to the absence of local level and culturally led food security actions. Crop production in the study sites is entirely dependent on rain and carried out using Indigenous farming knowledge. It has been practised traditionally with an animal-drawn *ard* plough called *maresha* in Amharic, but no effort has been made to capitalise on making this an integral part of tourism experience. Most local community members indicate that their farming is not adequately supported by technological advances and proper modern input sourcing. The respondents explained that local community members rarely adopt modern seeds and technologies because of the lack of knowledge and affordability reasons. Only a few households have access to irrigation from rivers or streams, and very few reported having hand-dug irrigation wells. As a result, the production of perennial crops such as fruits and vegetables is limited to the rainy seasons only.

Local community members also stressed the crucial livelihood roles of the possession of livestock. More specifically, the role oxen and horses play in households' livelihood undertakings such as farming, transportation, and means of income is acknowledged. Almost all participants indicated that livestock performance is not as productive as it used to be because of grazing land shortage and resource-use restrictions. Participants indicated that the concept of communal grazing lands is fast disappearing and a greater number of households used to rely on communal grazing land. Adding to this, a participant explained:

Livestock is not a matter of income and food; it is also a means for farming. Our life will be complicated unless something miracle happens. We are praying for our old good days where we used to drink milk like water.

This highlights the value of livestock holdings in terms of providing production and exchange entitlement to food and other necessary goods and services. In addition, its direct role as a food source, input for farming, and a means to earn income makes livestock vital to the local community's food security. The respondents also suggested that the shrinking performance of livestock livelihood activities directly or indirectly has influenced their households' food consumption and sociocultural practices. Butter represents one of the cultural identities as a cosmetic and a means of hair, skincare, and beauty treatment, which is no longer affordable because of the limited access to grazing land.

Indigenous Fit – Does Tourism Really Matter to Food Entitlement?

Local community members' responses to questions about what food insecurity means for them reveal a diverse range of opinions ranging from the lack of food at all to the absence or limitation of land and resources to perform traditional farming practices. While tourism was appreciated as one means of a livelihood diversification strategy, respondents noted that the income from tourism is insufficient to cover household food consumption and expenses. Most respondents don't think tourism can address food insecurity's root causes. Combined, insufficiencies in their own food production and their lack of capability to buy food are the two reasons to explain food insecurity. Many local peoples grow and consume food from their land, but they are also highly dependent on local markets. Participants' explanations demonstrate limited perspectives on tourism-supported food security. The most prominent explanation was around the relative importance of tourism earnings in assisting economic access to food. There is no strong contradiction between tourism food habits and culture-specific preferences, but explanations about nutrition are limited.

Tourism-development-induced changes in resource access and use are strongly contested issues impacting food security. Almost all respondents attest that their historical dependence on land and forest resources makes tourism sector resource-use issues more critical to their food security status. The ongoing demand for more productive farming and grazing land was seen as a challenge for tourism and park resources management, and these challenges are rooted in the inefficiency of the tourism sector in terms of its capacity to provide a viable alternative income or livelihood basis. Many from the local communities expressed their resentment over resource-use restrictions on grazing lands and firewood collection. A few local community members indicated that they used to collect firewood to generate income from selling that as well as for their household consumption, mainly for food preparation. Though it requires considerable effort regarding collection and transportation of the material to marketplaces, it was generally confirmed by the interviews that most landless households, as well as younger people, were predominantly engaged in firewood selling activities to supplement their household income and expenditure. A participant explained the change in resource access came about without communities' approval and its effect on subsistence activities to meet their food security requirements:

In the past, grazing land was free, and our access was not constrained. Now we do not have all the freedom to release our animals the way we used to because of the restriction from the government ... we do not have enough land to use it for grazing exclusively and there is no enough open grazing land [common pool]

As indicated in the above excerpt, restrictions that impact the community's livelihood and existence have been put in place by the government or the park staff with little consultation with the local community. Alternative means to address their grazing and firewood requirements have not been negotiated. The increased resource access needs of local communities were raised several times by interviewees, arguing that there was a sharp rise in the number of people struggling to access food resources. Participatory and resource-restrictive park management approaches in the area, along with other recurrent natural hazard events such as drought and rainfall variability, point to the possibility of increased risk of food insecurity.

Conclusions

This chapter contends that there are distinctive food security elements and matters for Indigenous communities related to food availability, access, utilisation, and stability. Promoting Indigenous food systems approaches and structural changes that systematically address the root causes of food insecurity is indispensable to protecting food systems (FAO, 2021). Tourism can draw from and impinges on many other activities and thus competes for the use of scarce resources. An emphasis on resource conservation and park management, assuming the economic value of tourism would more than compensate for the loss of access to traditional resources, seems to be the influencing factor in the relationship between tourism and food security in the study site. Food security concerns were not foremost for park and tourism officials. There seem to be no substitution behaviours available among community members in SMNP to increase the reliance on tourism for making a living. Instead, tourism's role in alleviating food insecurity is questioned. Tackling the root causes of food insecurity requires specific and culturally informed proactive actions. Food security actions and strategies that are rooted in non-Indigenous people's contexts are not always compatible with or supportive of the traditional food practices of Indigenous communities. There is a pressing need to alleviate tourism-induced restrictions over access to land and other resources. This holds particularly true for the landless population whose food security issues were largely hinged upon cash earnings dependent on local resources. Given the historically strong dependency on farming and livestock, any constraints on these Indigenous activities, such as those stemming from national park establishment and management, potentially bring food system instability and risk.

Using tourism as a tool to support Indigenous people's food security necessitates tourism planning frameworks that directly strengthen Indigenous people's food systems (Kuhnlein & Chotiboriboon, 2022) and can impact those already experiencing food insecurity disproportionally. The findings suggested that Indigenous communities have had minimal consultation and participation in planning and managing their resources and tourism. As a result, Indigenous communities do not appreciate tourism as a means of food security despite its small economic contribution. Tourism development initiatives also need to be inclusive and gain the full participation of Indigenous Peoples. Achieving Indigenous peoples' food security through tourism largely depends on how tourism intersects with access to and control over the natural resources used in the region. Tourism needs to be managed in a way that does not create resource-use conflict or endanger the traditional food systems. Planning and governance of tourism and food security based on traditional Indigenous knowledge and food systems by Indigenous peoples could help to establish and ensure more sustainable tourism and food practices. That approach seeks out potential improvements along with the food system, building Indigenous communities' capacity to produce adequate quantities and quality of traditional food to meet both their own dietary requirements and tourism sector demand. Building a tourism model that may assist in addressing structural issues is imperative to resolving food security challenges at the grassroots level. This necessitates Indigenous tourism-assisted food security frameworks, policies, and actions targeting Indigenous communities and their traditional food system.

References

Afenyo, E. A., & Amuquandoh, F. E. (2014). Who benefits from community-based ecotourism development? Insights from Tafi Atome, Ghana. *Tourism Planning & Development, 11*(2), 179–190.

Alexandratos, N. (1999). World food and agriculture: Outlook for the medium and longer term. *Proceedings of the National Academy of Sciences, 96*(11), 5908–5914.

Amhara Regional State Tourism and Parks Development Bureau. (2009). *Sustainable Tourism Networking Development Plan for North Gondar Administration Zone*. Bahir Dar, Ethiopia.

Ashley, C., & Roe, D. (2002). Making tourism work for the poor: Strategies and challenges in southern Africa. *Development Southern Africa, 19*(1), 61–82.

Bagelman, C. (2018). Unsettling food security: The role of young people in Indigenous food system revitalisation. *Children & Society, 32*(3), 219–232.

Barnett, J. (2011). Dangerous climate change in the Pacific Islands: Food production and food security. *Regional Environmental Change, 11*(1), 229–237.

Braun, V., & Clarke, V. (2006). Using thematic analysis in psychology. *Qualitative Research in Psychology, 3*(2), 77–101.

Burchi, F., & De Muro, P. (2012). *A human development and capability approach to food security: Conceptual framework and informational basis*. Rome: United Nations Development Programme, Regional Bureau for Africa. Retrieved from http://www.africa.undp.org/content/rba/en/home/library/working-papers/capability-approach-food-security.html

Burchi, F., & De Muro, P. (2016). From food availability to nutritional capabilities: Advancing food security analysis. *Food Policy, 60*, 10–19.

Candel, J. J. (2014). Food security governance: A systematic literature review. *Food Security, 6*(4), 585–601.

Carr, A. (2020). COVID-19, indigenous peoples and tourism: a view from New Zealand. *Tourism Geographies, 22*(3), 491–502.

Carter, K. N., Lanumata, T., Kruse, K., & Gorton, D. (2010). What are the determinants of food insecurity in New Zealand and does this differ for males and females? *Australian and New Zealand Journal of Public Health, 34*(6), 602–608.

Charlton, K. E. (2016). Food security, food systems and food sovereignty in the 21st century: A new paradigm required to meet Sustainable Development Goals. *Nutrition and Dietetics, 73*(1), 3–12.

Choi, H. C., & Sirakaya, E. (2006). Sustainability indicators for managing community tourism. *Tourism Management, 27*(6), 1274–1289.

Cidro, J., Adekunle, B., Peters, E., & Martens, T. (2015). Beyond food security: Understanding access to cultural food for urban Indigenous people in Winnipeg as Indigenous food sovereignty. *Canadian Journal of Urban Research, 24*(1), 24–43.

Coad, J., & Pedley, K. (2020). Nutrition in New Zealand: Can the past offer lessons for the present and guidance for the future? *Nutrients, 12*(11), 3433.

Coleman-Jensen, A. J. (2010). US food insecurity status: Toward a refined definition. *Social Indicators Research, 95*(2), 215–230.

Creswell, J. W. (2013). *Research design: Qualitative, quantitative, and mixed methods approaches.* Thousand Oaks: Sage publications.

Deaton, B. J., & Deaton, B. J. (2020). Food security and Canada's agricultural system challenged by COVID-19. *Canadian Journal of Agricultural Economics/Revue canadienne d'agroeconomie, 68*(2), 143–149.

Degarege, G. A. (2019). *Tourism, Livelihood Diversification and Food Security in Ethiopia.* University of Otago.

Degarege, G. A., & Lovelock, B. (2019). Sustainable tourism development and food security in Ethiopia: Policy-making and planning. *Tourism Planning & Development, 16*(2), 142–160.

Degarege, G. A., & Lovelock, B. (2021). Addressing zero-hunger through tourism? Food security outcomes from two tourism destinations in rural Ethiopia. *Tourism Management Perspectives, 39*, 100842.

Degarege, G. A., Lovelock, B., & Tucker, H. (2018). Empty bowls: Conceptualising the role of tourism in contributing to sustainable rural food security. *Journal of Sustainable Tourism.* doi:10.1080/09669582. 2018.1511719

Devereux, S. (1988). Entitlements, availability and famine: A revisionist view of Wollo, 1972–1974. *Food Policy, 13*(3), 270–282.

Devereux, S. (2001). Famine in Africa. In S. Devereux & S. Maxwell (Eds.), *Food security in sub-saharan Africa* (pp. 117–148). London: ITDG Publishing.

Devereux, S., & Maxwell, S. (2001). *Food security in sub-Saharan Africa.* London: ITDG Publishing.

Ericksen, P. J. (2008). Conceptualizing food systems for global environmental change research. *Global Environmental Change, 18*(1), 234–245.

Ethiopian Wildlife and Conservation Authority. (2013). Simien Mountains National Park History. Retrieved from http://www.ewca.gov.et/en/simien_mountain_national_park

Ethiopian Wildlife and Conservation Authority. (2016). Simien Mountains National Park. Retrieved from http://www.ewca.gov.et/en/simien_mountain_national_park

FAO, I., UNICEF, WFP and WHO. (2021). *The state of food security and nutrition in the world 2021. Transforming food systems for food security, improved nutrition and affordable healthy diets for all.* Rome: FAO.

Fitzpatrick, K. M., Harris, C., Drawve, G., & Willis, D. E. (2021). Assessing food insecurity among US adults during the COVID-19 pandemic. *Journal of Hunger & Environmental Nutrition, 16*(1), 1–18.

Food and Agriculture Organization. (2003). *Trade reforms and food security: Conceptualizing the linkages.* Rome: FAO.

Frankenberger, T. R., & McCaston, M. K. (1998). The household livelihood security concept. *Food Nutrition and Agriculture*, 30–35.

Gallegos, D., Ramsey, R., & Ong, K. W. (2014). Food insecurity: Is it an issue among tertiary students? *Higher Education, 67*(5), 497–510.

Gundersen, C., & Kreider, B. (2009). Bounding the effects of food insecurity on children's health outcomes. *Journal of Health Economics, 28*(5), 971–983.

Gundersen, C., & Ziliak, J. P. (2015). Food insecurity and health outcomes. *Health Affairs, 34*(11), 1830–1839.

Johnson, P. A. (2010). Realizing rural community-based tourism development: Prospects for social economy enterprises. *Journal of Rural and Community Development, 5*(1), 150–162.

Jones, A. D., Ngure, F. M., Pelto, G., & Young, S. L. (2013). What are we assessing when we measure food security? A compendium and review of current metrics. *Advances in Nutrition: An International Review Journal, 4*(5), 481–505.

Kim, K., Kim, S., & Park, C.-Y. (2020). *Food Security in Asia and the Pacific amid the COVID-19 Pandemic* Asian Development Bank, ADB Briefs, No. 139, doi.org/10.22617/BRF200176-2

Krystyna, S., & Philippa, R. (2020). Indigenous Peoples' Food Systems Hold the Key to Feeding Humanity. Retrieved from https://www.iied.org/indigenous-peoples-food-systems-hold-key-feeding-humanity#

Kuhnlein, H. V., & Chotiboriboon, S. (2022). Why and how to strengthen Indigenous peoples' food systems with examples from two unique Indigenous communities. *Frontiers in Sustainable Food Systems,* (6), 1–20.

Laborde, D., Martin, W., Swinnen, J., & Vos, R. (2020). COVID-19 risks to global food security. *Science, 369*(6503), 500–502.

Lapeyre, R. (2010). Community-based tourism as a sustainable solution to maximise impacts locally? The Tsiseb Conservancy case, Namibia. *Development Southern Africa, 27*(5), 757–772.

Maxwell, D. (1996). Measuring food insecurity: The frequency and severity of "coping strategies". *Food Policy, 21*(3), 291–303.

Maxwell, S. (1996). Food security: A post-modern perspective. *Food Policy, 21*(2), 155–170.

Maxwell, S., & Slater, R. (2003). Food policy old and new. *Development Policy Review, 21*(5–6), 531–553.

McDonald, B. L. (2010). *Food security.* Cambridge: Polity Press.

Ministry of Health (2019). *Household Food Insecurity among Children: New Zealand Health Survey.* Ministry of Health, Wellington, New Zealand

Mishra, K., & Rampal, J. (2020). The COVID-19 pandemic and food insecurity: A viewpoint on India. *World Development, 135,* 105068.

Misselhorn, A. A. (2005). What drives food insecurity in southern Africa? A meta-analysis of household economy studies. *Global Environmental Change, 15*(1), 33–43.

Myers, S. S., Smith, M. R., Guth, S., Golden, C. D., Vaitla, B., Mueller, N. D., . . . Huybers, P. (2017). Climate change and global food systems: Potential impacts on food security and undernutrition. *Annual Review of Public Health, 38*(1), 259–277.

Osmani, S. R. (1993). *The entitlement approach to famine: An assessment.* Helsinki: UNU World Institute for Development Economics Research.

Parnell, W. (2005). *Food security in New Zealand.* University of Otago

Pearce, D.G. (1992). Alternative tourism: Concepts, classifications, and questions. In V.L. Smith & Eadingtom, W.R. (Eds.), *Tourism Alternatives: Potentials and Problems in the Development of Tourism* (pp. 15–30). Chichester: Wiley.

Pérez-Escamilla, R. (2017). Food security and the 2015–2030 sustainable development goals: From human to planetary health: Perspectives and opinions. *Current Developments in Nutrition, 1*(7), e000513.

Santafe-Troncoso, V., & Loring, P. A. (2021). Indigenous food sovereignty and tourism: The Chakra Route in the Amazon region of Ecuador. *Journal of Sustainable Tourism, 29*(2–3), 392–411.

Scheyvens, R. (2002). *Tourism for development: Empowering communities.* London & New York: Prentice Hall.

Scheyvens, R. (2011). *Tourism and poverty.* New York: Routledge.

Scheyvens, R., Carr, A., Movono, A., Hughes, E., Higgins-Desbiolles, F., & Mika, J. P. (2021). Indigenous tourism and the sustainable development goals. *Annals of Tourism Research, 90,* 103260.

Sen, A. (1981a). Ingredients of famine analysis: Availability and entitlements. *The Quarterly Journal of Economics, 96*(3), 433–464.

Sen, A. (1981b). *Poverty and famines: An essay on entitlement and deprivation.* Oxford: Oxford University Press.

Stamoulis, K. G., & Zezza, A. (2003). *A conceptual framework for national agricultural, rural development, and food security strategies and policies.* Rome: Food and Agriculture Organization. Agricultural and Development Economics Division.

Tapsell, L. C., Probst, Y., Lawrence, M., Friel, S., Flood, V., McMahon, A., & Butler, R. (2011). Food and nutrition security in the Australia-New Zealand region: Impact of climate change. *Healthy Agriculture, Healthy Nutrition, Healthy People, 102,* 192–200.

Teddlie, C., & Tashakkori, A. (2009). *Foundations of mixed methods research: Integrating quantitative and qualitative approaches in the social and behavioral sciences.* Los Angeles: Sage Publications, Inc.

Telfer, D. J. (2015). The evolution of development theory and tourism. In R. Sharpley & D. Telfer (Eds.), *Tourism and development: Concepts and issues* (2nd ed., pp. 31–73). Bristol: Channel View Publications.

Tolkach, D., & King, B. (2015). Strengthening community-based tourism in a new resource-based island nation: Why and how? *Tourism Management, 48*, 386–398.

UN Committee on Economic Social and Cultural Rights (CESCR). (1999). General Comment No. 12: The Right to Adequate Food (Art. 11 of the Covenant). Retrieved from https://www.refworld.org/docid/4538838c11.html

United Nations. (2015). *Transforming our world: The 2030 agenda for sustainable development.* New York: UN. Retrieved from https://sdgs.un.org/2030agenda

Utter, J., Izumi, B. T., Denny, S., Fleming, T., & Clark, T. (2018). Rising food security concerns among New Zealand adolescents and association with health and wellbeing. *Kōtuitui: New Zealand Journal of Social Sciences Online, 13*(1), 29–38.

Zapata, M. J., Hall, C. M., Lindo, P., & Vanderschaeghe, M. (2011). Can community-based tourism contribute to development and poverty alleviation? Lessons from Nicaragua. *Current Issues in Tourism, 14*(8), 725–749.

22

INDIGENOUS KNOWLEDGE AS AN IMPORTANT CONTRIBUTION TO THE SUSTAINABILITY OF GEOTOURISM AND GEOPARKS

Martina Pásková, Taiwo Temitope Lasisi
and Abraham Cáceres Cabana

Introduction

Throughout history, human activities have often disrupted the delicate balance of the nature-earth ecosystem (Diamond, 1994). The consequences of human-induced environmental changes and overexploitation of natural resources have led to the collapse of civilizations. Various human endeavors, including tourism, have been associated with negative impacts on the socio-ecological landscape of host destinations (Alola et al., 2020; Wall, 2020; Pásková et al., 2021; Butler, 2019). To mitigate these negative effects and maximize the benefits of tourism, different forms of sustainable tourism, such as ecotourism, green tourism, and geotourism, have been proposed and explored in both practice and academic research (Weaver, 2006; Farsani, Coelho & Costa, 2012; Dowling, 2013; Dowling & Newsome, 2018; Pásková, 2022). Geotourism refers to... *"a form of natural area tourism that specifically focuses on geology and landscape. It promotes tourism to geosites and the conservation of geodiversity and an understanding of earth sciences through appreciation and learning. This is achieved through independent visits to geological features, use of geo-trails and viewpoints, guided tours, geo-activities, and patronage of geosite visitor centers"* (Newsome & Dowling, 2010: 4) and is increasingly becoming a mainstream research focus among eco-tourism researchers and practitioners alike.

The notion that geotourism combines the merits of geological and geographical perspectives with tourism development makes it appealing by providing opportunities for environmentally educative experiences that will result in sustainable and geologically-based benefits for local indigenous populations (Ólafsdóttir & Tverijonaite, 2018). From a destination management perspective, a geopark is described by McKeever and Zouros (2005: 274) as *"a unified area that advances the protection and use of geological heritage in a sustainable way, and promotes the economic well-being of the people who live there"*. Pásková (2018) argued that the use of Geoparks in tourism destinations offers destinations a channel to enhance the sustainability of the geotourism activities of such destinations.

UNESCO Geoparks like Colca y Volcanes de Andagua Geopark in Peru and Rio Coco Geopark in Nicaragua recognize the importance of Indigenous knowledge, which is passed

291

DOI: 10.4324/9781003230335-26

down through generations orally and shapes the characteristics of these destinations (Sen, 2005). Indigenous knowledge encompasses cultural, spiritual, environmental, social, and belief systems, providing a holistic understanding of local residents' lifestyles (Lauer, 2012). From an ethnoecology perspective, Indigenous knowledge involves the interaction, management, and perception of the environment by indigenous communities (Warburton & Martin, 1999). Scholars highlight the role of Indigenous knowledge in sustainability management, emphasizing its connection to ancestral knowledge for fostering sustainable destination development (Pásková, 2015, 2017).

Indigenous inhabitants and communities have knowledge not only of local flora and fauna as well as the functionality of the local ecosystem but also have skills based on practices and traditions handed down and utilized for generations, thanks to a rhythmic lifestyle mainly determined by natural processes (e.g., Gray, Parellada & Newing, 1998; Beltran & Phillips, 2000; Toledo, 2000; Nakashima & Roué, 2002; Borrini-Feyerabend, Kothari & Oviedo, 2004; Butler & Menzies, 2007; Bates et al., 2009; Gratani et al., 2014). Most of these skills are embedded in the diverse expressions of Indigenous spiritual customs, including superstitions, rituals, taboos, and legends – sacred periods or territories (e.g., Henige, 1999; Klubnikin et al., 2000; Cashman & Cronin, 2008; Németh & Cronin, 2009; Lauer, 2012; Samakov, 2015).

Most previous studies on indigenous knowledge and geotourism have focused on the impact of geotourism on indigenous identity (Pásková & Dowling, 2014). However, this chapter takes a different approach by using qualitative grounded theory to investigate the significance of Indigenous knowledge for geoparks and geotourism. It compares two UNESCO-designated geoparks, Colca y Volcanes de Andagua Geopark in Peru and Rio Coco Geopark in Nicaragua, utilizing various qualitative data collection methods such as interviews, discussions, analysis, and observations.

The Study

Research Design

This investigation is part of a larger project (see Pásková, 2018, 2017, 2015; Pásková & Hradecký, 2014; Pásková & Dowling, 2014) using people-centric, plant-based, post-cultural, holistic, and interdisciplinary approaches (Pásková, 2017).

The primary research objective was to explore the relationship between Indigenous knowledge and the sustainability of land use and natural resource utilization in the Geoparks. This chapter presents the findings of a comparative study aimed at identifying and comparing local Indigenous environmental knowledge and its potential to enhance geopark activities and geotourism sustainability. The study focused on answering the research question: "In what ways do the Andagua district (Peru) and El Apante (Nicaragua) communities differ in their utilization of Indigenous knowledge for sustainable geotourism management?"

Description of Research Localities

Andagua in the Colca and Volcanoes Andagua Geopark (CVAG)

According to oral history and some archaeological studies (Cáceres Cabana, 2019), the citadel of Antaymarca is where the first settlers of Andagua settled.

Since 1532, Peru has undergone economic, social, and political restructuring. These changes had a direct impact on the organization of the population, gradually eliminating some traditions

Figure 22.1 *"Pago a la Tierra"* ritual in the Peruvian Andes.
Photo: Pásková, M.

and customs, and forming a new idiosyncrasy where Western culture prevails. However, at the same time, the Inca culture was able to recover and the population of the district continued to carry out their activities such as *"Pago a la Tierra"* (payment to the Earth, see Figure 22.1), worshiping the sun, and the *"Mallquis"* (mummies), although many times in secret.

According to the last census of 2017, the district of Andagua had a population of 1,168 persons, who speak Quechua and Spanish. They are dedicated to farming, livestock, and a small sector of tourism activity. They cultivate broad beans, corn, and potatoes (*"papa liza"*), among other t bread products; the cultivation of alfalfa was largely replaced for reasons of improving dairy cattle. Andagua, and the annexes of San Isidro de Tauca and Soporo, generate approximately 2,500 litres of milk daily. There are a large number of families who are undertaking tourism with a view to the future since the district has great potential for development. It has a natural and cultural heritage, being part of the Colca and Andagua Volcanoes Geopark.

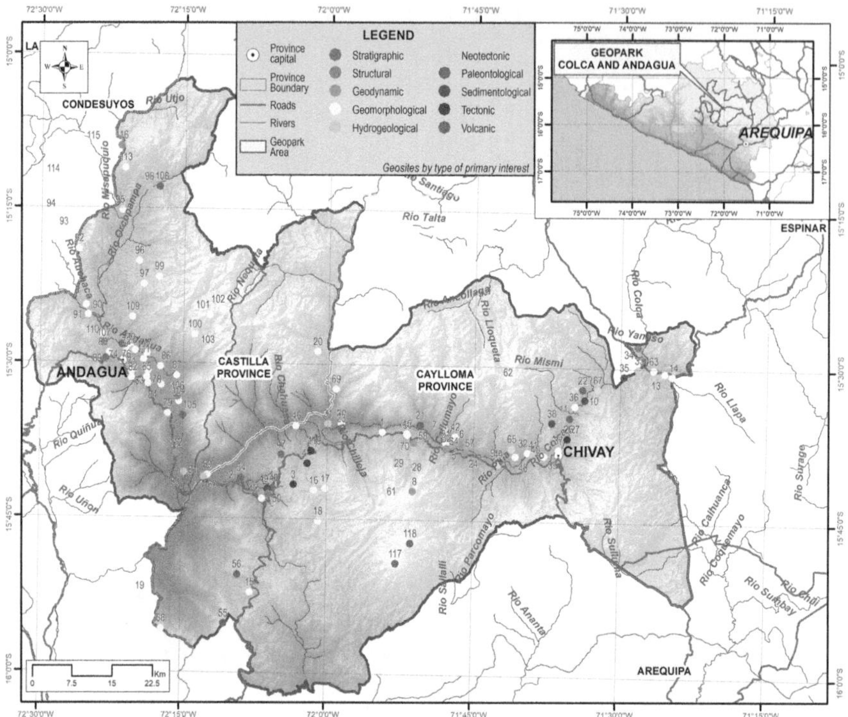

Figure 22.2 Map of the Colca and Volcanoes Andagua UNESCO Global Geopark, Peru.
Source: Pásková et al. (2021).

The district of Andagua represents 1 of the 14 districts of the Province of Castilla (see Figure 22.2) and the research locality of the Andagua community is shown in Figure 22.3. The Andagua district is part of the Valley of the Volcanoes, with a dry tundra climate, at an altitude of 3,000 meters above sea level to 4,500 meters above sea level. It has four annexed rural settlements, San Antonio, San Isidro de Taucca, Soporo, and Ccallhua being part of Alto Castilla, made up of six districts, Orcopampa, Choco, Ayo, Chachas, and Chilcaymarca, which make up the Colca and Andagua Volcanoes Geopark.

The topography of the district of Andagua is very varied and difficult to travel due to the high and very uneven hills, which present a cover of loose and moved blocks of volcanic materials.

The region to the west of the Andagua River is a *"puna"* surface that is over 4,400 meters a.s.l. high on which high peaks and summits stand out, such as the Coropuna of 6,494 metersa.s.l. and other points of more than 5,000 meters a.s.l. The rivers have deep valleys draining mostly to the Colca River; these tributaries have very deep channels with very steep flanks.

The Andagua jurisdiction boasts several extinct volcanoes, including the *"Mellizos"* (twins), Pucamauras, Pucamauras Chico, Canallamauras, Jechapita, Santa Rosa, Chilcayoc Grande, Chilcayoc Medio, Nina Mama, Huanacahuire, Quechapita, and Jenchaña (Pachari, 2020). Currently, the Andagua district relies on livestock, agriculture, and a small tourism sector, offering services such as food, lodging, and guidance. A small portion of the population works as professionals in education, banking, municipality staff, healthcare, and agriculture (Pachari, 2020). The area faces high unemployment rates, particularly among youth who lack training and employment opportunities.

Figure 22.3   Research spots indicated in the relief map of the Andagua locality, Peru.

Source: Google Earth and GPS Mapping (2022a).

The UNESCO designation as a Global Geopark provides a significant opportunity to improve the standard of living and address these challenges. Tourism in the heart of the Valley of the Volcanoes, including the Andagua district, has gained international recognition since 1932 through National Geographic's coverage (Cáceres Cabana, 2019). Although visitor numbers have been limited, adventure tourism has been the prevailing segment. The UNESCO Global Geopark designation has brought attention to the district, as it serves as a crucial passageway to other districts in the Valley of the Volcanoes.In terms of the destination life cycle concept, the period before the COVID-19 pandemic can be classified as the "involvement stage" (Butler, 1980). Residents adapted their family lodgings and restaurants, and some received training. Gradually, visitors started to come to the valley to experience the local customs, traditions, and geology, resulting in host families earning an additional income of 10% according to the local Visitor Registry. However, the Valley of the Volcanoes cannot be compared to the Colca Canyon, which is a more popular tourist destination within the Geopark and has already established itself nationally and internationally. Consequently, the Colca Canyon has received economic support for Geopark development.

Generally, with the economic support of AUTOCOLCA, the Destination Management Organization operating in the geopark area, tourism and related economic developments have been carried out in the geopark. In the post-pandemic economic reactivation, there are currently (2022–2023) very few visitors coming to the Andagua district compared to three years ago, resulting in a loss of income from tourism activities amongst the host families.

El Apante in the Rio Coco Geopark (RCG)

The Rio Coco Geopark is located near the Honduran border in Nicaragua's northwest and is rich in cultural and natural heritage as shown in Figure 22.4) the research locality of the El Apante community can be seen in Figure 22.5). Somoto Municipality, where the geopark seat is located, is the geopark's major focus. The establishment of the geopark spurred expectations and support for regional sustainable development and Indigenous cultural identity in the Rio Coco region, particularly among the Indigenous Chorotega population.

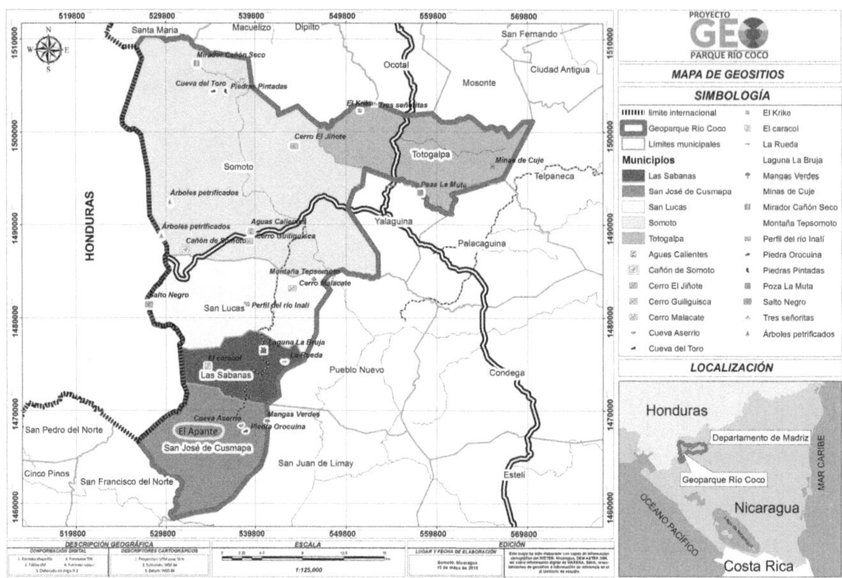

Figure 22.4   Map of the Rio Coco UNESCO Global Geopark, Nicaragua.

Source: Alcaldía de Somoto (2017).

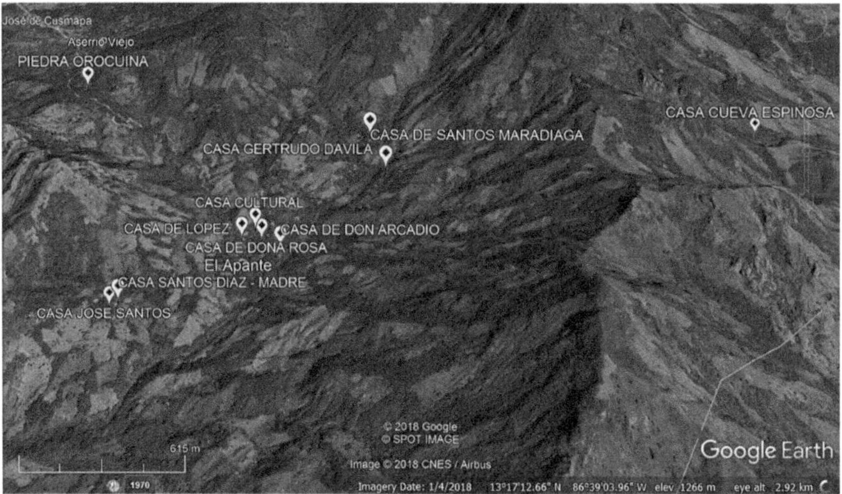

Figure 22.5   Research spots El Apante locality, Nicaragua.

Source: Google Earth and GPS Mapping (2022a).

Rio Coco's economy is primarily based on agriculture and livestock production, focusing on crops such as sorghum, upland rice, maize, beans, tubers like "malanga" and "cassava," as well as fruits and vegetables including onion, potato, sweet pepper, squash, watermelon, and "pipián." Cattle raising is the main livestock activity, followed by pig and poultry farming. To enhance the local economy, various initiatives have been introduced, specifically targeting geotourism.

These efforts have led to improvements in the livelihoods of the locals, the promotion of entrepreneurship, and the implementation of activities to enhance geotourism's economic impact. These activities involve raising awareness about geoparks, organizing workshops to create geopark routes with community members, training local guides, and establishing cooperatives that market their products under the geopark brand, showcasing the contributions of women in producing geopark-specific goods. El Apante community represents a part of the "Pueblo Indígena de San José de Cusmapa—Lugar Carrizal", one of the indigenous municipalities of the geopark. This community is self-organized by an Indigenous directive body called "Junta Indigena" or "Junta Directiva" and consults with the Council of Elders ("Consejo de Ancianos"). The Indigenous mayor of San José de Cusmapa described the organizational structure of the whole municipality as follows:

… our traditional organization that is already as a custom and as a tradition: Junta Directiva made up of seven members: the president, the vice president, the secretary, two members, two prosecutors and there is one of propaganda. But in addition to that, there is Consejo de Ancianos made up of seven members who have been presidents of the indigenous people. This is how we identify ourselves as the Chorotega indigenous people of San José de Cusmapa,….

The El Apante community's territory is divided into 4 sectors, delineated by 37 boundary stones known as "mojones." The Central Sector I comprise key community amenities such as an elementary school, the Indigenous House ("Casa Indigena"), a simple church, and a playground. Surrounding the community center are the other sectors: "Ojo de agua" (Sector II), "Mangas Verdes" (Sector III), and "Marañon" (Sector IV). Despite being only 4 kilometers away from San José de Cusmapa, the mountainous terrain poses challenges for accessing standard services such as healthcare and shopping, which are available in San José de Cusmapa. However, the community has addressed this by employing trained health brigade members ("brigadistas de salud"), local Indigenous healers ("curadores"), and three in-house elementary outlets in El Apante to provide necessary services (San José de Cusmapa Town Hall, 2014).

According to the research informants, El Apante is home to around 500 Indigenous individuals residing in 84 households, with living conditions characterized as basic and a majority falling under the category of extreme poverty (San José de Cusmapa Town Hall 2014). The area's natural environment is both rich and precarious, with volcanic and seismic activity, tropical cyclones, droughts, and landslides posing hazards. These circumstances contribute to the Indigenous peoples' strong connection to and reverence for nature. The focal group discussion revealed that most locals are farmers primarily cultivating "basic grains" (refer to Table 22.1), with approximately 5% engaged in cattle breeding and around 15% involved in coffee cultivation. Some community members sell their produce at the weekly fair in San José de Cusmapa, while others work in forestry, predominantly practicing silvopasture. Due to the challenging terrains at altitudes ranging from 1,000 to 1,600 meters above sea level, cultivation is not easily feasible, and the soils are prone to erosion and landslides.

Method

This research aimed to compare the potential of Indigenous knowledge for geotourism sustainability in two Latin-American UNESCO Global Geoparks. A qualitative research strategy, utilizing a grounded theory approach, was employed in two case studies. One of the authors carried out participant observation using the ethnographic technique whilst sharing living quarters in a home in the Andagua district (CVAG) for one week and the El Apante community's territory (RCG) for two weeks. Participant observation and semi-structured interviews were conducted, following a snowball technique for informant selection. The selection process was guided by the informant's level of Indigenous knowledge and geographic dispersion, with consultations made with the local "Junta Indigena" (see Table 22.1). The research progressed until data saturation was reached, and a field diary and digital audio recorder were used for documentation. The Indigenous house, Casa Indigena, served as a platform for participatory authentication during focus group discussions. The informants chosen were individuals with high levels of local Indigenous knowledge and significant responsibility. This rigorous research methodology ensured comprehensive data collection and authenticity of the findings (Strauss & Corbin, 1998; Gratani et al., 2014; Antonio et al., 2012).

All communication with the Indigenous individuals took place in Spanish, the official language. However, the residents also speak their respective Indigenous languages, resulting in the majority of local flora, fauna, and place names having Nahuatl names (RCG) or Quechua names (CVAG). Due to the authors' familiarity with the areas, there were no significant challenges in communicating with the informants and hosts. In the case of the Andagua district, interviews were conducted by one of the authors, who is a member of the local Indigenous community responsible for geotourism development. Prior permissions were obtained for photo documentation, audio recording, and the publication of respondent discourses. The study of the Colca and Volcanoes Andagua Geopark was based on primary research, while the Rio Coco Geopark study utilized secondary data from previous research conducted by one of the authors (Pásková, 2018). A comparative analysis was conducted based on the findings of the potential of Indigenous knowledge for geotourism sustainability for the two parks using the following criteria: local vegetation/rock/ clay types, living/ environmental conditions, source of knowledge, level of awareness of nature's value, the transmission of knowledge to the young generation, level and forms (approaches) of cognition, motivational, cognitive, and affective emotion, knowledge implementation, and potential of knowledge for geopark and geotourism sustainability. The criteria were prioritized based on knowledge gained from previous research (Pásková et al., 2021; Pásková, 2018, 2017, 2015; Pásková & Hradecký, 2014; Pásková & Dowling, 2014).

Findings

The comparison of the two case studies revealed similarities and differences (Table 22.1), but most notably the body of Indigenous knowledge is gradually diminishing in both territories. An elder male from RCG comments on this process as follows:

> … I think enough has been lost through carelessness. Now at this time, as we know so many things are coming. It is easier to buy in a pharmacy than to go looking for roots. Before, when we were so isolated, we were far from Somoto, we did not have transportation or access to communication routes. So, people knew a lot about medicine, and they kept their resources as reserves to take a remedy. People know little about this right now, and especially about plants ….

The main difference is that in the Rio Coco Geopark, the Indigenous language and typical Indigenous dresses are no longer used by local inhabitants although they publicly practice Indigenous rituals. One of the farmers (male) from RCG mentioned: *"A group from Cusmapa came and they did some rituals on the Orocuina stone, but we who farm here are not used to doing it. Those who did rituals in the past are gone, but currently, it is not practiced."* According to one of the

Table 22.1 Results of the comparative study

Criteria	Colca and Volcanoes Andagua Geopark	Rio Coco Geopark
Living / Environmental conditions	• Mountainous cultural landscape • Rural lifestyle based on traditional agricultural activities • Among the most typical plants belong Shaki Cactus, eucalypt, *yareta* (Azorella Compacta), alder, or *ayrampu* (Tunilla soehrensii) • Volcanic rocks are abundant there and used by local people ('*piedra laja*', *piedra pomez* ',' *piedra rustica*') • Indicated first problems with water contamination • Perceived impacts of climate change • Observed the disappearance of some plant and animal species • Perceived decrease in fertility of the soil (decreasing size of crops) • Agriculture is based mainly on raising life stock • Distance and size of *chacras* (agriculture parcels): 0,5 – 7 km from the house, approximately 10 ha • Cultivated crops: potatoes, corn, *habas* (broad beans), alfalfa, wheat, barley, quinoa • Collection of the *leña* (firewood) in the local forests	• The poor mountainous border region • Rural lifestyle based on traditional agricultural activities • The typical plants include the ceiba tree, *quapinol / algarrobo* (Hymenaea courbaril), or *matapalo* (Strangler fig) • Indicated problems with increasing plastics waste • Perceived impacts of climate change • Four species of pine trees (including endemic ones) • Tertiary volcanic formations (mainly andesite, dacite, or tuff) and sedimentary rocks • Sites on various hills with ancient pieces of ceramics and chiselled lithic material (obsidian, flint, etc.) - Indigenous ancestral settlements with possible ritual practices, these sites are archaeologically underresearched • Observed the disappearance of some plant and animal species • Perceived decrease in fertility of the soil (decreasing size of crops) • Distance and size of *chacras* (agriculture parcels): 0 – 8 km from the house, approximately 15 ha • Agriculture is based mainly on the cultivation of "basic grains": beans, corn, sorghum • Collection of the *leña* (firewood) in the local forests, there is a permanent lack of this resource in the region
Source of knowledge	• Ethnography (Takahashi Martínez 2012), historiography (Cáceres Cabana 2019), and archaeology (Menaker & Falcón Huayta, 2016, 2017) are researched and results disseminated at universities (to a limited extent to Indigenous communities • (Gran)parents and Elders (in decreasing extent), community, school (in a limited way) • Observation of the behavior of plants and animals • Direct observation of the movement of the moon, sun, and stars as well as interpretation of other cosmic phenomena	• Both archaeologic and anthropologic research are at the initial stage and results are not disseminated to Indigenous communities • (Grand)parents, Elders, and community (in decreasing extent), school (in a limited way) • Direct observation of the behaviour of plants and animals • Observation of the movement of the moon, sun, and stars as well as interpretation of other cosmic phenomena

(Continued)

Table 22.1 (Continued)

Criteria	Colca and Volcanoes Andagua Geopark	Rio Coco Geopark
Level of environmental awareness of nature's value	• Appreciation of natural processes and resources (without chemical contamination), a specific kind of social responsibility, and a specific perception of ecosystem services Indigeneity (Indigenous identity) perceived as a way of thinking and living • The key conception of *Pachamama* (Mother Earth) is a holistic perception of the world/cosmos, harmonic relation, and respect for the nature	• Indigeneity (Indigenous identity) perceived as a way of thinking and living • The key conception of *Pachamama* (Mother Earth) is a holistic perception of the world/cosmos, harmonic relation, and respect for the nature
Transmission of knowledge to the young generation	• There are changes in Indigenous knowledge transmission from personal to IT-based communication, both time and space shared across the generations are shrinking • Transmission through local schools is not always perceived as optimal • As a result, young people are not so interested (their interest is decreasing) in traditional knowledge and practices, they are retreating from wearing typical dresses such as *poncho* or *pollera* (a big one-piece skirt), and they are also not so eager to continue to use the Indigenous language (Inca language) *quechua*	• A certain group of young people feels Indigeneity is not appropriate for modern life, however, some youngsters are appreciating the possibility to learn from the Elders about Indigenous practices • The curricula in the local school do not include knowledge building and Indigenous identity • Indigenous knowledge is passed down entirely through the oral method. They do not have a practice of recording or organizing meetings about Indigenous knowledge.
Level and forms of cognition	• Lunar cycles as an indication of the timing of agricultural activities • Awareness of Indigenous origin, ancestors (Inca and Wari cultures), and related historical remains (first settlement Antaymarca citadel constructed by the Andagarunis tribe) • The local Indigenous peoples learn by doing and by observing the members of their family or the whole community, however, this way of learning is declining The new way of learning from the capacitation activities organized by the public bodies of nature protection	• Very limited knowledge of the ancestors and local Indigenous history, pride on the Maya origin, however obscurity on the concrete ethnic group (possibly *Chorotegas* or *Matagalpas*) • Very limited information and application regarding Indigenous knowledge in the local schools • The role of the Elders is still very important for both community and family life: they support the decision-making processes and provide emotional-spiritual guidance
Motivational, cognitive, and affective emotion	• Deeply rooted respect, continuation in rituals such as *Pago a la Tierra* (payment to the Earth), *tinkamientos al Apus* (thanksgiving to spirits of the mountains), *pichuhuira* (offerings to Mother Earth) when seeding • *Pachamama* (goddess of Earth) is perceived by local people as a basis of the existence of life and alimentation, they feel the obligation to care for nature	• A strong connection between the practical and spiritual life, religious aspects are part of daily life and work – knowledge and religion (syncretism) are mutually and narrowly integrated • The rituals connected to Mother Earth, Sun, or Moon are no more publicly practised.

Knowledge implementation	<ul><li>Collective ownership of the land and regarding decision-making, community ownership</li><li>Use of volcanic rocks for the construction of barriers, decorations, construction of ovens, and stoves, also for washing/pealing bodies, for cooking with stones in the earth</li><li>Construction of the house (basement) with stones and mud, the roof of *paja* straw (it takes approximately three months to build it)</li><li>Ceramics are no more produced in the Andagua district, just in two other localities of the geopark</li><li>The local people (mainly women) use to wear typical dresses such as *ponchos* (men) or *polleras* (women)</li><li>Clay is used for the production of *adobes* (bricks made of mud and straw)</li><li>In local agriculture, the rotation system, (potatoes/corn/barley/barley with alfalfa) is used, just organic fertilizers (*guano* or barnyard compost)</li><li>The role of the Elders is still very important for both community and family life: they support the decision-making processes and provide emotional-spiritual guidance</li><li>Use of volcanic rock for pavement of Andagua streets; for the building of *andenes* (terraces) at the slopes to improve conditions for the cultivation of local crops and demarcation of individual parcels</li><li>Exceptional use of volcanic material for creating *keritos* (Inca vases)</li><li>Currently, the tractor is used in farming, and no more *chaquitaclla* (foot plows originated from Inca times)</li><li>Seeds of *rocoto* (chili peppers) are inserted into the soil to eliminate /reduce plague</li><li>Collection and use of medical herbs for cooking, medicine, rituals</li><li>Decreasing the use of clay and ceramics</li><li>Partially, the agroforestry approach is applied</li><li>For the cultivation of local crops, the lunar cycles and astral phenomena are observed</li></ul>	<ul><li>Problems with the practical application of rights on the Indigenous land ownership and other rights recognized for *Pueblos indígenas* (Indigenous Communities).</li><li>Indigenous governing bodies still in use: *Junta directiva* (Board of Directors) as a decision-making body, *Consejo de Ancianos* (Council of Elders) as a supreme authority</li><li>Use of volcanic rocks for the construction of barriers, decorations, construction of ovens, stoves, weighing, grinding, *mojones* (boundary stones), also as washboards (tuff), whetstones or managers, etc.</li><li>Construction of the house with stones (basement) and mud, the roof of *paja* (straw), it takes approximately three months to build it</li><li>Clay is used for the production of bricks: *adobes* (clay and pine needles dried in the sun) and *ladrillos* (clay with cow dung burnt in the oven)</li><li>Local plants are used for the creation of various bowls and vases, baskets, bags, and mats. e.g. *jícaro* (Crescentia cujete), palm (leaves), and *tule* (Schoenoplectus acutus) used for *petate* (mat) production and erosion prevention</li><li>The sap of *quapinol / algarrobo* (Hymenaea courbaril) was and continues to be used for washing as a kind of soap</li><li>In local agriculture, the rotation system is used, preferentially without chemical fertilizers</li><li>The local people do not use to wear any typical dresses and do not speak the Indigenous language (except for their vernacular language when the majority of local plants, animals, and local toponyms have Nahuatl names</li><li>Collection and use of herbs for cooking and medicine</li><li>Intensive production of clay and ceramics, demanded by both local people and visitors</li><li>The agroforestry approach, mainly the silvopastoral practice, is applied</li><li>For the cultivation of local crops, the lunar cycles and astral phenomena are observed</li><li>Asking for permission when cutting the tree</li></ul>

(*Continued*)

Table 22.1 (Continued)

Criteria	Colca and Volcanoes Andagua Geopark	Rio Coco Geopark
Potential of knowledge for geopark and geotourism sustainability	• Continuation of traditional customs, dresses (*polleras, ponchos, sombreros*), cuisine, etc. • In the last years, the Indigenous language *Quecha* usage is supported by public authorities • Demonstration of applying knowledge of lunar cycles ○ Sharing of knowledge on the use of medical herbs: e.g. *muña* (Minthostachys setosa, Minthostachys mollis) to heal digestive problems and respiratory diseases, *marco / altamisa* (Ambrosia peruviana), *chacacoma* (Escallonia resinosa) with anticancer properties, *ajotillo* (Geranium filipes) for disinfection of tooths, ash from *huamantirca* (Aristeguietia discolor) serves for the fight against the plague • Storytelling: sharing of legends on local volcanos and their moon-like landscape, about '*sirenas*' of the Shankilay waterfalls • There is a local association of community tourism with efforts not to lose traditional customs, e.g. the '*Escarbo de acequias*' festivity • The key perceived natural attractions: 84 volcanos and lava phenomena, Shankilay waterfalls, Inca roads, Inca houses, Antaymarca citadel • A holistic approach to the presentation of the Earth's heritage - increased emphasis on linking the natural and spiritual values of the Earth's heritage, increased the authenticity of geotourism through the involvement of local Indigenous peoples • A few persons serve already as a guide for visitors • The typical plates in the typical oven from clay and stones with locally collected firewood (*leña*) • Souvenirs based on the local geology and history as *kerito*s (traditional Inca vases)	• Continuation in Indigenous cuisine, natural medicine, and various traditional activities • Sharing of knowledge on the use of herbs for ○ cuisine, e.g. *malanga* (Xanthosoma atrovirens) or *chayote* (Sechium edule), ○ medicine, e.g. *quebracho* (Aspidosperma) tree bark with antibiotic, antiseptic, anti-fever, and anti-inflammatory effects, the fruits of the *matasano* plant called also *zapote blanco* (Casimiroa edulis) with hypnotic and analgesic effects and *marango* (Moringa oleifera) for regeneration of organism and treating impotence problems ○ washing - as an alternative for soap, e.g. *quapinol* (Hymenaea courbaril) ○ farming - as an alternative for repellent, e.g. *hoja blanca* (Buddleia americana) • Preference for organic, diversified "family farming" over large monocultural farming with chemical treatment • A holistic approach to the protection, presentation, and interpretation of the Earth's heritage - increased emphasis on linking the natural and spiritual values of the Earth's heritage, increased the authenticity of geotourism through the involvement of Indigenous peoples, especially in the position of geoguides with traditional knowledge and skills • Storytelling regarding the key natural sites: sharing of legends on the Piedra Orocuina, e.g., about three small houses (large stones), on settlement of ancient peoples who had migrated from Guatemala and Honduras and spring of all rivers there • An innovation in geotourism through arts, meditative and medical programs based on geobotanical Indigenous knowledge.

Figure 22.6   Indigenous farming using local rock for stone walls in the Andagua community.
Photo: Pásková, M.

informants (female, CVAG), the cultural practices that they have "*are the use of typical clothing, traditional customs, Pago a la Tierra, speaking Quechua, ...*". The other female informant from CVAG added: "We *differ in the costumes that we wear polleras, sombreros more than all the women, and the belief in the stars, nature, the Apus, Coropuna volcano hills ...*". In spite of the fact that El Apante community (RCG) is a part of the Indigenous municipality "Pueblo Indígena de San José de Cusmapa" and it is self-organized by the "Junta Directiva" and "Consejo de Ancianos" (see Table 22.1), the local Indigenous peoples have problems with land ownership rights. The Andagua case (CVAG) shows a well-functioning community-based approach to land ownership used for individual farming (see Figure 22.6). One of the farmers (female) from CVAG described the situation as follows: "*Ownership of pasture land is not possible, but if we each have our land and that of the community where the animals graze, there is no private property. The peasant community is in charge through Junta Directiva if there is a jurisdiction and territory document...*".

In both instances, the Indigenous communities acquired knowledge of natural resource utilization through practical experience and observation within their families and communities. However, this traditional mode of learning is diminishing. As one of the farmers (female) from CVAG comments, the situation in schools does not represent an appropriate option:

> ... For example, I see that sometimes the mayor brings teachers and they are not doing it as they should, and I go and tell them that these things should not change them, I think that there have been changes in our culture, it is because there is no knowledge, they are not guided. They don't guide them, they don't ask, nobody guides young people ....

Currently, the modern ways of IT-based learning organized by public bodies such as local schools, government agencies, or town halls are emerging. However, the approach of local authorities is perceived as frustrating in some aspects. The public bodies for nature protection claim cooperation with the local Indigenous peoples; however, their input is not applied sufficiently and the value of the traditional Indigenous knowledge is underutilized in local opinion.

The results of the research show that the more remote and dispersed local communities are, the more Indigenous knowledge regarding nature and the traditional use of natural resources is preserved. As commented by an elderly female from the more remote part of Andagua district: *"The use of plants will always be and is part of the daily life of families and young people and children know the secrets for what they are used for, it is still transmitted from generation to generation…"*. Another female farmer from CVAG complete this picture: *"We try to transmit everything we have in culture, everything we do here, although most of them go to the city and forget about it."* However, the increasing pace of the modernization process in the core areas negatively influences the intensity and complexity of the intergenerational transfer and use of Indigenous knowledge in the peripheral areas. One of the participants of the focal group discussion (RCG) shared his experience: *"What can be done to recover what has been lost? There are things that we keep, some peasant people. For example, in my house, I have the grinding stone that was used to grind corn, coffee, and gourds even though I don't use them, the pinol, the comal, the clay pot, and in the communities, there are also taquezal houses, the adobe for the houses, the palm to cover. The indigenous house was made by youngsters of cement, it should have been made of palm leaves or like a small ranch, beautiful big, but modernism makes even the very traditions change"* (Figure 22.7). Most of the geopark's Indigenous knowledge of local nature and ways of using local natural resources points to the high relevance of this knowledge for the future sustainability of the geotourism development in the respective geopark territories. One of the farmers from RCG presented his opinion on this issue: *"If a tourist comes, we could show him how we live on the Earth. If a group of tourists comes, they could go to the community and stay in a house. It is possible because in the communities there are different crops and they could try natural things that do not harm the environment, such as natural fresh produce …"*. From this point of view, the crucial principles of the Indigenous knowledge are the use of natural raw materials that do not contain chemical substances which contaminate various components of the environment and the application of procedures which respect the dynamics of natural processes. One of the farmers (female) from CVAG underlined: *"Natural processes are respected, not using fertilizers, not using chemical products, other products that contradict nature, we still maintain as farmers that we are, we still maintain ancient traditions, not as much as in the city, practically in the city the land is made to produce the strength, here is still organic…"*.

Amongst the best preserved traditional environmental knowledge of the Indigenous communities in both geoparks is the use of different types of rocks and clay for various constructions and dyeing decoration and ceramics. An elder male from CVAG shared his living situation:

> I am an artisan and I am inclined to be a tourist guide and I take advantage of the volcanic stones, I am making little volcanoes and a kind of keritos for wine to use it to stain our land before I plant, ….

The core body of Indigenous knowledge is represented by the use of plants for healing and health protection. The elder women share their experiences as follows: *"I still like to use herbs, of course, pharmacy medicine cures instantly, but we still believe in our plants. For example, a holy herb al jatin, al marco, is a holy remedy for colic, shanki is a holy remedy for the liver, for colon cancer…"* (CVAG). *"The quebracho tree that when peeled there is a red shell that medicinal, to wash wounds, also for when the animals have pinta the quebracho is used…"* (RCG). *"There are threes forms of muña, one that has large leaves, another that has small leaves, and others that have smaller leaves. We find them in the fields, in the hills, and also in the highest part of the hill"* (CVAG). Another part of this Indigenous knowledge body relates to the production of wicker

Figure 22.7   Indigenous way of cooking and use of local clay for house and stove construction in the El Apante community.

Photo: Pásková, M.

furniture, cattle feeding, and food preparation (see Figure 22.7). An important role for local Indigenous communities is also the use and application of knowledge of the lunar cycle in agriculture and other rural activities. One of the GRC informants (male) explains:

The phases of the moon, the changes of the moon, were considered. To sow corn, it is not good to sow on a tender moon, because the corn cane remains tender, the cob is fine, but the cane bends. To cut down the trees, the term for the moon is sought, and this is still practiced.

Another GRC informant (male) describes the traditional way of treating parasite problems:

> For the issue of parasites the moon terms were used, my grandmother would tell my mom that boy has parasites, give him the dormilona root, when the moon is going to be tender, three days before the moon makes the change. You give the decoction for nine days. Yes, moon terms were used to deworm.

Specific cosmological knowledge and orientation in time and space were also observed. The central concept of Mother Earth represents the basic value framework of the lives of the original inhabitants, which leads them to perceive themselves as an integral part of nature. They expressed their link to Mother Earth in the following way: *"The Pachamama is the land where we were born, we are growing, we are living on this land, in all the villages of those we live is the Pachamama ..."* (male, CVAG). *"It's like we call it the Pachamama, it's like a god of the Earth, that land feeds us, reproduces food, that's why we respect the land ..."* (female, CVAG). *"I believe that the way of thinking and seeing the world and life of people who live in the countryside are always related to Pachamama nature..."* (male, CVAG). Based on the above, it can be argued that the use of Indigenous knowledge about nature in both geoparks supports the sustainable development of their territories including their geotourism activities. As commented by the local farmer (male, CVAG),

> ... it depends on local authorities, and the inculcation, above all, of local tourism association, that the custom not be lost, because if it were to be lost, tourists would no longer come because they are the ones who appreciate customs as dances and fiestas the most ....

Discussion and Conlusions

The results of this chapter have significant implications for geopark and geotourism stakeholders, as well as academic scholars in the field. By exploring the potential significance of Indigenous knowledge in the sustainable development of geotourism and geoparks, this study provides a comprehensive understanding of how Indigenous knowledge contributes to the sustainability of geotourism. The study's findings reveal that Indigenous knowledge encompasses not only traditional practices related to natural resources but also the spiritual dimension embedded in cosmology, particularly the concept of Mother Earth. This highlights the interconnection between traditional knowledge, spiritual beliefs, and the nature-earth ecosystem within a geopark destination. The study also emphasizes the importance of striking a balance between human needs and the environment, where both Indigenous knowledge systems and scientific knowledge systems play a vital role. Additionally, the comparative approach employed in this study expands the applicability of these findings beyond a specific geopark, providing valuable insights for geopark management worldwide. Overall, this research establishes a roadmap for incorporating Indigenous knowledge into the sustainability framework of geoparks, offering guidance for their future development towards sustainability. The Indigenous knowledge potential identified in both areas represents a great opportunity for both geopark management bodies and their geotourism development in Latin America. The continuation of traditional customs is endangered by the declining interest and motivation of the young Indigenous people, although the studied communities continue in the tradition of the application of natural medicine and enjoying the local Indigenous cuisine. One of the GRC Elders (male) attributes this decline to the general globalization process:

Relating health, there are plants, and each one is applied according to its procedure and need. We have many medicinal herbs, Bermuda grass is used to clean the kidneys, plantain as well, and swine grass to reduce inflammation and to burst boils. That has been the alternative before, but due to globalization it is now easier to go buy acetaminophen than to make tea ….

Documentation of Indigenous knowledge is essential for its preservation, but this can be somewhat contradictory and highly problematic. Indigenous knowledge is passed on from generation to generation exclusively by word of mouth, observation, and imitation (of verbal and nonverbal expression). Many Elders, as knowledge holders, are often illiterate, but generally recognized as experienced and wise: they prefer personal communication and demonstration to written or digital documentation in the intergenerational transfer of knowledge and skills. Ancestral knowledge is therefore not documented systematically by Indigenous peoples, and as in some other parts of the world the aforementioned processes of globalization, neo-colonialism, and related acculturation, (and in extreme cases, ethnocide or even genocide), have meant that such knowledge has been irretrievably lost. One of the CVAG informants (female) noted that her land *"has suffered mistreatment because previously there was gamonalism and the people have been dominated and it was a reason for people to be left behind in the education of everything…"*. Another informant from GRC shares her perspective:

Some people discriminate against us and say that we do not speak the ancient language and that we are not indigenous. It is not my fault that people from other places came and raped our women, and that our children are mestizos, but we recognize that in our blood we are indigenous, my mother and father are indigenous.

Field research on the system of knowledge and lifestyle of Indigenous peoples is necessary to adapt to this situation and support the need for its long-term character, requiring years of communication and consistent sharing of Indigenous life. Despite the fact that the scientific documentation of Indigenous knowledge as well as demonstrating Indigenous skills for research purposes is unnatural for common Indigenous persons, some, especially young people, seem to be interested in it. *"At school, they never told us about the history of our village. Our grandparents told us the story and the Council of Elders who was one of our best teachers, the old are living books because we learn from them. I have liked to read and investigate and I have asked the Elders. I am now the president of the Indigenous Youth Network"*, commented one of the local youths (male GRC).

With exceptions documented above, the young generation from both geoparks, are not very engaged in this transmission of traditional knowledge and skills, as they are building their knowledge and developing their skills increasingly in public schools. Elders from CVAG perceive this process in the following way: *"… it is perceived that young people do not want to learn the same things as us; they do not have the same indigenous identity as before, because they go out to the city and end up forgetting themselves"* (female). *"Relating to the use of herbs, many young people do not even believe what we say, what they believe the most is the pharmacy and health center, that's it…"* (female). *"Grandparents and grandmothers in each family always have an important role with the knowledge they have, these decisions are still respected, although in recent years it has been gradually lost"* (male). *"Knowledge is transmitted at fiestas as Pago a la Tierra, tilling the land is still maintained, only Quechua is being lost because young people speak little of this language ... "* (female). However, these educational institutions do not include material on Indigenous identity and Indigenous knowledge systems in their curricula. Indigenous children and young people living in the periphery of the region mostly can only continue to acquire and develop this knowledge and skills, primarily through observation, listening, and assistance to older people, especially their parents and grandparents. An elder female from RCG confided her feelings:

> First, it is necessary to prepare a group of people and young people. Those of us who still know some medicinal plants, try to make them known to young people because they don't know about them because it hasn't been followed up. I have difficulty and I'm old, I'm sick, but if the preparation is given at home I can participate even at the beginning, to give young people a light. I would like them to integrate and believe in natural medicine ....

However, the use of processes and materials that introduce pollution, along with the lack of waste management, points to the need to combine Indigenous knowledge with modern scientific knowledge in the application of sustainability concepts in geotourism management and the development of both geoparks.

Acknowledgement

We gratefully acknowledge the team of the International Geoscience Programme (IGCP) for their support in the implementation of the IGCP project No. 751 (Four Continents Connected through Playful Geoeducation), which enabled the work behind this chapter.

References

Alcaldía de Somoto (2017), '*Dosier de candidatura del proyecto de Geoparque Río Coco*', Archive of the Somoto Town Hall, Somoto, Nicaragua.

Alola, AA, Eluwole, KK, Alola, UV, Lasisi, TT & Avci, T (2020), 'Environmental quality and energy import dynamics: The tourism perspective of the Coastline Mediterranean countries (CMCs)', *Management of Environmental Quality: An International Journal*, vol. 31, no. 3, pp. 665–682.

Antonio, M, Kauffmann, M, Zamora Darsy, S, Vanegas Morgan, S, Soza Fuente, S, Puerta Chavarria, A, Forbes Brack, D, Watson, M & Calderon, R (2012), '*Manual de gestion cultural comunitaria Costa Caribe de Nicaragua: Herramientas de cultura y desarrollo 3*', https://unesdoc.unesco.org/ark:/48223/pf0000228333

Bates, P, Chiba, M, Kube, S & Nakashima, D (2009), '*Learning and Knowing in Indigenous Societies Today*', https://unesdoc.unesco.org/ark:/48223/pf0000180754

Beltran, J & Phillips, A (2000), '*Indigenous and Traditional Peoples and Protected Areas: Principles, Guidelines and Case Studies*', The World Conservation Union.

Borrini-Feyerabend, G, Kothari, A & Oviedo, G (2004), '*Indigenous and Local Communities and Local Areas: Towards Equity and Enhanced Conservation. Guidance on Policy and Practice for Co-managed Protected Areas and Community Conserved Areas*', International Union for Conservation of Nature, Gland, Switzerland.

Butler, RW (1980), 'The concept of a tourism area cycle of evolution: Implication for management of resources', *The Canadian Geographer / Le Géographe canadien*, vol. 24, no. 1, pp. 5–12, https://doi.org/10.1111/j.1541–0064.1980.tb00970.x

Butler, CF & Menzies, CR (2007), 'Traditional ecological knowledge and indigenous tourism', in *Tourism and Indigenous Peoples: Issues and Implications*; Butler, R., Hinch, T., Eds., Butterworth-Heinemann, Oxford, pp. 15–27.

Butler, RW (2019), 'Tourism carrying capacity research: A perspective article', *Tourism Review*, vol. 75, no. 1, pp. 207–211.

Cáceres Cabana, A (2019), '*Turismo rural Andagua historias de vida*', Asociacion de Turismo Andagua, Arequipa - Perú.

Cashman, KV & Cronin, SJ (2008), 'Welcoming a monster to the world: Myths, oral tradition, and modern societal response to volcanic disasters', *Journal of Volcanology and Geothermal Research*, vol. 176, no. 3, pp. 407–418.

Diamond, JM (1994), 'Ecological collapses of ancient civilizations: The golden age that never was', *Bulletin of the American Academy of Arts and Sciences*, vol. 47, no. 1, pp. 37–59.

Dowling, R & Newsome, D (2018), 'Geotourism: Definition, characteristics and international perspectives', in *Handbook of Geotourism*, Edward Elgar Cheltenham, Cheltenham, pp. 1–22.

Dowling, RK (2013), 'Global geotourism - An emerging form of sustainable tourism', *Czech Journal of Tourism*, vol. 2, no. 2, pp. 59–79.

Farsani, NT, Coelho, C & Costa, C (2012), 'Geotourism and geoparks as gateways to socio-cultural sustainability in Qeshm rural areas, Iran', *Asia Pacific Journal of Tourism Research*, vol. 17, no. 1, pp. 30–48.

Gratani, M, Bohensky, EL, Butler, JRA, Sutton, SG & Foale, S (2014), 'Experts' perspectives on the integration of indigenous knowledge and science in Wet Tropics natural resource management', *Australian Geographer*, vol. 45, no. 2, pp. 167–184.

Gray, A, Parellada, A & Newing, H (1998), '*From Principles to Practice: Indigenous Peoples and Biodiversity Conservation in Latin America: Proceedings of the Pucallpa Conference: Pucallpa, Perú, 17–20 March 1997*', IWGIA.

Henige, D (1999), 'Can a myth be astronomically dated? (Iroquois League, oral tradition, solar eclipse)', *American Indian Culture and Research Journal*, vol. 23, no. 4, pp. 127–157.

Klubnikin, K, Annett, C, Cherkasova, M, Shishin, M & Fotieva, I (2000), 'The sacred and the scientific: Traditional ecological knowledge in Siberian river conservation', *Ecological Applications*, vol. 10, no. 5, pp. 1296–1306.

Lauer, M (2012), 'Oral traditions or situated practices? Understanding how indigenous communities respond to environmental disasters', *Human Organization*, vol. 71, no. 2, pp. 176–187.

McKeever, PJ & Zouros, N (2005), 'Geoparks: Celebrating Earth heritage, sustaining local communities', *Episodes*, no. 28, pp. 274–278.

Nakashima, D & Roué, M (2002), 'Indigenous knowledge, peoples and sustainable practice', *Encyclopedia of Global Environmental Change*, vol. 5, pp. 314–324.

Németh, K & Cronin, SJ (2009), 'Volcanic structures and oral traditions of volcanism of Western Samoa (SW Pacific) and their implications for hazard education', *Journal of Volcanology and Geothermal Research*, vol. 186, no. 3–4, pp. 223–237.

Newsome, D & Dowling, R (2010), 'Setting an agenda for geotourism', *Geotourism: The Tourism of Geology and Landscape*, Goodfellow Publishers, Oxford, pp. 1–12.

Ólafsdóttir, R & Tverijonaite, E (2018), 'Geotourism: A systematic literature review', *Geosciences*, vol. 8, no. 7, pp. 234–250.

Pachari Rosello, JR (2020), '*Geoparque Valle de los Volcanes de Andagua*', Bachelor thesis, UNSA. Arequipa - Perú.

Pásková, M (2015), 'The potential of indigenous knowledge for Rio Coco Geopark Geotourism', *Procedia Earth and Planetary Science*, vol. 15, pp. 886–891.

Pásková, M (2017), 'Local and indigenous knowledge regarding the land use and use of other natural resources in the aspiring Rio Coco geopark', in *IOP Conference Series: Earth and Environmental Science*, p. 52018.

Pásková, M (2018), 'Can indigenous knowledge contribute to the sustainability management of the aspiring Rio Coco Geopark, Nicaragua?', *Geosciences*, vol. 8, no. 8, p. 277.

Pásková, M (2022), 'Geopark certification as an efficient form of sustainable management of a geotourism destination', in *Economics and Management of Geotourism*; Braga V., Duarte A., Marques C.S., Eds., Springer, Cham, pp. 65–85.

Pásková, M & Dowling, RK (2014), 'The usage of local and indigenous knowledge in the management of geotourism destinations', *14th International Multidisciplinary Scientific Geoconference SGEM*, pp. 805–812.

Pásková, M & Hradecký, P (2014), 'Aspiring geopark rio coco (Nicaragua)', in *SGEM2014 Conference Proceedings*, pp. 53–60.

Pásková, M, Wall, G, Zejda, D & Zelenka, J 2021, 'Tourism carrying capacity reconceptualization: Modelling and management of destinations', *Journal of Destination Marketing \& Management*, vol. 21, p. 100638. https://doi.org/10.1016/j.jdmm.2021.10063

Samakov, A (2015), 'Sacred sites: Opportunity for improving biocultural conservation and governance in Ysyk-Köl Biosphere Reserve, Kyrgyz Republic', https://www.communityconservation.net/ysyk-kol-kyrgyzstan/

San José de Cusmapa Town Hall (2014), 'Diagnostico socioeconomico subcuenca Rio Tapacalí', Madriz, Alianza por la Resiliencia: San José de Cusmapa, Nicaragua.

Sen, B (2005), 'Indigenous knowledge for development: Bringing research and practice together', *The International Information & Library Review*, vol. 37, no. 4, pp. 375–382.

Toledo, VM (2000), 'Indigenous peoples and biodiversity', in *Encyclopedia of Biodiversity*; Levin, S.A., Ed, Academic Press, Cambridge, MA.

Wall, G (2020), 'From carrying capacity to overtourism: A perspective article', *Tourism Review*, vol. 75, no. 1, pp. 212–215.

Warburton, H & Martin, A (1999), 'Local people's knowledge in natural resources research', in *Socio~economic Methodologies for Natural Resources Research*. Natural Resources Institute, Chatham, pp. 1–17.

Weaver, D (2006), '*Sustainable Tourism: Theory and Practice*', Elsevier Butterworth-Heinemann, Oxford.

23

REFLEXIVITY ON THE ESTABLISHMENT OF NATIONAL PARKS IN LIGHT OF THE CHAPTER BY MASON ET AL

Richard Butler

In her recent paper on deep reflexivity Crossley (2021) notes that discussing reflexivity in tourism studies is something which has come relatively late to that field of study. In the wider area of Indigenous studies, the term and its application is much better established. Citing Ateljevic et al. (2005) she goes on to state "In its broadest sense, reflexivity can be understood as the practice of reflecting on the conditions that enable and constrain the production of research in order to contextualize and qualify findings" (Crossley 2021, p. 207). In line with the approach to defining the term, this writer is focusing on the "positionality" aspect of writing and researching in the specific context of tourism and Indigenous peoples, and following Feighery (2006, pp. 270–271) who defined reflexivity as "the act of making oneself the object of one's own observation, in an attempt to bring to the fore, the assumptions embedded in one's perspectives and x of the world". In this specific instance, the issue of reflexivity was brought to my attention by the previous chapter by Mason et al in dealing with Indigenous issues stemming from the establishment of Banff National Park.

As a non-Indigenous person, certainly so in the Canadian situation in which I find myself, as a relatively new Canadian citizen (approximately half a century) and before that a citizen of a former colonising power in Canadian and other parts of the world, I have tried to be conscious of this position, particularly when researching and writing about other peoples and lands. In editing the two previous books with this title (Butler and Hinch 1996, 2007) both Hinch and I took particular care to avoid imposing our non-Indigenous backgrounds in the editing of those volumes, or in the chapters we contributed. In comparing those earlier volumes with this current one, however, one major difference emerges, and that is the fact that the Indigenous voice, as expressed by both Indigenous and non-Indigenous authors is much stronger and more unequivocally expressed in this current volume than in the earlier ones. That is, I think, a combination of an Indigenous co-editor who has been able to utilise personal contacts and knowledge to enlist many Indigenous contributors, and also the fact that there are far more Indigenous researchers, authors and entrepreneurs in the tourism field than in earlier years. This reflects the findings of Whitfield and Ruhanen (2016) who documented the rapid growth in academic work in Indigenous tourism, and the desire and ability of Indigenous individuals to make their views known through the academic (and other) literature and wider forms of media.

311

DOI: 10.4324/9781003230335-27

I do not write this piece with any sense of it being a "somewhat shamefaced confession of being white, male and middle class" (Feighery 2006, p. 277). I am not at all ashamed of being white, English, elderly and male, and I write this section to acknowledge that that background may mean that I am unaware of numerous issues and problems at both personal and global levels, in this instance in the context of tourism and Indigenous peoples. I have participated in tourism for eight decades and researched the subject for six decades, and one would hope, therefore, that I know a reasonable amount about tourism, both first-hand through experience as a tourist and second hand through research and study. I acknowledge that I have very limited experience of contact with Indigenous peoples and in turn, of their experiences with tourism. The chapter by Mason et al has revealed to me how little I (and most likely many others) knew about the impacts of the establishment of Banff and other western Canadian national parks on the Indigenous residents and users of those lands before the coming of the railway and through that, the creation of those parks.

Several of the chapters in this volume express strong emotions and viewpoints relating to the relationship between tourism and Indigenous peoples, from that of Higgins-Desbiolles in noting the rejection of the "Welcome to Country" theme, through the anticipated and actual use of Indigenous knowledge in recounting past abuses of Indigenous peoples and their resources to tourists as described by Jennings and Butler and Wilkinson, to the portrayal of unique and powerful Indigenous ways of living and operating by Mika et al, and Graci and Taylor, and the continued failure to understand and communicate appropriately with Indigenous peoples as discussed by Whitney-Gould. Reading these and the other chapters as a non-Indigenous person has enabled this writer to gain a better understanding of the attitudes and behaviours of Indigenous communities when facing the intrusion of tourism into their traditional lands and waters. In most cases, I was mostly unaware of the specific context being discussed and gained a great deal of knowledge. In this one particular case, however, I thought that I was well informed of the situation and context being discussed and it came as a considerable surprise to discover that my preconceptions, based on limited research but fairly extensive reading were far from complete and heavily reflected my past and present status.

The chapter in question as noted above, is the one by Mason et al, dealing with the establishment of Banff National Park (and other parks in western Canada) and the disregard of Indigenous rights and claims over access and activities when these parks were established. I found this chapter of particular personal interest as I had been involved in research and writing on Canadian and other national parks for several years and found the chapter somewhat puzzling and at odds with my recollections. In particular, I found it challenging that I had not encountered any previous discussion of the issues raised in that chapter and also thought it strange that publications on Banff and other western Canadian national parks with which I had been peripherally involved were not cited in the chapter. It seemed strange that neither set of writings cross-referenced each other, given the overlap between the subjects involved and thus, I decided to explore in more detail than might otherwise have been the case.

By way of context, I should note that I participated in two major national conferences on national parks (*Canadian National Parks Today and Tomorrow I and II*) held in Banff in 1968 and 1978 respectively. These were large conferences involving 94 presentations being made in total, with significant international participation, and I could not recall any mention of Indigenous peoples beyond one or two passing references to 'natives' and 'Indians', and certainly nothing I could recall about the issues raised by Mason et al. I went back to the volumes of proceedings (Nelson et al. 1968, 1979) to see if my memory was accurate or not and found that it had been mostly correct. Of the 43 papers in the first conference and 51 in the second, only 4 dealt primarily or even partly with Indigenous peoples and their relationships with national parks, past and what was then (1968 and 1978 respectively) present. There was one session on Indigenous use of the parks in the second proceedings and nothing beyond passing references in the first volumes. Even a paper titled "Man and His Environment the past 10,000 years An Approach to Park Interpretation" by Reeves (1968) only briefly mentioned the patterns of traditional Indigenous use, and then from a purely archaeological viewpoint, with no discussion of impacts from park establishment and management on the patterns and activities of Indigenous peoples.

Mair (1979), in his short Introduction to the session *Indigenous Use of national Parks and Related Reserves* commented on problems being encountered by the movement of people out of newly (1960s and 1970s) established national parks and stated that what he saw as the most significant change in views on this issue over the past ten years was "with the Indian and Inuit people.......We must accept and come to grips with this **newly** (emphasis added) understood reality or dimension" (pp. 343–344). The next paper by Freeman (1979), however, focused almost entirely on the value of Indigenous knowledge but in a purely ecological context for park management and did not deal with any negative social or cultural impacts on Indigenous populations through the establishment or management of either the old or the new national parks.

The only real discussion of Indigenous peoples and the effects of the creation of national parks on those peoples came in the paper by Hunt on legal aspects of the rights of Indigenous peoples in national and other parks in Canada. Hunt (1979, p. 608) noted

> It can be anticipated that these developments will give rise to new problems and issues. For one thing, the preservation of native rights in new parks may lead to pressures for the policy to be extended to older parks. If natives are to be involved in management decisions in a meaningful way, new institutions may be required.

In that statement, he correctly anticipated some of the innovations in park creation and focus discussed by Mason et al. but there was no ensuing discussion of these potential issues and problems, either in the paper or in subsequent questions and discussion from the floor of the meeting.

One should not be too critical of the organisers of these conferences for not paying more attention to Indigenous rights and the effects of park establishment on traditional activities and heritage. The primary arguments relating to National Parks in the academic and political contexts in the late 1960s and 1970s were focused on new park establishment (ten new parks had been established in Canada in the 1970s), and on the planning and management of parks (new and old), including in particular, the age-old argument between preservation and use of the parks (Nelson and Butler 1974). Local (non-Indigenous) residents of several of the new parks had been removed before or at the time of establishment, forcibly in at least one case (Kouchibuguac). The emphasis of many of the presenters at the meetings was strongly protectionist, e.g. "to require everything possible be done to safeguard the continued existence of each and every native species found in a park" (Beatty 1979, p. 628), except, presumably, the human species that is (my comment). In an open session, a comment from the floor noted that "It would be worthwhile to look in more depth at the human history elements of all of the parks" (Anon 1979, p. 781) but all the examples listed related to 19th century Caucasian artefacts and personnel.

Nowhere in the over 1,800 pages of the two proceedings is there any mention of Indigenous activities (other than hunting (trapping) and fishing) in the National Parks, such as gatherings, collecting, or other cultural activities, let alone occupying and living in the parks. This is bitterly ironic considering that the original establishment of national parks in the United States, where the concept was first put into practice, was heavily influenced by Romantic Age images of Indigenous peoples in those locations destined to become parks. Paintings, by artists such as Catlin, of Indigenous residents of the west reinforced that artist's views on establishing such parks:

> in his classic attire, galloping his wild horse…amid fleeting herds of elk and buffaloes. What a beautiful and thrilling specimen for America to preserve and hold up to her refined citizens and the world, in future ages. A nation's Park containing **man** (emphasis added) and beast, in all the wild freshness.
>
> (Catlin 1841, pp. 261–262)

As Nelson and Butler noted (1974, pp. 293–234)

> men such as Catlin definitely perceived the native people as part of the wilderness landscape, an idea that failed to take hold as the concept of wilderness evolved to the point where today it often seems devoid of any sign of man.

While such a rationale as that of Catlin and others owes much to the Romantic period of thought in the 19th century as Hall (2000, pp. 58–59) notes,

> It is significant that Catlin perceived that such parks should contain both nature and the North American Indian. The European advance across North America meant that either through disease, war or treaty, native peoples were pushed to the periphery of land which was regarded as suitable for development. Similarly, the first national parks were established in areas which were otherwise regarded as 'worthless' for economic development……The creation of national parks because they were, in one sense worthless, is a critical point to appreciate because in exactly the same way, native peoples were also expelled or forced to live on land that was otherwise worthless.

Similar arguments were made in Australia, New Zealand and other countries, where traditional occupiers (normally, but not exclusively, Indigenous) of land destined to be parks were removed and banned from the parks before or soon after they were established. In the Canadian context, the establishment of parks was a little different, although no more considerate of Indigenous rights, as the Canadian Pacific Railroad had shown how the "worthless" land in fact had real economic value potential in terms of tourism (Hart 1973). Such tourism was of a form in which Indigenous peoples were not anticipated or allowed to participate on their own terms and were at most, features of the parks to be looked at.

As other contributors to this volume have pointed out, the establishment of parks, reserves and other areas protected for primarily environmental preservation reasons have often created problems for Indigenous occupants of such areas by imposing restrictions on or prohibiting specific traditional activities such as hunting and gathering by what were the original users and residents of these lands and waters. This problem was not addressed very much at all in the two earlier books I have co-edited (Butler and Hinch 1996, 2007), although recognition of these problems did feature in a number of chapters in Butler and Boyd (1998), in particular those of Hall and Nepal. It is perhaps, a reflection of the changes which have occurred in the past quarter century whereby broader knowledge has broken the previous silence about the loss of rights, including those of access and resource extraction by Indigenous peoples resulting from a priority allocated by external agencies to environmental protection (and sometimes tourism) over the wishes and needs of the Indigenous peoples and others living in those areas.

Although the establishment of national and other types of parks is still occurring in many parts of the world, much more attention is being paid to residents of such proposed areas. Part of this stems from the more forcibly expressed views of Indigenous peoples and their supporters for their rights to take priority in terms of not just the establishment of such parks and reserves, but also on how these properties should be managed and how, if at all, tourism should be allowed to take place and to be developed. The lack of discussion of Indigenous peoples, their activities and their rights during the establishment of the national parks in western Canada is a reflection of the colonial attitudes of the times, whereby the preferences and in some cases, perceived needs of non-Indigenous populations took priority over the actual wishes and needs of Indigenous peoples, if these latter were even considered at all. That attitude has changed significantly for the better, although problems still exist, not least in the ways in which the establishment and development of tourism related activities and services are discussed, decided and implemented as Whitney-Gould notes so clearly.

Such developments do not explain the absence of discussion of Indigenous peoples in the two conferences discussed above. The reasons for holding the conferences were primarily to discuss environmental protection, new park establishment, and control over economic developments within existing parks, topics about which the interests of most of the academics and other speakers at the conferences were in agreement and that fact, coupled with the absence of Indigenous peoples in the organisation and participation in the conferences goes far to explain the omission of Indigenous issues. Knowing the individuals involved in organising the conferences I am confident that the absence of the issues raised by Mason et al. was based more on ignorance about the impacts of the establishment of the parks on the Indigenous residents and perhaps also a feeling that things had changed for the better over the intervening three-quarters of a century such that their relevance was less than the listed current problems being faced, rather than any deliberate attempt to silence or hide the issues noted in the chapter. My attempted reflexivity, however, having been prompted by the chapter by Mason et al., serves to remind me that there must be many other similar untold examples which remain unknown to many non-Indigenous people, even those like myself who had thought they had a good understanding of the situation.

References

Anon (1979) Comment from audience" in J.G. Nelson, R.D. Needham, S.H. Nelson, & R.C. Scace (eds.) *The Canadian National Parks today and tomorrow Conference II: Ten Years Later*, p. 781. Waterloo: University of Waterloo.

Ateljevic, I., Harris, C., Wilson, E., & Collins, F.L. (2005) Getting 'Entangled': Reflexivity and the 'Critical Turn' in Tourism Studies. *Tourism Recreation Research*, 30, 9–21.

Beatty, R.A. (1979) A Critique. In J.G.Nelson, R.D. Needham, S.H. Nelson & R.C. Scace (eds.) *The Canadian National Parks Today and Tomorrow Conference II: Ten Years Later*, pp. 627–630. Waterloo: University of Waterloo.

Butler, R.W., & Boyd, S. W. (1998) *Tourism and Mational Parks Issue and Implications.* Chichester: Wiley and Sons.

Butler, R., & Hinch, T. (Eds) (1996) *Tourism and Indigenous Peoples.* London: International Thompson Business Press.

Butler, R., & Hinch, T. (Eds) (2007) *Tourism and Indigenous Peoples Issues and Implications.* Amsterdam: Elsevier

Catlin, G. (1841) *Letters and Notes on the Manners, Customs and Conditions of the North American Indians.* New York: Wiley and Putnam.

Crossley, E. (2021) Deep Reflexivity in Tourism Research. *Tourism Geographies*, 23(1–2), 206–227.

Feighery, W. (2006) Reflexivity and Tourism Research: Telling An(other) Story. *Current Issues inTourism*, 9, 269–282.

Freeman, M.M.R. (1979) Traditional Land User as a Legitimate Source of Environmental Expertise. In J.G.Nelson, R.D. Needham, S.H. Nelson and R.C. Scace (eds.) *The Canadian National Parks Today and Tomorrow Conference II: Ten Years Later*, pp. 345–361. Waterloo: University of Waterloo

Hall, M.C. (2000) Tourism, National Parks and Aboriginal Peoples. In R.W. Butler and S.W. Boyd (eds.) *Tourism and National Parks Issues and Implications*, pp. 57–71. Chichester: John Wiley and Sons.

Hart, E.J. (1983) *The Selling of Canada: The C.P.R. and the Beginnings of Canadian Tourism.* Banff: Altitude Press.

Hunt, C.D. (1979) People and Parks Selected Legal Issues in Canada. In J.G.Nelson, R.D. Needham, S.H. Nelson & R.C. Scace (eds.) *The Canadian National Parks Today and Tomorrow Conference II: Ten Years Later*, pp. 605–626. Waterloo: University of Waterloo.

Mair, W.W. (1979) Introduction. In J.G.Nelson, R.D. Needham, S.H. Nelson and R.C. Scace (eds.) *The Canadian National Parks Today and Tomorrow Conference II: Ten Years Later*, pp. 343–344. Waterloo: University of Waterloo.

Nelson, J.G. & Butler, R.W. (1974). Recreation and the Environment. In I. Manners & M. Mikesell (eds.) *Perspectives on Environment*. Washington, DC: Association of American Geographers.

Nelson, J.G., & Scace, R. C. (1968) *The Canadian National Parks Today and Tomorrow*. Calgary: University of Calgary.

Nelson, J.G., Needham, R.D., Nelson, S.H., & Scace, R.C. (1979) *The Canadian National Parks Today and Tomorrow Conference II: Ten Years Later*. Waterloo: University of Waterloo.

Nepal, S. (2000) Tourism, National Parks and Local Communities. In R.W. Butler and S.W. Boyd (eds.) *Tourism and National Parks Issues and Implications*, pp. 73–94. Chichester: John Wiley and Sons.

Reeves, J. (1968) Man and his environment the last 10,000 years An approach to park interpretation. In J.G. Nelson, & R. C. Scace (eds.) *The Canadian National parks Today and Tomorrow*, pp. 121–126. Calgary: University of Calgary.

Whitford, M., & Ruhanen, L. (2016) Indigenous Tourism Research, Past Ghery 2005 should be Dated 2006 and Present: Where to from Here? *Journal of Sustainable Tourism* 24(8–9), 1080–1099.

SECTION 5

Indigenous Tourism Innovations and Developments

Increasingly Indigenous communities adopting innovative and distinctive ways to establish their offerings and to promote not only their existence, but the histories of their cultures. Different approaches to working with Indigenous communities are needed to ensure that external concepts and methods do not impinge or negatively affect Indigenous messages and controls on such developments. In some cases, it may take a considerable time to establish the confidence of local communities and to gain their support for development, and Bricker and Taukeinijkoro illustrate the efforts that can be required to gain local support for a shared development. In contrast, others have been undertaken independently by Indigenous entrepreneurs with a wider goal than simply developing tourism as Graci and Tayor reveal. In Northern Canada, the Indigenous population faces additional problems to those common to all such communities, as their geographical situation poses unique problems for food production, as well as for employment possibilities, and the establishment of Okpik Community Village represents a unique sustainable approach to development, with far wider benefits than just tourism sought for the local community. It represents a good example of an innovation combining tourism and other forms of development, and similar innovative examples are provided by Ara and Hoque with respect to the ways in which Indigenous handicrafts can assist local economies and the empowerment of Indigenous women in particular.

The incorporation of traditional Indigenous and ethnic arts and cultural features into tourism is a well-established element in Indigenous tourism but it can vary considerably in the ways in which that combination is developed. Ara and Hoque demonstrate such differences in marketing and promotion of the sale of handicrafts from local communities with varying effects upon Indigenous families. Similarly in Iran, where Indigenous peoples are regarded as ethnic groups, the artefacts that characterise the different peoples of that region reflect very different skills and approaches that each needs careful organisation and development to be able to maintain the necessary authenticity of the product in terms of its design and production. The step to using such handicrafts to develop political as well as economic strength is illustrated by Allanso and Novelli through their work with African artists in a variety of media, all working to utilise local materials and talent. It is clear that the education element that can be derived from the work of such artists is of tremendous

317

DOI: 10.4324/9781003230335-28

importance in stimulating and driving further development of Indigenous art and the successful introduction of such art to wider audiences, of which tourists are only one part.

In all of the above cases, one theme runs throughout, as embodied in the discussion and examples shown by Luthje et al., and that is the critical importance of deriving and ensuring the continuance of culturally sensitive policies and practices in tourism development. Without such principles, much of the innovation displayed in Indigenous tourism development can easily become distorted and divorced from the original cultural traditions, both physically and spiritually.

24

SOCIAL INNOVATION IN AN INDIGENOUS TOURISM DEVELOPMENT

Sonya Graci and Kylik Kisoun Taylor

Introduction

This chapter explores social innovation in regard to Indigenous tourism development in the Northwest Territories, (NWT) Canada. Tundra North Tours (TNT) in the Arctic has displayed resilience post- COVID-19 through the development of the Okpik Arctic Village (OAV) situated 16 kilometers North of Inuvik (NWT). Prior to COVID-19, this Indigenous-owned and operated tourism business was an exemplary example of sustainability and social innovation embracing the principles of Indigenous ways of living. TNT is examined to explore a best practice that is focused on authenticity, cultural preservation, sustainability, community engagement and empowerment, and social innovation. As an Indigenous entrepreneur, TNT has strived to find ways to ensure that not only is the business profitable, but that it contributes to the local community and to the empowerment of the Inuvialuit Peoples. The lessons learned from this case study can be used to support Indigenous tourism development that focuses on a Sustainable Livelihoods Approack SLA (Chaskin et al., 2001) within Canada and internationally.

Indigenous tourism in Canada is defined as tourism businesses "majority owned, operated and/or controlled by First Nations, Metis or Inuit peoples that can demonstrate a connection and responsibility to the local Indigenous community and traditional territory where the operation resides" (ITAC, 2017, p. 4). Butler and Hinch (2007) characterized the 'Indigenous tourism product' by examining the role of Indigenous societies in tourism and how they interact within the tourism framework. This 'Indigenous tourism product' includes activities in which "Indigenous peoples directly own or operate the tourism business or are indirectly involved by having their culture serve as the essence of the tourist attraction" (Butler & Hinch, 2007, p. 5). The Indigenous tourism product encapsulates a wide range of

> special events (dances, festivals, powwows), experiential tourism (guided hikes, cultural-interpretation programs, wildlife tourism, applied activities), arts and crafts, museums, historical recreations, restaurants, and accommodations, lodges, and resorts that celebrate Indigenous culture and are offered by, or located in, indigenous communities.
>
> (Gets & Jamieson, 1997 as cited in Lemelin et al., 2015, p. 318)

DOI: 10.4324/9781003230335-29

Tourism has been used as a form of participatory, community-based (or driven) development in Indigenous communities as it has the potential to address many of the economic, social, cultural and environmental challenges that these communities face (Graci, 2021; Colton &Whitney-Squire, 2010; Graci, 2010; Butler & Hinch, 2007; Zeppel, 2006; Colton, 2005; Zeppel, 2003; McGinley, 2003; Altman & Finlayson, 1993; Ingram, 1990). Benefits result from improved local economic wealth and increased community capacity, enabling community development and empowerment as well as an improved sense of responsibility as government reliance is reduced. Specific benefits include the preservation of natural and cultural heritage, increased education, training, and capabilities in business development and tourism, increased employment, economic diversification, improved infrastructure, enhanced environmental integrity, the sharing of Indigenous culture, the diminishment of existing social problems, and the allowance of traditional ways of living off of the land in a sustainable way.

Prior to the COVID-19 pandemic, the Indigenous tourism industry in Canada consisted of 1,900 Indigenous tourism businesses, employed over 40,000 workers, and contributed $1.9 billion of direct GDP to the Canadian economy (Conference Board of Canada & ITAC, 2021). Indigenous tourism was changing the economic landscape of Canada as the industry was substantially outpacing Canada's overall tourism activity in terms of both employment and GDP (ITAC, 2019). However, because of the restrictions on tourism and the resulting loss of business due to COVID-19, the Indigenous tourism sector in Canada has suffered immensely. The closing of the Canadian border to international travelers and strict limitations on domestic travel resulted in the temporary closure of many Indigenous tourism businesses, and it is estimated that around 800 of these businesses may never reopen (ITAC, 2022). Furthermore, Indigenous tourism lost $1.4 billion in direct GDP in 2020, resulting in the layoff of approximately 32,000 Indigenous tourism workers (ITAC, 2022). The Indigenous Tourism Association of Canada (ITAC) has provided many levels of support over the course of the pandemic to ensure that Indigenous tourism businesses would be able to be revived once tourism resumed. It is expected however that it will take a number of years to return to the successful pre-pandemic tourism levels (ITAC, 2022). Therefore, it is necessary for Indigenous tourism entrepreneurs to focus on developing businesses that are not only tourism focused but also incorporate a SLA that is both community focused and uses social innovation. The COVID-19 pandemic illustrated the importance of not only diversifying the tourism product but the need to incorporate social elements as well into tourism businesses. There is a greater focus on the need to attract and retain employees as well as giving back to the host communities. The principles of sustainability have always been in line with Indigenous values, however the pandemic has heightened this need amongst consumers and the search for authentic experiences has become more prominent.

Sustainable Livelihoods Approach and Social Innovation

At its core, the Sustainable Livelihoods Approach places at the forefront the needs of the community, participatory approaches to decision-making, access to resources, equitable distribution of economic benefits, local entrepreneurship, and economic diversity. The SLA is key to increasing community capacity as it not only demonstrates the value of community resources (i.e., cultural, social, natural) but also enables community members to develop the skills and competencies needed to leverage these resources for the development of their communities (Chaskin et al., 2001). A SLA enables community control through ownership of, and active participation in, the decision-making process, thereby improving the well-being of the community. This approach also works to significantly minimize economic leakage, allowing the earnings of businesses to remain primarily within the local economy (Al-Oun & Al-Homoud, 2008).

Social innovation involves the use of novel ideas, processes, strategies, procedures, methodologies, or instruments to better fulfill contemporary societal demands or to tackle social issues (Altinay et al., 2016). Individuals from underserved communities (i.e., those who are excluded by the wider society) are increasingly engaged in social innovation to fill the gaps left by ineffective governments, therefore inventing innovative strategies to relieve social issues and accelerate social change within their communities (Marr & Creelman, 2014). These individuals participate in a type of business known as "social entrepreneurship", which entails charging a market-changing innovation with a social purpose (Luke & Chu, 2013). As an underserved community, Canada's Indigenous population continues to be marginalized by mainstream Canadian society and is increasingly turning to social entrepreneurship for community development and self-determination (Pergelova et al., 2021). According to the Canadian Council for Aboriginal Business (CCAB, 2019), Indigenous entrepreneurs "alter traditional patterns of behavior by utilizing their resources in the pursuit of self-determination and economic sustainability via their entry into self-employment, forcing social change in the pursuit of opportunity beyond the cultural norms of their initial economic resources" (p. 11). Indigenous entrepreneurship is not just concerned with personal economic endeavors, rather Indigenous entrepreneurship and the associated enterprises are inclusive of economic, environmental, social, and cultural goals as integrated dimensions, thereby fulfilling a number of community-driven goals, such as protecting Indigenous land ownership and use, strengthening socio-economic conditions, and revitalizing tradition and culture (Maritz & Foley, 2018; O'Neill et al., 2015; Sengupta et al., 2015).

It is widely recognized that entrepreneurship and social innovation have great potential to lead social change and development, which is especially critical to Indigenous communities (Kline et al., 2014). Indigenous social entrepreneurs often strive to advance local development by establishing companies that prioritize community well-being, and research indicates that they are more likely to contribute to community causes and provide opportunities and support for disadvantaged people in their communities (Ngoasong & Kimbu, 2016; Richard, 2021). Empowering and supporting Indigenous peoples in entrepreneurship leads to more education and training opportunities for them, as well as for the wider community (Richard, 2021).

Entrepreneurship is often at the root of much societal growth and transformation, particularly within developing countries where institutions may be weak (Altinay et al., 2016; Ateljevic, 2009; Ngoasong & Kimbu, 2016), as well as in developed countries where government actions and inactions may lead to gaps in social welfare (Marr & Creelman, 2014). Traditionally, the concepts of social innovation and business (entrepreneurship) have been evaluated separately, but research posits that the two concepts can, and often do coexist (Ateljevic, 2009). This has led to the development of the concept of social entrepreneurship, defined as business ventures that are created and driven by the simultaneous pursuit of social and economic goals (Ateljevic, 2009; Ngoasong & Kimbu, 2016). Social entrepreneurship has great potential to lead social change and development (Kline et al., 2014; Ateljevic & Doorne, 2003). The concepts of the SLA and social entrepreneurship are exemplified in the case of TNT and the Okpik Arctic village.

Tundra North Tours

TNT is an Indigenous owned and operated tour operator that provides authentic Indigenous tourism experiences in Inuvik, Northwest Territories. TNT was founded by Kylik Kisoun Taylor, who is of Inuvialuit, Gwich'in, and Scandinavian descent. He grew up in Ontario, hunting, trapping and mushing dogs with his father. In 2003, Taylor made a permanent move to Inuvik, Northwest Territories, where he felt a deep connection to his Inuvialuit and Gwich'in roots, his family, the land

and his culture. His uncles and grandfather constantly took Kylik out in the bush where he learned the way of the land. They taught him the stories of his people and important cultural traditions and knowledge. In 2006, Kylik Kasoun Taylor started TNT with the dream of being able to share his love of the Arctic and all it has to offer with visitors from around the world (CCAB, 2022). He had a vision of preserving Indigenous culture, providing work opportunities for Indigenous peoples and connecting resources to Indigenous lands. Kylik Kisoun Taylor believes that tourism is a gateway for non-indigenous people to explore Indigenous culture (Taylor, 2018) and conducts adventure tours with an educational focus. TNT tourism has a low impact and consists of immersing tourists in the Arctic way of life. Tourists learn to build igloos, participate in traditional forms of food preparation, learn about Indigenous culture, and live on the land. TNT is focused on community integration; authenticity; building pride and cultural heritage, and sustainability, and employs 8 full time staff and approximately 50 temporary staff per year. The number of visitors TNT receives per year is approximately 300, and they mainly travel from British Columbia, Canada. The research that this chapter is based on has been developed through a number of conversations and a field visit to OAV. The authors have worked together on a number of projects over the years.

Community Integration

TNT has created an opportunity for Inuvialuit people to share their traditions through tourism whilst creating itineraries that maintain the safety and cultural integrity of the Indigenous community and provide a source of income for local residents seeking employment as a tour guide. Kylik Kisoun Taylor also believes that

> It is always easy to find a solution when it is an Indigenous owned company, because we are more community based and that helps us more in a financial way. Sharing culture and promoting culture is the biggest benefit for Tundra North Tours.
>
> (Taylor, 2018)

TNT inspires Inuvialuit youth and Indigenous youth across Canada to embrace their culture by showing that it is possible to be successful and make a living through sharing the Indigenous way of life. The support of Indigenous tourism is directly used to empower more community members.

Authenticity

Visitors experience authentic Indigenous cuisine, culture, and traditions as a result of the close-knit communities of the North. TNT incorporates cultural experiences that are based on the way of the land and the people residing on it. Tourists are able to discover the history of the burial site of Albert Johnson, known as the Mad Trapper of Rat River, and taste different Indigenous foods such as Muktuk (often made from the skin and blubber of the beluga whale) —a staple food of the Inuit diet as it contains healthy concentrations of vitamin C and D. Tourists on TNT trips can learn about the residential school system, hear stories from resident elders, and build an igloo in the winter (Figure 24.1).

Sustainability

TNT focuses on sustainability in their operations through low- impact tourism such as winter camping, composting toilets, harvesting and foraging, local manufacturing, and living off the grid.

Figure 24.1   Building an igloo in winter.

The tours include activities such as reindeer herding, ice fishing, snowshoeing, and other low impact activities. There is a focus on having guests immerse themselves in the local culture and participating in traditional activities such as igloo building and traditional food preparation. TNT is incorporating food sovereignty into their current business plan with the development of greenhouses, enhancing the local environment through irrigation and the addition of nutrients to the soil and farming. The purpose of their farm is to grow fresh produce and herbs that enhance the local culinary diversity of what is plentifully available, such as fish and game, for the local community and to assist in the creation of a sustainable food system in the Arctic region. Tourists can live on the land at the OAV, experience and immerse themselves in Canada's Arctic lifestyle.

Okpik Arctic Village

When COVID-19 hit in March 2020, TNT had been booked for the entire season. There was excitement that, after many years of hard work in the tourism industry, TNT was able to employ people at a wage rate that was above the normal standard of living, was able to provide tourists with an experience that was focused on cultural exchange and immersing themselves in the Inuvialiut way of living and ensure that the community benefited directly from tourism. COVID-19 however, stopped the tourism industry and TNT in its tracks, and tourism effectively ceased to exist in the Northwest Territories once travel restrictions came into force. It was at this moment that Kylik Kisoun Taylor decided to pursue an endeavor that was focused on social innovation as a means to not only stay afloat during COVID-19, but also to pivot his business to ensure future benefits for not only himself but also for his community. TNT has the rights to a large area of land and it was at this time that Kylik Kisoun Taylor decided to establish in that area an Arctic eco-village. The concept of an eco-village is not new and there are many examples worldwide of a community that is created based on shared work and values. The difference in the OAV is that it is Indigenously culture-based and located in the Arctic. Most Indigenous peoples of Canada have land and harvesting rights that allow them to access affordable land and to harvest its natural bounty. The need to grow produce and raise livestock is therefore not as important, cost- efficient, and culturally relevant to a sustainable food system. The OAV allowed Taylor to live off the land through creating a village that was self -sustaining. This would provide food sources, homes and employment to the local community. Due to COVID-19, TNT and Kylik Kisoun Taylor had to re-imagine tourism in the North. Since there was the potential to take a pause in TNT's tourism endeavors, it was an opportunity to identify current barriers to Indigenous tourism development in the North and how they could be addressed and re-oriented to better support the local community. It was at this point that the OAV was born.

Tundra North Tours wanted to create a model which alleviated the barriers to Indigenous community members working in tourism. Tourism wages traditionally have been very low compared to other resource based industries in the North and the cost of living is very high. According to Taylor (2022) "The North is one of the most difficult places in the world to do business, you know it is super remote, the costs are high, capacity is low, you know it's not an easy place". Many people are unable to receive child care in order to work. Food costs and food insecurity is extremely high as well as rent for those living in the town. The concept of the OAV was to have a place that was community-centric, thus ensuring a communal way of living that supported the need for food security, healthy living conditions, and assistance to family members. In addition to being a place that would attract tourists, this would be a place where people who were working with TNT could thrive and be able to provide affordable room and board as well as, in some cases, the opportunity to bring their children to work, and obtain training, support, language and a path to healing. In the future it is planned that homes will be built for elders (Taylor, 2022).

The OAV was developed as a social enterprise. As the cost of living is so high in the Northwest Territories, one way to support the community is to provide employee housing and other forms of support for employees and their families. Taylor's vision was to create a village that supported the traditional way of Inuvialiut and Gwich'in life including providing housing, culture-based employment for elders and employees, communal food production, a child-accommodating environment, and education in learning the Inuvialuit language. The village will not only be a place where tourists can come and share the traditional way of living in the North, but it will also have a wood mill, commercial fishery and greenhouses to grow food. This focus on food sovereignty is especially pertinent in the Northwest Territories as food costs are extremely high and culturally relevant food security is scarce. Access to fresh food is limited in the Northwest Territories. The OAV also provides access to language learning opportunities and cultural seminars that work with Elders to teach Indigenous youth about Inuvialiut and Gwich'in skills and culture. The development of the OAV will provide a safe and supportive space for the community. Taylor has plans for the OAV to become an incubator for future other culture and nature based related businesses to thrive. By providing support to new Indigenous businesses, the OAV can support the generation of products and services that will contribute to the purpose of the village. As Taylor states

The business incubator idea locally, I think works really well…. because we would be on the ground, actually operating the businesses and providing support on the ground. And because each community is so unique with everything such as land use agreements with different Indigenous groups, capacity issues, social issues, every community is so different. To have an incubator in the community with a team of people on the ground that can help with the day to day operations, as opposed to just the business planning and the marketing.

(Taylor, 2022)

Local businesses, such as honey production, for example, will lead to further food security for the village and also provide authentic experiences for tourists while supporting local Indigenous entrepreneurs.

The OAV would enhance the authenticity and opportunity of tourists to immerse themselves in the Inuvialuit and Gwich'in way of life. Tourists would be able to meet and spend time getting to know the people who live at the OAV. An elder was hired to teach cultural elements to the people of the village and to tourists. In addition to supporting community youth, each trip that goes out with TNT includes the opportunity for a young community member to join the trip. This ensures that the community has the opportunity to experience seeing their land.

Conclusion

As of 2022, the OAV has opened to tourists and the community alike. At the OAV, a sod house has been built, being the first model of sustainable housing for the village. The plan is to replicate this model of housing which is based on traditional methods and uses all reusable and recyclable materials. The OAV has also hosted many volunteers from platforms such as WOOOF throughout the summer of 2021 through 2022. Volunteers arrived from around the world to work on the OAV, creating a dynamic environment in which locals and tourists would be able to learn from each other.

This case study reveals that the Sustainable Livelihoods Approach and social innovation in Indigenous tourism development can be extremely beneficial for not only the tourism business but also for the local community as well. Tourism that is developed based on these principles focuses on ensuring the sustainability of the local community as well as the environment, and can serve as a model for tourism development worldwide. It also provides authentic experiences that can lead to an increase in education as well, as it is based on the concept of reconciliation. Local control and ownership lead to empowerment which in turn has led to the conservation of natural and cultural heritage. The OAV is a model for a sustainable way of life that embraces traditional knowledge.

References

Al-Oun, S., & Al-Homoud, M. (2008) 'The potential for developing community-based tourism among the Bedouins in the Badia of Jordan', *Journal of Heritage Tourism*, 3(1), pp. 36–54. https://doi.org/10.1080/1743873x.2008.9701249

Altinay, L., Sigala, M., & Waligo, V. (2016) 'Social value creation through tourism enterprise', *Tourism Management*, 54, pp. 404–417. https://doi.org/10.1016/j.tourman.2015.12.011

Altman, J.C., & Finlayson, J. (1993) 'Aborigines, tourism, and sustainable development', *The Journal of Tourism Studies*, 4(1), pp. 38–50.

Ateljevic, J. (2009) 'Tourism entrepreneurship and regional development: Example from New Zealand', *International Journal of Entrepreneurial Behavior & Research*, 15(3), pp. 282–308. https://doi.org/10.1108/13552550910957355

Ateljevic, I., & Doorne, S. (2003) 'Culture, economy and tourism commodities: Social relations of production and consumption', *Tourist Studies*, 3(2), pp. 123–141. https://doi.org/10.1177/1468797603041629

Butler, R., & Hinch, T. (2007) 'Introduction: Revisiting common ground', *Tourism and Indigenous Peoples: Issues and Implications*, pp. 2–12. Burlington: Elsevier. https://doi.org/10.1016/B978-0-7506-6446-2.50005-3

Canadian Council of Aboriginal Business. (2019) *Promise and Prosperity: 2016 Qualitative Research with Aboriginal Businesses in Canada*. Toronto: Canadian Council for Aboriginal Business.

Canadian Council of Aboriginal Business. (2022) https://www.ccab.com/events/kylik-kisoun-taylor-bio/

Chaskin, R.J., Prudence, B., Venkatesh, S., & Vidal, A. (2001) *Building Community Capacity*. Routledge. https://doi.org/10.4324/9781315081892

Colton, J.W. (2005) 'Indigenous tourism development in Northern Canada: Beyond economic incentives', *The Canadian Journal of Native Studies*, 15(1), pp. 185–206.

Colton, J.W., & Whitney-Squire, K. (2010) 'Exploring the relationship between Aboriginal tourism and community development', *Leisure*, 34(3), pp. 261–278. https://doi.org/10.1080/14927713.2010.521321

Conference Board of Canada & ITAC. (2021) *The Impact of COVID-19 on Canada's Indigenous Tourism Sector: Indigenous Tourism Sector: 2021 Update.* Vancouver, BC: Indigenous Tourism Association of Canada.

Graci, S. (2010) 'The potential for Aboriginal ecotourism in Ontario', *Geography Research Forum*, 30, pp. 33–148.

Graci, S. (2021) 'Indigenous ecotourism in Canada', *Routledge Handbook of Ecotourism*, p. 318. Routledge.

Ingram, P.K. (1990) *Indigenous Entrepreneurship and Tourism Development in the Cook Islands and Fiji: A Thesis Presented in Partial Fulfilment of the Requirements for the Degree of Doctor of Philosophy in Business Studies at Massey University*, Doctoral dissertation. Massey University.New Zealand

ITAC. (2017) *Indigenous Cultural Experiences Guide.* https://destinationindigenous.ca/blog/guide-indigenous-tourism-canada

ITAC. (2019) *Indigenous Cultural Experiences National Guidelines.* https://indigenoustourism.ca/corporate/national-guidelines/

ITAC. (2022) *The Indigenous Tourism Association of Canada Releases Its Revised Three-Year Strategic Plan.* Indigenous Tourism Association of Canada. https://indigenoustourism.ca/the-indigenous-tourism-association-of-canada-releases-its-revised-three-year-strategic-plan/

Kline, C., Shah, N., & Rubright, H. (2014) 'Applying the positive theory of social entrepreneurship to understand food entrepreneurs and their operations', *Tourism Planning & Development*, 11(3), pp. 330–342. https://doi.org/10.1080/21568316.2014.890126

Lemelin, R.H., Koster, R., & Youroukos, N. (2015) Tangible and intangible indicators of successful aboriginal tourism initiatives: A case study of two successful aboriginal tourism lodges in Northern Canada. *Tourism management*, 47, 318–328.

Luke, B., & Chu, V. (2013) 'Social enterprise versus social entrepreneurship: An examination of the 'why' and 'how' in pursuing social change', *International Small Business Journal: Researching Entrepreneurship*, 31(7), pp. 764–784. https://doi.org/10.1177/0266242612462598.

Maritz, A., & Foley, D. (2018) 'Expanding Australian indigenous entrepreneurship education ecosystems', *Administrative Sciences*, 8(2), p. 20. https://doi.org/10.3390/admsci8020020.

Marr, B., & Creelman, J. (2014) *Doing More with Less: Measuring, Analyzing and Improving Performance in the Not-for-profit and Government Sectors*, 2nd Edition. Palgrave Macmillan.

McGinley, K. (2003) 'Best practices: A planned approach to developing a sustainable aboriginal tourism industry in Mistissini', *Journal of Aboriginal Economic Development*, 3(2), pp. 12–19.

Ngoasong, M.Z., & Kimbu, A.N. (2016) 'Women as vectors of social entrepreneurship', *Annals of Tourism Research*, 60, pp. 63–79. https://doi.org/10.1016/j.annals.2016.06.002

O'Neill, B., Williams, P., Morten, K., Kunin, R., Gan, L., & Payer, B. (2015) *National Aboriginal Tourism Research Project 2015.* Aboriginal Tourism Canada. https://doi.org/10.1080/14724049.2019.1583754

Pergelova, A., Angulo-Ruiz, F., & Dana, L. (2021) 'The entrepreneurial quest for emancipation: Trade-offs, practices, and outcomes in an indigenous context', *Journal of Business Ethics*. https://doi.org/10.1007/s10551-021-04894-1

Richard, A. (2021) 'Mikwam Makwa Ikwe (Ice Bear Woman): A national needs analysis on Indigenous women's entrepreneurship', *Women Entrepreneurship Knowledge Hub*. Winnipeg, MB: I.H. Asper School of Business.

Sengupta, U., Vieta, M., & McMurtry, J.J. (2015) 'Indigenous communities and social enterprise in Canada', *Canadian Journal of Nonprofit and Social Economy Research*, 6(1), pp. 104–123. https://doi.org/10.22230/cjnser.2015v6n1a196

Taylor, K. (2018) Personal communication.

Taylor, K. (2022) Personal communication.

Zeppel, H. (2003) 'Ecotourism policy and Indigenous people in Australia', *Ecotourism Policy and Planning*, pp. 55–76. Cambridge: CABI Publishing.

Zeppel, H. (2006) *Indigenous Ecotourism: Sustainable Development and Management.* Cambridge: CABI Publishing.

25

PATHWAYS TO CULTURALLY SENSITIVE TOURISM POLICIES AND PRACTICES

Monika Lüthje, Emily Höckert and Outi Kugapi

Introduction

The Arctic is a vast area with a rich variety of Indigenous and other local cultures, histories, landscapes and political systems. Despite the differences, many Arctic regions face similar challenges, such as climate change, outmigration, long distances and limited opportunities for employment. Tourism plays an important role in many local economies, while its role is minor in others (Jóhannesson et al. 2022; Olsen et al. 2019). The COVID pandemic revealed both the fragility and importance of tourism in Arctic destinations and forced different actors to rethink the role of its future development. In many locations, tourism strategies are currently under revision. Nevertheless, despite the often-voiced need for more sustainable tourism futures (see for example, *Tourism Geographies*, 2020, 2 (3)), many tourism companies seem more eager to return their businesses to continuous growth.

Our exploration of the notion of cultural sensitivity in Arctic tourism began as a reaction to the long history of misrepresentation and misuse of Indigenous Sámi cultures by the tourism industry in our own country, Finland (see Kugapi et al. 2020a, 2020b; Saari et al. 2020; Sámediggi 2018). Stereotyping and cultural appropriation of Indigenous cultures have also taken place for a long time among tourism businesses elsewhere in the Arctic (Grimwood, Muldoon and Stevens 2019; Olsen et al. 2019; Viken, Höckert and Grimwood 2021). Together with a network of researchers, tourism entrepreneurs and other stakeholders including Sámi, Inuit and First Nation Indigenous peoples and the World Indigenous Tourism Alliance (WINTA), we wanted to contribute to changing the situation by starting the ARCTISEN project (Culturally Sensitive Tourism in the Arctic, EU's Northern Periphery and Arctic Programme). At the same time, our call for cultural sensitivity is not limited to Indigenous cultures or Arctic places but is relevant more broadly where different political organisations, rights, economic conditions, languages and worldviews are encountered (see also Viken et al. 2021).

Our work is founded on global-level policy papers, such as the United Nations Declaration on the Rights of Indigenous Peoples (UNDRIP 2007; see Higgins-Desbiolles 2007; Jamal 2019) and the Larrakia Declaration (2012), with a specific focus on Indigenous rights in tourism. Policies play a part in the Indigenous tourism system, as defined in Butler and Hinch (2007), where governments have a role in promoting economic and social Indigenous tourism development through

DOI: 10.4324/9781003230335-30

policy initiatives (see also Hall 2007). However, state institutions or development agencies that promote Indigenous tourism may use dissenting discourses whereby, for example, cultural aspects are overlooked or other aspects of sustainable development are prioritised over local cultures (de la Maza 2016; Warnholtz and Barkin 2018).

While sustainability seems to play a prominent role in many tourism policies, both in the Arctic and globally, the idea of sustainability is often understood only in environmental and economic terms, hence obscuring the cultural aspects of sustainability (see Carr, Ruhanen and Whitford 2016; Dangi and Jamal 2016; Soini and Dessein 2016). For instance, Whitford and Ruhanen's (2010) research on the development of Australia's policies for Indigenous tourism indicated that while these include a 'sustainable approach to Indigenous tourism development', both the socio-cultural and environmental aspects tend to be overlooked (Whitford and Ruhanen 2010: 491; see also de la Maza 2016). As a result, Indigenous tourism operators may choose not to operate with the notion of sustainability, although their values would be, at least partly, similar to the sustainability goals defined by the UN (Scheyvens et al. 2021).

It is also often unclear to what extent different upper-level policies are recognised and respected among tourism entrepreneurs in their daily business practices (see Fonseca and Carnicelli 2021). According to Whitford and Ruhanen (2010: 492), governments should take a more strategic approach to policy implementation, as 'Policies should provide a framework which draws upon Indigenous diversity in a consistent, collaborative, coordinated and integrated manner to provide the mechanisms to facilitate long-term sustainable Indigenous tourism'. In addition to calling for meaningful collaboration among different actors, Vargas et al. (2019) underlined the question of *how* social responsibilities should be implemented through policy frameworks.

This chapter joins a long line of discussions on sustainable policies and practices in Indigenous tourism settings (see Whitford, Bell and Watkins 2001; Whitford and Ruhanen 2016) viewed through the theoretical lenses of practice theory and cultural sensitivity. In order to enrich and deepen the prevailing discussions on sustainability in tourism, we approach the notion of cultural sensitivity as a respectful orientation to otherness based on recognition and reciprocity (Viken et al. 2021). We have found this approach to be helpful for addressing both the possibilities and challenges when a growing number of tourists, non-local tourism companies and seasonal workers are displaying an interest in the Arctic region (Olsen et al. 2019). Seasonal workers may be outsiders with little local knowledge, as in many places there is not enough local workforce. In the future, this is even more likely, as the COVID pandemic has forced many local tourism workers to switch to other jobs and it seems that they are not returning (Jóhannesson et al 2022; Macaulay, Bryce and Vainikka 2022). In the current situation (late 2022), it is timely to formulate more culturally sensitive tourism policies in the Arctic by highlighting the importance of local knowledge, as suggested in this chapter.

This chapter focuses on the role of a culturally sensitive orientation in tourism policies, asking what culturally sensitive policies would be like if Indigenous and other local tourism entrepreneurs that respect local cultures were able to play an important role in their formulation. We seek to answer this question by drawing on a wide range of stakeholder interviews and benchmarking in Arctic Finland, Sweden, Norway, Greenland and Canada. These voices bring to the policy table a diversity of lived experience and therefore relevant knowledge, priorities and perspectives about what works and what is meaningful (Head 2008: 9; see also Shove, Pantzar and Watson 2012: 145, 161–162).

By applying a practice-based method of analysis, the focus is on two bundles of culturally sensitive tourism practices that should be recognised and respected in policymaking. We understand policy as a set of regulations, rules, guidelines or development or promotion objectives and strategies that provide a framework within which decisions affecting tourism development and practices are taken within a destination. Policy defines 'the terms under which tourism operators must function' and 'provides a common direction and guidance for all tourism stakeholders within a destination' (Goeldner and Ritchie 2012: 326–327).

Before diving deeper into the Arctic context, we offer a short conceptual overview of practice theory and culturally sensitive orientation in tourism settings and describe our empirical material and analysis in more detail.

Practice Theory and Cultural Sensitivity

While cultural sensitivity has been, until recently, little-used and sparsely conceptualised in tourism contexts (Olsen et al. 2019; Read and Grimwood 2021; Viken et al. 2021), practice theories have been increasingly used within tourism studies over the last decade (see Bargeman and Richards 2020; De Souza Bisbo 2016; James, Ren and Halkier 2019; Lamers, van der Duim and Spaargaren 2017). Our theoretical framework brings together Shove et al.'s (2012) theory of social practices with the notion of culturally sensitive tourism, conceptualised by Viken et al. (2021). We approach these discussions with a relational lens; that is, instead of focusing on individuals as such, the theory of social practices and the cultural sensitivity framework both help to draw attention to situated activities and relations among groups of people (see Bargeman and Richards 2020: 1; Viken et al. 2021). Moreover, practice theorists see human agency and social structure as influencing each other in practices in reciprocal ways (Lamers et al. 2017: 55; see Shove et al. 2012: 3–4).

According to Shove et al. (2012), practices involve the active integration of three kinds of elements: materials, competences and meanings. By *materials*, they were referring to objects and what they are made of: infrastructure, tools, technologies, hardware, the body and other tangible physical entities. *Competences* refer to various forms of understanding and practical knowledge-ability, skill, know-how, or technique. By *meanings,* Shove et al. (2012) placed their focus on symbolic meanings, ideas and aspirations as well as the social and symbolic significance of participation in a practice at any one moment.

Shove et al. (2012) approached practices as performances and entities. Practices as performances involve the active integration of elements through the process of doing, whereas practices as entities are constituted through recurrent enactments of the practices as performances (see also Lamers et al. 2017: 56). In this chapter, we place our focus on the latter. We understand practices here as routinised 'doings' and 'sayings' performed by knowledgeable and capable human actors involving material objects and infrastructures (Lamers et al. 2017: 56).

In their research, Shove et al. (2012) drew special attention to social order, stability and change. According to them, practices are constantly in the process of formation, re-formation and deformation; they emerge when connections between elements are made and change when new elements are added to them or when existing elements are combined in new ways. Practices survive only if there are practitioners who are willing and able to integrate their elements and, in doing so, keep them alive. When connections between elements are broken, practices disappear. Shove et al. (2012) also argued that the promotion of more sustainable ways of life should be rooted in an understanding of the elements of practices and the connective tissue that holds them together.

Although Shove et al. (2012: 136) stated that 'links and connections between elements and practices are rooted in past inequalities and constitutive of similar patterns in the future', they also advocated active policy measures that promote transitions in undesirable practices. They emphasised that it is impossible to control social practices because they are emergent and their development unpredictable, but it is nevertheless possible to increase the chances that more desirable practices emerge or survive instead of less desirable ones (Shove et al. 2012: 146, 163; see also Lamers et al. 2017: 59–60).

Our research suggests that by bringing the discussions on social practices together with the notion of cultural sensitivity, we can better understand how inclusive, legitimate and effective sustainable tourism policies can be formulated. The recent article 'Cultural Sensitivity: Engaging Difference in Tourism' by Viken et al. (2021) advanced a conceptual understanding of this notion by drawing on collective efforts by the ARCTISEN project (see Olsen et al. 2019). The article extends recurrent concerns associated with how Indigenous cultures are enfolded in various land and water governance, development and research contexts (Lehtola 2015; Tuhiwai Smith 2012).

The notion of cultural sensitivity can be seen as a highly relevant approach to contexts in which different languages, rights, political organisations, influences, economic conditions and worldviews are encountered. Diverse complaints within tourism discourse often boil down to the stereotype of tourism actors being driven by insensitivity—that is, by a self-centred attitude and by treating others as means to their ends (see Dinhopl and Gretzel 2016; Höckert 2018). These kinds of orientations to tourism and tourism encounters *per se* make it difficult to envision more respectful and caring futures (Biddle and Swee 2012; Jamal 2019; Smith 2011). To move beyond stereotyping, appropriating and assimilating tourism practices, it is important to explore the possibilities of culturally sensitive orientations.

To advance the conceptualisation of cultural sensitivity in tourism, Viken et al. (2021) built on the phenomenologically informed model of intercultural sensitivity developed by Milton Bennett (1986) that sees cultural sensitivity as a way of relating to cultural diversity. Bennett's model helps to disentangle and clarify the various ways people experience cultural differences and attach meanings to them. In its simplest form, sensitivity was understood by Bennett as the ability to sense and engage with difference—that is, as one's relation to otherness. Cultural sensitivity is hence approached here as a subjective orientation towards otherness that simultaneously shapes and is shaped by, different kinds of social processes, narratives and practices, including those associated with tourism.

Bennett (1986) suggested that developing cultural sensitivity requires not only knowledge and competence but also self-reflection that leads to new understanding, awareness and attitudes. Bennett further submitted that the more that individuals can recognise and give non-evaluative meanings to cultural nuances and different perspectives, the more they hold culturally sensitive perceptions. Thus, enhancing intercultural sensitivity implies not only the acquisition of new skills but also the development of one's own consciousness. His model hinges on a conceptual split between what he labelsethnocentric and ethnorelative experiences of difference. While the ethnocentric realm involves essentialising difference and detaching us from others, the ethnorelative realm is based on openness to diversity.

To understand cultural sensitivity in the tourism context, it becomes vital to recognise the fundamental ontological and epistemological differences between ethnocentric and ethnorelative orientations (Viken et al. 2021). Viken, Höckert and Grimwood's (2021) conceptual framework of cultural sensitivity, illustrated in Figure 25.1, builds on Bennett's thoughts and previous research that addressed authorities', tourists' and tourism companies' practices with Indigenous groups. While the ethnocentric circle encompasses a more self-centred and dualistic approach to cultures (see also Chambers and Buzinde 2015), the idea of ethnorelativity calls for openness towards other ways of being and knowing.

In order to break the negative circle of ethnocentric ways of being and knowing, Viken et al. (2021) drew attention to the ethnorelative notions of recognition, respect and reciprocity. These

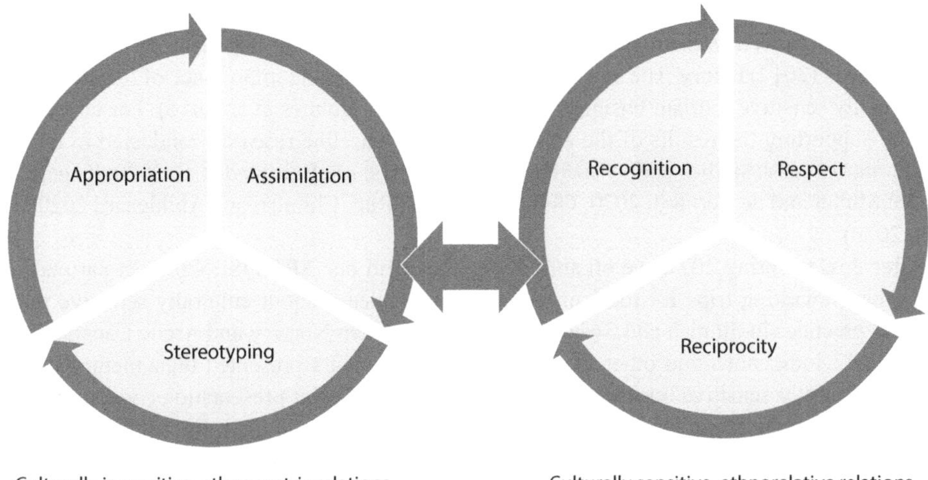

Figure 25.1   A conceptualisation of cultural sensitivity in tourism.

Source: Viken et al. (2021: 5)

values can also be seen as founding elements in the declarations that seek to secure the rights of Indigenous peoples (see Larrakia Declaration 2012; UNDRIP 2007). As the late Johnny Edmonds of WINTAsummarised, 'cultural sensitivity is about respect', and tourism is an area in which this is particularly important (personal communication, ARCTISEN Kick-off, 28 November 2018). Instead of categorising different actors or policies as culturally insensitive or sensitive as such, cultural sensitivity can be seen as a personal and professional social disposition and practice that fosters inclusive intercultural relationships. This means that culturally sensitive processes are shaped by different actors and are always socially and locally situated. While tourism practices focus on the processes of doing and saying, the notion of cultural sensitivity can be understood as a reflexive position and orientation that shapes these processes. In other words, the motives and strategies behind different kinds of practices can be understood in terms of ethnocentric and ethnorelative orientations towards otherness.

In the following analysis and discussion, we provide examples of respect, recognition and reciprocity, both as an orientation to otherness and as a practice among different actors and stakeholders, and suggest how the findings could be applied to future tourism policies.

Empirical Materials and Methods

Our empirical materials consist of interviews and notes from benchmarking trips conducted in 2019–2021. The primary purpose of the interviews was to co-create video material for an open-access online course on culturally sensitive tourism directed to tourism entrepreneurs (see Abilgaard et al. 2022; ARCTISEN online course). In the videos, 17 Indigenous and other local tourism entrepreneurs and other relevant tourism and cultural sector stakeholders, such as Destination Management Organisation managers, in Finnish and Swedish Lapland, Greenland and Northern Norway reflect on their own practices in terms of cultural sensitivity and suggest how, in their view, culturally sensitive tourism practices could be enhanced further.

The video interviews were transcribed and analysed, first by data-driven thematic content analysis to summarise their content, which focused on two themes: collaboration and authenticity.

We followed this with a practice-based analysis, whereby we categorised the doings described in the interviews as well as the materials, competences and meanings they consisted of, drawing on Shove et al.'s (2012) theory. The analysis drew our attention to the importance of local knowledge in culturally sensitive tourism business practices (see also Holmes et al. 2016). For contextualising and supporting the results of the analysis, we used baseline research conducted in the same four countries (desk studies and 99 stakeholder interviews; see Brattland et al. 2020; Kugapi et al. 2020b; Müller and de Bernardi 2020; Olsen et al. 2019; Ren, Chimirri and Abildgaard 2020; Saari et al. 2020).

After this, in spring 2021, we organised, together with our ARCTISEN project partners, four online benchmarking trips for tourism entrepreneurs to learn about culturally sensitive tourism business practices in Finnish and Swedish Lapland, Northern Norway and Arctic Canada. During the trips, 13 Indigenous and other local tourism entrepreneurs presented their methods of conducting culturally sensitive business with the help of PowerPoint presentations, videos and open discussions with the trip participants. Notes were made during the trips. Interestingly, the entrepreneurs described in their company presentations similar practices to those of the entrepreneurs in the video interviews, which is further support for our findings.

Culturally Sensitive Practices in Arctic Tourism

We now move on to present our analysis of two bundles of practices that tourism companies utilise in their businesses, according to our empirical material; that is, practices that are carried out in the same geographical location and are loosely connected to each other (see Shove et al. 2012: 81–91). The first bundle consists of reciprocal practices that enhance collaboration, and the second consists of respectful practices related to authenticity. The connecting element between the two bundles is local knowledge, which also requires and deserves recognition.

Reciprocal Practices of Collaboration

Instead of approaching reciprocity as the mutual exchange of gifts, we understand it as a caring relationship that produces affective or ethical values, such as friendship, trust and mutual understanding (Kuokkanen 2007; Sabourin 2013: 306). For instance, in the Sámi cultures, general reciprocity is central to an old tradition called *verdde* (Svensson and Viken 2017), which refers to a person who you can trust. The notion of reciprocity can also help us explore the possibilities of co-producing new tourism knowledge and practices (Höckert 2018; Ren, van der Duim and Jóhannesson 2018; Viken et al. 2021).

Based on the interviews, planning tourism business activities together with Indigenous and other local peoples plays an important role in the local tourism entrepreneurs' practice of collaboration (see also Brattland et al. 2020: 9; Ren et al. 2020: 22). The aim of such collaboration is to plan activities so that they do not disturb the locals by adapting the activities to life within the local community. According to the interviewees, tourism entrepreneurs who act with a culturally sensitive orientation check in advance that their planned business activities do not disturb the local people's lives, discuss and negotiate with them and search for solutions together with them. Recognition of different actors, mutual respect and open discussion enable consideration of others' practices, problem-solving and negotiation of various future development options (see also Kugapi et al. 2020b: 10; Müller and de Bernardi 2020: 20–24, 27).

For instance, reciprocal practices of collaboration are crucial when it comes to questions of land use. The aforementioned Larrakia Declaration (2012) underlines the importance of respecting 'customary law and lore, land and water, traditional knowledge, traditional cultural expressions' and cultural heritage in tourism decisions. According to the interviewees, a recurrent respectful practice involves tourism entrepreneurs checking daily in advance that the nature excursions they have planned for their clients do not disturb reindeer herding (see also Kugapi et al. 2020b: 15–16; Müller and de Bernardi 2020: 21, 25). Similarly, it might be necessary for entrepreneurs to check, for instance, that a planned tourist excursion does not visit a location where locals are hunting. Both reindeer herding and hunting are part of local cultures respected by tourism entrepreneurs who act in a culturally sensitive way.

The interviewees also emphasised the practice of planning business activities so that they would be of benefit to as many locals as possible and correspond to their needs. This includes, for example, designing tourism products that can be produced in collaboration with (other) local entrepreneurs. In Northern Norway and Greenland, there have been initiatives where local needs form the basis of all tourism planning instead of having tourist needs as the starting point of tourism development. In this kind of community-centred way of developing tourism (Higgins-Desbiolles et al. 2019), DMOs and tourism companies search for the kinds of clients who the local people wish to welcome as visitors to their home villages or towns (see also Ren et al. 2020: 21, 26). The idea is that certain kinds of tourism development attract particular types of tourists (see also Kugapi et al. 2020b: 24; Müller and de Bernadi 2020: 24). Recent Indigenous case studies in the USA and Australia have also described similar tourism business planning practices whereby businesses are planned to serve or be of benefit to Indigenous cultures (Scherrer 2020; Swanson and DeVereaux 2017).

In sum, when different Indigenous and other local perspectives are involved in tourism business planning, it is more probable that the result presents local cultural diversity better (Brattland et al. 2020: 18, 27).

Respectful Practices of Authenticity

Based on our empirical material, the practice bundle of respecting authenticity consists of being true to oneself and telling the truth to tourists. Being true to oneself is existential authenticity (Wang 1999), which in tourism research is normally connected to tourists who find their true selves by travelling. In culturally sensitive tourism business practices, it is, however, the entrepreneurs and their employees who are true to themselves. This means that the entrepreneur plans their business so that it fits their values and interests and searches for employees and clients who share similar values and interests. For example, if the entrepreneur is passionate about making handicrafts, they offer handicraft travel packages for tourists interested in them. Values-based businesses have also been found to be common among lifestyle entrepreneurs whose values may be in many respects similar to the dimensions of cultural sensitivity (e.g., reciprocity, collaboration, trust and communality) (see Ateljevic and Doorne 2000; Hurst et al. 2020; Kugapi et al. 2020a; Viken et al. 2021).

According to the interviewees, being true to oneself also means that the entrepreneurs' and their employees' culture is in what they do in their own lives and that they do not try to be something that in reality they are not. For example, they do not try to be traditional representatives of Indigenous or other local cultures if these are not their true selves, which can be seen as countering the culturally insensitive practices of stereotyping and cultural appropriation taking place in Arctic tourism. According to the interviewees, it is enough to tell tourists about one's own life in the place they are visiting. Nothing more exotic is needed (see also Olsen et al. 2019: 38–39). Moreover, the interviews indicate the importance of collaboration with Indigenous people when non-Indigenous entrepreneurs wish to tell tourists about Indigenous cultures. From the point of view of cultural sensitivity, the best practice is that those to whom a particular culture belongs—and who know best what can be shared with outsiders—are the ones who recount it to tourists (Brattland et al. 2020: 18–9, 27–29; see also Müller and de Bernardi 2020: 20).

The entrepreneurs' answers demonstrate how different kinds of stories can be used to create both connections between different actors and also as tools for the ongoing processes of self-determination (Read and Grimwood 2021), epistemic decolonisation and reconciliation (Kramvig and Førde 2020). It can be said that sharing true stories with one's guests is connected to being true to oneself. In the interviews, there is the idea both in true stories and in one's true self of an objective truth, but these truths can also be seen as being socially and culturally constructed in the local community (see Wang 1999). In other words, one's true self and true stories are constructed in interactions with others and are always in a state of becoming.

According to the interviewees, telling the truth means telling the truth about one's products (e.g. not saying that they are local or Indigenous if they are not), discussing things that one knows, telling true stories about one's own life and the lives of one's family, country and people and providing facts that are correct. The last mentioned is important, among others, to the Indigenous Sámi, about whom the tourism industry has disseminated much incorrect information for decades, especially in Finland (Kugapi et al. 2020b; Sámediggi 2018; see also Müller and de Bernardi 2020: 19). According to some of the Indigenous interviewees, due to the colonisation and modernisation processes, Indigenous tourism entrepreneurs may have to study, learn, revitalise or even reinvent their cultural traditions in order to be able to present them to tourists and narrate correct information about them (see also Kramvig and Førde 2020; Olsen et al. 2019: 39).

Based on the interviews, telling the truth to tourists also includes telling them the kinds of behaviour that is appropriate or inappropriate in the local culture(s) and nature. Many NGOs, DMOs and local communities have written guidelines to inform visitors of appropriate behaviour in the Arctic (see Kugapi et al. 2020b: 24, 27; Read forthcoming; Ren et al. 2020: 14–15, 22). The culturally sensitive way to create these local policies is to do it in collaboration with the local people (see Chapter 11, editors' note), focusing on their needs, rights, cultural norms and rules (see Read forthcoming). This is a way to involve the local communities in tourism governance and, more specifically, in defining what cultural sensitivity means in their home villages or towns on a practical level of tourist behaviour (Brattland et al. 2020: 28).

Recognition of Local Knowledge

Based on the empirical material, local knowledge plays an essential role in both the practices of collaboration and authenticity (see also Brattland et al. 2020: 14, 26, 28; Holmes et al. 2016; Kugapi et al. 2020b: 31; Müller and de Bernardi 2020: 30; Olsen et al. 2019; Ren et al. 2020: 22). This is a connecting element between the two practice bundles, and here it means knowledge of and understanding the practices of local people and the rhythms of local nature, both of which are intertwined (see also Valkonen and Valkonen 2018: 18–19). It is necessary not only for collaboration and telling the truth to tourists but also for being true to oneself. Local knowledge is for a large part experiential; it comes from living in a local place and community—and it is what makes the local people local (Ingold and Kurttila 2000: 194; Valkonen and Valkonen 2018: 12, 16).

Belonging to the local community creates an interdependence between its members, where everyone is in relation to the local values. This can be seen as the core of culturally sensitive tourism (Brattland et al. 2020: 27). Local knowledge may be inherited knowledge acquired by living in the same place even for several generations, and it can also be Indigenous knowledge. Based on our empirical material, we see that at its deepest, it stems from existential insideness in the local place: the place is experienced as full of meaning without deliberate and self-conscious reflection, while the person is an accepted part of the place and it is part of them (Relph 1976: 55). In that sense, it is also connected to existential authenticity and being true to one's self. In previous research on tourism entrepreneurship, local knowledge has been studied as a source of innovation and as an enabler of authenticity and differentiation, which gives a competitive advantage in the market (see Dias et al. 2020); however, besides these market-oriented perspectives, it is also an important enabler of cultural sensitivity and, as such, it can be seen as the connecting tissue that holds the practices of collaboration and authenticity together (see Shove et al. 2012: 36). This is important to recognise when designing and implementing culturally sensitive tourism policies.

Conclusion: Pathways to Culturally Sensitive Tourism Policies

Cultural sensitivity offers a novel framework for developing tourism policies, services and businesses and for approaching tourism encounters in ways that can enhance recognition, respect and reciprocity towards otherness in Arctic tourism and beyond. Our analysis suggests that, in order to become more sustainable, tourism policymaking should on the one hand, opt for a culturally sensitive lens and on the other, draw attention to the competence elements of tourism business practices.

According to Shove et al. (2012), practices may be transformed by changing their elements. Based on our analysis, we suggest that policies should aim to change the competence elements of culturally insensitive tourism business practices in the Arctic, which, in turn, would change their material and meaning elements to more culturally sensitive ones. As local knowledge is essential in culturally sensitive tourism business practices, policies should ensure that it is involved in the business planning and operations of tourism companies. To achieve this, tourist numbers may, for example, be limited by local people's capacity and willingness to be tourism entrepreneurs and workers. On the other hand, outsider companies might be obliged to collaborate with local communities or employ local workers. Policies could also be formed to promote tourism entrepreneurs and workers' understanding of cultural sensitivity as a concept and practice. These actors, among others, would gain from studying (e.g. via a self-study online course) the basics of cultural sensitivity in Arctic tourism, and those coming from outside the region could be required to pass a test before being allowed to operate there. Passing the test would not automatically imply that the person would act in a culturally sensitive way, but if the course and test were carefully designed, it would be probable that they would result in changed practices.

As our empirical material shows, the interviewed and benchmarked tourism entrepreneurs followed a culturally sensitive tourism policy in practice. That is, they had a practical understanding of and were knowledgeable about culturally sensitive tourism business. On the other hand, upper-level policies may be unfamiliar to them or difficult to follow. An illustrative example of this is the creation of the 'Principles for Responsible and Ethically Sustainable Sámi Tourism' (Sámediggi 2018), which were drafted by the Sámi Parliament in Finland following a participatory approach involving Sámi tourism entrepreneurs and other Sámi stakeholders. However, after the participatory phase, the parliament made the principles stricter, which resulted in some Sámi tourism entrepreneurs complaining that it did not understand what doing business was like and had imposed requirements that were impossible to follow in practice (Kugapi et al. 2020b; Suomi 2020).

Policies are needed, but how can they be made more inclusive, legitimate and effective? As Kilpinen (2022) stated, policies do not do anything by themselves; instead, they are brought to life by humans. Kilpinen (2022) also suggested that policies should not only be implemented but also be humanised to peoples' lives by inclusion and dialogue—in other words, by creating a

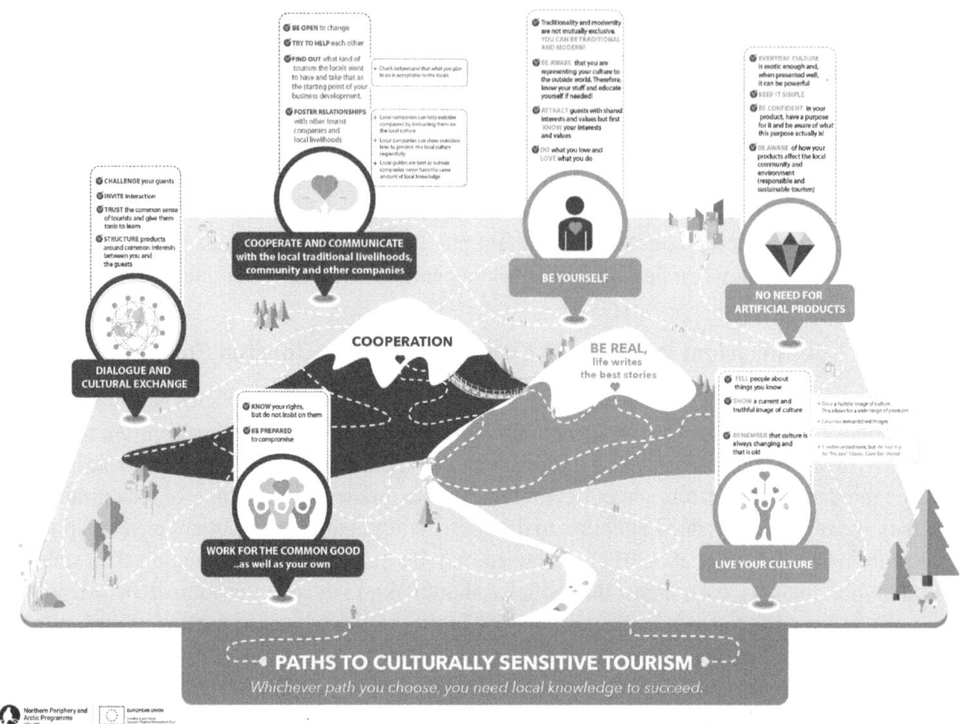

Figure 25.2 Roadmap for culturally sensitive tourism business.

Source: ARCTISEN project.

common ground for discussion and participation in the process at hand. An example of this kind of policymaking is the roadmap designed by the ARCTISEN project for tourism companies (see Figure 25.2) based on the interview material presented in this chapter. The roadmap gives guidance on how to conduct culturally sensitive tourism business in practice. It can be seen as a policy framework and could be taken as a model when drafting tourism policies (see also Warnholtz and Barkin 2018).

Respecting and recognising local entrepreneurs' and other local stakeholders' practices, knowledge and participation while formulating tourism policies could make those policies more approachable and effective. Various participatory methods could be utilised for fostering stakeholder engagement in the policymaking process. For example, art-based and service design methods may foster recognition, respect and reciprocity among different participants (see Li et al. forthcoming). We acknowledge that participatory approaches have their challenges that need to be taken into consideration (see Höckert et al. 2021; Vernon et al. 2005) but see them as important for formulating tourism policies that enhance culturally sensitive, sustainable tourism futures. Arctic tourism would be quite different if, for example, the practices of collaboration and authenticity described in this chapter were widely adopted in the region.

References

Abilgaard, M.S., Höckert, E., Kugapi, O., Lüthje, M. and Ren, C. (2022) "Enhancing Culturally Sensitive Tourism in an Online learning Environment," in J.R. Edelheim, M. Joppe and J. Flaherty (eds.) *Teaching*

Tourism: Innovative, Values-based Learning for Transformational Practices, Cheltenham: Edward Elgar Publishing, 157–159.

ARCTISEN online course (n.d.) *Cultural Sensitivity in Arctic Tourism*, viewed 27 March 2022, <https://blogi. eoppimispalvelut.fi/learnculturaltourism/>

Ateljevic, I. and Doorne, S. (2000) "'Staying within the Fence': Lifestyle Entrepreneurship in Tourism," *Journal of Sustainable Tourism* 8(5), 378–392.

Bargeman, B. and Richards, G. (2020) "A New Approach to Understanding Tourism Practices," *Annals of Tourism Research* 84, 102988.

Bennett, M.J. (1986) "A Developmental Approach to Training for Intercultural Sensitivity," *International Journal of Intercultural Relations* 10(2), 179–196.

Biddle, N. and Swee, H. (2012) "The Relationship between Wellbeing and Indigenous Land, Language and Culture in Australia," *Australian Geographer* 43(3), 215–232.

Brattland, C., Jæger, K., Olsen, K., Dunfjell, O., Elle, M. and Viken, A. (2020) *Cultural Sensitivity and Tourism. Report from Northern Norway*, Rovaniemi: Publications of the Multidimensional Tourism Institute.

Butler, R.W. and Hinch, T. (2007) *Tourism and Indigenous Peoples: Issues and Implications*, London: Butterworth Heinemann.

Carr, A., Ruhanen, L. and Whitford, M. (2016) "Indigenous Peoples and Tourism: The Challenges and Opportunities for Sustainable Tourism," *Journal of Sustainable Tourism* 24(8–9), 1067–1079.

Chambers, D., and Buzinde, C. (2015) "Tourism and Decolonisation: Locating Research and Self," *Annals of Tourism Research*, 51, 1–16.

Dangi, T. and Jamal, T. (2016) "An Integrated Approach to Sustainable Community-based Tourism," *Sustainability* 8, 475.

de la Maza, F. (2016) "State Conceptions of Indigenous Tourism in Chile," *Annals of Tourism Research* 56, 80–95.

de Souza Bispo, M. (2016) "Tourism as Practice," *Annals of Tourism Research* 61, 170–179.

Dias, A., Silva, G. M., Patuleia, M. and González-Rodriguez, M. R. (2020) "Developing Sustainable Business Models: Local Knowledge Acquisition and Tourism Lifestyle Entrepreneurship," *Journal of Sustainable Tourism*, ahead of print, 1–20.

Dinhopl, A. and Gretzel, U. (2016) "Selfie-taking as Touristic Looking," *Annals of Tourism Research* 57, 126–139.

Fonseca, A.P. and Carnicelli, S. (2021) "Corporate Social Responsibility and Sustainability in a Hospitality Family Business," *Sustainability* 13, 7091.

Goeldner, C.R. and Ritchie, J.R.B. (2012) *Tourism. Principles, Practices, Philosophies*, Hoboken: John Wiley & Sons.

Grimwood, B.S.R., Muldoon, M.L. and Stevens, Z.M. (2019) "Settler Colonialism, Indigenous Cultures, and the Promotional Landscape of Tourism in Ontario, Canada's 'Near North'," *Journal of Heritage Tourism* 14(3), 233–248.

Hall, M. (2007) "Politics, Power and Indigenous Tourism," in R. Butler and T. Hinch (eds.) *Tourism and Indigenous Peoples: Issues and Implications*, London: Routledge, 305–318.

Head, B.W. (2008) "Three Lenses of Evidence-based Policy," *The Australian Journal of Public Administration* 67(1), 1–11.

Higgins-Desbiolles, F. (2007) "Taming Tourism: Indigenous Rights as a Check to Unbridled Tourism," in P.M. Burns and M. Novelli (eds.) *Tourism and Politics: Global Frameworks and Local Realities*, Amsterdam: Elsevier, 83–107.

———, Carnicelli, S., Krolikowski, C., Wijesinghe, G. and Boluk, K. (2019) "Degrowing Tourism: Rethinking Tourism," *Journal of Sustainable Tourism* 27(12), 1926–1944.

———, Trevorrow, G. and Sparrow, S. (2014) "The Coorong Wilderness Lodge: A Case Study of Planning Failures in Indigenous Tourism," *Tourism Management* 44, 46–57.

Hinch, T. and Butler, R. (2009) "Indigenous Tourism," *Tourism Analysis* 14(1), 15–27.

Höckert, E. (2018) *Negotiating Hospitality. Ethics of Tourism Development in the Nicaraguan Highlands*, London: Routledge.

——— Höckert, E., Kugapi, O. and Lüthje, M. (2021) "Hosts and Guests in Participatory Development," in A. Harju-Myllyaho and S. Jutila (eds.) *The Future of Inclusive Tourism*, Bristol: Channel View Publications, 15–32.

Holmes, A.P., Grimwood, B.S.R., King, L.J. and the Lutsel K'e Dene First Nation (2016) "Creating an Indigenized Visitor Code of Conduct: The Development of Denesoline Self-determination for Sustainable Tourism," *Journal of Sustainable Tourism* 24(8–9), 1177–1193.

Hurst, C. E., Grimwood, B. S. R., Lemelin, R. H. and Stinson, M. J. (2020) "Conceptualizing Cultural Sensitivity in Tourism: A Systematic Literature Review," *Tourism Recreation Research* 46(4), 500–515.

Ingold, T. and Kurttila, T. (2000) "Perceiving the Environment in Finnish Lapland," *Body & Society* 6(3–4), 183–196.

Jamal, T. (2019) *Justice and Ethics in Tourism*, London: Routledge.

James, L., Ren, C. and Halkier, H. (eds.) (2019) *Theories of Practice in Tourism*, London: Routledge.

Jóhannesson, G.T., Welling, J., Müller, D.K., Lundmark, L., Nilsson, R.O., de la Barre, S., Granås, B., Kvidal-Røvik, T., Rantala, O., Tervo-Kankare, K. and Maher, P. (2022) *Arctic Tourism in Times of Change. Uncertain Futures – From Overtourism to Re-Starting Tourism*, TemaNord 2022:516, Copenhagen: Nordic Council of Ministers.

Kilpinen, P. (2022) *Inhimillinen strategia*, Helsinki: Alma Talent.

Kramvig, B. and Førde, A. (2020) "Stories of Reconciliation Enacted in the Everyday Lives of Sámi Tourism Entrepreneurs," *Acta Borealia* 37(1–2), 27–42.

Kugapi, O., Huhmarniemi, M. and Laivamaa, L. (2020a) "A Potential Treasure for Tourism – Crafts as Employment and a Cultural Experience Service in the Nordic North," in A. Walmsley, K. Åberg, P. Blinnikka and GT. Johannesson (eds.) *Tourism Employment in Nordic Countries*, London: Palgrave, 77–99.

———Kugapi, O., Höckert, E., Lüthje, M., Mazzullo, N. and Saari, R. (2020b) *Toward Culturally Sensitive Tourism: Report from Finnish Lapland*, Rovaniemi: Publications of the Multidimensional Tourism Institute.

Kuokkanen, R. (2007) *Reshaping the University: Responsibilities, Indigenous Epistemes, and the Logic of the Gift.* Vancouver: UBC Press.

Lamers, M., van der Duim, R. and Spaargaren, G. (2017) "The Relevance of Practice Theories for Tourism Research," *Annals of Tourism Research* 62, 54–63.

Larrakia Declaration (2012) *The Larrakia Declaration on the Development of Indigenous Tourism*, viewed 30 March 2022, <https://www.winta.org/_files/ugd/24cdc6_a47b5d589b2d40ae9139c36a9ee5589b.pdf>

Lehtola V. P. (2015) "Sámi Histories, Colonialism, and Finland," *Arctic Anthropology*, 52, 22–36.

Li, H., Lüthje, M., Björn, E. and Miettinen, S. (forthcoming) "Fostering Stakeholder Engagement in Sustainable Cultural Tourism Development in a Nature-Based Site: A Case Study on Using Methodological Layering of Art-Based Methods," in A. Mandic and S.K. Walia (eds.), *Routledge Handbook of Nature-Based Tourism Development*, London: Routledge.

Macaulay, B., Bryce, R. and Vainikka, V. (2022) *Impacts of COVID on the Development of Ethical Tourism*, Ethical Tourism Recovery in Arctic Tourism (ETRAC) project, viewed 31 March 2022, <https://etrac.interreg-npa.eu/resources/>

Müller, D. K. and de Bernardi, C. (2020) *Reflections on Culturally Sensitive Tourism: The Case of Sweden*, Rovaniemi: Publications of the Multidimensional Tourism Institute.

Olsen, K.O., Abildgaard, M.S., Brattland, C., Chimirri, D., de Bernardi, C., Edmonds, J., Grimwood, B.S.R., Hurst, C.E., Höckert, E., Jaeger, K., Kugapi, O., Lemelin, R.H., Lüthje, M., Mazzullo, N., Müller, D.K., Ren, C., Saari, R., Ugwuegbula, L. and Viken, A. (2019) *Looking at Arctic Tourism through the Lens of Cultural Sensitivity. ARCTISEN – A Transnational Baseline Report*, Rovaniemi: Publications of the Multidimensional Tourism Institute.

Read, J.B. (forthcoming) "Have We (Un)Intentionally Made Indigenous Law a Tourists' Code? A Research Note," *International Journal of Tourism Policy* 12(4), 443–450.

———, and Grimwood, B.S.R. (2021) "Ecotourism Development through Culturally Sensitive Universalism," in D.A. Fennell (ed.) *Routledge Handbook of Ecotourism*, London: Routledge, 103–116.

Relph, E. (1976) *Place and Placelessness*, London: Pion.

Ren, C., Chimirri, D. and Abildgaard, M.S. (2020) *Toward Culturally Sensitive Tourism: Report from Greenland*, Rovaniemi: Publications of the Multidimensional Tourism Institute.

———, Jóhannesson, G.T., Kramvig, B., Pashkevich, A. and Höckert, E. (2021) "20 Years of Research on Arctic and Indigenous Cultures in Nordic Tourism: A Review and Future Research Agenda," *Scandinavian Journal of Hospitality and Tourism* 21(1), 111–121.

———, van der Duim, R. and G.T. Jóhannesson (2018) "Co-creation of Tourism Knowledge," in C. Ren, G.T. Jóhannesson, R. and van der Duim (eds.) *Co-creating Tourism Research: Towards Collaborative Ways of Knowing*, London: Routledge, 1–10.

Saari, R., Höckert, E., Lüthje, M., Kugapi, O. and Mazzullo, N. (2020) "Cultural Sensitivity in Sámi tourism: A Systematic Literature Review in the Finnish Context," *Finnish Journal of Tourism Research* 16(1): 93–110.

Sabourin, E. (2013) "Education, Gift, Reciprocity: A Preliminary Discussion," *International Journal of Lifelong Education* 32, 301–317.

Sámediggi (2018) *Principles for Responsible and Ethically Sustainable Sámi Tourism*, viewed 27 March 2022, <https://www.samediggi.fi/ethical-guidelines-for-sami-tourism/?lang=en>

Scherrer, P. (2020) "Tourism to Serve Culture: The Evolution of an Aboriginal Tourism Business Model in Australia," *Tourism Review* 75, 663–680.

Scheyvens, R., Carr, A., Movono, A., Hughes, E., Higgins-Desbiolles, F. and Mika, J.P. (2021) "Indigenous Tourism and the Sustainable Development Goals," *Annals of Tourism Research* 90, 103260.

Shove, E., Pantzar, M. and Watson, M. (2012) *The Dynamics of Social Practice. Everyday Life and How It Changes*, London: Sage.

Smith, M. (2011) "Après moi le deluge: Ethics, Empire, and the Biopolitics of Last Chance Tourism," in H. Lemelin, J. Dawson and E.J. Stewart (eds.) *Last Chance Tourism. Adapting Tourism Opportunities in a Changing World*, London: Routledge, 153–167.

Soini, K. and Dessein, J. (2016) "Culture-Sustainability Relation: Towards a Conceptual Framework," *Sustainability* 8, 167.

Suomi, K. (2020) *Vastuullinen paikallisuus saamelaismatkailun eettisen ohjeistuksen näkökulmasta*, lecture notes, Vastuullinen paikallisuus, Inarijärven aarteet, delivered 2 September 2020.

Svensson, G. and Viken, A. (2017) "Respect in the Girdnu: The Sámi Verde Institution and Tourism in Northern Norway," in A. Viken and D. Müller (eds.) *Tourism and Indigeneity in the Arctic*, Bristol: Channel View Publications, 261–280.

Swanson, K. and DeVereaux, C. (2017) "A Theoretical Framework for Sustaining Culture: Culturally Sustainable Entrepreneurship," *Annals of Tourism Research* 62, 78–88.

Tuhiwai Smith, L. (2012) *Decolonizing Methodologies: Research and Indigenous Peoples*. London: Zed Books.

UNDRIP (2007) *United Nations Declaration on the Rights of Indigenous Peoples*, viewed 30 March 2022, <https://www.winta.org/_files/ugd/24cdc6_b35c62570725412895cc05eb5ceff95c.pdf>

Valkonen, J. and Valkonen, S. (2018) "On Local Knowledge," in T.H. Eriksen, S. Valkonen and J. Valkonen (eds.) *Knowing from the Indigenous North: Sámi Approaches to History, Politics and Belonging*, London: Routledge, 12–26.

Vargas, V.R., Lawthom, R., Prowse, A., Randles, S. and Tzoulas, K. (2019) "Implications of Vertical Policy Integration for Sustainable Development Implementation in Higher Education Institutions," *Journal of Cleaner Production* 235, 733–740.

Vernon, J., Essex, S., Pinder, D. and Curry, K. (2005) "Collaborative Policymaking. Local Sustainable Projects," *Annals of Tourism Research* 32, 325–345.

Viken, A., Höckert, E. and Grimwood, B.S.R. (2021) "Cultural Sensitivity: Engaging Difference in Tourism," *Annals of Tourism Research* 89, 103223.

Wang, N. (1999) "Rethinking Authenticity in Tourism Experience," *Annals of Tourism Research* 26, 349–370.

Warnholtz, G. and Barkin, D. (2017) "Development for Whom? Tourism Used as a Social Intervention for the Development of Indigenous/Rural Communities in Natural Protected Areas," in I. Borges de Lima and V.T. King (eds.) *Tourism and Ethnodevelopment: Inclusion, Empowerment and Self-determination*, New York: Routledge, 49–65.

Whitford, M., Bell, B. and Watkins, M. (2001) "Indigenous Tourism Policy in Australia: 25 Years of Rhetoric and Economic Rationalism," *Current Issues in Tourism* 4(2–4), 151–181.

———— and Ruhanen, L. (2010) Australian Indigenous Tourism Policy: Practical and Sustainable Policies?" *Journal of Sustainable Tourism* 18(4), 475–496.

———— and Ruhanen, L. (2016) "Indigenous Tourism Research, Past and Present: Where to from Here?" *Journal of Sustainable Tourism* 24(8–9), 1080–1099.

26

THE UPPER NAVUA
CONSERVATION AREA

Reflections on Ecotourism and Community

Kelly Bricker and Kasimiro Taukeinikoro

Introduction

Fiji is a South Pacific Island nation of approximately 330 islands covering a region of 1.3 million km^2 lying nearly 1,900 miles north of Sydney, and just over 3,000 miles southwest of Honolulu. The total land area of Fiji is approximately 18,333 kilometers, which is slightly larger than Hawaii and its population is 935,974. Nearly one-third (110) of the 332 islands are inhabited, the largest being Viti Levu which is divided north to south by a mountain range with several peaks above 3,000 feet (900 meters) including Tomanivi, at 4,344 feet (1,324 meters) the highest point in Fiji. It is home to 70% of the population of the country, of whom 59.9% live in urban areas, and 44.1% in rural communities.[1] Fiji is primarily dependent on tourism and subsistence agriculture, with tourism as the principal economic driver and major source of economic exchange.

This chapter discusses the introduction of ecotourism into the rural hinterland of Fiji and the issues faced by the two villages and the tour operators involved. The area concerned (Figure 26.1) is in the Highlands of Viti Levu, at a point where two river systems, the Wainikoroiluva ('Luva) and the upper reaches of the Navua Gorge, flow through dense tropical rainforest. The focus is especially on the development of whitewater recreation, and how it is perceived in the villages of Nakavika and Nabukelevu, which have hitherto relied primarily on subsistence farming and, in the case of Nabukelevu, the sale of timber.

Ecotourism in Fiji

The tourism industry in Fiji has grown substantially over the past 50 years. During 1999, more than 409,955 overseas visitors spent F$600 million in foreign exchange (Fiji Visitors Bureau, 1999), compared to 2019, when visitor arrivals reached a new high of 894,389, a 2.8% increase from 2018.[2] Tourism is directly or indirectly responsible for over 40,000 jobs in the formal sector, where it is the most important business.

The recognition of "ecotourism" was relatively new when this project began in 1998. In February 1999, the government adopted an ecotourism policy, *Ecotourism and Village-based*

DOI: 10.4324/9781003230335-31

340

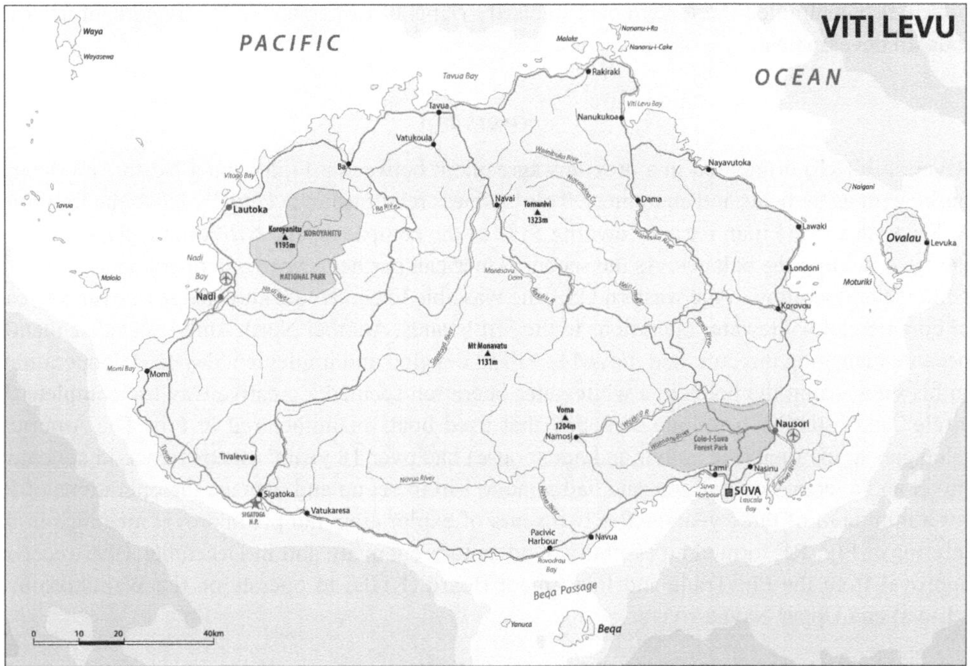

Figure 26.1   Highlands of Viti Levu, the largest island in Fiji.

Source: Bricker (author).

Tourism: A Policy and Strategy for Fiji (Ministry of Tourism and Transport, 1999), and the definition of ecotourism was:

> A form of nature-based tourism which involves responsible travel to relatively undeveloped areas to foster an appreciation of nature and local cultures, while conserving the physical and social environment, respecting the aspirations and traditions of those who are visited, and improving the welfare of the local people.

The policy incorporated five main principles. It recognized that ecotourism should complement, but not compete with more conventional tourism, that tourism should take second place to conservation, and at times, may be banned or restricted in areas considered especially vulnerable, and that successful implementation of the policy can be achieved only through social cooperation. In addition, information on village-based tourism should be centrally available and regularly monitored, and successful implementation of the policy would be based on the existence of strong and effective institutions.

Among other matters covered in the policy document was the need to develop an awareness of the physical and cultural environment. At the time Rivers Fiji began its operations, most tourism development had occurred in the islands and major urban areas, while the policy was also focused on rural areas and the need to enhance the quality of life in more remote villages, advocating fiscal, and other incentives to assist village-based projects and the development of necessary skills. The policy also committed the government to clarify rights of ownership and to avoid conflicts over tourism development.

Rivers Fiji

Rivers Fiji (RF) originated in a previous agreement between a Fijian and a North American to build whitewater boats and equipment. This business relationship eventually led to the formation of RF, with a sole Fijian partner (owning 51% of the company), and two Americans, resident in the USA holding the balance. As the second American partner owned and operated a successful whitewater operation in the western USA, he was able to contribute knowledge of over 50 years of commercial whitewater operations to the Fiji Islands. Another North American subsequently, became managing director, and moved to Fiji to develop and implement day-to-day operations. In his view, using the boats in a whitewater operation seemed a creative way to "complete the circle" and establish a unique company that used boats manufactured in Fiji. The American managers at the time (the author and her spouse) had over 18 years' of experience in adventure travel and overseas operations, and had gone to Fiji to set up and manage the operations of RF for a minimum of three years. After two years of exploration and pre approval meetings in the interior of Fiji, RF formed an ecotourism adventure company and in December 1997 received approval from the Fiji Trade and Investment Board (FTIB) to operate on the Wainikoroiluva ('Luva) and Upper Navua rivers.

The Role of Government

Permission from the FTIB is required before overseas investors can form a business in Fiji, a process that can take some time. As FTIB was unwilling to allow another business to operate, if it competed with or resembled any current tourism programs in the Navua River region, several enquiries were made to confirm no similar operation existed in either Namosi or Serua Province.

It was also necessary for RF managers to discuss the project with landowning patrilineages (*mataqali*) who controlled access to the rivers, and the company consulted with all the villages along each corridor. At the request of the provincial leaders (*Roko Tuis*) of each Provincial District office (Namosi and Serua), a second series of meetings with the villages of Nakavika and Nabuke-levu occurred after FTIB approval. The primary focus of these meetings was to secure signatures of approval by relevant *mataqali* for RF to begin business. The fact that these meetings were held again was somewhat confusing to each village, for they had assumed the first round of approvals was adequate.

At the time, RF also discussed the project with the Fiji Visitors' Bureau (FVB), the Department of Tourism, the Ministry of the Environment (including the Department of Forestry), local tour operators, and the World Wildlife Fund for Nature (WWF). In general, government bodies were unaware of the economic and ecotourism potential of whitewater recreation, but they supported the project, primarily because it would take tourism opportunities to remote areas.

Once established, RF continued to liaise with government entities, provincial offices, and land-owners and found an especially enthusiastic ally in the FVB. It was at the behest of the FVB that a meeting was held in July 1998 with several ministers and directors from the Departments of Envi-ronment, Tourism and Transport, Fijian Affairs, and Women and Culture, at which RF outlined a proposal for developing Fiji's hinterland, an area not included in the traditional sun, sand, and sea (3S) tourism for which Fiji tends to be better known. At the meeting, RF also explained the need for a conservation area on the Upper Navua River corridor.

During the meeting, which was a turning point in RF's attempts to establish its operation in Fiji, government officials emphasized the importance of continued dialogue with, and involvement of, local communities. Importantly, after more than 2.5 years of developing the RF ecotourism project, the suggestion was first made that the iTaukei Land Trust Board (TLTB) (then, the NLTB), a parastatal organization set up in 1940 to administer native land for the benefit of Fijian owners, should be involved, (see www.itaukeiaffiars.gov.fj). Accordingly, the TLTB states the following:

No iTaukei land shall be dealt with by way of lease or license under the provisions of the Act, unless the Board is satisfied that the land proposed to be made the subject of such a lease or license is not being beneficially occupied by the iTaukei owners and is not likely during the currency of such lease or license to be required by the iTaukei owners for their use, maintenance and support.

(TLTB, 2023: 2)

In retrospect, the omission of the TLTB from previous discussions had been fundamental. At the time of the establishment of the company, more than 80% of land in Fiji was communally owned by *mataqali*, and as such, is reserved for their special use and cannot be leased without their consent (FTIB, 1998). It was because of this that RF was advised to reach formal agreements through TLTB with *mataqali* in the areas they wished to use, to maintain long-term access to the river. This type of legal agreement, it was felt, was one way to protect and conserve the Upper Navua River corridor's biodiversity and provide consistent social and economic benefits to communities. Hence, a formal agreement was established as a "lease for conservation."

Local Communities and Landowners

The land along the Upper Navua River corridor is native land, and before anyone outside the landowning group could use it (commercially or otherwise), permission by the *mataqali* or other kinship groups would have to be obtained. When whitewater specialists from America visited Fiji to explore the rivers and assess their commercial potential, they therefore had to be introduced through traditional welcoming ceremonies (*sevu sevu*). Such meetings involved sitting with community elders, discussing the project and the fees to be paid to gain access to the river. Crucially for RF, any future investment required villagers' agreement on several points. First, RF should have long-term access to the river corridor for commercial whitewater activities (kayaking and rafting), and camping activities. Second, local communities, if they wish, would participate in running and guiding the operation; and, finally, the *mataqali* would support RF in conserving the river and the surrounding area. After these discussions, the villagers agreed, and the project continued.

Partnerships and Agreements: The Lease for Conservation

RF wanted an equitable partnership with the government and local landowners to enable the business to remain viable and productive for many years and bring benefits to local people and the conservation of the river corridor. At the same time, the structure and interests of villages and the needs of the communities varied, and arrangements with *mataqali* also differed according to circumstance. Agreements cannot be "rubber-stamped." In the case of the Upper Navua River, for example, formal leases were necessary to protect RF's investment in roads, bridges, and product development, as well as from commercial mining and logging encroachment. Arrangements with the village of Nakavika were arrived at during the welcoming meetings and accepted through less legalistic, but "shared understanding" by RF, the village, and the Provincial Council. This process involved the presentation of a *tabua* (whale's tooth) by RF to the village of Nakavika and associated *mataqali*. Traditionally, presented to distinguished guests, or exchanged at weddings, funerals, and birthdays, the *tabua* is also presented at the conclusion of agreements and contracts.

The Upper Navua River

This element of the operation, described here as the Upper Navua River project, and based just outside the village of Nabukelevu, was created through agreements involving the TLTB (which managed the lease on behalf of the landowning *mataqali*), the villagers, and RF.

Commercial logging and mining threats to the pristine river corridor were then, and are still ongoing. As a result, RF decided to proceed with a long-term lease for conservation, meant to conserve the river corridor and protect benefits for local communities, and the quality of the river experience. Such a lease was unusual, as the lessee was not to be engaged in extracting resources, or a typical tourism lease, which was building a resort project. While approvals from the *mataqali* were quickly obtained, prolonged negotiations with the TLTB delayed the start of the company's operation by 18 months.

According to TLTB, the delays occurred because of the peculiar nature of the lease request. It was finally issued in November 1999, 18 months after RF had been advised to approach TLTB and 22 months after the FTIB had given permission for RF to begin its operation. The lease of the road, which was for 50 years, involved a one-off payment to TLTB, which was distributed, after the deduction of administrative costs and other·fees, to three *mataqali*. In addition, RF was to pay an annual fee, in return for which it could operate whitewater programs and have exclusive use of the road.

At the time of writing, the law of Fiji permits logging within 30 meter of waterways, and roads in the whitewater rafting area sometimes come within 2 kilometer of the river's edge. It was discovered that a proposal existed for a road to the river to extract gravel for other logging roads in the area, and the proposed extraction site was directly opposite one of the largest waterfalls on the river and in the middle of RF's proposed operation. Such threats from the logging industry led RF to seek the establishment of a linear biosphere reserve in future lease agreements.

The concept behind the lease for conservation was to protect biodiversity, while benefitting local communities. There was also a safety concern. The proposed reserve would considerably reduce the risks to RF's clients. There would also be long-term benefits for both RF and the *mataqali*, as it would protect the canyon from felled trees and other debris from logging, enhance the tourist experience, ensure the corridor was kept pristine, enable RF and landowners to evaluate the nature and growth of ecotourism in the area, and provide jobs for rangers and guides. The concept of a protected linear biosphere reserve had been approved by the relevant *mataqali* along the river and the proposal was finally accepted by the TLTB in November 2000.

The Wainikoroiluva ('Luva) River

RF engaged in another arrangement with *mataqali* of the village of Nakavika, where the Wainikoroiluva ('Luva) River is located. The TLTB was not involved, yet agreements still required signatures of approval from heads of all the relevant *mataqali*. While many discussions centered on the initial project, subsequent meetings focused on payments for daily operations. Community meetings were held to outline a plan of action and proposed activities operated by RF. The villagers preferred to retain the existing agreements with RF, arrived at by "traditional" negotiations.

Therefore, all parties, including the Provincial council, the Head of the Nakavika village tribe, and associated *mataqali,* signed documents confirming RF as the sole operator for the sections of the river outlined in their program.

Once support from the villagers and the Provincial office had been secured, RF felt confident about investing further in training guides, developing trails, building thatched huts (bures), marketing, and program expansion. Indeed, after nearly two years of operation, the company expanded its operations to include a two-day whitewater experience and overnight camp. However, soon after agreement with the villagers had been reached, RF's position was challenged by another operator interested in the same stretch of the river. Unfortunately, the Chief who organized the original agreement, passed away, and the original agreement was not honored. RF decided to move its operations to a different section of the river, the Middle Navua as a result. As part of the original agreement, fees were paid to landowners for areas incorporated into the program of activities, for disembarking tourists at the village (put-in and take-out fees), for the use of sites for lunch, for a trek to a waterfall, and for take-out areas (where RF ends the trip). In the end, RF stopped operations in Nakavika, as the program was splitting the village, causing arguments amongst various *mataqali,* and where funds were channeled was debated. The original agreement had been based on funds directly benefitting the entire community, yet the situation no longer appeared to be built on this premise. Based on consultation with the guides and managers from Nakavika, RF ended its program on the Wainikoroiluva.

Monitoring the RF Operation

At the time of writing, one-day trips continue to run at least. Three times a week on the Upper Navua. The value of the agreements lies not in the paper on which they are written, but in the relationships between the tour operator and the villagers. If they disapproved of the actions of RF, even with the most formal of agreements, villagers could always resort to roadblocks and hinder the operation in the beginning of the relationship. However, it was agreed that liaison between RF and the communities was a critical position, in ensuring consistent communication, and building trust. Building such relationships takes time and understanding on the part of the operator and the villagers. In several instances, for example, villagers had to educate the tour operators about the culture of the village, and the villagers in turn needed to be informed about the ways of tourism and tourists.

One method of furthering this process of mutual understanding was the use of focus groups in both villages, and these were started at the end of 1998. In Nakavika, they were set up six months after commercial operations were under way, while on the Upper Navua Gorge, they commenced with the first training trips from the village of Nabukelevu. Initial meetings weld in both villages to seek permission, discuss the purpose of the study, introduce research assistants, and schedule meetings, and someone from each village was recruited to assist in scheduling and organizing visiting groups. After discussions with the chiefs, the *Turaga Ni Koro* (the village speaker and representative of the District Council), elders, and research assistants, it was decided to split focus groups by gender and age. Men and women were divided, and two age groups, covering the range 18–30 and 31 years and after, were set up. The local assistant in each village recruited volunteers from each *mataqali* and four focus groups were formed, numbering from 10 to 16 people, from each village, totaling 45 in Nakavika and 26 in Nabukelevu.

Interviews in the local Fijian dialect were conducted by two research assistants from the University of the South Pacific, and groups met twice a week at times convenient to the villagers, in the evenings, and at weekends. Discussions were recorded and later translated into English. The results show how villagers perceived tourism, its costs, and benefits, how they viewed what they had to offer tourists, and their perception of other issues that follow from opening the region to tourism (Bricker, 1999).

Perceived Village Attractions and Attitudes

The results of these focus groups revealed that both villages felt that the traditional Fijian way of life was unique, and would be appreciated by tourists. As one man remarked, *"tourists come to experience our way of life, behavior, and warm hospitality of which we are renowned for."* Another villager discussed the importance of showing visitors "their culture":

> For those of us who live in the village [pause] there exists the proper and genuine Fijian way of life or norms and behavior which could be shown to tourists. For example, the mekes [dances] - these are traditional genuine mekes, which is unique here, and one that we learn which could be shown to tourists. Another example is the preparation of the traditional yaqona ceremony. The tourists also have their own unique culture and ceremonies, which we often see on TV So we can also show them our culture and ceremonies.

At the time of writing, guests do not necessarily visit villages. The villagers discussed this and referred to guests coming to see their way of life:

> Another thought is for the tourists to come to the village and experience the Fijian way of life. They get to taste traditional Fijian delicacies and come to know some Fijian herbal medicine, which might be beneficial to them. They [tourists] can't understand or appreciate our culture and traditions if they just arrive and depart from our village. They must spend time with us. What I'm trying to say here is that they must take some treasured memories with them when they depart from here … They can't do that if they don't spend time with us!

Nabukelevu villagers also believed it was important to share the Fijian "way of life" with tourists and implicitly to recognize cultural differences:

> Here in Nabukelevu, our Fijian way of life is very much intact - that is, love and respect for all, good interpersonal relationships. When the tourists visit our main urban centers, it is still like the life they live in their country of origin. These tourists want to see something different, experience, taste, touch and feel something different. They want to come to villages for their holidays - to experience something different.

In addition to whitewater rafting and kayaking, villagers listed several activities they felt would be appropriate to offer guests. Examples included visiting caves and ancient fortresses, birdwatching, hiking, or trekking, dances and preparation of traditional foods, handicrafts, storytelling, myths, legends, and horse riding. Villagers also considered their natural surroundings to be an "attraction" for tourists:

> Another aspect of their visit is that in their countries they are surrounded by white concrete jungle. When they come to our place they see the mountain ranges, breath of fresh air, the unbounded and pristine forest where they will realize that it has been untouched since the creation of the world.

Up to the time of writing, river trips have been RF's primary focus, but there have been occasions when visitors have paid fees to stay overnight in the village, lodging with acquaintances of RF staff. In response, the company set up a camp area near the river on land belonging to one of the *mataqali*. Villagers expressed an interest in developing the site: they plan to provide tents for accommodation and traditional meals cooked in underground ovens (*lovo*).

Perceived Benefits of Tourism

Villagers in Nakavika and Nabukelevu identified several communal and individual benefits that could result from tourism development.

Increased revenues were clearly important. For example, at the time of writing, RF pays a fee for every guest traveling down the Upper Navua River. This is divided amongst every *mataqali* along the river corridor. From the information given at the focus groups, funds have been used to build schools, improve the village facilities (for example, by upgrading footpaths), and develop spin-off businesses. As one community member reported:

> …with the development of tourism, the potential is there to achieve greater things, such as going into new business ventures, trucks, and shops for example. Most importantly, is our quest for giving our children the very best education.

Other village groups also benefit directly and indirectly from the tourist dollar. Rugby teams, youth groups, and school groups have worked on individual projects to earn cash for fund-raising projects. Projects identified by communities, such as healthcare and building improved water systems have been facilitated by RF and partnerships wishing to assist communities in the rural highlands.

Villagers recognized that ecotourism was a way for them to improve their standard of living:

> I think the proceeds we receive from tourism will dramatically transform our village to a better one in terms of the structure, fittings and fixtures and design of our homes. We all know it is only money that will generate development, nothing else. Therefore, in terms of the future development, I think it will move from good to better.

Another remarked, "*I welcome the development of tourism to our village because we will I be able to further develop our school, our village, and our standard of living.*"

Further employment and training were seen to derive directly from tourism. At the outset of the project, RF worked with *mataqali* to rotate opportunities for employment as whitewater guides. In the beginning, meetings were held with village elders to discuss guidelines, job descriptions, and anticipated days of work per month. Then the elders selected individuals on a rotational basis (between *mataqali*) for training, a decision which, according to one of the guides, brought the *mataqali* and the community more "together". Those interested and ultimately selected, receive a four-month whitewater guiding course, which includes swift water rescue, and in-service training related to food preparation, customer service, and ongoing training with the Red Cross cardio-pulmonary resuscitation first aid training courses. As noted later, the transfer of these trainings remains beneficial for professional development, as well as during crises such as the COVID-19 pandemic.

The *mataqali* were pleased with the employment of their young men, and the guides took pride in their role. Although the new work opportunities required them to manage their time to meet the demands of the job, families, and friends, which clearly prompted a process of adjustment, their kin, and other villagers considered the results to be positive, suggesting that guides spent less time around the kava bowl (which involves communal drinking of a Fijian non-alcoholic brew made from kava, (*Piper methysticum*)) and were thus more reliable and responsible. For their part, the guides realized that their opportunities to socialize were reduced, and sometimes they even had to miss a game of rugby (which is hugely popular throughout Fiji). However, they liked earning money because it allowed them to visit relatives, buy goods, and gain access to preventive medical care. Villagers also acknowledged that the guides spoke English more proficiently.

According to some villagers, contact with tourists had involved them in a more general learning experience, and focus groups at the start commonly stressed the benefits of cultural exchange. As one group member remarked:

> I really appreciate the tourists' visits. I often ask them where they come from and it fascinates me just to look at them, especially their smiling faces. When I see all these tourists from all over the world it's like I'm living in their place of origin.

Indeed, because of meeting tourists, some individuals expressed interest in one day, themselves, traveling overseas:

> I know that these tourists come from developed countries whereby they have good jobs and enjoy high salaries. If they need some people to be employed in their home country, then it would be nice for us to go and work there.

Villagers also felt there was an opportunity to learn from other cultures:

> I think tourists will bring a lot of good things and changes to our community. We will learn from them - new ways of doing things, perhaps.

Finally, villagers felt that tourism could benefit their children and future generations because it was sustainable. This view is summarized by a young man from Nabukelevu, who had experienced considerable logging over the last ten years:

> Tourism is very different from other industries such as the logging industry or the fishing industry. The logging and the fishing industry will take away from our land our resources, in a way depriving us of our resources. Whereas for tourism, tourists would not take away any of our resources. They will just come, watch, and experience our resources that also include our culture, and us the people. When the logging company departed from here, they departed with our forests. The land is stripped bare. When the fishing company departed from our waters, they departed with our marine life and our waters are scarce or even extinct with marine life. Our resources must be safeguarded so that our future generations can also utilize them.

Clearly, villagers from the highlands approached tourism with a positive outlook. According to Taukeinikoro (1999), close ties among the villagers, and the monitoring of all developments by RF, created a sense of trust and expectation in the benefits of tourism. The pace of development was slow in the beginning, and communities took time to adjust, and because guides work generally on a part-time basis, they are still able to attend to village and family needs. Finally, as the guide's role requires them to interpret surroundings familiar to them, they develop confidence in their presentation and pride of place in their home environment.

The Perceived Costs of Tourism Development

Residents were also aware, though, that tourism could have negative impacts. They felt that tourists might disrupt village life and offend local dress codes, and that tourism could lead to an unequal distribution of wealth. They clearly recognized that the regular routines of village life would be put under some pressure by tourism and emphasized the need for discussions within the communities over tourist activities, the time and the duration of visits, and the rules that tourists should follow. As one male villager remarked:

> [With] further development of tourism in our village we have to meet and discuss the days that the tourists should visit our village so that it does not clash with other village activities. We welcome the development of tourism to our village; we just have to decide the time and day of their visit.

It was felt that tourists should not visit the villages every day, and commonly argued that Sunday "should be free … because it is a day of worship", and "Monday is a day devoted to community work". When RF first started in Nabukelevu, Mondays and Wednesdays were set up as training days for the guides. However, the village then levied a fee on the guides, whose time spent on the river took them away from their communal obligations. As a result, training activities were quickly transferred from Monday to Friday.

Community members required advance notice of tourist visits and expected to greet them with the traditional ceremony (*sevu sevu*). As one villager remarked, "People do not like tourists who walk into the village unannounced or they [tourists) do not have the courtesy to ask for permission from the *Turaga Ni Koro*".

For most villagers, rules regarding dress and behavior are important.

> Tourism would be a problem if the tourists just roam around our area and that there are no guidelines or rules for them to follow. If we [in Nabukelevu village] stick to our way of life and the tourists follow what they are supposed to do and how they are supposed to behave, then I do not see any problems.

It was generally felt that young people might imitate tourists and thus come into conflict with "the Fijian way of life".

> I would like to emphasize the point of dressing to developers- to wear clothing suitable or proper for the village community and behavior that they display not to influence the young children or youth that may lure them away from the norms, values, morals that they were brought up with and which holds society together.

It was also believed that such problems could be overcome by establishing rules for visitors to the village, asking them to wear a sulu (sarong), and if necessary, to avoid actions that might influence young people unduly.

Finally, villagers were concerned that incomes derived from tourism should be equally distributed. As one remarked:

> Tourism is the source of income for tour guides and for the village. Money has both good and bad advantages and is seen to be the root of all evil. With increasing wealth within the mataqali and village, there comes a time when there will be disharmony [and] this disharmony will lead to the barring of tourists to the village and the eventual closure of such enterprise.

Mataqali are compensated whenever their land is used by tourists, and as indicated earlier, cash obtained from porter work is shared. Equity was achieved by careful and equal distribution of training opportunities and employment to all the *mataqali* involved along the river corridor, as designated by the lease agreement.

Infrastructure and Access

Mataqali were aware that the development of tourism necessitated a basic level of infrastructure, and that access to the area was important if the ecotourism venture was to succeed. When the physical situation of the villages is understood, such concerns clearly have some validity.

The road providing access to villages near the Upper Navua River is approximately 12 kilometer from the Queen's Highway (the main road which, along with the King's Highway, encircles the island), and its condition is often poor, as it is used periodically to transport timber. It normally takes about an hour to reach the landing point on the river, but occasional landslides and severe mud conditions can make access difficult. Villagers felt that if tourism was to succeed, government assistance was essential, especially with roads, water, schools, and clinics:

> Government should improve our road, tar-seal our road because our road is very rugged. It could hinder the coming of more tourists to our village.
>
> Our village needs the improvement of our water supply system. We have discussed and did some fund raising and were able to raise $12,000. Government contributed $32,000 for this project. As of today, nothing has materialized, and we have not contacted Government's Water Supply Department - thus the delay.

Many villagers were especially concerned about schools and health care:

> Government should improve the condition of our school in terms of building, textbooks, and school equipment all in the effort to improve the education standard of our children.
>
> We need to improve the condition and facilities of our Dispensary.

Reflecting Forward...

It has been suggested that "alternative" tourism that develops slowly, and leaves control with local people is less likely to have negative implications (Butler, 1991), and this case study indicates that planning for a slow start, attention to education and training, maximizing local benefits, and continued evaluation and feedback are crucial for ecotourism to be successfully introduced into a rural area that has no previous experience of tourism.

Planning for a Gradual Start

Clearly, delays that prevent any growth at all are commercially unacceptable, but research has demonstrated that when growth occurs gradually, some negative social impacts are minimized. (Butler, 1991; Stevens, 1991). The speed at which a project develops depends ultimately on the interest and understanding of the local community. An unusual drought during the first year of operation meant that the RF project grew slowly in the beginning. However, this allowed both the village and RF to adjust the training and timing of the operation. It also enabled villagers to assess the nature of tourism and the benefits it brought to the community, and to make plans for future developments-it provided a solid foundation from which to grow.

Education and Training

While the human impact is considered one of the most destructive forces within protected areas, advocates of ecotourism claim that education can minimize tourism's negative consequences (Pietri, 1992). Pre-departure information can facilitate more effective management of tourists in ecotourism programs, and field personnel, especially, have a crucial role to play. "Tour operators and guides must be particularly aware of, and sensitive to the areas they promote because they develop the itineraries, and are responsible for their clients" (Ziffer, 1989: 31). Wight (1993: 3) suggested that ecotourism should involve "education among all parties – local communities, government, non-governmental organizations, industry, and tourists (before, during, and after the trip)" (emphasis in the original).

Throughout its history, RF has taken an active role in educating the government, local communities, youth, and other non-governmental entities about whitewater programs that they have developed. Visitors are provided with pre-departure fact sheets, while the training manual used in the company's whitewater school has been distributed to guides, some of whom have attended eco tourism workshops, subsequently disseminating the information in their village. Videos and social media on the river project and ecotourism, and information on the proposed conservation area have been a part of meetings with landowners and government officials; RF produced an interpretive guide for its area of operations. The company believes that this type of activity has enormous positive effects on the overall development of the programs.

While the COVID pandemic was likely the most challenging time in the history of the company, throughout its history and evolution, Rivers Fiji has been involved in activities that support its local communities, and build resiliency and sustainability into its DNA. For example, the company has engaged in numerous efforts to support ecotourism and the sustainability of tourism within the country including:

- Outreach and conservation awareness programs in partnership with Nature Fiji, for youth and families along the Upper Navua Conservation Area (UNCA);
- Awareness programs for our UNCA through village visitation, workshops, and annual rafting invitations with Mareqeti Viti to assist interpreting the unique ecosystem of the UNCA;
- Employment and training opportunities for young men and women in whitewater rafting and guiding;
- Assist communities in improving their freshwater systems and other infrastructure needs;
- Organizing foreign doctors and local doctors for annual village medical clinics and improved access to healthcare;
- Co-Founder of the Duavata Sustainable Tourism Collective; and,
- Conducting swift water training for cyclone events and associated flooding.

The company is also working on a succession plan, training the next generation of managers, guides, and owners for the future. The National Government has used the company as a model for ecotourism and conservation. For example, as an outcome of this ecotourism enterprise,

- RF created the 1st lease for conservation, over 615 hectares of privately owned/native lease conservation area known as the Upper Navua Conservation Area (UNCA) supported by tourism dollars;
- The UNCA became Fiji's first and only Ramsar Site, (wetland of international importance) in 2006;
- RF is a member of the National Wetlands Steering Committee;
- RF has been invited on many occasions to governmental or NGO's meeting either as RF or Duavata;
- RF was invited to the RAMSAR Convention meeting in Japan in November of 2017.

Maximizing Local Benefits

It is commonly accepted that ecotourism can provide long-term benefits "to the resource, to the local community, and to the industry (e.g., benefits may be conservation, scientific, social, cultural, or economic)" (Wight, 1993: 3). However, although RF's company's philosophy is to "leave no trace" of its operations, the introduction of a new type of tourism product has caused managers some concern. As the founding Managing Director commented:

> Once we open these corridors up to the world, we open them up to the good and the bad. As a result, we believe RF has an inherent responsibility to protect and conserve the resources in which we operate (February 3, 1998).

Rivers Fiji's initial mission, to encourage conservation, worked well to create a lease for conservation and the subsequent biodiversity "buffer zone" on the Upper Navua Conservation Area. This was the first of its kind in Fiji, where a private entity leases an area simply to protect it from extractive use, and preserve the pristine quality of the experience, thus addressing economic, socio-political, and environmental sustainability of the project.

Thus far, villagers have combined tourism with farming, and the logging activities of one company also continue. As the focus group data indicated 24 years ago, they believed that tourism could bring additional economic and other benefits to their communities, and these were likely to increase in the future. Women's groups, sports organizations, and youth groups all expressed interest and now engage in special projects and ways of earning funds, and the women's groups also wanted to develop arts and crafts for sale to visitors. During the COVID-19 pandemic, the fact that many employees retained their connection to their farms, provided a level of resiliency, during the worst tourism downturn in the history of RF.

Conclusion

This paper argues that partnerships with rural communities can succeed if ecotourism operators are knowledgeable and support the principles of ecotourism. Can ecotourism work if one company takes responsibility – however willingly – for environmental conservation and guardianship when it is simultaneously involved in running a business for profit? In fact, several resorts in Fiji take responsibility for the island environments that they lease or own, improving them and protecting them from future damage. This should be of no surprise, for in nature-based tourism the environment is the key business asset. If it is damaged, the business is at risk. At the same time, in Fiji the long-term health of the environment is usually the responsibility of the *mataqali* and, with or without outside assistance, only the landowners can steer a sustainable course of action.

If Fiji is to continue with sustainable tourism development successfully, it must evaluate, regulate, and monitor competition and control the type of use that rural areas can potentially support. Guidelines for tourism operations in natural areas must be developed, with inputs from all stakeholders (including mataqali), and should be followed by a comprehensive plan to monitor all ecotourist activities on a regular basis. Whitewater recreation, for example, requires extensive knowledge and skill to operate safely, but successful operations are always likely to be imitated, often without respect for the environment and safety procedures implemented in the original product development. As this case study shows, organizations such as the TLTB must be able to evaluate ecotourism projects and advise the government according to recognized criteria of best practice. If this occurs, joint and effective responsibility becomes possible, and all is not then left to the good will and social conscience of the tour operator.

Finally, it can be concluded that partnerships in ecotourism cannot be based solely on good working relationships at the local level. They must also be supported by government and regional entities that monitor and regulate tourism. In the end, sustainability can be achieved only if all stakeholders cooperate in the endeavor and thread a common interest in sustainability into all new developments.

Acknowledgements

This case study is the result of the hard work and dedication of numerous individuals. The authors would like to thank the local village partners such as the villages of Nabukelevu and Wainadiro who also part of the Rivers Fiji rafting operation. A special mention to the landowning units or *mataqali*, and traditional landowners of the Upper Navua Conservation Area. They are *mataqali* Navau, Vunitavola, Sauturaga, Vunimoli, Naviaraki Korovisilou, Naviaraki Nabukelevu, Ketenatukani, Cawanisa and Sauturaga Nabukelevu. Their knowledge, vision, and enthusiasm for ecotourism development in the rural highlands forms the basis of this case. Vinaka vakalevu is also due to the RF Management and staff, 100% Fijian operated, who make the magic happen for guests day in and day out, to the research assistants at the University of the South Pacific who transcribed and interpreted their dialect, and for the hard work and vision of George Wendt, may George eternally rest in peace, Nate and Kelly Bricker, Colin Philip, for his vision and conservation ethic, may Colin eternally rest in peace, and Glenn Lewman of Sotar. And to one founding member of the RF team, who so aptly stated, "conservation is something we have to wake up to every day, and say, we have to protect this place, it is a fight that is never over."

Notes

1 Ibid.
2 Fiji Bureau of Statistics.

References

Bricker, K. (1999). *Sustainable Rural Tourism Development: A Case Study of Ecotourism in the Highlands of Fiji.* (Unpublished manuscript), University of the South Pacific, Suva, Fiji.
Butler, R.W. (1991). Tourism, environment, and sustainable development. *Environmental Conservation* 18(3): 201–209.

Duavata Sustainable Tourism Collective. (2023). Retrieved from https://www.duavatasustainabletourism. org/, Accessed January 23, 2023.

Fiji Visitors Bureau. (1999). *1999 Fiji International Survey*. Ministry of Tourism and Transport: Suva, Fiji.

FTIB. (1998). *An Investor's Guide to Fiji. Fiji Trade and Investment Board*, Suva, Fiji.

Harrison, D. (Ed.) (1999). *Ecotourism and Village-based Tourism: A Policy and Strategy for Fiji.* Ministry of Tourism and Transport: Suva, Fiji.

iTaukei Land Trust Board (TLTB). (2023). Retrieved from https://www.tltb.com.fj/About-Us/Governing-Legislation, January 2, 2023.

Mara-Nailatikau, A.I. (1999). *Minister of Tourism Address to the Forum*. The 1999 Fiji Tourism Forum, Nadi, Fiji.

Pietri, O.A. (1992) *The Ecotourism Dilemma*. The Ecotourism Society, p. 3.

Stevens, S.F. (1991). Sherpas, tourism, and cultural change in Nepal's Mount Everest region. *Journal of Cultural Geography* 12(1): 39–58.

Taukeinikoro, K. (1999). *Nakavika: A Guide's View of Tourism.* (Unpublished manuscript) University of the South Pacific, Suva, Fiji.

Wight, P. (1993). Ecotourism: ethics or eco-sell? *Journal of Travel Research* 31(Winter): 3–9.

Ziffer, K. (1989). *Ecotourism: The Uneasy Alliance*. Conservation International and Ernst and Young, Washington, DC.

27

INDIGENOUS HANDICRAFTS BASED DOMESTIC TOURISM IN BANGLADESH

Easnin Ara and Md Ariful Hoque

Introduction

Moving beyond mere "sun, sand and sea destinations" (Sinclair, 2003, p. 141) tourists have been revealing their interest in first-hand and authentic experiences of local communities' cultures, ethnicity, livelihood, and history (Butler & Hinch; 2007; Smith, 2016). Particularly different elements of Indigenous ethnic cultures (e.g. handicrafts, dances, songs, festivals, foods) have received significant attention in this regard. Such a growing interest of tourists towards Indigenous cultures and lifestyles paves the way for Indigenous people's involvement in tourism in different contexts. Indigenous people's involvement in tourism has long been emphasised by different governments and development-oriented organisations primarily as a tool for Indigenous socio-economic development through the promotion and commercialisation of their distinct cultural elements.

Within this process, handicrafts have received tourists' attention, not only being seen as symbolic objects and/or souvenirs of the memory of their visit, but also being representative of a distinct ethnic culture (Swanson & Timothy, 2012; Husa, 2019). Studies can be found addressing the handicrafts-tourism nexus primarily from the humanistic perspective, focusing on the roles of different human actors (e.g., tourists, entrepreneurs, artisans), and such a predominant humanistic focus also calls for understanding the role of non-human actors in tourism (Ara et al., 2022; Ren, 2011). Various tourism development studies, including those focused on the Indigenous culture-based tourism, have tended to focus on the appeal of this type of tourism to international tourists. However, our study aims to understand the role of handicrafts, particularly Indigenous handicrafts, in developing domestic tourism in Bangladesh. Along with its already developed nature-based tourism, Bangladesh – a home of many small ethnic groups – is currently observing a growing interest for cultural tourism, focusing on the unique culture and life-styles of the different Indigenous ethnic groups. Moreover, there is an explicit policy emphasis by the country's government in terms of developing tourism that prioritises ethnic cultures and crafts. Among the different small ethnic groups in Bangladesh,the Manipuri and Bawm communities have particularly received popularity through their traditional handicrafts-centric tourism involvement. We adopted the lens of ANT to understand the enacting capacity of non-human actors, handicrafts in particular, in terms of domestic tourism development. A qualitative case study approach was applied with multiple modes of data collection to gather empirical evidence in this regard.

355

DOI: 10.4324/9781003230335-32

Tourism and Indigenous People

Indigenous people's involvement in tourism is no longer a new phenomenon. The shift from mass tourism to alternative forms of tourism has acted as a catalyst in this regard. Within the broader domain of cultural tourism (Smith, 2016), the involvement of Indigenous people in tourism has been termed 'Indigenous tourism', a phrase which is also often interchangeably used with tribal tourism, aboriginal tourism, ethnic tourism, and anthropological tourism in different contexts (see Zeppel, 2006; Sofield & Li, 2007) due to its anthropological roots (Weaver, 2010). The basis of such a form of tourism is often the distinct cultural elements of different Indigenous ethnic groups.

The potential for developing tourism centring on Indigenous culture has further brought to the fore the attention of different stakeholders (e.g. Indigenous communities, governments, aid agencies, and non-governmental organisations (NGOs)) to consider this form of tourism as a tool for such communities' development. This is especially so in many developing areas, where Indigenous people often suffer from multiple deprivations (e.g. economic and socio-cultural), and where tourism involvement has been promoted in terms of addressing their poor conditions (Zeppel, 2006; Kennedy & Dornan, 2009). While economic benefits primarily in terms of creating employment and earning opportunities have been portrayed as the key motivators for Indigenous involvement in tourism, several socio-cultural benefits (e.g. preservation of traditional cultural elements, reinforcing indigenous identity, reconciliation between indigenous and non-indigenous people) have also been advocated (see Cassel & Maureira, 2015; Higgins-Desbiolles, 2003; Hoque et al., 2020; Whitford & Ruhanen, 2014).

It is clear, however, that things are not as straightforward as they may seem initially. There exist a number of controversies in terms of the involvement of Indigenous people in tourism, including the frequent limited participation and voice of Indigenous people, the dominance of non-Indigenous people, the loss of traditional lands and displacement; issues of traditional gender roles, and leakage of benefits, as reported (see Coria & Calfucura, 2012; Hipwell, 2007; Ishii, 2012; Melubo & Carr, 2019; Phommavong & Sörensson, 2014; Zeppel, 2006). Moreover, such tourism has often been developed with external support (e.g. aid agencies and NGOs) in many cases - especially in developing contexts - primarily focusing on international tourists; a specific emphasis that has been criticised as a narrow focus in this regard (Pawson et al., 2017). These issues further highlight the need for understanding the role of Indigenous culture as a focus for domestic tourism development. Moving forward with this theme, among the different cultural elements of Indigenous ethnic communities, the next section explores the handicrafts-tourism nexus.

Traditional Handicrafts in Tourism

Handicrafts have their origins as objects for daily requirements or more broadly to meet the practical needs of different communities in their day- to- day life (Swanson & Timothy, 2012; Islam & Carlsen, 2016). Within this role, handicrafts have also become a platform by which different Indigenous communities are able to express and maintain their cultural traditions while producing those items that help to promote their ethnic identity as well (Cohen, 2016; Novelli & Tisch-Rottensteiner, 2012; Smith, 2016). Henceforth, handicrafts and traditions are perceived to be entwined with each other, through the historical and cultural expressions of the artistic producers (Crainic, 2014). Schaffer (2011) argues that handicrafts fulfil the role of a platform, representing the time and history when they took form. Linking tradition with handicrafts, McCarthy (2014), and Swanson and Timothy (2012) observe that handicrafts get their meaning from their connection with place and time. Furthermore, handicrafts have been considered as a combination of both objects and concepts (Kendall, 2014; Margolis-Pineo, 2016; Risatti, 2007). On the one hand, the exposure of handicrafts as objects emphasises the function or the materials through which they are produced such as wood, clay, glass, metal, fiber, etc. On the other hand, the notion of handicrafts as a concept emphasises the tradition or essence relating to the object. Hence, the meaning of indigenous handicrafts is embedded equally in the traditional (what something is) and formal (how it looks) aspects of a particular distinctive ethnic group.

Acknowledging and extending this view Stecher (2014) suggests that handicrafts need to describe the work of particular community members who have specialised technical skills in producing objects that reveal the expressive and aesthetic essence of their tradition, and which not only serve a practical function but also have ceremonial and spiritual connotations. Hence, the traditions of a particular group/community are entangled as a core essence for Indigenous handicrafts where tradition bears a sense of historical and cultural values (Cohen, 2016; Crainic, 2014; Shuzhong, & Prott, 2013) .Such connections of handicrafts with different groups/communities have established their relationship with tourism through tourists' growing interest in experiencing cultural elements of different distinct communities (e.g. Indigenous and/or ethnic) (Smith, 2016; Xu et al., 2014; Yang & Wall, 2014). In this regard, handicrafts can be found within the 12 cultural components (e.g. history, music, handicrafts, work, dress, architecture, language, religion, education, customs, gastronomy and leisure activities) identified by Ritchie and Zins (1978) which could reveal unique expressions of ethnic groups, and create appeal for tourists too. In addition, Smith (1996) emphasised the 4 H's – heritage, habitat, history and handicrafts – as essential factors for developing Indigenous tourism which in turn illustrates the importance of handcrafts for tourism development.

Within this panorama of tourism, Indigenous and/or ethnic handicrafts generally find their position as a communicator between host and guest – people differing in their lifestyles, cultures and habitats – (Chang et al., 2008; Yan, 2019) reflecting the entanglement of handicrafts with tourism (Ara et al., 2022). Tourism invokes memories, and souvenirs in the form of handicrafts of a host community or country fulfil this need through a combination of ethnic and economic viability (Chang et al., 2008; Husa, 2019; Xu et al., 2014). Adding to this, 'Place' has also been considered an important form of viability that validates the linkage between traditional handicrafts and tourism. In this regard, several researchers acknowledge handicrafts (in the form of souvenirs) as a thematic identifier of the place visited (see Hume, 2014; Peters, 2011; Swanson & Timothy, 2012; Timothy, 2005).

While an entwined relationship between tourism (development) and handicrafts seemingly has been observed from the extant literature, the current study aims to clarify and reconfirm that relationship adhering to a non-anthropocentric approach in the context of two Indigenous ethnic communities in Bangladesh.

Understanding Non-human Agency – Actor Network Theory (ANT)

The prime focus of our study was to understand the agency of a non-human actor, in particular indigenous handicrafts, calling for a theoretical lens that could facilitate the understanding of the handicrafts' influence over other actors to develop and/or promote tourism (in this case, domestic tourism). In this regard, ANT has been adopted as the theoretical underpinning because of its novel ways of understanding that the ability to 'act' not only resides in humans but also in non-human actors (Ara et al., 2022; Beard et al., 2016; Jóhannesson et al., 2015; Ren, 2011). Henceforth, the notion of networking or 'actor network' (Middelveld, 2012) has been considered. ANT asserts that a network is the outcome of a series of actions which bring different entities (both human and non-human) together through various interactions associated with certain goals (Ren, 2011;

Van der Duim et al., 2014). This practice of networking was found helpful in terms of identifying the relevant actors within the handicrafts-tourism nexus, while providing an understanding of the relational capacity of handicrafts to bring those actors together.

Bangladesh Context

In Bangladesh, Indigenous people are generally known as *Adivasi* – a Bengali term. However, the identity of Indigenous people in Bangladesh is a contested phenomenon (Barkat, 2016). Instead of being recognised as 'Indigenous', they are referred as 'ethnic sects and communities', 'tribes', and/or 'minor races' (Kapaeeng Foundation 2013). Also, there is controversy over the number of Indigenous groups in Bangladesh (Minj & Khakshi, 2015). While the focus in this chapter is not on the issues related to that debate, we feel it is important to provide a broader understanding of the context. Bangladesh's Indigenous communities reside both in hill tracts (Chittagong Hill Tracts - CHT) and in plains lands. Each Indigenous group is unique in terms of lifestyle and culture expressed through their traditional crafts, celebrations, customs and rituals, languages, costumes, dances, songs, and food that significantly differ from the mainstream Bengali population (D'Souza, 2013; Hoque, 2020). Such cultural diversity has made these communities significant tourist attractions in Bangladesh. Consequently, several Indigenous communities have become involved in tourism utilising their different cultural characteristics, primarily handicrafts. Furthermore, while the country records relatively insignificant numbers of international tourists currently, it has been experiencing a boom in domestic tourism in the last decade (Ara, 2020; Fabricius et al., 2017). For this research into Indigenous handicrafts, two different Indigenous ethnic communities – one from the hill tracts and the other from plains land – were chosen as case studies to explore the agency of handicrafts in Indigenous tourism.

Both the settings presented here are part of two larger research projects, where research in the first setting was conducted by Hoque and in the other by Ara. Fieldwork in the research settings was conducted in 2017 and 2018 respectively. The first group of people – Faruk Para Bawm community, resides in the south-eastern part of the country, in Bandarban, CHT. The other participants, from the Manipuri community, reside in the northeastern part of the country in Moulvibazar, Sylhet. While the two Indigenous communities differ in terms of their ethnicity (Bawm and Manipuri) and location, they complement each other by sharing aspirations for domestic tourism development centred on their traditional handicrafts. Alongside their various cultural attractions (including dance, songs, festivals, and food), both communities are well-known particularly for their textile handicrafts with unique designs and traditional weaving techniques (See Figures 27.1 and 27.2).

In both cases, women are primarily involved with the handicraft production. For example, Bawm women are famous for their backstrap weaving technique and produce a range of artefacts, while Manipuri women are experts in handloom weaving and highly renowned for weaving 'Manipuri *Sari*' and other traditional Manipuri items (e.g. *Fanek-Fidup*, scarf) with a unique Manipuri motif – *Moirang*. Though both groups used to weave these items for their personal use, they now consider the production and selling of these to tourists as an opportunity to complement their traditional livelihood (subsistence farming), following tourists' growing interest in their traditional handicraft items.

Fieldwork

A qualitative interpretative approach was adopted for this study which facilitated an in-depth understanding of the research settings. In this regard, multiple data collection methods were employed, including semi-structured interviews, focus group discussions (FGD), and observations, in order to

Figure 27.1   Traditional backstrap loom used by Bawm people.

gather empirical evidence from the relevant actors (community people, out-of-community people, institutional representatives, and tourists) involved in the handicrafts-tourism nexus in both settings. Research participants in both settings were selected following a purposive snowball sampling technique (Bakas, 2017). In study site one, 29 semi structured interviews with different participant groups (community participants including weavers, sellers and other tourism entrepreneurs; institutional participants including government and non-government organisations; and other relevant participants e.g. private tourism operators) along with an FGD with the community participants. Later, in study site two, 41 semi-structured interviews with relevant participant groups (community participants including weavers, sellers and middlemen; out-of-community tourism entrepreneurs including sellers, tour operators, and tour guides; institutional personnel including government and NGO officials; and tourists) were conducted. Also, one FGD was conducted with the weavers in order to obtain a better understanding of the research phenomenon. The researchers' observations in both settings were incorporated. Thematic analysis (Braun & Clarke, 2006) was utilised to make sense of the collected data, outlining the role played by indigenous handicrafts in domestic tourism development in Bangladesh. The following sections document the key findings based on several broader themes: demand-supply interaction, modification of handicrafts, community control, and external facilitation and the comments and interpretations are based on the responses and comments of the interviewees.

Figure 27.2   Traditional Manipuri Handloom.

Bawm Handicrafts and Tourism Development

Faruk Para (a Bawm village) is adjacent to one of the most popular tourism destinations – Shoilo Propat (a mountain waterfall) – in Bandarban. This village, being the home of an Indigenous community, often attracts tourists. They visit the area to experience the Bawm culture and particularly to buy the Bawm handicrafts. Reflecting on this, one of the community members (Diam) commented: "… *now a massive number of tourists come here … and want to buy these (handicrafts)*". Such demand is further connected to the tourists' interest in the unique Bawm weaving technique, indicating the handicrafts' agency in relation to attracting and developing tourism. In this context, the presence of domestic tourists is highly visible. Restrictions on foreigners' travel in the region were commented upon by a research participant (Simon) in this regard: "… *mainly the Bengali people visit here. You could hardly see any foreign tourist here! … It is very complex.*" Foreign travellers need to follow a range of formalities when visiting this region. Such restrictions are primarily associated with the government's suspicious attitude towards foreign travellers in relation to the socio-political complexities in this region.

The community's friendly attitude towards tourists was found to be a key contributor to the growing tourist demand for handicrafts. Being an economically marginal community, these people have considered tourist demand as an opportunity for generating employment and earning, by producing and selling their traditional handicrafts in small shops at Sholilo Propat. One of the participants (Shamol) mentioned: "*In the beginning, there were only one or two shops, but now you will see there are about 20 to 30 shops"*. Subsequently, the community has developed a community-based approach to meeting tourists' demand for handicrafts. In almost all the families, women were especially involved in producing and selling handicrafts. Within this community-based system, a group of women weaves the items inside the village, another group buys these items wholesale from the weavers and finally sell those items to tourists in small shops at Shoilo Propat. Importantly, the community controls the process from creation to sale, and does not allow any outsiders to sell handicrafts at Shoilo Propat.

Nevertheless, handicrafts have been modified in terms of items, colour, and shape to meet the growing tourist demand. For example, along with traditional clothing, they now weave and sell various other small-sized items (e.g. shawls, mufflers, placemats, bags, and towels) in response to tourists' preferences and convenience, as reflected in one of the weaver's (Alema) comment: *"... tourists love our woven items and often ask for items such as muffler, small bags that are easy to carry. So we weave these too"*. It is important to mention here that both the community members and tourists do not consider the inclusion of different items as a threat to the authenticity of handicrafts, as those items are woven using the traditional looms following the traditional Bawm weaving technique. Tourist preferences were noted by many participants including Van (a handicrafts seller): *"When tourists come they usually ask whether the items are woven by us [Bawm people]. We reply, 'of course'! ... We are trying to bring more beautiful designs [using traditional weaving process] to attract more tourists."*

Another important finding was the involvement of external facilitators, primarily development-oriented organisations. For example, the United Nations Development Programme (UNDP) and different NGOs were found to be facilitating the community in terms of its tourism involvement (based on the traditional weaving and handicrafts) through capacity building (e.g. training and business capacity development) and creating access to finance (e.g. financial grants, microcredit, and co-operative credit union). One of the NGO participants (Abdul) noted: *"By taking money from us, what do they do? They weave the shawls. During the winter they especially do this task as this is a tourism destination. During the summer, they grow pineapple and mangoes in their orchards"*. This quote reflects the potential of handicrafts in bringing the community and NGOs together and the agency of handicrafts in creating an alternative livelihood opportunity for the community during the high tourist season (December to March). This statement reflects the limitation of the seasonality of demand in terms of handicrafts-based domestic tourism in this particular context.

Manipuri Handicrafts and Tourism Engagement

Sreemangal – one of the seven *Upazillas* (sub-districts) of Moulvibazar district – has had increased tourism demand relating to different ethnic groups with their unique lifestyles and cultures. Among the different ethnic groups, Manipuri peoples' involvement in tourism is highly visible in this context, primarily through their handicrafts following the use of unique Manipuri motif – *'Moirag'* (a specific and artistic design) – and their traditional weaving technique. A tourist (Asif) opined: *"... you surely know that Manipuri handicrafts are very well recognised, mainly their handloom weaving ... My wife also likes their unique design a lot. Whenever we visit Sreemangal, we buy something (Manipuri handicrafts) every time"*. This statement reflects the ability of Manipuri handicrafts to attract tourists and gain their attention. Such attention has subsequently contributed to the development of several handicrafts markets (encompassing a number of shops in each market) in different parts of Sreemangal (e.g. Ramnagar, Radhanagar, Jalalia road). A significant tourist flow throughout the year was reported by the research participants and while both domestic and international tourists visit this community, the level of domestic tourists is greatest, as exemplified by a handicrafts seller (Surjo): *"Mostly people from our country visit here. International tourists also come, but not that much often (emphasised). I mean they do not come regularly ..."*. Within this setting, handicraft related activities (both weaving and selling) were found to be performed both by community (Manipuri) and out-of-community (Bengali) people. The involvement of out-of-community people was driven by the growing tourists' demand for Manipuri handicrafts. While such out-of-community involvement was not appreciated by the Manipuri people, no initiative was taken by the community to restrict business within their community. Also, the differing involvements of both men and women were apparent in the weaving-selling practices of these handicrafts, whereby women were primarily involved in weaving and men were mostly involved in selling.

Changes in handicrafts were identified as a result of tourists' demanding low priced ethnic souvenirs. To meet such demand, handicrafts were found to have been changed in terms of materials used (from cotton yarn to synthetic yarn), weaving techniques (from handloom weaving to powerloom weaving), and the types of items woven and sold (from traditional Manipuri items e.g. *Fanek-Fidup* to other diversified non-Manipuri items, e.g. showpieces, different Bengali attire, toys, bedsheets and pillow cases). Interestingly, neither the tourists nor the community residents appeared to be concerned regarding such changes and/or the inclusion of such non-Manipuri items. On the one hand, the community residents have accepted such diversification because of the increased earning prospects from handicraft-centred tourism. On the other hand, tourists appear happy to be able to get Manipuri handicrafts at a low price. In this regard, a community seller (Jatinmoy) stated: *"You know, not all tourists are same! Most of them do not want to spend much [emphasized]. Actually, they like cheaper items. That's why we have to keep all these [both Manipuri and non-Manipuri items]"*.

Furthermore, handicrafts were found to gain the attention of different government and non-governmental organisations relating to the promotion and protection of Manipuri handicrafts. In this regard, the Bangladesh Handloom Board (BHB) – a government organisation – has been facilitating the weavers through its micro-credit and small and medium enterprise (SME) loan programmes, and training. Such government facilitation was confirmed by a representative of BHB (Rahmat): *"We provide training to hundred weavers every year ... Training is provided on weaving, and colouring the yarn."* Also, several NGOs were found to facilitate both the weavers and sellers of Manipouri handicrafts by providing micro-credit to them. In addition, two NGOs – Centre for Community Development Assistance (CCDA) and Ethnic Community Development Organisation (ECDO) – have been actively promoting and selling (collecting from the local weavers and sellers) Manipuri handicrafts in places beyond Sreemangal following the growing interest in handicrafts as noted by NGO participant (Jaman):

We do this (selling) for their promotion and they [shop-based sellers] are very happy, you know! They need not to do anything but their items are sold and they are getting money along with free promotion as those receive mass exposure.

Thus, both individuals and institutions are working together to capitalise on the handicrafts-tourism potential.

Conclusion: Non-human Agency, Indigenous Handicrafts and Tourism Development

Utilising ANT's notion of networking, this chapter has examined the central agency of indigenous handicrafts (a non-human entity) in domestic tourism development in two different settings in

Bangladesh. Handicrafts were found to gain such agency within the relational networking of weaving, selling, and promotion and protection practices utilised by the communities involved. This reflected the handicrafts' influence over others – both human and non-human entities – relating to tourism. In those networking practices, human actors have different roles (e.g. weavers, sellers, tourists) and non-human actors (besides the handicrafts) include traditional weaving techniques and looms, designs used in weaving, and institutions (both government and NGOs). All these actors are drawn together by the relational capacity of handicrafts caused by the handicrafts' ethnic, economic, and place viability associating with tourism. Despite the difference in ethnicity and location, findings from both communities illustrate the development of Indigenous handicrafts-based tourism, depicting similarities in some cases with dissimilarities in other instances as shown in the following table (Table 27.1).

Both examples bring to the fore the demand-supply interaction as the reflector of the agency of handicrafts in developing domestic tourism. Addressing the earlier studies (e.g. Chang et al., 2008; Hoque et al., 2020; Husa, 2019), the findings indicate an increasing demand by tourists for unique cultural elements (in this case handicrafts) of Indigenous people, in turn motivated by a number of actors (e.g. community and out-of-community people, institutions, tourists) involved in tourism development. Within these interactions, handicrafts were found to have been changed through the inclusion of different items in both settings following the preferences and demands of tourists'. However, the main difference is over whether to include traditional weaving-based items and/or other sourced handicrafts that have no connection to the community's culture. While the Bawm community has introduced different small-sized woven items relating to tourists' demand for convenience, the Manipuri people have introduced a range of diversified items (including non-Manipuri items) moving beyond their own woven and/or produced items to meet tourists' demand for low-priced souvenirs. Although earlier studies raised the question of 'authenticity' for the commodification of traditional cultural elements (Cohen, 2016; Novelli & Tisch-Rottensteiner 2012), our findings differ in that no such concern was found in any of the settings. Arguably, this absence of concern for authenticity could be connected to the capitalisation of the economic agency of handicrafts both for the community (generating income) and tourists (low-priced items). Within this demand-supply interaction, the two settings differ in terms of the impact of seasonality. While the first setting is influenced by seasonality which is quite common in tourism (Tao & Wall,

Table 27.1 Summary of the findings highlighting similarities and dissimilarities in two case studies

Key aspects	Bawm community	Manipuri community
Handicrafts – centred tourism involvement	Present	Present
Multiple stakeholders' presence	Present	Present
Involvement of women and men	Women in both weaving and selling	Women in weaving and men in selling
Modification and/or change in handicrafts	Present (Due to tourists' demand for smaller items)	Present (Due to tourists' demand for low-priced items)
Inclusion of non-woven/sourced handicrafts	Absent	Present
Community control	Present	Absent
Concern for authenticity	Absent	Absent
Seasonality	Present	Absent

2009), the Manipuri handicrafts were found to attract tourists throughout the year, affirming their agency in contributing to the development of domestic tourism.

These two communities' contexts further provide important insights regarding gender discourse. Both communities feature the high involvement of women in capitalising on the economic agency of handicrafts relating to tourism. Notably, in the first setting, all the activities – from weaving to selling – are performed by the Bawm women. The second setting differs in this regard as weaving is performed by women and selling is primarily by men, enabling a mutual earning prospect. Importantly, such involvement of women in the handicrafts-tourism businesses did not reflect any evidence of the breaking down of traditional gender roles, which contradicts some earlier studies that found the involvement of women in tourism as problematic in terms of the gender division of labour and power dynamics (Ishii, 2012; Phommavong & Sörensson, 2014). Additionally, this study's findings reveal how the economic prospect of selling handicrafts to tourists can reduce gender inequality within the community, which in turn has the potential to contribute to Indigenous communities' economic (e.g. employment and earning) and social empowerment (e.g. enhanced social capital) at least to some extent.

The settings and processes examined also demonstrate the capacity of handicrafts in linking the communities with different external facilitators (both government and non-governmental organisations) followed by the promotion and protection of indigenous handicrafts for each community's development. Other literature also noted the involvement of external facilitators in Indigenous culture-based tourism (see Zeppel, 2006; Kennedy & Dornan, 2009). Nonetheless, the two settings differ in terms of community control relating to handicrafts-based tourism initiatives. While community control is considered as a key aspect of Indigenous tourism (Hinch & Butler, 2007), the Manipuri setting reflects an absence of the perceived need for such control through their indifference to outsider involvement in this regard. This raises a concern – namely, whether such community indifference in terms of control may be problematic for the sustainable development of Indigenous handicraft-centred tourism in the long run and any immediate resolution of this concern is unclear. The two Bangladeshi communities' experiences extend our understanding of Indigenous tourism beyond the conventional anthropocentric view by illustrating how handicrafts can play a central role in developing Indigenous-culture-based tourism.

Acknowledgement

This chapter originates from the doctoral research projects of the authors. We sincerely acknowledge the Bawm and Manipuri community members and all other research participants whose valuable insights have shaped this chapter.

References

Ara, E. (2020). *Handicrafts-enacted tourism realities in Bangladesh* (Unpublished doctoral thesis). University of Otago, Dunedin, New Zealand.

Ara, E., Tucker, H., & Coetzee, W. (2022). Handicrafts-enacted: Emplacing non-human agency in ethnic tourism. *Journal of Hospitality and Tourism Management, 50*, 345–354. https://doi.org/10.1016/j.jhtm.2022.01.008

Bakas, F. E. (2017). A beautiful mess: Reciprocity and positionality in gender and tourism research. *Journal of Hospitality and Tourism Management, 33*, 126–133.

Barkat, A. (2016). *Political Economy of Unpeopling of Indigenous Peoples: The Case of Bangladesh*. Dhaka: Muktobuddhi Prokashani.

Beard, L., Scarles, C., & Tribe, J. (2016). Mess and method: Using ANT in tourism research. *Annals of Tourism Research, 60*, 97–110. https://doi.org/10.1016/j.annals.2016.06.005

Braun, V., & Clarke, V. (2006). Using thematic analysis in psychology. *Qualitative Research in Psychology, 3*(2), 77–101.

Butler, R., & Hinch, T. (Eds.). (2007). *Tourism and Indigenous Peoples: Issues and Implications.* London: Routledge.

Cassel, S. H., & Maureira, T. M. (2015). Performing identity and culture in indigenous tourism – A study of indigenous communities in Québec, Canada. *Journal of Tourism and Cultural Change, 15*(1), 1–14.

Chang, J., Wall, G., & Chang, C. L. (2008). Perception of the authenticity of Atayal woven handicrafts in Wulai, Taiwan. *Journal of Hospitality & Leisure Marketing, 16*(4), 385–409.

Cohen, E. (2016). Ethnic tourism in mainland Southeast Asia: The state of the art. *Tourism Recreation Research, 41*(3), 232–245.

Coria, J., & Calfucura, E. (2012). Ecotourism and the development of indigenous communities: The good, the bad, and the ugly. *Ecological Economics, 73*, 47–55.

Crainic, N. (2014). The meaning of tradition. In D. Mishkova, M. Turda, & B. Trencsényi (Eds.), *Discourses of collective identity in central and southeast Europe 1770–1945* (pp. 299–305). New York: Central European University Press.

D'Souza, B. J. (2013). Struggle of indigenous peoples for cultural and political rights. In P. Gain (Ed.), *Bangladesh: Land, forest and forest people* (pp. 227–232). Dhaka: Society for Environment and Human Development (SEHD).

Fabricius, M., Raj Joshi, S., Kumari Lama, A., Choudhury, D., Nue, M. M., Chakma, S., Rasul, G., Uddin, K., & Shahidul Alam, M. (2017). *Tourism destination management plan for the Bandarban hill district, Bangladesh (2017–2027).* Kathmandu: International Centre for Integrated Mountain Development.

Higgins-Desbiolles, F. (2003). Reconciliation tourism: Tourism healing divided societies! *Tourism Recreation Research, 28*(3), 35–44.

Hinch, T., & Butler, R. (2007). Introduction: Revisiting common ground. In R. Butler & T. Hinch (Eds.), *Tourism and Indigenous Peoples: Issues and implications* (pp. 1–12). London: Routledge.

Hipwell, W. T. (2007). Taiwan aboriginal ecotourism: Tanayiku Natural Ecology Park. *Annals of Tourism Research, 34*(4), 876–897.

Hoque, M. A. (2020). *Community-based indigenous tourism, NGOs and poverty in Bangladesh* (Unpublished doctoral thesis). University of Otago, Dunedin, New Zealand.

Hoque, M. A., Lovelock, B., & Carr, A. (2020). Alleviating Indigenous poverty through tourism: The role of NGOs. *Journal of Sustainable Tourism*, 1–18. https://doi.org/10.1080/09669582.2020.1860070

Hume, D. L. (2014). *Tourism art and souvenirs: The material culture of tourism.* Abingdon: Routledge.

Husa, L. C. (2019). The 'souvenirization' and 'touristification' of material culture in Thailand: Mutual constructions of 'otherness' in the tourism and souvenir industries. *Journal of Heritage Tourism, 15*(3), 279–293.

Ishii, K. (2012). The impact of ethnic tourism on hill tribes in Thailand. *Annals of Tourism Research, 39*(1), 290–310.

Islam, M. F., & Carlsen, J. (2016). Indigenous communities, tourism development and extreme poverty alleviation in rural Bangladesh. *Tourism Economics, 22*(3), 645–654.

Jóhannesson, G. T., Ren, C., & van der Duim, R. (2015). *Tourism encounters and controversies: Ontological politics of tourism development.* Surrey: Ashgate Publishing Ltd.

Kapaeeng Foundatoin. (2013). *Human rights report 2012 on indigenous peoples in Bangladesh.* Dhaka: Kapaeeng Foundation.

Kendall, L. (2014). In tangible traces and material things: The performance of heritage handicraft. *Acta Koreana, 17*(2), 537–555.

Kennedy, K., & Dornan, D. (2009). An overview: Tourism non-governmental organizations and poverty reduction in developing countries. *Asia Pacific Journal of Tourism Research, 14*(2), 183–200.

Margolis-Pineo, S. (2016). Crafted: Objects in flux. *The Journal of Modern Craft, 9*(1), 83–89.

McCarthy, C. (2014). To foster and encourage the study and practice of Mauri arts and crafts: Indigenous material culture, colonial arts and crafts and New Zealand museums. In J. Helland, B. Lemire, & A. Buis (Eds.), *Craft, community and the material culture of place and politics, 19th-20th century* (pp. 59–81). Surrey: Ashgate Publishing Ltd.

Melubo, K., & Carr, A. (2019). Developing indigenous tourism in the bomas: Critiquing issues from within the Maasai community in Tanzania. *Journal of Heritage Tourism, 14*(3), 219–232.

Middelveld, S. (2012). *Coral reefs in Wakatoby National Park Indonesia: Insights from Actor-Network Theory* (Unpublished master's thesis). Wageningen University, Netherlands.

Minj, A., & Khakshi, J. W. (2015). *Mapping BRAC development activities relating to indigenous peoples.* Dhaka: Integrated Development Programme, BRAC.

Novelli, M., & Tisch-Rottensteiner, A. (2012). Authenticity versus development: Tourism to the hill tribes of Thailand. In O. Moufakkir & P. M. Burns (Eds.), *Controversies in tourism* (pp. 54–72). Wallingford: CABI.

Pawson, S., D'Arcy, P., & Richardson, S. (2017). The value of community-based tourism in Banteay Chhmar, Cambodia. *Tourism Geographies, 19*(3), 378–397.

Peters, K. (2011). Negotiating the 'place' and 'placement' of banal tourist souvenirs in the home. *Tourism Geographies, 13*(2), 234–256.

Phommavong, S., & Sörensson, E. (2014). Ethnic tourism in Lao PDR: Gendered divisions of labour in community-based tourism for poverty reduction. *Current Issues in Tourism, 17*(4), 350–362.

Ren, C. (2011). Non-human agency, radical ontology and tourism realities. *Annals of Tourism Research, 38*(3), 858–881.

Risatti, H. (2007). *A theory of craft: Function and aesthetic expression.* Chapel Hill: University of North Carolina Press.

Ritchie, J. R. B., & Zins, M. (1978). Culture as determinant of the attractiveness of a tourism region. *Annals of Tourism Research, 5*(2), 252–267.

Schaffer, T. (2011). *Novel craft: Victorian domestic handicraft and nineteenth-century fiction.* Oxford: Oxford University Press.

Shuzhong, H., & Prott, L. (2013). Survival, revival and continuance: The Menglian weaving revival project. *International Journal of Cultural Property, 20*(2), 201–219.

Sinclair, D. (2003). Developing indigenous tourism: Challenges for the Guianas. *International Journal of Contemporary Hospitality Management, 15*(3), 140–146.

Smith, V. L. (1996). Indigenous tourism: The four Hs. In R. Butler & T. Hinch (Eds.), *Tourism and Indigenous Peoples* (pp. 283–307). London: International Thomson Business Press.

Smith, M. K. (2016). *Issues in cultural tourism studies* (3rd ed.). London: Routledge.

Sofield, T., & Li, F. M. S. (2007). Indigenous minorities of China and effects of tourism. In R. Butler & T. Hinch (Eds.), *Tourism and indigenous peoples: Issues and implications* (pp. 265–280). Oxford: Butterworth-Heinemann.

Stecher, A. de. (2014). Souvenir art, collectable craft, cultural heritage: The Wendat (Huron) of Wendake, Quebec. In J. Helland, B. Lemire, & A. Buis (Eds.), *Craft, community and the material culture of place and politics, 19th-20th Century* (pp. 37–57). Surrey: Ashgate Publishing Ltd.

Swanson, K. K., & Timothy, D. J. (2012). Souvenirs: Icons of meaning, commercialization and commoditization. *Tourism Management, 33*(3), 489–499.

Tao, T. C. H., & Wall, G. (2009). Tourism as a sustainable livelihood strategy. *Tourism Management, 30*(1), 90–98.

Timothy, D. J. (2005). *Shopping tourism, retailing, and leisure.* Clevedon: Channel View Publications.

van der Duim, R., Ampumuza, C., Ahebwa, W.M. (2014). Gorilla tourism in Bwindi impenetrable national park, Uganda: An actor-network perspective. *Society & Natural Resources, 27*(6), 588–601.

Weaver, D. (2010). Indigenous tourism stages and their implications for sustainability. *Journal of Sustainable Tourism, 18*(1), 43–60.

Whitford, M., & Ruhanen, L. (2014). Indigenous tourism businesses: An exploratory study of business owners' perceptions of drivers and inhibitors. *Tourism Recreation Research, 39*(2), 149–168.

Xu, H., Zhouyuan, T., & Sisi, N. (2014). The impact of cultural tourism on the innovation of ethnic handicraft production in Dali, China. *Asia Pacific World, 5*(2), 82–100.

Yan, S. (2019). The art of working in hair: Hair jewellery and ornamental handiwork in Victorian Britain. *The Journal of Modern Craft, 12*(2), 123–139.

Yang, L., & Wall, G. (2014). *Planning for ethnic tourism.* Burlington: Ashgate Publishing.

Zeppel, H. (2006). *Indigenous ecotourism: Sustainable development and management.* Wallingford: CABI.

28

INDIGENOUS TOURISM IN IRAN

Atefeh Ahmadi Dehrashid and Zahed Ghaderi

Introduction

The unique and rich ethnic groups that live in Iran have great potential to be involved in, and to benefit from Indigenous tourism, but only very limited attention has been paid to date to the issues of Indigenous tourism in Iran, despite its significant capabilities. Scholarly works have more focused on ecotourism, cultural tourism, nature-based tourism, and other forms of tourism (Ghaderi et al., 2022b), with the result that Indigenous tourism has remained virtually untouched and unexplored. Therefore, the purpose of this chapter is to introduce and identify Indigenous tourism locations and attractions in Iran, the qualification of tourism infrastructure and Indigenous tourism companies, possible methods of promoting Indigenous culture, note the current participation of Indigenous people in the development of tourism, and the role of supportive and managerial structures in developing Indigenous tourism.

Indigenous Tourism

Indigenous tourism is a very complex social phenomenon, and due to the great variety of its programs and goals, it has mental dimensions, and generates all kinds of social, economic, and cultural communications (Hinch and Butler, 2009; Sofield, 1993; Weaver, 2010). Currently, the rapid globalization process in cultural trends has created a sense of a need for, and an increasing tendency to stabilize national and local identities, especially among developing countries (Mika and Scheyvens, 2022). In addition, the development of electronic communication flows and media on a global scale, have facilitated the access of more people to culture, and at the same time, the consequences of globalization have increased interest in distant cultures and Indigenous tourism (Fan et al., 2020; Scheyvens et al., 2021; Sinclair, 2003). Indigenization of tourism is the adaptation of tourism to the environment and cultural space of each place. In this approach, there is a two-way flow in tourism; on the one hand, Indigenous tourism includes actions and reactions of tourists, and on the other hand, impacts the size, scope, and functional dimensions of tourism (Whitford and Ruhanen, 2016).

Today, the approach to tourism in the global economy is an important parameter in the process of local economy dynamics because it can also create a continuous process of income and

DOI: 10.4324/9781003230335-33

employment opportunities at tourist destinations. The economic restructuring affected by the growing demand for income and employment tends to consume the environment in the form of economic and cultural spaces (Ryan and Aicken, 2005; Butler et al., 2012). One of the ways of using the environment is Indigenous tourism, which creates numerous economic benefits by preparing the environment for the consumption of tourists (Butler and Hinch, 2007). Therefore, specifying tourism is important for economic dynamism on a small and local scale (O'Gorman et al., 2007).

In the context of tourists' actions and reactions, there is a concept of behavioral adaptation that occurs in the framework of the interaction between hosts and guests. Visual media plays an important role in the development of Indigenous tourism because of the trust that people have in them more than other information sources (Cassel and De Bernardi, 2021). The media, by showing natural beauties, has an impact on tourists' decisions to travel and choose a travel destination. It also has caused the consolidation of individualistic identity, and the change of interests and choices in the tourism experience. This has led to the emergence of a new type of tourism, which is the opposite of mass tourism, and that is individual and family travel to ethnic destinations, which is increasing over time. Normally, such tourists are aware of their duties such as the needs, and sensitivities of the host community and respect them. In this regard, tourism planners in destinations attempt to attract certain types of tourists according to the market trend and host-guest interaction and try to provide the necessary conditions for Indigenous tourism development (Kelly-Holmes and Pietikäinen, 2014).

In the World Tourism Organization report "Tourism: Vision 2020", it is predicted that Indigenous and cultural tourism will be one of the five key tourism markets in the future (UNWTO, 2002), and such growth represents an increasing challenge for managing tourism in cultural and local settings. Given that the presence of tourists affects the daily life of local residents, topics such as the culture and identity of the destination people have been the focus of researchers and cultural institutions in societies with high cultural and heritage tourism capacities. Indigenous tourism, as an alternative or replacement to mass tourism, offers a model for local development in which the original characteristics and unique components of the cultural identity of the destination are used to distinguish a specific tourist destination from others (Butler and Hinch, 2007). In the process of developing Indigenous and cultural tourism, the spiritual aspects of the Indigenous culture as cultural heritage are basic parameters that are highly important (McIntosh and Ryan, 2007).

The basis of Indigenous tourism includes cultural assets such as local narratives, cultural heritage, lifestyle and specific identities of a community, and cultural values and attitudes (Hunter, 2011). In the UNESCO report titled "Our Creative Diversity", culture is defined as a general collection of spiritual, material, intellectual, and emotional assets that are characterized as a society or social group, and it includes creative expression (oral history, language, literature, arts, and handicrafts), and social actions (patterns of social interaction) and they are shown in material, or built forms such as sites, buildings, historical urban centers (Ruhanen and Whitford, 2019), landscapes, art, objects, and handicrafts (Scheyvens et al., 2021). In this way, culture, which is one of the main foundations of Indigenous tourism can be defined as a "living identity". Indigenous tourism is the emergence of human interests to satisfy the curiosity about how people live in the environment around them, as well as visiting the physical aspects of their lives, which are presented in jobs and professions, music, literature, dance, food, drama, handicrafts, language, and religious rituals (Butler and Menzies, 2007; De la Maza, 2016). Cultural heritage is a manifestation and expression of people's culture and history; it gives the experience of tourists a local and global character, and is specified by cultural contact and composition (Carr et al., 2016; Kunasekaran et al., 2017). Smith (2015, p. 125) has provided the following definition of Indigenous and cultural tourism

"The visitor's passive, active and interactive contact with cultures and societies through which one can earn new experiences with educational, creative, and fun matters". Similarly, World Tourism Organization (UNWTO, 2002) considers Indigenous tourism as tourists' journeys to visit Indigenous community and their cultural assets including performing arts, folklore, festivals and events, historical sites, nature, and culture.

Indigenous Tourism in Iran

Having many diverse ethnic communities throughout the country, Iran has great potential for indigenous tourism. However, while having a multi-ethnic society including Turks, Arabs, Baloch, Kurds, and Lures is a significant tourism resource, these ethnic communities do not have unique distinct human rights and self-governance in Iran (UNPO, 2012). The government, for many reasons, does not grant them the right to self-governance in order to maintain Iran's internal security; and the concept of "indigenous" or "ethnic" is integrated within the framework of the concept of national unity with an emphasis on Iranian identity. Needless to say, the people of these regions have been suffering from the lack of cultural, religious, economic, social, and political rights for a long time (Malakoutikhah, 2020). They have been

> systematically excluded from many benefits in investment and development related to natural resource wealth, despite the fact that the principle of self-determination that all peoples are allowed to freely determine their political status and freely pursue their economic, social and cultural development, is enshrined in the International Covenant on Economic, Social and Cultural Rights, and in the Iranian constitution
>
> (UNPO, 2012, p. 6)

The current ruling regime of Iran has declined to call these native people "indigenous", but rather officials prefer to call them minorities. Nevertheless, these communities, informally are Indigenous groups that were the first inhabitants of the Iranian plateau.

However, despite having such potentiality, this kind of tourism has been ignored by decision-makers and tourism planners in Iran (O'Gorman et al., 2007; Ghaderi, 2011). This ignorance, deliberately or unintentionally, is despite the fact that most of these Indigenous communities are living under severe poverty, and thus tourism has the potential to make a significant contribution to poverty alleviation through creating employment opportunities and income for local communities (Ghaderi and Henderson, 2012). In addition, tourism can act as a strategy for the social-cultural integration of Indigenous communities at both local and national levels where local communities desire this. Therefore, in order to highlight the significance of Indigenous tourism in Iran, the following examples are presented as indigenous tourism corridors linking prominent elements and attractions in ethnic destinations.

Locations and Potential for Indigenous Tourism in Ethnic-cultural Areas of Iran

Iranian ethnicities include the Kurdish, Lures, Turkmen, Turks, Baloch, and Arabs. Most Iranian ethnic groups are settled in the border areas of Iran and have cultural-religious links with neighboring countries. For example, Baloch's with the southeastern region of Pakistan, Turkmen in the north east of Iran in the neighborhood of Turkmenistan, Azeri in the north west borders with Azerbaijan and Turkey, Arabs in the west and southwestern border, and finally Kurds in the west

Table 28.1 Largest Iranian ethnic groups

Ethnic groups	(%)	Location
Azari	13	East and West Azarbaijan, Zanjan, Ardabil and Qazvin
Kurd	6.9	West Azerbaijan, Kurdistan, Kermanshah, Ilam, Khorasan
Lur	4	Lurestan, Kohkoliveyeh and Boyer Ahmad, Fars, Khuzestan
Arab	3	Khuzestan
Baloch	2	Sistan va Baluchistan
Turkmen	1.1	Golestan

Source: Compiled by the authors.

and north west of the country in the vicinity of Iraq and Turkey. The locations of these groups are divided by political borders in most cases.

The largest Iranian ethnic groups are Azeri. The population of Azari is about 13% of the Iranian total, Turkmen are about 1.1%, and Kurds are around 6.9%. The population of Arabs in the southwest of Iran is estimated to be 3% and the Baloch ethnic group in the southeast of the country is about 2% of the total population (Iran's Center for Statistics, 2022), (see Table 28.1). The potential development for Indigenous tourism in Iran is mainly in the ethnic-cultural corridors in these areas, especially in the border regions, where every ethnicity has unique customs, cultural events, special dialects, traditional clothes, handicrafts, and ethnic arts (Music, Dance, and Drama) (Chianeh et al., 2019; Torabi et al., 2019; Zamani-Farahani and Henderson, 2010).

Ethnic-cultural Regions of Azaris

This group is the second-largest ethnic group and the largest Indigenous group in Iran. Their communities are scattered in the northwestern part of Iran, in the provinces of West Azerbaijan, East Azerbaijan, Ardebil, Zanjan, and Qazvin (Center for Statistics, 2022), and it has expanded to Hamedan and Gilan. Azaris are distinguished by their dialect and other cultural events, but in terms of religion and ritual ceremonies, they have similarities with the majority of the nation. In the field of literature, Bayati is considered to be the most common form of folk poetry or the oral literature of the Azeri people. The most famous music of Azariabjani is Aashiq. As well as important types of Azerbaijani music, we can note, the Azerbaijani performance art of Tar, Balaban, Naqara, Baghlama and Garman. The most important handicrafts and local artifacts are carpets, weaving, needlework, pottery and ceramics, and woodcarving (Figure 28.1)

Ethnic-cultural Regions of Kurds

Kurds are the second largest Indigenous group in Iran. The general opinion is that Kurds are Iranian and Indo-European who settled in the Kurdistan region about four thousand years ago (Ghasemi, 1984). The Kurdish language consists of different dialects such as *"Kermanji"*, *"Sorani"*, *"Zarza"*, *"Gourani"* and *"Kalhor"*, and in terms of the structure and roots, it is one of the Iranian branches of Indo-European languages. In terms of religion, Kurds are Muslim (from both Sunni and Shiite sects). However, among the Kurds, there are also other religions represented such as Christian, Jewish, Bahai, Zoroastrian, and Yarsan, in small numbers. The Kurds, in addition to

Figure 28.1   Azari Pottery dishes.

Figure 28.2   Palangan, the stair-stepped village and tourism destination in Kurdistan, Iran.

having common customs with the Persians, also have their special traditions including marriages, religious ceremonies and cultural events, local products, and food (Ghaderi et al., 2018). It should also be mentioned that the architectural style of stair-stepped in some parts of Kurdistan due to the geographical and mountainous situations has added local and indigenous attractiveness. The Indigenous people of Kurdistan have unique ethnic arts (music and special dance), and according to UNESCO, the city of Sanandaj, the center of Kurdistan, has been registered as the third creative music city in the world in 2019. SomeIts most important attractions and cultural events (Figures 28.2–28.4) include celebrating *Nowruz* in rural areas of Kurdistan which attracts hundreds of thousands visitors each year, and this is one of the most important Indigenous tourism activities performed by local people, especially in the village of Palangan in Hawraman region. In the past decade, many tour companies have organized different travel packages to Kurdistan, where Indigenous people are living. Local people in some rural and urban areas such as Hawraman, Palangan, Hajij, and Marivan have chosen to becomerelatively integrated into tourism activities (Ghaderi et al., 2018).

Ethnic-cultural Regions of Arabs

Iranian Ahwazi Arabs are settled in the central and southwestern parts of the Khuzestan province. The inhabitants of Khuzestan were originally Arabs in terms of language and ethnicity. In addition to having some common features with the Persians in terms of religious affairs and national events, this ethnicity also has distinctive characteristics such as language, dress, and cultural features. The coffee ceremony of Arabs has been registered as an intangible heritage at the national level. The

Figure 28.3 Kurdish dance.

Figure 28.4 Nowruz festival celebration Palangan.

Dalleh Coffee is one of the cultural objects and tools of Arab people, which has always been used as a cultural symbol in official ceremonies.

Another ancient custom of this people is *Gargiyan*, which is performed every year on the 15th of Ramadan among the Arabs of Khuzestan. In terms of music, *Alvaniyeh* music is the name of one of the styles or melodies of Khuzestan Arabic music. The name is taken from the name of its singer, Alwan Al-Shou'i, who specialized in playing this style and is considered one of the spiritual heritages of the Arabs of Iran. The most important handicrafts and products of this ethnicity include *Kapu* weaving (made from the stems of palm trees), weaving mats, local carpets and rugs, firewood, metalwork, and leather products. The combination of both the tangible and intangible heritage of this ethnicity has created an impressive. Indigenous tourism corridor in this part of the country. Nevertheless, Iranian Arabs have not been appropriately involved in tourism activities due to the sensitivity of their culture, and the reluctance of the authorities to engage Indigenous Arabs in decision-making and planning affairs.

Ethnic-cultural Regions of Baloch

The indigenous Baloch peoples live in an ethnic corridor located in a geographical area that includes southeastern Iran, southwestern Afghanistan, and eastern Pakistan. Iran's Baluchistan has a spatial and geographical link with the Baluchistan state in the Federal Republic of Pakistan and the Baloch region of Afghanistan. This ethnic group has Balochi clothing, customs and arts

Figure 28.5   Local clothes of Baloch ethnic peoples.

Figure 28.6   Local clothes of Baloch ethnic peoples.

(Baloch dances and music), and other special Indigenous cultural products and local foods that distinguishes it from other ethnic groups (Figures 28.5 and 28.6).

One of the most famous features of these people is their genuine hospitality. In addition, in the field of music, this ethnicity has unique folklore such as *Lylo* (lalaee), *Mutak* (lament), *Amba* (songs of sailors), *Nazink* (women's song in a joyous ceremony), *Ghazal* melodic by dervishes, Divine mentions of *Gawati* and *Mald* (music therapy ceremonies), *Saut* (happy wedding songs), *Liko, Shep Taki* (birthday songs), and *Zahiruk* (Romantic songs). Baloch handicrafts are made with natural tools and include. mat weaving, pottery making, needlework, Baluchi embroidery, coin embroidery, and cream embroidery which are among the most important arts of both Baloch women and men. *Kalpurgan* pottery is universally famous, and the village of Kalpurgan, the world's only living pottery museum, is a UNESCO world heritage site. This handicraft was made by Baloch women alone without a pottery wheel. Among other cultural works and beautiful and delicate artistic products of Baloch women is needlework. Balochi needlework has a seal of authenticity from UNESCO and is also registered in the list of the spiritual heritage of Iran.

Several national and local travel companies organize package tours to visit Baloch Indigenous destinations. However, safety and security is a major issue in this region, due to its proximity to Pakistan and Afghanistan. Except for a few well-prepared destinations in Zahedan, the center of Sistan and Baluchistan, as well as the Chabahar port, infrastructure and facilities are not provided, and the Indigenous people have not been integrated into tourism programs due to the lack of awareness, limited capacity of building programs, as well as other cultural barriers.

Figure 28.7   Local clothes of Turkmen ethnic.

Figure 28.8   Turkmen ethnic handicrafts (carpet and wearing felt art).

Ethnic-cultural Regions of Turkmen

Turkmen are scattered in many territories. Apart from Turkmenistan, which is their main habitat, and Iran, they also live in neighboring countries like China, Afghanistan, Turkey, and Iraq. Iranian Turkmen live in the land between the Atrak and Qarasu rivers, and in the North Khorasan province (Quchan, Bojnoord, and Sarakhs). The territory of the Turkmen people begins from this region located to the east of the Mazandaran Sea and extends toward Central Asia. Most of Iran's Turkmen live in Golestan province, in the cities of Gonbad, Bandar Turkmen, and Maravetepe. These people have ethnic-cultural potential and attractions such as Turkmen arts (dance and music), handicrafts, clothing, special language and customs (religious and ritual), and their local foods (Mosammam et al., 2016). (Figure 28.7)

Living in the vicinity of the Golestan National Park, Turkmen people have participated in projects to conserve their natural and cultural heritage including being involved in tourism activities as well. However, as tourism has not been well integrated into their livelihood, they gain only limited benefits from tourism (Figure 28.8).

Ethnic-cultural Regions of Lurs

Lur is the name of an Iranian ethnic group, living in the west and southwest of Iran. This ethnic minority has two main branches: the big Lur and the small Lur. Big Lurs mostly live in Kohgiluyeh and Boyer Ahmad, Fars, Bushehr, Chaharmahal, and Bakhtiari. They are also scattered in some

Figure 28.9 Local clothes of Lur peoples.

Source: Atefeh Ahmadi Dehrashid.

parts of Isfahan, Khuzestan, and Lurestan. Small Lurs live in Lurestan and Pushtkoh (Ilam), and some parts of southern Kermanshah, Hamadan, and Andimshek in Khuzestan province. Among the attractions of Indigenous tourism, Luri heritage music has a high potential and is divided into seven parts in terms of content and theme, including epic, romantic, mourning, work, humorous, religious, and introducing nature. Another potential of Indigenous tourism of this ethnicity is *Luri* dance, such as *Chupi* performed with drums and fiddles, *Sama* dance, and harvesting dance, The special features of the Lurs ethnic group such as language, customs, clothing, art (music and dance), local foods, and their special cultural events distinguish this ethnicity from others (Vafadari, 2008). (Figure 28.9). Indigenous tourism in Lurestan has a huge potential due to the abundance of both natural and cultural heritage of Lurs. Local and national travel companies currently arrange various package tours to this region to visit the Indigenous people.

Indigenous Bio-cultural Heritage in Iran and Indigenous Tourism Development

There is a two-way relationship between tourism and culture; as culture is considered an attribute for tourism development, and tourism also plays an important role in cultural development (Richards, 2021). Similarly, the development of sustainable tourism could have also a significant impact on local culture such as strengthening the cultural and artistic heritage of local communities, thus revitalizing customs and traditions, ethnic arts, lifestyle, local economic activities recovery, and advocating for neglected architectural styles. By introducing and specifying the culture of the destination community, tourism can both increase people's awareness of local customs and traditions and could revive local arts and Indigenous industries (Bonet, 2011). Tourism planning should pay attention to the ethnic-cultural diversity of a multicultural society like Iran (Ward-Perkins et al., 2019). Previous studies have considered four dimensions of authentic Indigenous cultural topics in Indigenous tourism (Csapo, 2012; Du Cros and McKercher, 2020).

Socio-economic and Infrastructure Dimensions

This dimension focuses on the effects of ethnic tourism development on job opportunities, increasing life standard levels, improving the quality of ethnic products, improving communication and public services, improving welfare and recreational facilities, reducing moral and social problems, and diminishing family concepts (Girard and Nijkamp, 2009; Noonan and Rizzo, 2017). The findings of these studies show that ethno-cultural diversity has a positive effect on strengthening social dimensions, ethnic authenticity, cultural protection, and cultural norms. Further, the development of Indigenous tourism has resulted in the preservation of the cultural capital of

Indigenous communities (Ghaderi et al., 2020; Seyfi and Hall, 2018), while the presence of tourists in ethno-cultural areas with tourism potential has increased cultural pride and the desire to display cultural values, symbols, and cultural customs among host societies, and it affects the active participation of the local community (Alipour et al., 2013; Khodadadi, 2016). In this regard, one of the obstacles in Iran is the lack of tourism services and infrastructure, especially in the border areas of Iran, which are often remote and inaccessible. Most of the target villages of tourism in the ethnic-cultural areas do not have suitable road networks and communication infrastructure and have poor conditions in terms of welfare-recreational and health services which need to be taken into consideration in Indigenous tourism development planning (Ghaderi et al., 2018; Taleghani et al., 2014).

The Participation of Local People in Indigenous Tourism Development

Today, residents as one of the main actors of tourism, play a very important role in the development of tourism in tourist destinations. Their participation in tourism development programs and policies in ethnic and cultural areas will depend on their level of acceptance of the effects and consequences of tourism among them. One of the primary aspects of local participation in the tourism sector is hospitality. Hospitality as the behavioral pattern of the host society is an intangible element of tourism, which can have a positive impact on the mental image of the tourism culture of the host society (Richards, 2021). In this regard, the content analysis of studies published on Indigenous tourism in Iran indicates that people's participation in the development of sustainable indigenous tourism has three aspects: social-cultural participation, economic participation, and environmental participation. Research shows that the development of Indigenous tourism among destination communities in Iran leads to an increase in the sense of belonging, revival of local customs, raising the level of cultural awareness, and breadth of the opinion of the host community (Vazin and Sheikhi, 2021).

In the economic dimension, the level of participation of local people depends on the level of their benefits from economic activities related to tourism such as job creation, income generation, and development of the local economy. If locals gain tangible economic benefits from tourism, they most probably will have enough motivation to participate in developing Indigenous tourism. Regarding environmental participation, empirical studies show that Indigenous peoples' environmental awareness has increased in recent years (Ghaderi et al., 2022b), and they are sensitive to their natural heritage. This awareness has improved their participation in conserving their environment. Attracting greater participation of local people in Indigenous tourism development programs in Iran requires capacity building, expansion of educational-promotional programs, formation, strengthening of associations, and involving local people in decision-making and policy-making affairs.

Community Poverty and the Economic Sustainability of Indigenous Tourism

The unbalanced levels of development in the provinces of Iran reflect the centralized planning system which relies on the Center- Periphery model, the qualification of development levels, and social welfare indicators in ethnic-cultural areas are lower than in other parts of the country. Also, the level of poverty, according to the latest report of the Iran Statistics Center (2022), in ethnic-cultural areas with a high potential for Indigenous tourism is higher than in other areas (Aghmaei and Lawell, 2022; Enami et al., 2019; Farzanegan and Habibpour, 2017). According to the same census, about 26 million of the population of 80 million are below the poverty line and are often located in ethnic-cultural border areas. Therefore, it is necessary to take a strategic planning approach for Indigenous tourism development and provide conditions for developmenting Indigenous tourism and reducing the level of poverty in these areas (Goodwin, 2007; Nooripoor et al., 2021; Zarandian et al., 2016). In this regard, pro-poor tourism is designed for the development of tourism to increase the net profit for the poor from the tourism sector, and ensure that tourism growth helps reduce poverty. This type of tourism seeks to provide special economic, social, environmental, and cultural benefits to the poor community (Harrison, 2008).

The aims of this type of tourism are to increase local employment and improve the active participation of local people in the decision-making process and development programs. However, the role of local tourism providers in poverty reduction in Indigenous areas in Iran is not significant (Torabi et al., 2019). Surveying the conditions of local tourism companies in the ethnic-cultural areas of Iran, and the role that they can play in reducing poverty reveals that these companies are very few in number with limited resources and responsibilities (Ghaderi et al., 2018). Also, some of them are facing severe challenges such as low quality of service delivery and performance, structural issues at the micro level, a shortage of professional employees, and at the macro level, issues such as an excessive expansion of the public sector, currency fluctuations, inefficient business policies, lack of economic stability, and lack of suitable physical infrastructure in the tourism sector (Seyfi and Hall, 2018). Considering that local people are intended to be the main beneficiaries of Indigenous tourism, they need to be fully involved and integrated into Indigenous tourism programs for the long-term and sustainable economic development.

Managerial and Supporting Infrastructure (Plans, Policies, and Programmes)

The review of documents, politics, and development plans of Iran's tourism sector shows that in the planning for the years to 2025, sustainability (Veicy, 2018), comprehensiveness (comprehensive social justice for all social classes), balanced development, and the use of historical, cultural, and natural heritage have been emphasized. Moreover, in the country's cultural engineering map, there is a direct reference to Islamic-cultural management and the institutionalization of tourism patterns, development of religious tourism, and expansion of foreign tourism in religious, pilgrimage, scientific, and cultural sites. Reviewing the documents, policies, and executive programs in the tourism sector, especially related to Indigenous tourism in Iran, shows that the tourism development prospects have not been fully implemented. The success of many of these programs requires careful implementation, including legal and protective requirements, facilitation for creating jobs in the Indigenous tourism sector, provision of infrastructure based on prioritization plans and needs' assessment, political support, and provision of required financial resources. Also, participation of the Indigenous community in special tourism areas, the localization of tourism policies in accordance with the ethnic-cultural context of the tourism target areas, education and promotion of a healthy and sustainable tourism culture, and monitoring of the implementation of programs are all required

The first step in attracting tourists to an area with tourism potential is usually by introducing and promoting the tourist attractions to the potential market. In Iran, marketing Indigenous tourism and other forms of tourism have been performed in a traditional way including television advertising, installation of information banners, and setting up of information centers. This method does not work well in the global digital world where the use of advanced technologies is vital for any success in competitive markets.

The success of any tourism program requires the full support and participation of the host, and the local communities at different levels (Ghaderi et al., 2022b). Local people, particularly in Indigenous communities, have to be considered and incorporated in the development of tourism, in decision-making and policy`-making, so that they can participate in the implementation of programs, and become well acquainted with the problems, obstacles, and benefits of this industry (Yang and Wall, 2016).

Conclusion

Iran is the birthplace of one of the richest civilizations in history and has many archaeological, historical, and cultural attractions for tourism. If planned well, the diversity that has arisen from the ancient Iranian culture and civilization would contribute to a series of corridors of Indigenous tourism (Butler et al., 2012; Ghaderi et al., 2022a). In addition, the development of Indigenous tourism can have a positive effect on the prosperity of the local economy (Heydari Chianeh et al., 2018; Khodadadi, 2016) as well as the conservation of the local environment and cultural heritage. Nevertheless, as this study has shown, Indigenous communities have not been integrated in the tourism planning and development of Iran, and local residents have received only very limited benefits from tourism. Several reasons can be cited to explain this situation. First, the centralized structure of the government and the lack of sufficient and necessary attention to the border areas, where the majority of Indigenous people live, have not enabled tourism to succeed (Ghaderi and Henderson, 2012). Second, due to the lack of facilities, infrastructure, and communication networks in the areas where Indigenous people live, access to these sites is often difficult, and local people are unable to provide tourism services. Third, the Indigenous people do not have enough knowledge and information about tourism (Sinclair, 2003), thus they have only limited participation in existing tourism programs. Finally, tourism stakeholders are reluctant to initiate investments and tour companies have made only limited efforts to promote Indigenous tourism destinations among national and international markets (Curtin and Bird, 2022).

References

Aghmai M and Lawell C-YCL. (2022) Energy, economic growth, inequality, and poverty in Iran. *The Singapore Economic Review* 67: 733–754.

Alipour H, Kilic H and Zamani N. (2013) The untapped potential of sustainable domestic tourism in Iran. *Anatolia* 24: 468–483.

Bonet L. (2011) Cultural tourism. *A Handbook of Cultural Economics, Second Edition.* Edward Elgar Publishing, 166–171.

Butler R and Hinch T. (2007) *Tourism and Indigenous Peoples: Issues and Implications*, Routledge.

Butler CF and Menzies CR. (2007) Traditional ecological knowledge and Indigenous tourism. InButler R & Hinch T. (Eds.) *Tourism and Indigenous Peoples: Issues and Implications*, Routledge, 33–45.

Butler R, O'Gorman KD and Prentice R. (2012) Destination appraisal for European cultural tourism to Iran. *International Journal of Tourism Research* 14: 323–338.

Carr A., Ruhanen L and Whitford M. (2016) Indigenous peoples and tourism: the challenges and opportunities for sustainable tourism. *Journal of Sustainable Tourism* 24: 1067–1079.

Cassel SH and De Bernardi C. (2021) Visual representations of Indigenous tourism places in social media. *Tourism Culture & Communication* 21: 95–108.

Center for Statistics. (2022) Iran's population, Retrieved on 20 August from www.amar.ir

Chianeh RH, Rezatab Azgoomi SK and Kian B. (2019) Tourism in Iran. *Experiencing Persian Heritage.* Emerald Publishing Limited, 11–25.

Csapo J. (2012) The role and importance of cultural tourism in modern tourism industry. *Strategies for Tourism Industry-micro and Macro Perspectives* 10: 201–212.

Curtin N and Bird S. (2022) "We are reconciliators": When Indigenous tourism begins with agency. *Journal of Sustainable Tourism* 30: 461–481.

De la Maza F. (2016) State conceptions of indigenous tourism in Chile. *Annals of Tourism Research* 56: 80–95.

Du Cros H and McKercher B. (2020) *Cultural Tourism*: Routledge.

Enami A, Lustig N and Taqdiri A. (2019) Fiscal policy, inequality, and poverty in Iran: assessing the impact and effectiveness of taxes and transfers. *Middle East Development Journal* 11: 49–74.

Fan KHF, Chang T and Ng SL. (2020) The Batek's dilemma on Indigenous tourism. *Annals of Tourism Research* 83: 102948.

Farzanegan MR and Habibpour MM. (2017) Resource rents distribution, income inequality and poverty in Iran. *Energy Economics* 66: 35–42.

Ghaderi Z. (2011) Domestic tourism in Iran. *Anatolia* 22: 278–281.

Ghaderi Z, Abooali G and Henderson J. (2018) Community capacity building for tourism in a heritage village: the case of Hawraman Takht in Iran. *Journal of Sustainable Tourism* 26: 537–550.

Ghaderi Z and Henderson JC. (2012) Sustainable rural tourism in Iran: a perspective from Hawraman Village. *Tourism Management Perspectives* 2–3: 47–54.

Ghaderi Z, Michael Hall C, Scott N, et al. (2020) Islamic beliefs and host-guest relationships in Iran. *International Journal of Hospitality Management* 90: 102603.

Ghaderi Z, Michael Hall MC and Ryan C. (2022a) Overtourism, residents and Iranian rural villages: voices from a developing country. *Journal of Outdoor Recreation and Tourism* 37: 100487.

Ghaderi Z, Shahabi E, Fennell D, et al. (2022b) Increasing community environmental awareness, participation in conservation, and livelihood enhancement through tourism. *Local Environment* 27: 605–621.

Ghasemi, R. (1984) *Kurds and Their Ethnic Unity*: Amirkabir, 1363. (In Persian).

Girard LF and Nijkamp P. (2009) *Cultural Tourism and Sustainable Local Development*: Ashgate Publishing, Ltd.

Goodwin H. (2007) Indigenous tourism and poverty reduction. *Tourism and Indigenous Peoples*. Routledge, 84–94.

Harrison, D. (2008) Pro-poor tourism: a critique. *Third World Quarterly* 29(5): 851–868.

Heydari Chianeh R, Del Chiappa G and Ghasemi V. (2018) Cultural and religious tourism development in Iran: prospects and challenges. *Anatolia* 29: 204–214.

Hinch T and Butler R. (2009) Indigenous tourism. *Tourism Analysis* 14: 15–27.

Hunter WC. (2011) Rukai indigenous tourism: representations, cultural identity and Q method. *Tourism Management* 32: 335–348.

Iran's Center for Statistics. (2022) Iran's latest census by province. Retrieved on 5 October from www.amar.ir

Kelly-Holmes H and Pietikäinen S. (2014) Commodifying Sámi culture in an indigenous tourism site. *Journal of Sociolinguistics* 18: 518–538.

Khodadadi M. (2016) Challenges and opportunities for tourism development in Iran: perspectives of Iranian tourism suppliers. *Tourism Management Perspectives* 19: 90–92.

Kunasekaran P, Gill SS, Ramachandran S, et al. (2017) Measuring sustainable Indigenous tourism indicators: a case of Mah Meri Ethnic Group in Carey Island, Malaysia. *Sustainability* 9: 1256.

Malakoutikhah, Z. (2020) Iran: sponsoring or combating terrorism?*Studies in Conflict & Terrorism* 43(10): 913–939.

McIntosh AJ and Ryan C. (2007) The market perspective of indigenous tourism: opportunities for business development. *Tourism and Indigenous Peoples*. Routledge, London, 91–101.

Mika JP and Scheyvens RA. (2022) Te Awa Tupua: peace, justice and sustainability through Indigenous tourism. *Journal of Sustainable Tourism* 30: 637–657.

Mosammam HM, Sarrafi M, Nia JT, et al. (2016) Typology of the ecotourism development approach and an evaluation from the sustainability view: the case of Mazandaran Province, Iran. *Tourism Management Perspectives* 18: 168–178.

Noonan DS and Rizzo I. (2017) Economics of cultural tourism: issues and perspectives. *Journal of Cultural Economics* 41: 95–107.

Nooripoor M, Khosrowjerdi M, Rastegari H, et al. (2021) The role of tourism in rural development: evidence from Iran. *GeoJournal* 86: 1705–1719.

O'Gorman K, McLellan LR and Baum T. (2007) Chapter 18- Tourism in Iran: central control and indigeneity. In: Butler R and Hinch T (eds) *Tourism and Indigenous Peoples*. Butterworth-Heinemann, 251–264.

Richards G. (2021) *Rethinking Cultural Tourism*: Edward Elgar Publishing.

Ruhanen L and Whitford M. (2019) Cultural heritage and Indigenous tourism. *Journal of Heritage Tourism* 14(3): 179–191..

Ryan C and Aicken M. (2005) *Indigenous Tourism: The Commodification and Management of Culture*: Elsevier.

Scheyvens R, Carr A, Movono A, Hughes E, Higgins-Desbiolles F and Mika JP. (2021) Indigenous tourism and the sustainable development goals. *Annals of Tourism Research* 90: 103260.

Seyfi S and Hall CM. (2018) Tourism in Iran: an introduction. *Tourism in Iran.* Routledge, 3–37.

Sinclair D. (2003) Developing indigenous tourism: challenges for the Guianas. *International Journal of Contemporary Hospitality Management* 15: 140–146.

Smith MK. (2015) *Issues in Cultural Tourism Studies*: Routledge.

Sofield TH. (1993) Indigenous tourism development. *Annals of Tourism Research* 20: 729–750.

Taleghani GR, Ghafary A, Asgharpour SE, et al. (2014) An investigation of the barriers related to tourism industry development in Iran. *Procedia - Social and Behavioral Sciences* 120: 772–778.

Torabi Z-A, Rezvani MR and Badri SA. (2019) Pro-poor tourism in Iran: the case of three selected villages in Shahrud. *Anatolia* 30: 368–378.

Unrepresented Nations and Peoples Organization [UNPO]. (2012) UNPO Report Outlines Threats to Indigenous Peoples' Cultural, Economic, and Social Rights in Iran. Retrieved on 14 October 2022 from https://unpo.org/article/18957

UNWTO. (2002) Tourism 2020 Vision Vol. 7 Global Forecast and Profiles of Market Segments (English version) (online only). https://www.e-unwto.org/doi/abs/10.18111/9789284404667

Vafadari A. (2008) Visitor management, the development of sustainable cultural tourism and local community participation at Chogha Zanbil, Iran. *Conservation and Management of Archaeological Sites* 10: 264–304.

Vazin N and Sheikhi AR. (2021) Cultural impacts of tourism development in Iran from the perspective of experts. *Journal of Tourism of Culture* 1(3): 21–28. doi:10.22034/toc.2020.241748.1012

Veicy H. (2018) The study of tourism industry in national basic laws of Islamic Republic of Iran. *Strategic Studies of Public Policy* 7: 93–112.

Ward-Perkins D, Beckmann C and Ellis J. (2019) *Tourism Routes and Trails: Theory and Practice*: CABI.

Weaver D. (2010) Indigenous tourism stages and their implications for sustainability. *Journal of Sustainable Tourism* 18: 43–60.

Wei L, Qian J and Zhu H. (2021) Rethinking indigenous people as tourists: modernity, cosmopolitanism, and the re-invention of indigeneity. *Annals of Tourism Research* 89: 103200.

Whitford M and Ruhanen L. (2016) Indigenous tourism research, past and present: where to from here? *Journal of Sustainable Tourism* 24: 1080–1099.

Yang L and Wall G. (2016) *Planning for Ethnic Tourism*: Routledge.

Zamani-Farahani H and Henderson JC. (2010) Islamic tourism and managing tourism development in Islamic societies: the cases of Iran and Saudi Arabia. *International Journal of Tourism Research* 12: 79–89.

Zarandian N, Shalbafian A, Ryan C, et al. (2016) Islamic pro-poor and volunteer tourism —the impacts on tourists: a case study of Shabake Talayedaran Jihad, Teheran —a research note. *Tourism Management Perspectives* 19: 165–169.

29

CONTEMPORARY ARTS AND INDIGENOUS ARTS-BASED TOURISM IN WEST AFRICA

Clive Allanso and Marina Novelli

Introduction

The international art scene has been known for promoting 'cultural diplomacy' (Guerzion, 2019) by taking the work of Indigenous artists to an international arena and creating what could be defined as a new niche tourism segment and leading to greater and more meaningful cultural exchanges, facilitating a better cross-cultural understanding, and overcoming conventional and limited images and narratives (Novelli et al., 2022).

This chapter provides a helicopter view of core issues that emerged from empirical research undertaken, between January and May 2022, with various key stakeholders within the contemporary arts value chain in Nigeria and Ghana. The research sets out to create an understanding of the socio-economic impact of contemporary arts on sustainable (Indigenous tourism) development. The research employed a Rapid Situation Analysis (Koutra, 2010) approach to allow adaptability within a complex research environment, and included an array of artists, curators, gallery owners, arts foundations, small indigenous businesses, and quasi-governmental organisations. The case study methodology was based on two study settings, (Nigeria and Ghana) with empirical research involving key stakeholders from three geographically diverse regions in Nigeria (Lagos, Port-Harcourt and Abuja) and two in Ghana (Accra and Aburi). Forty semi-structured interviews, four focus groups and multiple overt participant observations and informal discussions were undertaken across the contemporary arts value chain. Ethnographic and participatory methods, including narrative and visual methods, were also utilised throughout the research fieldwork, with visits to arts markets, events, art galleries and exhibitions playing an important role in enabling the researcher (and lead author of this chapter) to gain a clearer understanding of the study settings, particularly in respect to Indigenous interpretation and meaning of local traditional symbols and materials as well as cultural norms.

The research highlights that Indigenous artists are increasingly becoming a reckoning voice for their respective communities and environments. and using their arts as a vehicle to highlight the plight of specific causes and create new images of Indigenous arts. This has allowed them to

Contemporary Arts and Indigenous Peoples

Arts tourism is identified as a growing field of inquiry in tourism studies and includes any activity that requires travel to see art, with a fascinating relationship between art travel, tourism

381

DOI: 10.4324/9781003230335-34

and individual or community development through to the potential to create 'touristic desire' and 'places as tourist destinations' (Novelli et al., 2022, p. 232). Novelli et al. (2022) identify the opportunities that the emerging Artscape provides for networking, collaborations and exchanges of ideas and the potential of using art to change narratives through creative infrastructures that include art exhibitions, biennales, etc. and to open up opportunities for creatives to get together and build on shared values in a supportive environment. These new patterns of mobility are the potential building blocks of a greater narrative that can be used to develop the shortcomings in the overall tourism strategy and infrastructure for developing nations in West Africa.

Contemporary Arts across West Africa involves a vast number of established and emerging people and organisations, from privately owned galleries like Alexis Galleries, Terra Kulture and Nike Art Gallery in Lagos (Nigeria), and Gallery 1957, Kuenyehia Trust for Contemporary Art and, Anoff Gallery in Accra, to showrooms and centres dedicated to art, culture, and heritage, through to national museums and cultural centres. working to promote arts and (art-based) tourism. However, these efforts are limited and take place in silos, with very few collaborative ventures taking place. 'Tourism has been identified as a catalyst for economic growth for Africa; however, tourism is such a multi-sector industry that many development challenges constrain Africa's tourism potential' (Novelli et al., 2021, p. 35). Within this emerging growth sector rests the steady rise of the arts sector, significantly contributing towards African nations' GDP through the increasing activities at biennales, art fairs, arts and craft markets, and a leading number of exhibitions held in venues in major cities across West Africa. Although there are very few empirical studies on the growth and impact of art-based tourism in Africa, primarily due to e-commerce barriers that view tourism as 'a non-essential activity' (Weigert, 2019, p. 1167), but also to do with the complexities arising in defining the constituent parts of 'measurement', not only of what, but by whom and for what purpose. It is generally accepted that there are growing markets for arts, trinkets and knick-knacks and travel memorabilia, with Indigenous tourism being identified as 'lucrative business, enriching some tribal budgets and profiting individual craftsmen and guides' (Butler and Hinch, 1996, p. 238) enabling an increase in tourism as well as developing links to other Indigenous offerings in the areas of music, fashion and film industry, showcasing indigenous talents on a national and globally evolving arena, but recently skewered with a twist towards mass production of neo-modern tourism inspired art.

In the 1980s, there was a prolific rise in privately owned galleries and museums across sub-Sahara Africa. They range in size and sophistication but commonly their role and function continue to be modelled on those of the economic sales route initiated by the West and conglomerate organisations. The owners were found to be amongst the celebrity entrepreneurs and unabashedly work hard to promote artists with 'great fanfare' and entrepreneurship appropriate to the art itself (Findlay, 2012). The artists themselves also play a significant role in the valuation process, as the ontological presence allows direct contact between artist and collector which often adds to the price of the work (Findlay, 2012), with international collectors recognising a large emerging market in developing economies for contemporary African art (Robertson, 2016).

Since Nigeria's independence in 1960, the Lagos art world has continued to evolve, becoming one of 21st-century Africa's fastest-growing art economies (Castellote and Okwuosa, 2020). Lagos has progressively developed most of the elements that characterise the social system generally described as an 'art world.' In recent years there has been a rise in the growth of the contemporary art scene across West Africa. Subsequently, there has been a steady increase, demand and consumption in both the local and international markets across Africa for indigenous contemporary art, fashion, and music of West African origin.

The awakened consumer consciousness led to an increased demand for artefacts and a proliferation of emerging artists jumping onto the copy-cat bandwagon to produce items to meet the increased demands. These were ably assisted by the increasing practice of established artists who were moving towards developing their art to meet the demands of sustainability and community development and progressively looking at their communities and environments and pulling into their repertoire concepts such as bricolage.

Since the 1990s, African contemporary arts has been experiencing a significant growth attracting international, regional and local interest. Artists have been reflecting on their localities and environments and began to use raw materials as assemblage, objet truvés or forms of bricolage, introducing materials that were used in everyday society. Artists such as El Anatsui (Ghana/ Nigeria) portrayed material from the immediate environment, while others like Abdoulaye Konaté incorporated body wear into their art (Njami et al., 2005). Increasingly, emerging artists continuously draw heavily upon this style and manner in an attempt to potentially attract the kind of attention and benefits that established artists have long worked hard to secure.

During the same period, there was an amplified outcry for people and organisations to become socially aware of the need to protect the environment through sustainable practices and to promote innovation, entrepreneurship, and outputs that had less impact on the environment and specifically provided correlated benefits for the Indigenous people and their communities. Such developments presumably, and at times unconsciously, were addressing the United Nations Millennium Development Goals (MDG's) and subsequently the Sustainable Development Goals' (SDGs) priorities. Therein lies a complexity within the micro-niche segment of indigenous arts-based tourism across West Africa, predominantly to do with how efforts are implemented and monitored through culturally diverse and complex relationships (Novelli et al., 2022) leading to local sustainable development.

In describing African artists, Njami identified their productivity in line with 'noble sense of bricolage' as a relationship between 'matter and 'know-how'' as the mechanism or tool to gaze upon in order to feel or understand contemporary African art (Njami et al., 2005, p. 23). The art of entrepreneurial bricolage, the type that is associated with recycling, upcycling, and tourism-led marketing can be seen across Africa. In Ghana, sacks that were designed to carry and transport flour, cocoa, and charcoal, are given a multiplicity of functions beyond their designed for transportation requirement. The fishermen use them to create sails and their uses are found in the forms of pillowcases, dishcloths, lining for other materials/high-status garments, shirts, shorts and shoulder bags (Nicklin and Salmons, 2002). These bricolage techniques inspired arts forms that can be seen across numerous arts galleries and biennales (i.e. Ibrahim Mahama's work exhibited at the 2019 Venice Biennale, Gerald Chukwuma's art installation at the Grand Palais Ephémérè in Paris 2022) as well as souvenir shops and stalls mushrooming all over the streets and open spaces across most West African nations, in response to the increasing tourist market demand.

The increased use of locally resourced material in the production of West African contemporary arts continued to be highlighted by artists like Dili Humphrey, who is also known by his nickname 'professor junkman' or the 'junkman from Africa'. He is widely known as a contemporary artist who practices painting, performance art and sculpture using recycled materials and assemblage techniques from articles collected from the urban environment and used to reflect the waste and commercialisation of commodity and destruction of the environment. His use of 'forms of waste' and 'junk art, is used 'in depicting people of different classes, men, women, young and old' (Sowole, 2011). Countries across West Africa host a diverse and eclectic number of art organisations and institutions, varying in size and composure, featuring ownership by Indigenous and foreign nationals, collaborative or joint ventures and a varying combination of them. In Ghana for instance, these range from the Ghana National Museum to the Foundation for Contemporary Art, to numerous galleries like Gallery 1957, Nubuke Foundation, the ANO Centre for Cultural Research, Kuenyehia Trust for Contemporary Art, Green Butterfly community initiative and many other emerging organisations.

Similarly, in Nigeria, the African Artists Foundation, Centre for Contemporary Art-Lagos, Alexis Galleries, Terra Kulture and other establishments like Moeshen Gallery in Abuja and the arthouse complex of Diseye Tantua, aptly described as D-Artist SHRINE in Port-Harcourt provide fertile grounds for a growing arts-based tourism scene. However, these efforts are limited and fragmented and largely driven by the efforts of individuals or community champions valiantly working towards raising the profile and, the talents, and the heritage of their communities. The struggle to gain national or artistic acclaim is often impeded by national lack of knowledge or cohesion on the value that these enterprises can bring to the global south, primarily because there are a large number of people who still hold the notion that growth is to be found in the international market with less 'rosy' interest and support from West African people and organisations (Ruiz, 2016, p. 1).

Concurrently, the introduction of recyclable material enabled artists to portray local landscapes and voices and speak out about issues affecting their environment, and to raise other matters of concern such as inequality, poverty, and health. In a study of 23 diverse entrepreneurs by Stinchfield et al. (2013, p. 1) 'bricolage' is identified as a 'unique form of entrepreneurial behaviour' by looking at the integration of materials, responsiveness to changing market conditions and methodologies applied by entrepreneurs to set goals, identify opportunities, and secure resources to exploit those opportunities. The outcomes of that study suggested that motivations were not always linked to the traditional economic models or driven by profit maximization.

It should, however, be noted that there is insufficient literature and understanding of how these entrepreneurial activities or modalities are designed, operate or perform over time against the backdrop of the communities they are said to support or portray. In her study, Swigert-Gacheru (2011, p. 158) argues that the phenomenon of the 'African Renaissance' shows that emerging cultural practices highlighted struggles within the creative sector, mainly around economic and socio-political topics and thus forming a 'Jua Kali Ingenuity' which could be seen as an informal artisan movement being developed through the notion and introduction of bricolage and innovative entrepreneurship into 'junk art' and also through the resourcefulness of local creatives being entrepreneurial and being able to make a flour sack into any number of objects from artworks to bags, trousers and pillowcases, or figurines built of rags and wastes (Swigert-Gacheru, 2011, p. 131; Nicklin and Salmons, 2002, p. 87; Sowole, 2011). This concept is widely acknowledged across West Africa as art assemblage or repurposing of disused materials.

Historically, art of African origin has been inextricably linked to colonialism, with artists either trying to draw on the past or otherwise working towards moving away from the past. The notion of what constitutes art from or of Africa continues to be viewed through a Western gaze that depicts this form of art as tribal or primitive. Willett (1993) disputes this view and asserts that some aspects of African arts, namely woodwork and sculptures, are a 'highly developed and sophisticated art form with thousands of years of history behind' them (Willett, 1993, p. 27). Contemporary West African artists are now breaking this mould and increasingly using their art as a vehicle for changing the narrative and the gaze.

Gerald Chukwuma, from Nigeria, is one of these artists whose talents make use of wood panels with multiple artistic effects (see Figure 29.1) and styles such as burning, painting and chisel work and are based on the indigenous culture and heritage of traditional Uli and Nsibidi symbols linked to historical practices of traditional Nsukka art. Quarshie Tornu (see Figure 29.2) and Diseye Tantua similarly make use of the artist's bricolage expressionism. These artists regularly draw

Figure 29.1   Showcasing bricolage at 1-54, Contemporary African Art Fair, Somerset House, London, 2021 February.

upon the concepts of bricolage by making use of materials locally sourced by Indigenous people and directly tie into providing the communities that they work with an active platform to engage, discuss, and raise the issues and challenges they face, from pollution to poverty, lack of education or other basic infrastructures.

What is described here in terms of history, indigenous traditions, arts practices, and the interesting bricolage concept are in themselves aspects that have proved to be of interest to audiences exploring the West African region, either by choice as arts enthusiasts and/or experts, or by chance as they happen to visit a locality and become exposed to some of artistic expressions discussed.

From Contemporary Arts to Indigenous Arts-based Tourism

Carr et al. (2016, p. 1068) identify Indigenous tourism as 'commonly viewed as a means of facilitating socio-economic benefits to Indigenous individuals, communities and host regions'. Recent case studies on contemporary African art and its socio-economic impact on society in West Africa (Nigeria and Ghana) in 2022, highlighted a general consensus that it would be difficult to produce any accurate statistics around art-based tourism or to ascertain the true picture of the impact and input for Indigenous people. This is particularly highlighted as a capacity and capability issue, as many of the relevant nations did not collect empirical data around tourism as it is still not seen to

Figure 29.2   Fieldwork in Ghana with Quarshie Tornu (Allanso, own collection 2022).

be an economic asset. There are significant strides being made in a number of African nations that have begun to recognise the value of tourism and its impacts on Indigenous people, in terms of elevation from poverty, links to sustainability, increased entrepreneurship and overall community welfare and development. However, these still are reliant on being driven by the tourist gaze or notion of what constitutes the tourism industry, with Africa being primarily associated with safari, wild animals, rural and traditional living in huts, further complicating any advancement on what constitutes contemporary Indigenous tourism and sustainability. This makes it difficult deciding where to draw the line between opportunity and striking an equilibrium between managing and benefiting the continent and its Indigenous people through initiatives that are responsible and sustainable.

The normal media of communication is language, which across Africa has been used by colonisers as a weapon of 'colonial domination' (Fasan, 2015, p. 7), designed to be as inclusive as it is exclusive and thus offering a separatist and elitist barrier to indigenous people and communities. Artists have increasingly used visual applications to have their voices portrayed or heard in order to enable a greater narrative between indigenous people and their communities as well as those who govern or visit them. In giving the Indigenous agency and a voice, Emi Faloughi's entrance at the 2022 biennale, Grand Palais Ephémérè in Paris, showcased the international interest in traditional representation of the feminine creative art sector arising from West Africa and brought it into the international arena. Faloughi's work with recycled wood and traditional Benin bronze casting, represents the growing interest of the vicissitude and intersection between ancient traditional work skills and post-modernity in tourism development. As an emerging African artist, Faloughi draws upon her multi-cultural heritage and works to combine her passion for working with wood, to the production of bronze art and electrification. She links the Indigenous Benin artisans to the modern world through pieces that draw upon bricolage, to portray Indigenous people in their daily lives, from their labours in the workplace to their food, recreation, and family traditions (see Figure 29.3). Faloughi highlights that by taking her art to the international setting, a new narrative

Figure 29.3 Ayo players, a traditional game played by everyone (Courtesy of Emi Faloughi, Artists own collection 2022).

is created that dispels the myth and notion of African art as being inferior, and allows for the narrative of the Indigenous people to be discussed at a global level. This notion continues to be valid as it is widely acknowledged that the growth of Indigenous art from Africa requires an international market, as there is insufficient support from Indigenous people and organisations (Ruiz, 2016).

Kelechi Chinwendu is another emerging artist who packs a punch when it comes to being creative, dynamic and using her art to tantalise the receiver and take them on an incredible journey that combines history, myth, facts, cultural and heritage. Her art is based on traditional Igbo methods and is direct and sensual, she challenges the notion of gender and sexuality and draws on heritage and oral histories to put back into perspective ancient traditions and values distorted by colonialism and foreign religions. Kelechi's works are thus reflecting and putting into perspective the Nigerian notion of 'material and spiritual legacy [which] …are well incorporated into the living patterns of the people, which cannot be separated from their spiritual philosophies' and in what Uduji et al. describe as being 'prised objects for the encouragement and protection of tourism' (Uduji et al., 2020, p. 123).

Kasfir (2000) suggests that 'artistic production of all kinds needs to be viewed through the lens of the culture producing it' (Kasfir, 2000, p. 105). Kelechi's work allows for expressionism in deep-routed traditional beliefs but yet evolving to a present-day independence that allows freedom of expression, being and equality. Her works are deep-routed in ancient methodologies which she uses as a storyboard to bridge the past to the present and places subjective topics at the core of her expressions. Kelechi's work draws the recipient into a magical world of thought and reflection and gives voice to subject matter that might traditionally be seen as taboo, bringing this into perspective. It is easy to see how such creative structures are becoming more appealing to the tourist gaze, offering deep insight and connectivity to subject matters like gender roles and feminine power and strength, sex, and sexuality. She declared that the growing interests of tourists and tourism-related industries have encouraged her to continue to use her art as a mechanism for education and information exchange. In an interview, as part of a case study in February 2022, during her residency at the African Artists Foundation in Lagos, Nigeria, Kelechi also commented that the emerging art industries across Africa were enabling greater engagement with indigenous peoples and with foreigners and gave greater exposure to the issues and challenges faced by those who were normally invisible.

Community Based Indigenous Initiatives

In Ghana, 'Green butterfly' (see Figure 29.4) is a bi-monthly activity that captures the true spirit of the Indigenous people through their arts, crafts, and food. They have created a community-based shared space which increasingly attracts both indigenous tourists and those from faraway lands. This multiplicity of individual artisans, crafters, caterers, and small businesses allows for greater integration and assimilation of the Ghanian culture and heritage, drawing upon the concept of 'shop local' to be shared and enjoyed, whist at the same time contributing to the social and economic growth and regeneration of localised neighbourhoods across Accra. The bi-monthly open-air sessions increasingly pull together indigenous talents across a spectrum of fields and attract a diverse audience of local residents, tourists and travellers, into a single venue to share visual arts, authentic cuisine, sustainable sourced localised products and a vibrant escapism from the monotony of life.

Nana Osei Kwadwo identified that Serge Attukwei Clottey created the 360LA project to enable inclusive community participation and to create a safe space for experience and knowledge (Novelli et al., 2021). Embedded within this concept are the community-led initiatives that are increasingly encouraging governments across West Africa to not only engage with the arts sector but to see the links towards tourism and Indigenous wellbeing brought about by local community engagement

Figure 29.4   Green Butterfly initiative, May 2022, Accra Ghana (Allanso, own collection 2022).

that is linked to the UN Agenda 2030 for sustainable development. This encourages tangible and equitable benefits to communities through tourism yield and the impact on government organisations that need to develop initiatives to recognise the benefits and income that tourism can bring to a country and at the grassroot level, to people in their communities (Novelli et al., 2021, p. 18).

The 360LA project actively uses the concepts of bricolage and sustainability (see Figure 29.5) in their creative art function and also serves as the spirit of Indigenous community cohesion through the engagement of a diverse range of people within the La community. This venture gives jobs and security to a large number of youths, and they can be seen working and sharing l29.ife experiences daily as they undertake learning their new skill sets. Each one of these workers will tell you about their other roles and means of income but those often are subject to contract or commissions, while their role at the 360LA project guarantees not only a steady income stream but brings safety in numbers, comradeship, and exposure to wider markets. The duality and increased need to be innovative to survive is not new to them nor unique to one part of Africa. In fact, in Bindura town in Zimbabwe, young people survive the hostile competitive environment by adopting multiple roles and agency for survival (Kabonga et al., 2021). The 360LA project in Ghana is open to visitors and tourists and the artists will happily engage and take visitors on a tour of their world, drawing upon the resources they use, how and where they find them. They explain how their struggle and plight are being rewarded through the uptake of environmentally sustainable practices and how these are slowly impacting on improving the community they live and work in, which subsequently leads to more tourists and better prospects.

Covid-19 has left a scar where there was already a bruised area around sustainability and survival, the difficulties imposed by Covid-19 have been seen in and around the markets across West Africa. An increased use of up/recycled materials, an increase in people actively understanding the economic value of discarded items, and seeing how indigenous artisans are creatively using such material, not just as sources of income, but also to generate awareness and contribute to the tourism development arena are clearly evident. This phenomenon has also led to an increase in interest and consumption by Indigenous people for the art that was being locally sourced and produced and saw a 'draw of people to visit places in proximity to their residence' (Novelli et al., 2022, p. 232).

The Yibo Koko, Séki Dance Troupe and Indigenous Voices

The Séki dance troupe is taking the histories, heritage and culture of the grass roots and using dance as a method of giving a voice to Indigenous people. The dance troupe transforms oral histories through dance, to highlight the issues and concerns of the Indigenous people in rural areas, highlighting the problemmes of colonialism through to modernity, drawing on symbolism, traditional apparel, and attitudes. This moderate but highly influential group of artisans is drawn from diverse rural backgrounds, giving hope to under-represented people in society and also affording them opportunities to learn, develop and grow, by providing education and skills training in music, song writing, dance, and sustainable community living.

Yibo Koko, founder of the Séki dance troupe, by background is an artist, business entrepreneur, innovative creator, and leading figure in local and national arts for Nigeria. Koko has created a multi-cultural hub in the heart of the city of Port-Harcourt, from where his artistic productions are increasingly attracting interest both from the Indigenous people of the Niger Delta and foreign travellers. His workshop increasingly draws Indigenous people from far and wide, into a safe haven where cultural and bygone traditional art forms are rejuvenated and used to inform, educate, and change negative narratives, remove stereotypes, and promote community spirit and wellbeing. The open-air rehearsals can be heard booming around the neighbourhood and passers-by

Figure 29.5   Serge Attukwei-Clottey's installation, Accra Ghana (Novelli, own collection 2018).

and onlookers are often found peering althrough the fences into the garden to catch a glimpse of rehearsals and a snapshot of new productions. He is also a cultural ambassador for tourism and has taken the Séki dance troupe to an international level, exporting the Nigerian performance arts across the globe and generating and inspiring the Nigerian tourism product. His creation of Séki performance art, as a major tourism product reflects the culture and value of 'traditional/Indigenous theatre [and] includes performances that have mimetic characteristic, some elements of conflict, which are derived from the people's cultural heritage' (Bell-Gam, 2009, p. 11).

The Séki programme is centred around rejuvenating society through education and knowledge exchange; by bringing jobs to Indigenous people; promoting social awareness, and changing attitudes towards key issues like sustainability, through the collection of discarded material like paper, plastics and wood for conversion into creative art pieces. However, the issues are debateable around whether art or art tourism can lead Indigenous people out of poverty. There are still debates being held around the attitude and beliefs about how tourism impacts and benefits communities, which suggests that to some degree Indigenous people might use such developments as alternative tourism 'defined as development that is less commercialised and consistent with the natural, social, and community values of a host community' (Gursoy et al., 2010, p. 381), to escape abject poverty, but it is not indicative of the change required to move their communities towards fulfilled sustainability or to becoming completely self-sufficient. This research showed that efforts of community champions continue to encourage Indigenous people's sustainability and development, albeit on the immediate and small scale, acts as a voice to political and governmental institutions towards enabling greater engagement and productivity for diverse communities, and this is supported by Gursoy et al. (2010) through their study of mass and alternative tourism, that finds that the implications for Indigenous people must consider their attitude through examination of very complex factors.

Other Contemporary Artists and Media

Kasfir (2000) also highlights the complexities involved in establishing and identifying art that is produced by traditional Indigenous cultural practices through 'traditional genres' versus that which is produced for the 'gallery circuit' and therefore constitute nothing more than 'overgrown local workshops' (Kasfir, 2000, p. 105). This complexity extends to the Indigenous people's relationships between themselves and socio-economic encounters they have with the value chain production from raw material to delivery to customers which generates the income they desire. Post-colonialism has seen a rise in innovation and entrepreneurship among Indigenous people, some of whom shift between producing art that is cultural, traditional, and historical, towards that which is produced for mass consumerism and driven purely by financial gain. The economic factors driving such changes are often seen as leading to skill loss, duplication, and repetition of same objects in the markets, destroying creativity and historical integrity, amongst many other concerns and increasingly affecting the socio-economic power struggle brought about by modernity, where individuals who do not agree with the 'power brokers face ostracism of other penalties' (Butler and Hinch, 1996, p. 293). In addition to this is globalised pressures placed on artisans directly through concepts such sustainability.

Despite the numerous attempts and advances by artists, entrepreneurs, non-government & government organisations, and international collaborators, to promote culture, heritage, art or art-tourism as a media for engagement and connectivity between Indigenous people and their communities in West Africa (and to the wider world beyond their territory) there continues to be a high culture of disconnect and distrust between people and agencies. In Nigeria for instance, this is often associated with the term '419', a reference to the police penal code system that highlights levels of fraud linked to 'circumvention of systems' (Smith, 2001, p. 808), and has earned the nation a reputation for high-level fraud, all of which hampers progress and development. Primarily this is because there are no proper structures, laws or even codes of conduct that can be applied across all of the creative art sectors, or even as a code of conduct across Indigenous West African countries.

The impact of poor governance leaves a number of small industries and enterprises that are trying to flourish, at the mercy of those who exploit people and the market for profit and who often are a hinderance to those who need to learn how to position themselves correctly to be viewed favourably by these sources of finance providers (Abor and Biekpe, 2007). However, technological advances have allowed innovative practices to be established which can circumvent poor governance and allow Indigenous enterprise to be developed. Key factors include solar power and mobile phone use and affordability, which bring connectivity to the internet and enabling Indigenous people to access global markets. Boateng et al. posit that as of 2012 Nigeria had more than 100 million subscribers and that if harnessed properly, its impact could become significant for micro-enterprises (Boateng et al., 2014, p. 32).

Support and mentorship and developmental opportunities are some other important factors affecting indigenous people. In an interview conducted in February 2022 with Mathew Oyedele, curator at the Alexis Galleries, Lagos, Nigeria), it was identified that curators were working hard to change the narrative and improve the socio-economic prospects for emerging artists. Oyedele highlighted the joint efforts between curators and galleries in promoting and establishing residences that included workshops, discussion groups, and networking opportunities for artists to not only develop their sown style and techniques but to learn how to best position their business and to encourage collaborative working, sustainable practice and much more. Oyedele's use of social media, the 'Arts Discourse web blogs, continuously explores emerging contemporary art works and events and he uses his knowledge of the struggles that artists and art institutions face to promote a dialogue for youth in finding their voice (Nwakunor, 2022). Similarly, the international award-winning author on tourism in Nigeria, Pelu Awofeso, strives to link the heritage and culture of the people of Nigeria with the diverse tourism potential for Indigenous peoples. Awofeso (2005) illustrates the significance of Indigenous domestic tourism and draws upon the multifaceted and diverse cultures of 250 ethnic nationalities that can be found within the internal borders of Nigeria (Awofeso, 2005).

Film, Music and Fashion

Nollywood has emerged as the third biggest movie and entertainment industry and closely rivals productivity alongside those of Hollywood and Bollywood. Nollywood currently produces numerous works that showcase Nigerian and other West African cultures, heritage and art on a global scale. The phenomenal uptake of Afro-beat music is shown by it significantly charting across the globe with non-Indigenous people being found singing along to 'pidgin' English verse and rhyme or using African words and terminology in their daily discourse. The explosion of African print and textile across the fashion industry has been driven by a vast and assorted array and neo-blend of traditional materials and designs and patterns based on Indigenous heritage and portrayed by the stars on screen.

Yolanda Okereke-Fubara is an internationally acclaimed and award-winning costume designer who uses her knowledge of traditional formal and informal apparel to bridge the gap between the masses and the global world, highlighting community issues and providing employment and developmental opportunities for under privileged youths in society. In an interview in January 2022, she highlighted how sourcing materials led her to working with Indigenous people across many different tribes in Nigeria and across West Africa. She draws on their localised heritage and culture to inspire creativity in design and to establish local community-led initiatives and apprenticeships, helping children and young adults to benefit from her entrepreneurial and creative skills. She identified a number of initiatives where she has encouraged collaboration and partnership working with local businesses when creating on-set bespoke designs to fit in with period dramas and film production in rural areas.

Conclusion: The Potential of Indigenous Arts-based Tourism in West Africa

The global recognition, growth and demand for creative artefacts that support innovative and entrepreneurial talent and tourism development and therefore Indigenous entrepreneurship, continues to expand through the use of social media, technological advances, and the dynamic influences of Nollywood. Increasingly there is more awareness of the contribution that art-tourism can bring to nations' economies across West Africa. While some nations like Nigeria are still in their infancy in tapping into this modality, other nations like Ghana and Senegal have already taken major strides in developing practical and financial initiatives to enable growth linked to tourism. These include increased governmental support for cultural and heritage works, imposing a tourist tax or levy system to enable further development of the emerging market, and even drastically changing and simplifying the ease of travel through swift processing of tourism-related visits.

However, there continue to be significant gaps in joined-up thinking and mechanisms that promulgate policy or govern mutually benefited growth across individuals, private industry and governmental agencies. Conversations with an array of people in the art, governance, marketing business indicate that 'Trust within a partnership' is very hard to establish when there are very few or no standards, frameworks or laws to promote good governance or support failing ventures. This is perpetuated by failing leadership and the lack of integrated ventures towards developing tourism that would directly benefit the Indigenous people. The lack of such initiatives is primarily driven by the view that the economic benefits of establishing an arts-based tourism market is disproportionate to the benefits, and that the arts do not generate as much income as mass tourism (Butler and Hinch, 1996) and benefits would be disproportionate to the required costs of infrastructure to develop this market segment. All in all, there might still a long way to secure African contemporary arts as a widely pursued tourism experience, but what this research has showed is that the foundations are being laid for an Indigenous tourism segment of interest to an increasing number of visitors.

References

Abor, J. & Biekpe, N. (2007). Corporate governance, ownership structure and performance of SMEs in Ghana: Implications for financing opportunities. *Corporate Governance (Bradford)*, 7, 288–300.

Awofeso, P. (2005). *Nigerian festivals: The famous and not so famous*, Nigeria, Homestead Pub.

Bell-Gam, H. L.(2009). Strategies for the development of tourism and theatre industries in Nigeria: Rivers state perspective. *Creative Artist: A Journal of Theatre and Media Studies*, 3, 11.

Boateng, R., Hinson, R., Galadima, R. & Olumide, L. (2014). Preliminary insights into the influence of mobile phones in micro-trading activities of market women in Nigeria. *Information Development*, 30(1), 32–50.

Butler, R. & Hinch, T. (1996). *Tourism and Indigenous Peoples*, London, International Thomson Business Press.

Carr, A., Ruhanen, L. & Whitford, M. (2016). Indigenous peoples and tourism: The challenges and opportunities for sustainable tourism. *Journal of Sustainable Tourism*, 24, 1067–1079.

Castellote, J. & Okwuosa, T. (2020). Lagos art world: The emergence of an artistic hub on the global art periphery. *African Studies Review*, 63, 170–196.

Fasan, R. (2015). 'Wetin dey happen?': Wazobia, popular arts, and nationhood. *Journal of African Cultural Studies*, 27, 7–19.

Findlay, M. (2012). *The value of art: Money, power, beauty*, London; Munich, Prestel.

Guerzion, G. (2019). La Biennale e il rilancio della cultural diplomacy. *Atribune*. https://www.artribune.com/arti-visive/2019/05/biennale-musei-cultural-diplomacy/ [Accessed 28 October 2022].

Gursoy, D., Chi, C. G. & Dyer, P. (2010). Locals' attitudes toward mass and alternative tourism: The case of Sunshine Coast, Australia. *Journal of Travel Research*, 49(3), 381–394.

Kabonga, I., Zvokuomba, K., Nyagadza, B. & Dube, E. (2021). Swimming against the tide: Young informal traders' survival strategies in a competitive business environment in Zimbabwe. *Youth & Society*, 55(2), 280–299.

Kasfir, S. L. (2000). *Contemporary African art*, London, Thames & Hudson.

Koutra, C. (2010). Rapid situation analysis: A hybrid, multi-methods, qualitative, participatory approach to researching tourism development phenomena. *Journal of Sustainable Tourism*, 18, 1015–1033.

Nicklin, K. & Salmons, J. (2002). Contemporary and commercial art in Ghana: Recycling text, image and material. *Journal of Museum Ethnography*, 14, 84–93.

Njami, S., Duran, L., Museum Kunst, P. & Hayward, G. (2005). *Africa remix: Contemporary art of a continent*, London, Hayward Gallery Publishing.

Novelli, M., Ampong, E. A. & Ribeiro, M. A. (2021). *Routledge handbook of tourism in Africa*, London, Routledge.

Novelli, M., Cheer, J. M., Dolezal, C., Jones, A. & Milano, C. (2022). *Handbook of niche tourism*, Cheltenham, Edward Elgar Publishing.

Nwakunor, G. A. (2022). Alexis Galleries… Contemporary Africa in new perspective. *The Guardian (Nigeria)*, 12 October 2022. https://guardian.ng/art/alexis-galleries-connecting-africa-in-new-perspective/ [Accessed 28 October 2022].

Robertson, I. (2016). *Understanding art markets: Inside the world of art and business*, London, Routledge.

Ruiz, C. (2016). The art of Ghana. https://www.ft.com/content/a02d29c8-341a-11e6-ad39-3fee5ffe5b5b [Accessed 28 October 2022].

Smith, D. J. (2001). Ritual killing, 419, and fast wealth: Inequality and the popular imagination in southeastern Nigeria. *American Ethnologist*, 28(4), 803–826.

Sowole, T. (2011). Going to 'Baboon House'? Try Professor Junkman. *A-Arts News Reciew Interview* [Online]. https://www.africanartswithtaj.com/2011/09/dilomprizulike-aka-junkman. [Accessed 15 September 2021].

Stinchfield, B. T., Nelson, R. E. & Wood, M. S. (2013). Learning from Levi-Strauss' legacy: Art, craft, engineering, bricolage, and brokerage in entrepreneurship. *Entrepreneurship Theory and Practice*, 37, 889–921.

Swigert-Gacheru, M. (2011). Globalizing East African culture: From Junk to jua kali art. *Perspectives on Global Development and Technology*, 10, 127–142.

Uduji, J. I., Okolo-Obasi, E. N. & Asongu, S. A. (2020). Sustaining cultural tourism through higher female participation in Nigeria: The role of corporate social responsibility in oil host communities. *The International Journal of Tourism Research*, 22, 120–143.

Weigert, M. (2019). Jumia travel in Africa: Expanding the boundaries of the online travel agency business model. *Tourism Review (Association internationale d'experts scientifiques du tourisme)*, 74, 1167–1178.

Willett, F. (1993). *African art: An introduction*, London, Thames and Hudson.

30

DON'T WORRY. WE HAVE YOUR BEST INTERESTS AT HEART

Kelly Whitney-Gould

Introduction

This chapter examines the need for thoughtful language when working with Indigenous peoples involved in the visitor industry. To borrow a phrase, words matter, and they can be wielded wisely or not. It matters because the outcomes will foster collaboration or dissension. Indigenous peoples around the world have been forced to endure the loss of their languages, their lands, and their identities as a result of colonialism, neocolonialism, institutional racism, and the politics of power—legitimized through language. As a means of unpacking this issue, three commonly used industry words and phrases will be examined: benefit, resource, and service, in the context of Haida Gwaii. The islands of Haida Gwaii lie some 80 kilometres off the northwest coast of British Columbia, Canada. They comprise the unceded traditional territory of the Haida people, who have a vibrant heritage born of the land and watersand kept alive across countless generations. Their experience of Haida Gwaii is fully expressed through their oral and material traditions of language as history, song, dance, design, artwork, naming practices, and ways of being. Speakers will tell you that their 'language comes from the land.'

This statement suggests that it is the land that creates language, which in turn informs tangible and intangible material culture (Whitney-Squire 2016). Broadly constructed, this rests in contrast to those who would argue that it is culture that informs language. The two perspectives are quite different. While there are cognitive and biological functions of language, based on arguments of nature versus nurture, the Haida believe language holds them within a relationship: land and people—people and land. One does not exist without the other, generation to generation: past, present, and future. Put another way, *language speaks the ancestor*—they are connected or relational across time and place.

The Importance of Language

The loss of Indigenous languages globally is quite simply a tragedy. The argument that languages are fluid and evolving, while sufficiently true, belies the fact that many languages have not been lost because of natural shifts in societal or cultural norms; rather, they were systematically removed. In the case of the Haida and many Indigenous Peoples in Canada and elsewhere around the world,

395

DOI: 10.4324/9781003230335-35

languages were lost due to colonialism, disease, systemic racism, and bad policy. In Canada, the residential school system not only forced children to stop speaking their own languages, but removed them from their homes, families, and lands. The naked aggression of the church and state was couched in terms of *having their best interests at heart.*

For those unaware of this tragic history, the first residential schools opened in Canada in the 1880s with the intention of *educating the Indians* (Long and Dickason 1996). The following statements offer some insight into the mindset of the period:

> For the girl who has earned only the rudiments of reading, writing ... [she] will be worth vastly more as mistress of a log cabin than one given years of study.
> (Department of Indian Affairs and Methodist Church Pamphlet 1906; cited in Long and Dickason 1996)

> Of all the things government did, the one thing that did the most harm was their education policies; removing children from the home and family; away from their language and culture ... we call it cultural genocide.
> (Long and Dickason 1996)

> For colonization to be fully effective ... the conquered will only submit to the theft of everything they hold when they can be convinced that it is done for their own good.
> (Manual and Posluns 1974)

Parental consent was not required (Paul 2000). Children were forcibly removed from their homes and families and placed in educational institutions, often hundreds of miles from home. Though shocking, the last residential school in Canada closed only in 1996—ending over 100 years of re-education policy and practice (Haida Gwaii Museum). The recent discovery of dozens of unmarked graves at several of these church-run institutions has reopened old wounds and memories of the catastrophic abuse that had resulted (CBC News 2022).

Not to belabour the point, but at the turn of the 20th century, the population of the Haida had decreased to fewer than 600 people from an estimated 'tens of thousands' (CHN 2022). Once thriving villages were wiped out and abandoned, with survivors eventually relocating to one of two remaining sites: Skidegate (HlGaagilda) and Old Massett (Gaw) (Steedman and Collison 2011). At one time, several dialects were spoken across the islands and southern Alaska. Today, only two are spoken in Haida Gwaii: Xaayda kil (southern dialect) and Xaad kil (northern dialect). A third dialect is spoken in Alaska. And these, by a small and diminishing number of fluent Elders. The Haida remain committed to restoring their language, despite the catastrophic odds against their ability to achieve that outcome.

Personal Background

My name is Kelly Whitney-Gould. I was born in central British Columbia and my ancestors hailed from western Europe, several generations ago. I grew up in the Yukon Territory and my personal values are grounded in principles of egalitarianism. Yet, whether I am aware of it or not, these values are also framed and informed within a history of societal norms, for example, hard work, duty to family, individualism, and representative government. Knowing where I'm from says a great deal about who I am as a person and a great deal more about my values—expressed not only through my actions, but in the words that I speak and write.

My work with the Haida began in 2010 when I arrived in Skidegate to undertake masters research exploring the relationship between aboriginal ecotourism and community-based development. I spent two months volunteering with the Haida Heritage Centre at K̲ay Llnagaay and departed with a growing interest in supporting the expanded use of local Indigenous languages in tourism settings. To that end, my doctoral work at the University of Otago in Aotearoa/New Zealand (2011–2014) examined tourism and language revitalization in Haida Gwaii. Returning permanently to the islands in 2014, my postdoctoral research (2014–2016) focused on improving local language opportunities and a community assessment of language projects.

Through education and experience, I now see my personal values as mouldable and open to sensible discussion, as opposed to the irrefutable absolutes I once thought them. Said's (1979) ground-breaking work, *Orientalism*, describes the process of 'creating the other' where we label and organize ideas, cultures, and peoples based on our idea of what they are or are not, creating. a self-referring system of statements and biases that at once authorizes and validates beliefs and assumptions—camouflaged as knowledge. Importantly, Said (1979: 5–6) qualifies this by observing that (1) there is always some degree of corresponding reality and (2) it is a mistake to assume that these myths and lies have no teeth.

My point is that the words I speak and write carry a lot of baggage. There is no escaping this and it is not a matter of casting blame or fault. It is a matter of acknowledgement and action. Having laid a foundation, the purpose of this chapter is to examine the need for thoughtful language when working with Indigenous Peoples involved in the visitor industry. As stated above, words matter, and they can be wielded wisely or not. It matters because the outcomes will foster collaboration or dissention in advancing culturally appropriate, community-centric development. Or not, as the community chooses.

Local Visitor Context

Indigenous Peoples around the world have been forced to endure the loss of their languages, their lands, and their identities because of colonialism, neo-colonialism, institutional racism, and the politics of power—legitimized through language. The same can and has been said about tourism (Hall and Lew 2014; Jaworski and Thurlow 2013; Johnston, 2000). It is my hope that what follows will spark thoughtful discussion about the damage and lost potential that occurs when we fail to understand the weight and meaning that our words carry for Indigenous Peoples. The best practices listed at the end of this chapter offer pragmatic suggestions aimed at tourism planners, visitor associations, destination management, consultants, and government agencies who seek to work effectively with their Indigenous counterparts.

The ruggedness and remote coastal location of the islands and the unique and beautiful culture of the Haida people have attracted visitors for many years. Many people, myself included, have visited, and fallen in love with this magical place and never left. When people ask me to describe the islands, I tell them, 'It is soft here,' meaning the air, the land, and the forest hold a gentleness that I have never experienced elsewhere. Life is quiet on the islands. There are no shopping malls, no movie theatres, and a single set of streetlights near the ferry terminal.

Queen Charlotte (Daajing.giids) is the largest community on island and serves as a hub for the provision of most services and amenities. The population of the islands is small—approximately 4,000 with a land mass similar in size to that of Hawai'i Island. Graham Island lies to the north and Moresby Island to the south. These land masses are split approximately in the middle by Skidegate Inlet, providing direct water access to the west coast. In addition to the communities already noted, there are also Masset, Sandspit, Port Clements, and Tlell.

Tracking visitor numbers is a relatively haphazard process; however, an exit survey completed in 2019 by the Misty Islands Economic Development Society (MIEDS 2019) suggests that approximately 32,000 people made the sojourn annually pre-Covid. Of these, approximately 5,500 touch down briefly (enroute to sport fishing lodges), as do a similar number of business travellers. For those who do spend time and money on the islands, the economic value approximates $15.8 million (CAN).

These figures are relatively small given the potential; however, tourism has been kept to a minimum for several reasons: level of trip planning required; high cost of travel; high cost of amenities and services; limited accommodation; short visitor season; and limited road access into remote areas. While most visitors rate their experience at 82% positive (summer) the reality of spending time on the islands presents some very real challenges (MIEDS 2019). The following hypothetical visitor statement says a great deal: 'I am lost on a logging road and have not done my laundry in two weeks.' I jest, but only partially. It is easy and dangerous to get lost on the logging roads and it is difficult to find a laundromat with working machines.

Consider that repairing a machine requires it to be shipped off the island at great cost to the business. If you get lost or injured on one of the 1000s of kilometres of old logging roads, you will have to self-rescue unless you carry the appropriate emergency response equipment. While the raw challenge of this will attract some visitors, the majority are mainly interested in visiting friends and family (14%), family vacation (15%), sightseeing nature and wildlife (14%), Haida cultural tourism (13%), and visiting the Gwaii Haanas National Park Reserve and Haida Heritage Site (11%) (MIEDS 2019).

One of the main challenges to building the islands' visitor industry is to understand that it is not simply a matter of getting more people to this destination. Currently, islanders and many local businesses would be hard-pressed to accommodate more visitors and there is a vocal contingent of residents who are not overly keen on the industry for myriad reasons. Most problematic is that the infrastructure to support an expanded industry is either non-existent, crumbling, or requires significant long-term investment. Key to this is understanding that the land itself is gridlocked within overlapping rights of private/corporate ownership (logging), land title negotiations, and land use planning that has much of the islands and surrounding oceans designated as protected areas, conservancies, or ecological reserve areas.

The potential for tourism development on the islands is significant, a fact long recognized by the federal, provincial, regional, and local governments. It is to this end that a steady stream of consultants, tourism planners, and destination management organizations have sought to make sense of how best to proceed. Armed with their reports, statistics, and visions of helping the islands' communities they deploy weapons of strategic planning, community development, and stakeholder engagement to prepare, polish, and sell the experience of Haida Gwaii. The dilemma is ideological. On one hand, you have a hidden neoliberal agenda that pushes for bigger, better badder due to the primarily economic benefits that can be had (Amoamo et al. 2018; Duffy 2015; Keul 2014; Pavlović and Knežević 2017; Wearing and Darcy 2011). Conversely, you have Indigenous communities pushing for realized forms of sustainability (Bunten 2010; Hillmer-Pegram 2016; Loomis 2000; McIntosh et al. 2004; Salazar 2011). In many places around the world, Indigenous and non-indigenous communities are seeking restrictions on growth or advocating for degrowth (Hall 2009; Higgins-Desbiolles et al. 2019).

Since moving to Haida Gwaii permanently in 2014, I have had the privilege of sitting on the Council of the Haida Nations' Tourism Committee as a Consultant Advisor. I currently serve as a Councillor for the Village of Port Clements and chair the local Vibrant Community Commission. This has provided some insight into the machinations and challenges of supporting tourism development on the islands. So too, my work on the island and with several provincial universities has given me the opportunity to work directly with Indigenous students, communities, entrepreneurs, and small businesses. All of whom, in their own way, seek to empower, build capacity, and create meaningful opportunities for their people.

The Language Issue

Moving back to the 'language' part of this discussion, gaining practical insight into this issue is critical to creating a safer space in which to work with Indigenous communities. The need for practical examples is required because, as Bhabha (1994: 336) puts it, 'Even when there is a resolution of meaning, there is a problem of its performativity.' My pragmatic interpretation of this is to say that on some level we know what the problem is; what we lack is the ability to recognize how and where we transgress. In other words, the ability to view our words and actions through the eyes of another's experience. Hollinshead and Suleman (2017: 6) raise this issue, calling for greater consideration of how *power* operates within the industry 'to variously make, demake, and remake particular peoples, places, pasts, and presents.' To this end, their argument and call to action is that 'more empathetic forms of dialogue with the decolonizing world' (Ibid) are desperately required. I couldn't agree more.

Some of the earliest strategic planning meetings I attended were with small groups of locals coming together to discuss the challenges of tourism development. It is a complex hodgepodge of local businesses, local government, and local associations seeking ways to work together respectfully. However, as Bhabha (1994) suggests, these discussions take place within a particular and continuing narrative of place, which reflects the dilemma of belonging. On Haida Gwaii, this equates to those seeking ethical reparation because they know they belong and those who insist they belong because they are there. To Bhabha's point, whilst the broader objectives of the meetings are in general terms benign, they mask myriad underlying narratives, which are highly complex.

One of the ways islanders have learned to manage these narratives is through the use of a simple meeting tool called a 'parking lot.' Typically, it is used to note things that need to be remembered or followed up on after the meeting. On Haida Gwaii, the *parking lot* has taken on a more important role by giving a voice to those affected—an acknowledgment of the underlying narratives. It was at such a meeting that I first witnessed participants attempt to sort through the underlying complexities of tourism development. There were the usual side issues, for example, externally controlled service providers such as BC Ferries or Air Canada. While a few private vessels make their way to the islands each year; general off-island visitor access is restricted to air travel and ferry service. These systems are expensive to maintain and operate—creating a bottleneck wherein high prices and low volumes cancel each other out in terms of growth opportunities.

Different Values and Meanings

However, other issues were deeply embedded within the fabric of this people and this place: rights of access, respect for the land, necessity of balance, and the connectedness of everything. The gist being, that tourism should reflect this place—the Haida version of authenticity and not visitors' expectations. So too did the use of the local language arise. Not only in terms of reclaiming placenames and the urgent need to make the language visible and accessible (Whitney-Squire 2016), but the *lingua franca* of the industry came under scrutiny—as considered from the perspective and aspirations of the Haida.

Table 30.1 Excerpt / parking lot discussion

Problem word	Suggested alternatives
Authentic	Genuine
Culture	Life ways
Resource	Quality
Benefits	Values
Use	Work with
Service	Experience
Other	Us / we / our

Critically, many Haida see tourism as a means of transitioning through or adapting to the present realities of modernity, meaning that tourism is not the goal or objective; rather, it is a means to a different end—a different future reality. This perspective fundamentally situates people and places at the centre or locus of community planning and decision-making. In contrast, the centre or locus of decision-making for development, via tourism, remains solidly entrenched within the purview of economic growth. Herein, the underlying challenge is that the only value being measured is that of dollars and cents. There may be indirect values to be had, but the objective by-and-large is more tourism, more visitors, and more money. As a result, the words we use to justify *growth and development* rest on very different foundations.

In recognition of some of these dynamics, the concept of community-based tourism development was intended to shift some of the power to the community, by offering a measure of acknowledgment that communities have legitimate rights and needs that supersede those of tourists. It offered communities a greater voice in terms of how much and what forms the industry should appropriately take. However, community-based tourism has its detractors, such as Blackstock (2005) who notes critical failures with the approach. The assumption is that (1) development is desirable, (2) consensus is possible, and (3) communities have control. As a village councillor in a very small community, I would tend to agree with Blackstock's assessment even 17 years later.

Given the schism between measures of economic worth and those of the intrinsic values of Indigenous places and people, the community-based tourism approach is a strong miss! While there are no doubt exceptions, the majority do not achieve the gains they seek, as the need for economic growth inevitably overwhelms the values and aspirations of the community. It has been my observation that one of the fundamental stumbling blocks to achieving greater reciprocity and balance rests in understanding the power exerted over Indigenous Peoples in the form of industry jargon. My awareness of the issue grew as a result of the number of words and phrases that increasingly came under fire from residents, students, entrepreneurs, and small business owners. Table 30.1 offers a partial list of problem words, taken from an original 'parking lot' sheet, followed by the preferred alternatives discussed at that time.

Some of the above words have come under scrutiny previously; for example, Wang (1999) provides a detailed overview of the history and complexities associated with the word *authenticity*. See Rickly (2022) for an in-depth analysis and historical review of the subject matter. Grillo (2003) examines the term *culture*, which is often framed as essentialist or concerned with the loss of identity. Similarly, Zygadlo et al. (2003) explored the need for representations of *living culture* that embodies both traditional and contemporary expressions. Wearing and Darcy (2011) examine the process of *othering* in the visitor industry, describing it as cultural cannibalism. Ben Sherman (2017), Chair of the World Indigenous Tourism Alliance (WINTA), routinely shared a selection of related concepts that undermine Indigenous life ways, for example, competition, aggression, and materialism. The overarching theme of these articles and presentations is similar to that of Hollinshead and Suleman (2017: 8), who articulate the need for a 'word gene bank' which is more appropriately suited to a broader and grander purpose within academia and the visitor industry than that which is currently being deployed.

Outside of tourism studies, the Indigenous meanings of certain words and concepts are also receiving greater attention. Downey et al. (2022) conducted a literature review of the meaning of *water* in Australian river communities. Of the 1900 articles reviewed, only 26 examined Indigenous cultural understandings of the word. Of great interest is that the authors note that even fewer articles examine non-indigenous cultural connections to water. Their conclusion was that those authors 'have not problematized non-indigenous and Indigenous cultural connections to water in the same way.' (op cit 9) The implication is that neither set of values / meanings is being examined overly well. In the field of education, Lipe (2019) draws on a number of scholars to compare the underlying values of western science and that of Indigenous knowledge systems for the purpose of building common understandings. While broadly conceptualized, the objective is to unpack the ways in which knowledge systems are constructed and therefore misunderstood. Prest and Goble (2021: 24) explore the challenge of 'conveying the culturally constructed meanings of local Indigenous musics and the worldviews they manifest…,' amidst the hegemony of the English language and dominant colonial discourses.

In addition to the terms discussed above (authenticity / culture / othering), 'parking lot' participants raised the following three words (benefits, resource, and service) repeatedly. With the exception of the word *service*, which is discussed in the sense of elitism by Taum (2010), these terms have not been given careful examination in the literature within the context of the present discussion. To demonstrate the concern, a closer examination of the non-indigenous etymology of these words is warranted. The following excerpts were obtained from the Online Etymology Dictionary (2022). Meanings and concepts not relevant to this discussion were excluded; underlined words are the author's; years denote [c. century] when following a number and [c. circa] when preceding a number.

Benefit

- *financial* support … to which one is *entitled* late [c. 1890]
- good or noble deed; *helpful* or friendly action [late 15 c.]
- *advantage, profit*; beneficial thing [late 14 c.]
- raise money for some *deserving unfortunate person* or charitable cause' [c. 1680]

Resource

- any means of supplying a *want or deficiency* [c. 1610]
- possibility of *aid or assistance* (often negative) [c. 1690]
- the meaning *expedient, device*, shift also from [c. 1690]
- a country's wealth, *means of raising money* and supplies [c. 1779]

Service

- condition of *slavery*; act of serving, occupation of an *attendant servant;* [11–12 c.]
- assistance, *help*; a helpful act; *provision* of food; sequence of dishes served [13 c.]
- state of being bound to *undertake tasks; at someone's direction* [mid 13 c.]
- service or employment in court or administration [13 c.]

Table 30.2 Expression of core values

<u>X</u>aayda kil / <u>X</u>aad kil	Meaning / core value
Yahguudan	• respect for oneself
Yahguudang.ngaay	• respect for all living things
	• fit for respect
Tll Yahda	• making it right
Isda Yahda	• did it right
Isda ad diigaa isda	• giving / sharing
	• reciprocity
	• reciprocal relationships

Source: <u>K</u>ii'iljuus (Barb Wilson) (2009), Ministry of Education (2017), SHIP <u>X</u>aayda Kil Glossary (2016) and Steedman and (Jisgang) Collison (2011).

The etymological history of these words is not offered up as evidence of a particular perspective. Rather, it has been provided as evidence as to the underlying cause of confusion depending upon one's worldview, heritage, and experience. I would suggest, however, that the underlined words and concepts specifically work to undermine Indigenous values and aspirations. In contrast, Table 30.2 examines three concepts / phrases in the Haida language that express some of the core values that underpin the experience and knowledge of this people and these place.

The expression of the above values rests in stark contrast to those expressed by words and phrases used routinely in the tourism industry. While one list does not necessarily fully exclude the other (*to Said's point that there is always some degree of corresponding reality*), based on the conversations and discussions held over the course of many workshops and community meetings, the following statements broadly reflect the perspectives of those present.

Service	We are not carriers of water [we are a proud people].
Resource	Our culture is not your resource [as in cultural resources].
Benefit	A financial transaction [not our measure of success / value].

The ideological schism caused by the use of these words is deeply felt and the alternatives noted in Table 30.1 (above) suggest that other word choices would hold greater meaning and relevance to the lifeways of the Haida people. However, even when the use of certain words, for example *tourist* or *industry*, was challenged by the Haida in the midst of strategic planning sessions, their concerns and suggestions were met with *that is the industry norm*—a dismissal of any discussion or consideration of the alternatives.

An Alternative Approach

In an effort to examine these and other words in greater detail, it was to this end in 2019 that a research proposal was submitted to the Social Sciences and Research Council of Canada. While the research was never completed due to the onset of the covid-19 pandemic, the following excerpt from the original application explains the premise:

This research seeks to examine the embedded values of words and phrases routinely used in the tourism industry that often conflict or contrast with Indigenous worldviews. The purpose of the study is to identify commonly held misunderstandings that may affect the involvement and participation of Indigenous Peoples. The iterative nature of the study will enable rich descriptions for thematic analysis and offer culturally appropriate alternatives to encourage greater respect and transmission of Indigenous values within the tourism and hospitality industry.

(Source: Author)

The significance of the research was to provide evidence of Indigenous perspectives and experiences and a performative basis on which to acknowledge and instill local cultural values within a dominant and globalized industry. The objective was to challenge accepted norms within the industry using a form of discourse analysis which already acknowledges that the industry is rife with contested and embedded representations of power and identity constructions that favour western values and understandings (Higgins-Desbiolles 2009, 2022; Hillmer-Pegram 2016; Meo-Sewabu and Walsh-Tapiata 2012). Working in collaboration with WINTA, it was anticipated that this research would:

- develop alternative, culturally appropriate value expressions
- provide opportunities to test alternative value expressions in tourism settings
- encourage deeper cross-cultural learning in education and training in tourism
- demonstrate a commitment to the process of conciliation, diversity, and inclusion

Left largely to their own devices, the Haida and islands residents have, over time, built a small eclectic group of local businesses that welcome guests to their shores. There are no corporate resorts and there are no large cruise ships. There would be little support for that level of incursion into life on the islands. The Haida have put their support behind boutique ecolodges, museums, cultural centers, boat tours into Gwaii Haanas, coffee shops, local dining experiences, fishing charters, guided nature hikes, and educational talks, to list a few services available to visitors. Island residents have developed a range of unique bed and breakfasts, guided ecotours into the parks, an assortment of souvenirs and other shops, and a hodgepodge mix of community events. Municipalities have broadly worked to improve on islands transportation, marine harbours, and supported local trail development. However, nothing is ever simple and one could put everything noted above into a big bowl and give it a good mix up. The result would give one a much better sense of reality.

Into this walks an unsuspecting consultant, destination planner, tourism advocate, or government agency with a set of externally oriented deliverables and the challenges are readily apparent. Hollinshead (2009: 553) rightly argues that engagement with Indigenous Peoples is not predictable and the exchange of ideas and actions is messy. Further, he suggests that tourism specialists are not able to 'gauge the indigenous interest in outside/introduced objects,' meaning that they do not have the tools to understand or even imagine the potentialities. My interpretation of this is to argue congruently that the very words and jargon used so routinely within the tourism industry and academia broadly is one of the key stumbling blocks to creating a realized space for emergent belonging (Bhabha 1994; Hollinshead 1998). Critically, Wearing and McDonald (2002: 205) argue that striking a balance 'between old forms of knowledge and the new…lie[s] in the hands of the communities themselves.' Further, the authors note that underpinning effective practices is the need for a new 'language of management.'

Building on this, the use of *community-centric* as a means of orienting the concept of development to the needs and aspirations of the community is to understand that Indigenous scholars are increasingly advocating for tourism development informed by customary practices, cultural

403

values, and holistic worldviews (Zygadlo et al. 2003). Spiller et al. (2011: 153) suggest 'multi-dimensional approaches based on an ethic of care' as does Amoamo et al. (2018: 491–478) in terms of 'value exchange and collective agency,' noting as well the need for a 'language of economic diversity.' Carr et al. (2016: 1076) offer an edited series of articles broadly focused on a growing body of work aimed at shifting 'sustainable tourism away from its western developed-world roots to embrace other scenarios.' This body of work strongly suggests a need for Indigenous-centred tourism suited to the aspirations of people and place. The objective therein is to think about how visitorship can meet the needs and aspirations of the community as opposed to the status quo wherein communities are altered and changed to meet the needs of the industry ... *due to the benefits to be had.*

Going back to the 'unsuspecting consultant,' I have read innumerable reports, strategies, and plans and listened to a litany of cringe-worthy statements made by people with the best of intentions. The sum total of this body of work can be found collecting dust on shelves everywhere. It is my view that the following best practices are underpinned by a need for (1) greater sensitivity in the words we so easily resort to and (2) culturally appropriate investigations into the linguafranca of the industry.

In summary:

1　Invest in and plan for the linguistic and cultural safety of participants.
2　Use a 'parking lot' strategy to respect the experience of those present.
3　Request and discuss words and phrases that are considered problematic.
4　Ask for alternatives and use this to inform discussions and the content of reports.
5　Incorporate local Indigenous place names and relevant terminology in written materials.
6　Avoid using old reports to build background information. They are rife with errors and stereotypes.
7　Adopt principles of conduct (e.g., the Larrakia Declaration).
8　Do not set the conditions for planning and deliverables in advance.

Ethical research, good policy, and thoughtful engagement on the part of destination development planning and management would go a long way to providing safe spaces in which Indigenous Peoples and communities are empowered to draw on local knowledge and local experience. The need for sensitive, co-constructed words and phrases is critical to fostering collaboration; such that we have the potential to advance the evolution of culturally appropriate language to support sustainable practices.

Conclusion

Living and working on Haida Gwaii has demonstrated that the challenge of moving toward community-centric Indigenous tourism faces three major obstacles on the part of off-islanders: (1) lack of detailed localized knowledge, (2) lack of recognition of local expertise, and (3) lack of accountability. The industry, on the whole, could do a great deal more to effectively raise awareness of these problems among those working in management, those tasked with gathering information, and those working behind the scenes. Beyond the elusive promise of development, there are the ethics of simply doing good work and engaging thoughtfully and meaningfully with Indigenous Peoples. Not easily done. Everyone is busy. Herein lies the challenge of accountability. It is never anyone's fault and intended kindnesses are misplaced due to a lack of knowledge. A vicious and repeating cycle becomes ensnared in the politics of neoliberal agendas and the need for growth.

When it is said to an Indigenous community, *don't worry, we have your best interests at heart,* that is a lie. The statement may be heartfelt, but it is unlikely that the speakers are prepared to back it up with effective or meaningful action. Here too, accountability for misspoken words is a massive concern because there is no recourse and little discussion. Herein, I would argue for a progressive approach where we work to get the language right in order to effect change—possibly creating new words that garner co-created expressions of values and meanings.

Our words are the building blocks of understanding, of sharing, of relationships, and of trust. Building awareness of the inherent nature and meaning of tired and overused words and phrases in the industry can only provide us with better options. It will also provide an opportunity to gain insight into the perspectives of those we seek to work with ethically and constructively. It can only enrich our perspectives, demonstrate meaningful respect for myriad ways of being in this world, and help us to find grander narratives in which to cooperate. Clearly, we need to reinvent the wheel.

References

Amoamo, M., Ruckstuhl, K. and Ruwhiu, D. (2018) "Balancing Indigenous Values through Diverse Economies: A Case Study of Māori Ecotourism," *Tourism Planning & Development*, 15(5), pp. 478–495.

Bhabha, H. K. (1994) *The Location of Culture*. New York: Routledge.

Blackstock, K. (2005) "A Critical Look at Community Based Tourism," *Community Development Journal*, 40(1), pp. 39–49.

Bunten, A. C. (2010) "More like Ourselves: Indigenous Capitalism through Tourism," *The American Indian Quarterly*, 34(3), pp. 285–311.

Carr, A., Ruhanen, L. and Whitford, M. (2016) "Indigenous Peoples and Tourism: The Challenges and Opportunities for Sustainable Tourism," *Journal of Sustainable Tourism*, 24(8–9), pp. 1067–1079. doi: 10.1080/09669582.2016.1206112.

CBC News (2022) 60 Minutes Anderson Cooper. "Canada's Unmarked Graves: How Residential Schools Carried out 'Cultural Genocide' against Indigenous Children. https://www.cbsnews.com/news/canada-residential-schools-unmarked-graves-indigenous-children-60-minutes-2022-02-06/

CHN (2022) Council of the Haida Nation. "History of the Haida Nation." Retrieved from https://www.haidanation.ca/haida-nation/

Downey, H., Spelten, E, Holmes, K. and Van Vuuren, J. (2022) "A Rapid Review of Recreational, Cultural, and Environmental Meanings of Water for Australian River Communities," *Society & Natural Resources*, Feb2022, P1, 35(5), 556–574.

Duffy, R. (2015) "Nature-Based Tourism and Neoliberalism: Concealing Contradictions," *Tourism Geographies*, 17(4), pp. 529–543.

Grillo, R. D. (2003) "Cultural Essentialism and Cultural Anxiety," *Anthropological Theory*, 3(2), pp. 157–173.

Hall C. M. (2009) "Degrowing Tourism: Décroissance, Sustainable Consumption and Steady-State Tourism," *Anatolia*, 20(1), pp. 46–61.

Hall, C. M. and Lew, A. (2014) "Speaking Heritage Language, Identity and Tourism." In A. Lew (Ed.), *The Wiley and Blackwell Companion to Tourism* (pp. 418–432). Hoboken, NJ: John Wiley and Sons.

Higgins-Desbiolles, F. (2009) *Capitalist Globalisation, Corporate Tourism and Their Alternatives*. New York: Nova Science.

Higgins-Desbiolles, F. (2022) "The Ongoingness of Imperialism: The Problem of Tourism Dependency and the Promise of Radical Equality," *Annals of Tourism Research*, 94. doi: 10.1016/j.annals.2022.103382.

Higgins-Desbiolles, F. et al. (2019) "Degrowing Tourism: Rethinking Tourism," *Journal of Sustainable Tourism*, 27(12), pp. 1926–1944.

Hillmer-Pegram, K. (2016) "Integrating Indigenous Values with Capitalism through Tourism: Alaskan Experiences and Outstanding Issues," *Journal of Sustainable Tourism*, 24(8–9), pp. 1194–1210.

Hollinshead, K. (1998) "Tourism, Hybridity, and Ambiguity: The Relevance Of," *Journal of Leisure Research*, 30(1), pp. 121–156.

Hollinshead, K. (2009) "Tradition and the Declarative Reach of Tourism: Recognizing Transitionality—The Articulation of Dynamic Aboriginal Being: A Critique of Swain's Overview of Continuity and Adaptation in 'Australian' Aboriginal Reality, with Relevance for th e Discerning Presentation of Mutable/in-between Peoples Elsewhere," *Tourism Analysis*, 14(4), pp. 537–555.

Hollinshead, K. and Suleman, R. (2017) "Time for Fluid Acumen: A Call for Improved Tourism Studies Dialogue with the Decolonizing World," *Tourism, Culture & Communication*, 17(1), pp. 61–74.

Jaworski, A. and Thurlow, C. (2013) "Language and the Globalizing Habitus of Tourism: Towards a Sociolinguistics of Fleeting Relationships." In N. Coupland (Ed.), *The Handbook of Language and Globalization* (pp. 264–286). West Sussex: Wiley Blackwell.

Johnston, A. (2000) "Indigenous Peoples and Ecotourism: Bringing Indigenous Knowledge and Rights into the Sustainability Equation," *Tourism Recreation Research*, 25(2), pp. 89–96.

Keul, A. (2014) "Tourism Neoliberalism and the Swamp as Enterprise," *Area*, 46(3), pp. 235–241.

Kii'iljuus (Barb Wilson) in F. Brown, F. and Brown, Y. K. (Eds.). (2009) *Staying the Course, Staying Alive – Coastal First Nations Fundamental Truths: Biodiversity, Stewardship and Sustainability*. Victoria: Biodiversity BC.

Lipe, D. (2019) "Indigenous Knowledge Systems: Missing Link in Scientific Worldviews." In H. Tomlins-Jahnke, S. Styers, S. Lilly and D. Zinga (Eds.), *Indigenous Education: New Directions in Theory and Practice* (pp. 453–481). Edmonton: UA Press.

Long, D. A. and Dickason, O. P. (1996) *Visions of the Heart: Canadian Aboriginal Issues* (1st ed.). Toronto: Harcourt Canada.

Loomis, T. M. (2000) "Indigenous Populations and Sustainable Development: Building on Indigenous Approaches to Holistic, Self-Determined Development," *World Development*, 28(5), pp. 893–910.

Manual, G. and Posluns, M. (1974) *The Fourth World: An Indian Reality*. New York: Macmillan.

McIntosh, A., Zygadlo, F. K. and Matunga, H. (2004) "Rethinking Māori Tourism," *Asia Pacific Journal of Tourism Research*, 9(4), pp. 331–352. doi: 10.1080/1094166042000311237.

Meo-Sewabu, L. and Walsh-Tapiata, W. (2012) "Global Declaration and Village Discourses: Social Policy and Indigenous Wellbeing," *AlterNative: An International Journal of Indigenous Peoples*, 8(3), pp. 305–317.

Ministry of Education (2017) X̱aayda Kil / X̱aad Kil Grades 5 to 12: Integrated Resource Package. School District No. 50 Haida Gwaii. Retrieved from https://www.sd50.bc.ca/wp-content/uploads/2020/09/XK-IRP-merged-10-Jan-2017-SK.pdf

Misty Island Economic Development Society (2019) *Go Haida Gwaii 2019 Visitor Survey: Final Report*. https://www.mieds.ca/wp-content/uploads/2020/05/Haida-Gwaii-Final-Report-May-20-2020.pdf

Online Etymology Dictionary (2001–2022) *Online Etymology Dictionary*. Retrieved from https://www.etymonline.com/

Paul, D. N. (2000) *We Were Not the Savages*. Halifax: Fernwood Publishing.

Pavlović, D. and Knežević, M. (2017) "Is Contemporary Tourism Only a Neoliberal Manipulation?" *Acta Economica Et Turistica*, 3(1), pp. 1–98. doi: 10.1515/aet-2017-0007.

Prest, A. and Goble, J. S. (2021) "Language, Music, and Revitalizing Indigeneity: Effecting Cultural Restoration and Ecological Balance Via Music Education," *Philosophy of Music Education Review*, 29(1), pp. 24–46.

Rickly, J. M. (2022) "A Review of Authenticity Research in Tourism: Launching the Annals of Tourism Research Curated Collection on Authenticity," *Annals of Tourism Research*, 92. doi: 10.1016/j.annals.2021.103349.

Said, E. W. (1979) *Orientalism* (1st Vintage books edn.). New York: Vintage Books.

Salazar, N. B. (2011) "Community-Based Cultural Tourism: Issues, Threats and Opportunities," *Journal of Sustainable Tourism*, 20(1), pp. 9–9.

Sherman, B. (2017) Leadership for Worldwide Sustainable Indigenous Tourism. *Conference: Sustainable Indigenous Tourism*. Ryerson University and Vancouver Island University, Nanaimo, BC.

SHIP Xaayda Kil Glossary (2016) Elders of Skidegate and Skidegate Haida Immersion Program. ISBN: 978-0-9940525-6-8.

Spiller, C. et al. (2011) "Relational Well-Being and Wealth: Māori Businesses and an Ethic of Care," *Journal of Business Ethics*, 98(1), pp. 153–169. doi: 10.1007/s10551-010-0540-z.

Steedman, S. and (Jisgang) Collison, N. (2011) *That Which Makes Us Haida: The Haida Language*. Kay Llnagaay: Haida Gwaii Museum Press.

Taum, R. (2010) "Tourism." In C. Howes and J. Osorio (Eds.), *The Value of Hawai'i: Knowing the Past, Shaping the Future* (pp. 31–38). Honolulu: University of Hawai'i Press.

Wang, N. (1999) "Rethinking Authenticity in Tourism Experience," *Annals of Tourism Research*, 26(2), pp. 349–370. doi: 10.1016/S0160–7383(98)00103-0.

Wearing, S. and Darcy, S. (2011) "Inclusion of the 'othered' in Tourism," *Cosmopolitan Civil Societies: An Interdisciplinary Journal*, 3(2), pp. 18–34.

Wearing, S. and McDonald, M. (2002) "The Development of Community-Based Tourism: Re-Thinking the Relationship between Tour Operators and Development Agents As Intermediaries in Rural and Isolated Area Communities," *Journal of Sustainable Tourism*, 10(3), pp. 191–206.

Whitney-Squire, K. (2016) "Sustaining Local Language Relationships through Indigenous Community-Based Tourism Initiatives," *Journal of Sustainable Tourism*, 24(8–9), pp. 1156–1176. doi: 10.1080/09669582.2015.1091466.

Zygadlo, F. K., McIntosh, A., Matunga, H. P., Fairweather, J. and Simmons, D. (2003) *Māori Tourism: Concepts, Characteristics and Definitions*, Canterbury, NZ: Tourism Recreation Research and Education Centre, Lincoln University.

Indigenous Involvement in Planning Development

In this final section of the volume, chapters explore the scale, role and problems associated with Indigenous involvement in the planning and development of tourism. As has been noted in several earlier chapters, Indigenous communities have achieved a much greater role in the decision-making and policy formulation related to the introduction and expansion of tourism in Indigenous communities. This larger role has brought with it a number of problems, including convincing residents that their voices will be heard and their wishes accepted in terms of scale and type of development, and being given the necessary detailed information about the impacts that any such development would bring about in their communities. Of particular concern in this regard is communication between Indigenous peoples, their representatives and those external, mostly non-Indigenous, representatives of governments, agencies and private individuals and corporations proposing developments. One is reminded of the late President Reagan's probably apocryphal comment that the scariest eight words in the English language were "We're from the government here to help you" in this context. Difficulties arising from being spoken to as opposed to speaking with were noted by Hollinshead in an earlier chapter, and this issue is highlighted by Whitney-Gold in this section in discussing the different terminologies used by Indigenous locals and external visitors. Not only do the same words often imply different meanings, but in some cases, there are no equivalent terms in English (or presumably other 'colonial' languages) to some terms in Indigenous languages. The whole issue of communication is of paramount importance if appropriate and correct information is to be exchanged and the results of development transparent.

LeMoigne discusses the way local grassroots organisations have been able to gain some degree of empowerment in Latin America and make valuable inputs into decisions affecting Indigenous regions and peoples. It is clear, from earlier chapters of Calcufuro and Paskova et al. that not all issues have been resolved in Latin America, and that despite the progress made, some age old problems relating to land allocations, movement of people, alienation of rights and limits on opportunities still exist and need to be resolved. The larger scale organisations LeMoigne discusses represent progress in these areas however, despite local scale individual issues remain extant. The creation of large, national scale Indigenous organisations has been a necessary step toward gaining both recognition and increased empowerment, and the Indigenous Tourism Association of Canada is one example of how such organisations can secure funding to advise and assist grassroots

DOI: 10.4324/9781003230335-36

developments by Indigenous groups to develop in appropriate ways with guarantees of quality and common messages for tourists. Ensuring levels of quality at a national scale is important in helping to overcome the aversions to Indigenous offerings noted by Holder in her chapter, and common messages about the history of colonisation, subjugation and cultural genocide can help with the process of reconciliation and the gaining of deserved respect for such communities (as noted by Butler and Wilkinson, and Jenning). Whitford et al. end this section on another positive note in demonstrating the successful participation of Indigenous groups with government agencies in producing the First Nations Tourism Action Plan for Queensland. They note the process involved in gaining cooperation between the different parties and how this has resulted in an Action Plan to develop appropriately the spectacular Indigenous tourism opportunities that exist in the state of Queensland. This example can work as another example of how cooperation with and recognition of Indigenous voices can result in development that hopefully meets the needs and wishes of the Indigenous peoples of such areas.

31

INDIGENOUS TOURISM ASSOCIATION OF CANADA (ITAC)

Richard Butler

Abstract

This chapter is composed of excerpts from Indigenous Tourism Association of Canada (ITAC) publications with permission from ITAC. ITAC is not alone in representing Indigenous tourism operators, but it provides a good example of what can be achieved by cooperation and integration amongst Indigenous peoples at a national level. It provides advice, organization, and assistance to Indigenous groups involved in tourism to enable them to survive in tourism in Canada and appropriately represent Indigenous culture. ITAC works with national and provincial governments in the formulation of policies related to Indigenous tourism development and in the development of programmes to assist the growth of Indigenous tourism and its recovery after the COVID-19 pandemic. It provides definitions of the terms involved, guarantees authenticity and quality to tourists and runs programmes to improve business efficiency and increase benefits to Indigenous communities. Except where shown in Italics, the following text has been drawn from ITAC's publications and websites in 2023. Some changes in the order of material only have been made.

Introduction

The Indigenous Tourism Association of Canada (ITAC) is a national non-profit Indigenous tourism industry organization established in 2015. ITAC is the lead organization tasked with growing and promoting the Indigenous tourism industry across the country. Inspired by a vision for a thriving Indigenous tourism economy sharing authentic, memorable and enriching experiences, ITAC develops relationships with groups and regions with similar mandates to enable collective support, product development, and promotion and marketing of authentic Indigenous tourism businesses in a respectful protocol.

Its mission is to provide leadership in the development and marketing of authentic Indigenous tourism experiences through innovative partnerships.

Its vision is a thriving Indigenous tourism economy sharing authentic, memorable and enriching experiences.

DOI: 10.4324/9781003230335-37

The members of ITAC represent Indigenous-owned and Indigenous-controlled tourism businesses from across the country. ITAC's Board of Directors includes representatives of Indigenous-owned and Indigenous-controlled organizations from every province and territory.

Since 2015, ITAC has strived to support and strengthen Indigenous tourism experiences throughout Canada. It is now widely recognized as a global leader in Indigenous tourism development and marketing. Its reputation is built upon its successes, but also the strength of its partnerships – at the federal, provincial and territorial levels and also with the Indigenous communities across the country who have chosen to welcome visitors to their lands. ITAC focuses on creating partnerships between associations, organizations, government departments and industry leaders from across Canada to support the growth of Indigenous tourism in Canada and address the demand for the development and marketing of authentic Indigenous experiences. It has an established membership process that enables Indigenous tourism industry partners to engage with and show support for Indigenous tourism.

Indigenous Peoples in Canada

In Canada, the term "Indigenous" is an umbrella term that describes the original inhabitants. "First Nations" are peoples who have historically lived between the Atlantic and Pacific Oceans, south of the Arctic. The "Inuit" have historically lived in Canada's Far North. The "Métis" are descendants of the historical joining of First Nations peoples and European settlers. Each group of the Indigenous peoples in Canada has their own rich and vibrant culture largely influenced by the environment in which they traditionally lived. While Indigenous is a term of preference to some people, it is always best practice to use the nation or home community. When in doubt, ask the person or supplier how they self-describe.

There are more than 60 distinct Indigenous languages spoken in Canada from 12 different language families, more than 630 First Nation communities, over 50 Inuit communities and approximately 600,000 Métis people across Canada, and around half of the Indigenous population in Canada live in cities. There are approximately 2 million Indigenous people in Canada today, about 5.3% of the Canadian population.

ITAC has endorsed the following definitions specific to Indigenous tourism, as they resulted from previous national and extensive consultation of industry, Elders and community.

Indigenous Tourism – all tourism businesses majority-owned (51%), operated and/or controlled by First Nations, Métis or Inuit peoples that can demonstrate a connection and responsibility to the local Indigenous community and traditional territory where the operation resides.

Indigenous Cultural Tourism – meets the Indigenous tourism criteria and in addition a significant portion of the experience incorporates Indigenous culture in a manner that is appropriate, respectful and true to the Indigenous culture being portrayed. The authenticity is ensured through the active involvement of Indigenous people in the development and delivery of the experience. There are tourism businesses that are neither majority-owned nor operated by Indigenous people who offer "**Indigenous tourism experiences.**" Authentic Indigenous cultural tourism is by Indigenous people, not about Indigenous people.

Role of ITAC

The purpose of the ITAC is to improve the socio-economic situation of Indigenous people within the ten provinces and three territories of Canada. It does that through the provision of the following services to Indigenous tourism operators and communities, or those looking to start a cultural tourism business:

- Economic development advisory services
- Conferences

- Professional development training and workshops
- Industry statistics and information.

The mission of ITAC is to provide leadership in the development and marketing of authentic Indigenous tourism experiences through innovative partnerships and to enable a thriving Indigenous tourism economy sharing authentic, memorable and enriching experiences. It represents Indigenous-owned and Indigenous-controlled tourism businesses from across the country, and its Board of Directors includes representatives of Indigenous-owned and Indigenous-controlled organizations from every province and territory. ITAC develops relationships with other groups and/or regions with similar mandates, uniting the Indigenous tourism industry in Canada. It will work to enable collective support, promotion and marketing of authentic Indigenous cultural tourism businesses in a respectful protocol.

ITAC is proud to present its members with a suite of digital marketing tools, resources and courses to support the development and literacy of online marketing. Consumers research and purchase mainly through online channels, and ITAC wants to ensure that Indigenous tourism experiences and products are easily found and bookable. The *Original Mark of Excellence* assures visitors of a quality tourism experience. It is displayed by businesses that offer products and services that are truly authentic and that have been accredited by the ITAC. Any business that earns the accreditation is demonstrating that it has the right standards in place to deliver a quality experience to visitors. These standards are set by Indigenous tourism operators for Indigenous tourism operators, and the programme enables ITAC to apply consistent and fair criteria to any Indigenous business seeking formal recognition as being "Market Ready."

Action Plan 2022–2023

The ITAC has launched its 2022–2023 Action Plan for the continued resurgence of Indigenous tourism across Canada. This Action Plan supports the sustainable restoration of Indigenous tourism in Canada as the sector leader to support annual growth in jobs, marketing, GDP contributions and the creation of new Indigenous tourism businesses.

ITAC's 2022–2023 Action Plan focuses on a four-pillar approach to business that includes:

- **Leadership:** Increasing positive industry awareness, advocacy and accountability to gain market respect, establish funder confidence and build member support. This includes generating own-source revenues through investment in the International Indigenous Tourism Conference, led by ITAC.
- **Partnerships:** Uniting the Indigenous tourism industry in Canada, bringing together members and key industry organizations to collaborate and maximize results. Key to ITAC's success will be leveraging resources with federal, provincial and regional partners through investments in provincial and territory Indigenous tourism organizations.
- **Development:** Encouraging product development and investment from non-Indigenous provincial and territorial partners in Indigenous tourism and destination development, to ensure the industry remains competitive internationally. Tourism HR Canada will play a key role with ITAC's training and labour strategies.
- **Marketing:** Promoting Indigenous tourism and its positive community and cultural impacts with high-value, targeted marketing as well as through research, media and sales efforts. This includes marketing investments of $30 million over three years with ITAC leveraging an additional $18 million in support and matching funds from dynamic industry partnerships.

Complementing the four-pillar methodology are three priorities for ITAC. First, inspiring ITAC member businesses across the country to rebuild or refocus their business offerings through targeted business support, education and training. Next, leveraging partnership opportunities and investments with provincial and territorial Indigenous tourism organizations to maintain their membership and infrastructure. Lastly, by strengthening ITAC as the national leader and advocate for Indigenous tourism operators, stabilizing funding through the federal government, partners and members.

We believe the full recovery of our industry is possible by 2025, and we have set aggressive goals to reach those levels. ITAC will continue to leverage strong domestic and global demand for exceptional, export-ready Indigenous tourism experiences and invest in a strategy to recover Indigenous tourism jobs. And these expected 21,000 new jobs will return $204 million to the federal government over three years through CPP and EI contributions, and federal income tax. ITAC has achieved strong results throughout the COVID-19 pandemic. This success would not have been possible without the ongoing support of our funding partners and the Government of Canada.

The success of past ITAC strategies has been a result of ITAC's efforts to support member businesses in becoming more market- and export-ready and ITAC's building of a network of provincial and territorial partners across Canada. This coordinated approach to development and marketing activities is led by ITAC as the national voice for industry leadership and advocacy. Over the past seven years, the sector has seen unprecedented growth in Indigenous tourism offerings, resulting in new job creation and an increased contribution to GDP – up from $1.4 billion to nearly $2 billion. Additionally, from 2016 to 2019, ITAC supported a 100% increase in the number of market-ready and export-ready Indigenous tourism businesses in Canada.

ITAC's strategic recovery plan supports the sustainable restoration of Indigenous tourism in Canada as the sector leader in annual growth in jobs, GDP contributions and the creation of new tourism businesses. As the Indigenous tourism industry in Canada recovers from the shock of the COVID-19 pandemic, ITAC has set its sights on moving forward in the most adaptable and sustainable manner possible. ITAC understands that this undertaking will present challenges, both known and unknown, and is prepared to face them head on. It has forged national partnerships that include 11 Destination Canada, Tourism HR Canada, WestJet, Parks Canada and the Tourism Industry Association of Canada. The input of these partners is invaluable as ITAC works with them to rebuild the industry in Canada. ITAC has achieved strong results throughout the COVID-19 pandemic. This success would not have been possible without the ongoing support of our funding partners and the Government of Canada. To achieve our 2022–2025 targets for the Indigenous tourism industry in Canada, ITAC's recovery plan requires a $65 million investment over three years.

In 2021, we began to see a slow recovery from the impacts of COVID-19 on the Indigenous tourism sector nationally. A strong Indigenous tourism industry is vitally important to the visitor economy in Canada, providing significant economic growth and stability across urban and rural communities. Additionally, Indigenous tourism operators across the country provide unparalleled opportunities for reconciliation and education for non-Indigenous people coast-to-coast-to-coast, while supporting the cultural and economic vitality of Indigenous nations and communities. Prior to the COVID-19 pandemic, tourism was one of the fastest-growing industries in the world. As one of Canada's largest industries, tourism was responsible for $105 billion in GDP, provided 1 in 10 Canadian jobs – approximately 1.9 million – and was made up of 225,000 small- and medium-sized businesses across the country. ITAC is proud to be seen as a leader amongst many global nations committed to also growing their Indigenous tourism industries.

Following three consecutive years of positive growth, international travel to and from Canada declined by 73%, from 96.8 million travellers in 2019 to 25.9 million in 2020, primarily because of the COVID-19 pandemic. Additionally, travellers to Canada from both the United States and overseas countries were down by 93% in December 2020. This of course impacted ITAC members, with many struggling to survive. The year 2021 saw modest improvements to those numbers, and there is a reason for optimism as provinces and territories begin moving towards final phases in vaccination distribution plans. Vaccines have been increasingly administered to the general population and will continue into the summer.

This hope and optimism is shared by ITAC as we experience the resiliency of our members and destination partners. November and December of 2021 saw an increase in unemployment in the tourism industry; however, it's important to note that this is a typical trend year over year. Public health measures are changing in several provinces, and it is, therefore, possible that the previous employment momentum seen in 2021 will re-emerge as public health protections are eased across the country. The unemployment rate in the tourism sector in January 2022 was 11.9%, which is a significant decrease from the same time in 2021 when the unemployment rate stood at 18.6% as reported by Tourism HR Canada. With Tourism HR Canada's help, ITAC will be investing in training and labour strategies, with the knowledge that an investment in Indigenous tourism will employ more Indigenous workers than an investment in any other sector. Also, as borders and public health measures continue to open, it is expected that the Indigenous tourism industry will help lead the economic recovery of the tourism industry in Canada and provide valuable human resources to fill important tourism roles across the country.

ITAC is focused on continuing to ensure stability and recovery of the Indigenous tourism economy in 2022 and beyond. Through our Build Back Better strategy (https://indigenoustourism.ca/wp-content/uploads/2022/01/ITAC-Building-Back-Better-2022-2025.pdf), presented to over 300 national delegates during the National Indigenous Tourism Conference, ITAC will continue to promote authentic, sustainable and culturally rich Indigenous tourism experiences. It is more important than ever to continue to showcase Canada as a premier Indigenous tourism destination while growing and strengthening our resilient membership base in Canada. Our advocacy work will also continue as we strive to educate Canadians, provincial, territorial and federal governments and the tourism industry on the multiple benefits of a thriving Indigenous tourism industry, from coast-to-coast-to-coast.

As the Indigenous tourism industry in Canada recovers from the COVID-19 pandemic, ITAC has set targets to move forward while being adaptable and sustainable. Recognizing that this undertaking will present known and unknown challenges, ITAC is prepared to face them head on using proven strategies and funding resources.

Three Main Priorities of the 2022–2023 Action Plan

2022–2025   Targets

The 2022–2023 Action Plan works towards our three-year strategy goals and priorities: achieving contributions towards Canada's GDP, growing the Indigenous employee workforce back to 2019 numbers and maintaining our membership year over year. ITAC's revised targets for 2025 are to return to pre-COVID-19 levels. With the required investment of $65 million, ITAC believes it will hit the following targets by the end of 2025: grow our Industry to $1.9 billion in direct GDP contributions; grow to 1,900 Indigenous tourism businesses; and grow to 40,000 Indigenous tourism employees.

2022–2023   Recovery to Resiliency Actions

As ITAC continues to manage the effects of the COVID-19 pandemic, it is shifting its focus to restarting the Indigenous tourism industry and the promotion of Indigenous destinations that are open and ready to welcome visitors. ITAC will ensure that its members have the training and resources they need to meet regional health and safety requirements and will continue to work with Destination Canada and provincial and territorial partners on the launch of domestic marketing campaigns. Collectively, this work supports the sustainable recovery of the sector and will prepare ITAC's members for renewed promotion to international markets. The objective (being) is to safely engage with members to restart the Indigenous tourism industry with a focus on domestic travel while continuing to build capacity amongst ITAC's membership and providing targeted business support where required.

Leadership activities include being an advocate for the industry at the federal level to ensure continued, easy and fair access to federal support and to expand participation and representation in national-level tourism organizations.

Partnership activities are to support the ongoing recovery of the Indigenous tourism industry, to increase financial support for the provincial and territorial Indigenous tourism associations, to host an in-person International Indigenous Tourism Conference in 2023 and to grow ITAC membership across Canada.

Development activities are to implement the Original Accreditation Program and align business support to improve products or services; to deliver support to help businesses meet accreditation programme market readiness standards; to support the implementation of public health and safety requirements in the "new normal" with a focus on future-proofing the industry; and to implement and share provincial and territorial reopening and response plans. In order to create a strategy to recover jobs and increase Indigenous engagement in the tourism industry by implementing ITAC's job recovery strategy, ITAC will launch an Indigenous Tourism Innovation Lab to help future-proof ITAC member businesses; create Provincial and Territorial Readiness Toolkits to support destination development; create a directory of online training resources for members; develop the Indigenous Culinary Ambassador network; and collaborate with educational institutions to deliver Indigenous tourism and culinary programmes.

Marketing activities are to deliver virtual marketing campaigns and public service announcements to keep Indigenous tourism top of mind in the aspirational travel planning of consumers; to collaborate with media and public relations partners to monitor sentiment and respond to opportunities as the nation shifts out of lockdown; and to organize and manage sales missions and roadshows for key markets. Other elements are to create niche content to promote the various sectors of the industry; grow the Indigenous pavilion at Rendez-Vous Canada and other trade and media events; implement targeted domestic and international marketing campaigns using The Original Original brand; deliver one-on-one coaching and financial support to members for digital sales and marketing training; expand travel trade and media networks in key markets; increase investments in partnerships with TripAdvisor, Airbnb and other online travel agencies; build on the brand awareness of The Original through Indigenous channels; optimize the user experience of Destination Indigenous websites and social media channels to showcase and promote ITAC members; create content with industry partners to support media, trade and marketing campaigns; and deliver influencer and storytelling campaigns.

Required Investment: The 2022–2023 Action Plan is the first step in the implementation of ITAC's three-year strategy goals and priorities. The required investment of $21,276,100 will support ITAC's work towards returning to, and then moving beyond, pre-COVID-19 industry statistics for direct GDP contributions, the number of Indigenous tourism businesses and the number of Indigenous tourism employees.

Deliverables Are to Be in Four Areas

Leadership: Increasing positive industry awareness, advocacy and accountability in order to establish funder confidence and build support for ITAC and its members and ITAC's provincial and territorial Indigenous partners.

Partnerships: Uniting the Indigenous tourism industry in Canada by bringing together members and aligning and leveraging relationships with provinces and territories, and creating national partnerships to support Indigenous tourism.

Development: Working in partnership with provincial and territorial Indigenous tourism associations to encourage member development and to grow investment in Indigenous tourism, and further development of provincial and territorial Indigenous tourism networks.

Marketing: Promoting Indigenous tourism businesses and the positive community and cultural impacts their operations have; conducting high-value, targeted marketing as well as research, media and sales efforts; leveraging marketing resources with provincial and territorial organizations to maximize investment; and raising awareness of ITAC as a national and international brand.

The Original Original

The ITAC launched its new campaign, The Original Original, on Indigenous People's Day, June 21, 2021. The campaign aims to educate travellers, modernize their perception of Indigenous experiences and rebuild the industry, which was disproportionately devastated by the pandemic.

A key component of The Original Original is a new brand mark (Figure 31.1) that will help travellers better identify and book experiences from Indigenous-owned tourism businesses across Canada.

"The Original Original campaign is a reflection of our communities as they really are: diverse, authentic, empowered and current" (Keith Henry, president and CEO of ITAC).

> Our greater mandate at ITAC is to leverage tourism to help support the revitalization and broader understanding of Indigenous culture in a way that contributes positively to Indigenous communities. The Original Original mark supports this mandate by helping travellers better distinguish and support authentic businesses, and lift our voices.
>
> (Indigenous Tourism Association of Canada (adobe.com)

The Original Original mark identifies tourism businesses that have been vetted by ITAC using four key criteria: the business is at least 51 per cent Indigenous-owned, it's a 31business that embraces the values of Indigenous tourism, it offers a market or export-ready experience, and is an ITAC member.

The Original Original mark artwork aims to explore the ethos of this very concept by placing two-letter "Os" within each other, representing the world, as well as the cycle of life. At the centre of these circles is a fire symbol that possesses a single flame, but is divided into three parts. This distinction represents each of the three groups of Indigenous peoples in Canada: First Nations, Métis and Inuit. Through this branded mark, ITAC aims to further develop widespread recognition of authentic Indigenous experiences across the country. (*These can be accessed via the following websites.*)

DestinationIndigenous.ca is ITAC's newly launched digital experience platform. Travellers can check out the platform to find out what is open now to discover. Be sure to look for The

Figure 31.1   The Original Original.

Original Original branded mark to ensure the experience is an authentic one. Find authentic Indigenous vacation packages and enjoy once-in-a-lifetime experiences alongside legendary hospitality with Indigenous vacation packages.

indigenouscuisine.ca lists Indigenous culinary experiences, restaurants and recipes from across Canada. For a list of experiences and storytellers, download ITAC's Culinary Directory.

buyauthentic.ca features Indigenous artisans and makers across the country that can introduce you and your audience to authentic Indigenous arts and crafts.

canadianpowwows.ca lists powwows and guest protocols. Experience our living cultures at a powerful gathering of Indigenous celebration.

Appendix

ITAC has recently begun a new study to enable increased participation by Indigenous women in tourism as described below.

The Northern WE in Tourism Study

In addition to the Building Back Better programme outlined above, another specific initiative is *The Northern WE in Tourism Study.* (https://indigenoustourism.ca/programs-services/the-northern-we-in-tourism-study)

The Purpose of the Study

Working alongside women in the North to identify pathways to culturally responsive training, resources and wrap-around services by exploring the journey of best practices in each community, we will make informed recommendations for investment. The Northern WE in Tourism team will be conducting interviews with organizations that identify as providing training, wrap-around support, emotional support, financing/funding, networking opportunities and resources to women entrepreneurs in tourism in the North. The Northern WE project is funded by the Government of Canada under the *Future Skills* programme.

ITAC, in partnership with allied researchers, is creating sustainable livelihoods for Indigenous women entrepreneurs in tourism in the Canadian North. Information gathered through The Northern WE in Tourism Study will result in increased collaboration between women entrepreneurs in the North to identify pathways to sustainable livelihoods. Ensuring women have access to the resources, services and networks required to support them in building sustainable livelihoods through tourism will increase well-being and strengthen destination development in the Northwest Territories, Yukon, Nunavut.

The Northern WE in Tourism study provides ITAC with insights, data and means to fix a broken system. This partnership will apply Indigenous knowledge to build a systems response based on the recommendations and relationships developed in this study. Connecting Indigenous women to networks that link them to the social determinants of health through tourism will improve personal well-being and enhance the well-being and economic success of their home communities.

References

Indigenous Tourism Association of Canada (adobe.com)
https://indigenoustourism.ca/programs-services/the-northern-we-in-tourism-study/
https://indigenoustourism.ca/wp-content/uploads/2022/01/ITAC-Building-Back-Better-2022-2025.pdf

32

INDIGENOUS TOURISM INTERNATIONAL FRAMEWORK, RIGHTS, AND EMPOWERMENT OF GRASSROOTS ORGANIZATIONS

Latin America and the Chilean Case

Jean-Philippe Le Moigne

Introduction Indigenous Tourism in the Americas

Indigenous tourism has experienced significant expansion in recent decades. Since the first steps taken by Canada, Australia, and New Zealand, it now occupies an important place on the global scale, getting the respect of Indigenous communities, government agencies, the tourism industry (tour operators and tourism agencies), and non-governmental organizations. In Latin America and the Caribbean the development of Indigenous tourism is no exception. Significant movements can be considered to have been started internationally in the region by two key events. The first was an event developed in San José de Costa Rica by the International Labor Organization (ILO) through its program REDTURS (Sustainable Tourism Network in Latin America), from which the Otavalo Declaration emerged. This is a document developed by the participants in the event, establishing "the need to promote genuine expressions of the cultures of Indigenous peoples as a source of differentiation and competitiveness in the national tourist offer" and two years later the same program released the Declaration of San Jose, which mentioned the importance of "the self-management of tourism and the leading role of communities in the planning, development and supervision of activities in their territories".

The second relevant event occurred in 2002 with the Indigenous Tourism Forum celebrated in Oaxaca Mexico during the International Year of Ecotourism declared by international organizations linked to the United Nations through the UNWTO. The Ecotourism World Summit held in Quebec City (Canada) delivered a Declaration about Ecotourism development (UNWTO, 2002), the same year the World Summit on Sustainable Development was held in Johannesburg in South Africa (United Nations, 2002), ratifying the compromize on sustainable development after the Rio Summit in 1992. These two initiatives in the same year showed the world that humans must make key changes in their behaviors aiming to protect biodiversity.

419

DOI: 10.4324/9781003230335-38

Even with this significant movement towards "sustainability", Indigenous peoples working in tourism felt that ecotourism was a green-washing movement that was giving more benefits to non-Indigenous peoples than supporting Indigenous entrepreneurs and Indigenous lands. This can be seen in the declaration made by Indigenous peoples in the Indigenous Tourism Forum in Mexico, where the Indigenous participants explicitly expressed their view that the sustainability and ecotourism movement had left them out and that ecotourism has been used by the tourism industry to do greenwashing (Rethinking Tourism Project, 2002). In the same document, the Indigenous voices were raized for the first time in invoking Indigenous rights in the development of tourism.

Holding these events in Mexico was no coincidence, since the movement as such had already begun there a couple of years before and the United Nations had already become aware of this, creating in 2000 the REDTURS project (Sustainable Tourism Network among Indigenous and Rural Communities in Latin America) which was financed directly by the ILO. This organization had begun working with different organizations in a first step (2000–2002), mainly in the countries of Bolivia, Brazil, Costa Rica, Ecuador, Guatemala, and Peru, which was to be expanded to other countries of the continent in the coming years, until its disappearance related to the end of funding by ILO. REDTURS had the purpose of "accompanying the communities in the processes of reflection, search for solutions and application of strategies to face the challenges of the globalized tourism market, enhancing its strengths and overcoming its shortcomings" (ILO, 2006: 7), thus making it possible to combat poverty in rural areas.

This ILO project can be considered to be the initiative that really triggered the development of Indigenous tourism in the region, regardless of the fact that Indigenous tourism was not called as such. In most of the countries of the American continent at that time (and until approximately 2015) when talking about tourism of Indigenous communities, this was designated as rural tourism, community tourism, community-based tourism, original tourism, or even solidarity tourism, but it was hardly ever heard of as Indigenous tourism. There were some exceptions such as in Mexico, where Indigenous tourism began to be talked about at the beginning of the 21st century, but only by Indigenous organizations and not from government agencies.

In the years in which REDTURS was active in the area, the ILO recognized that Indigenous peoples shared a common experience of marginalization and discrimination and that the non-recognition of the rights of Indigenous peoples to their lands, territories and natural resources affected these peoples who were dedicated to their traditional activities. The project also allowed those involved to recognize problems faced by this type of enterprize, such as the lack of access to credit and marketing resources (still present today). It is no coincidence that this project was carried out by ILO, since it is this organization that published the first international document that deals with the protection of the rights of Indigenous peoples, through the famous convention on Indigenous and tribal peoples (ILO, 1989).

Although this project was important in the region, there are other issues that are relevant to consider in understanding the evolution of Indigenous tourism in the region and the relationship between the grassroots organizations and the many Government agencies.

The International Spectrum (International Laws and Technical Documents)

Twenty-one documents linked to Indigenous rights have appeared worldwide between 1948 and 2020 (see Figure 32.1), including international declarations, conventions, and technical documents about human rights, Indigenous peoples, tourism, and sustainability. This aspect of Indigenous tourism has not been well documented, but authors such Whitford and Ruhanen (2016: 4) recognize that the international Indigenous legal framework is a key element in Indigenous tourism development. In reality, Indigenous tourism cannot be conceived without considering Indigenous rights, notwithstanding the fact that some people argue that Indigenous tourism is just another form of tourism, because it is impossible not to recognize that discussion is about tourism developed by native peoples, therefore their cultures, territories, and rights are at stake. In this context Baleva (2019) offers an interesting overview about the implications of tourism and Indigenous rights land, showing through examples around the world, how tourism has caused "eviction from traditional lands, overuse of habitat related to increased tourist demand and the destruction of habitat to create tourism infrastructure" (Baleva, 2019: 3).

The international Indigenous rights movement started after the Second World War with the Declaration of Human Rights in 1948. In 1989, the convention No. 169 of the ILO was released, initiating and revealing a path towards the protection of Indigenous rights (ILO, 1989: 2). Later in 2007, a milestone was achieved with the United Nations Declaration on the Rights of Indigenous Peoples, known as UNDRIP. All these documents played a role in Indigenous tourism that must be considered in the planning process of tourism.

At the level of the American continent, it is appropriate to also consider the American Declaration on the Rights of Indigenous Peoples (ADRIP) (OAS, 2016), which connects with Indigenous tourism directly through UNDRIP, and which was the base for the Larrakia Declaration (discussed in the following section). The Larrakia Declaration is a document created in 2012 by Indigenous people that established the principles of Respect, Protection, Empowerment, Consultation, Business, and Community to guide the development of Indigenous tourism globally (PATA, 2015: 15, 63–67).

Regarding the technical documents, during the last few years, the United Nations has released many publications, of which the Sustainable Development Goals 2030 (UN, 2015), the Global Code of Ethics (UNWTO, 2001), and recommendations on sustainable development of Indigenous tourism (UNWTO, 2019) are notable. These technical documents (some of them refer directly to Indigenous tourism, others to Indigenous businesses in general or to how to work with Indigenous peoples) have created a clear route to follow by giving examples and methodologies. The emergence of international technical documents is extensive, and the list is long with more than 27 laws, declarations, and publications that influence the development of Indigenous tourism directly or indirectly. Figure 32.1 shows the evolution of these documents through the years, and it is possible to see that since 2016 there has been an exponential growth linked to an international interest in Indigenous rights and Indigenous tourism.

All of these documents are important, and the following (Table 32.1) is a list of those that have supported the development of Indigenous tourism in Latin America.

The Success of Indigenous Tourism Organizations in the International Scene

It is no secret that Indigenous tourism has increased in importance on the international scene, probably as an answer to the advancement of international laws and technical documents. One of the key elements linked to this Indigenous tourism "revolution" is the creation of grassroots organizations.

At the international level, the World Indigenous Tourism Alliance (WINTA) emerged as a global umbrella organization implementing the United Nations Declaration on the Rights of Indigenous Peoples (UNDRIP) through tourism worldwide by way of the application of the *Larrakia Declaration* principles, the key document written up by the first Pacific Asia Indigenous Tourism Conference (PAITC) held in Darwin Australia, in the land of Larrakia Aboriginal people. WINTA

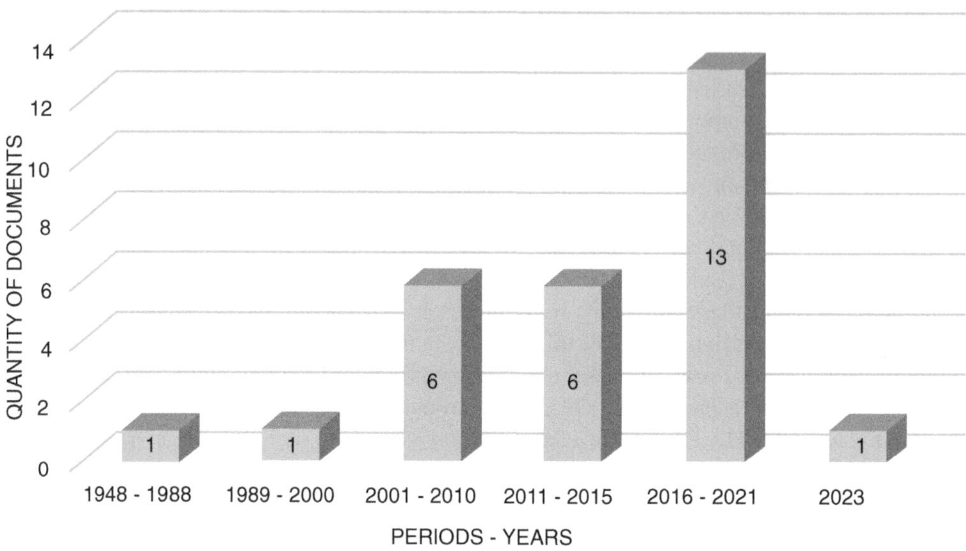

Figure 32.1   Evolution of declarations, conventions and technical documents related to Indigenous tourism development.

Source: Author, J. P. Le Moigne, Indigenous Tourism Alliance (WINTA).

Table 32.1 International laws and technical documents affecting Indigenous tourism development

International documents	Year
The Universal Declaration of Human Rights (United Nations)	1948
Convention 169 on Indigenous and tribal peoples in independent countries. International Labor Organization (ILO)	1989
United Nations Declaration on the Rights of Indigenous Peoples (UN)	2007
Larrakia Declaration on the development of Indigenous Tourism (UNWTO - WINTA)	2012
American Declaration on the Rights of Indigenous Peoples (OAS)	2016

Technical documents	Year
Indigenous Human Rights in the Asia-Pacific Region. Review, Analysis and Check-list (PATA)	2015
Indigenous peoples and the travel industry. Global good practice guidelines (GAdventures)	2017
Protect and Promote Your Culture: a Practical Guide to Intellectual Property for Indigenous Peoples and Local Communities (WIPO)	2017
National Guidelines: Developing Authentic Indigenous Experiences in Canada (ITAC)	2018
Recommendations of the World Committee on Tourism Ethics on Sustainable Development of Indigenous Tourism (UNWTO)	2019
UNWTO Inclusive Recovery Guide Sociocultural Impacts of COVID-19 Issue 4: Indigenous Communities (UNWTO)	2021
Compendium of Good Practices in Indigenous Tourism – Regional Focus on the Americas (UNWTO – WINTA)	2023

was implemented with the aim of creating and supporting an international network of individuals and Indigenous groups dedicated to the development of tourism. Its primary goal is to support cooperation and the implementation of strategies for the development of Indigenous tourism. It does this by working with the tourism industry and government agencies linked to tourism and Indigenous development in different countries to promote partnerships and strengthen respect for Indigenous wisdom, values, and knowledge, and by supporting Indigenous communities and Indigenous tourism organizations (see www.winta.org). One of the first publications that appeared to guide Indigenous tourism technically was published by the Pacific Asia Travel Association (PATA) and WINTA in 2015. Entitled *Indigenous Tourism & Human Rights in Asia & the Pacific Region* (PATA, 2015), the document offers a checklist built upon core aspects of the Larrakia framework, identifying their relevance to four stakeholder groups (the Indigenous communities, private sector tourism developers working with or from Indigenous communities, public sector authorities at a local or national level that govern tourism, and non-government agencies that advocate or support responsible tourism development with Indigenous peoples.

It is important to highlight that WINTA has had a close working relationship with the United Nations World Tourism Organization (UNWTO) since 2012, after the recognition of the *Larrakia Declaration* through partnership work for the development of relevant technical publications that have guided and established parameters for the development of Indigenous tourism worldwide. These include the recommendations for the development of Indigenous tourism (UNWTO, 2019), which for the first time provides recommendations for the different actors that are involved with tourism activities, including tour operators and travel agencies, in the design of Indigenous tourism products, the distribution of benefits for Indigenous communities, and contributions to the empowerment of Indigenous communities. It also includes recommendations for tour guides, Indigenous communities, and tourists. During the COVID pandemic, WINTA and UNWTO, with the support of other organizations, published a document titled *"Inclusive Recovery Guide Socio-cultural Impacts of COVID-19 Issue 4: Indigenous Communities"* aiming to support Indigenous communities in the recovery process after the tremendous damage caused by COVID-19 and the resulting travel restrictions.

WINTA is also the custodian of two important international events, the World Indigenous Tourism Summit (WITS), and the PAITC. PAITC has met in Australia (Darwin, 2012), Canada (Vancouver, 2015), and Chile (2021), and WITS has been held in Aotearoa New Zealand (2018), Perth, Australia (2023) and Kaohsiung, Taiwan (2024). These events are very relevant because they create an opportunity for dialogue between leaders and organizations of Indigenous Tourism so as to develop a joint vision of the future and further empowerment related to the development of Indigenous Tourism worldwide.

Before explaining the actual situation of Indigenous tourism in Latin America I would like to acknowledge the Indigenous tourism organizations that are leading the movement at the worldwide level and that have gained recognition because of their professional performance, Indigenous tourism research, and the creation of tools and methodologies. Considering the parameters mentioned it is essential to include in this list the Indigenous Tourism Association of Canada (ITAC), the Western Australia Indigenous Tourism Operators Council (WAITOC), NZ Māori Tourism, and the American Indian Alaska Native Tourism Association (AIANTA). All of these are leading the Indigenous tourism development in their respective countries and have gained respect within their relevant Government agencies and tourism industries. Table 32.2 lists these organizations.

Readers should pay particular attention to the case of ITAC in Canada, which has shown that it is possible to have an Indigenous leadership in charge of tourist activity at the national

Table 32.2 Leading Indigenous tourism organizations

Name	Abbreviation	Country	Website
World Indigenous Tourism Alliance	WINTA	Global	www.winta.org
Indigenous Tourism Association of Canada	ITAC	Canada	www.Indigenoustourism.ca
Western Australia Indigenous Tourism Operators Council	WAITOC	Australia	www.waitoc.com
Māori Tourism New Zealand	MTNZ	New Zealand	www.maoritourism.co.nz
American Indian Alaska Native Tourism Association	AIANTA	United States	www.aianta.org

level, assisting with building governance of tourism activities in Indigenous territories through joint work with provincial and territorial Indigenous tourism organizations. In addition, ITAC has taken giant steps in terms of working with partners in the Canadian tourism industry and especially with government agencies who have believed in the organization mainly because of the professionalism demonstrated and the results obtained from a constant research process regarding the contribution of Indigenous tourism to the Canadian economy. The organization is currently carrying out a marketing campaign that has won international awards and is also working on reconciliation through tourism activity [see also Chapter 31 on ITAC's COVID recovery program].

Indigenous Tourism Governance in Latin America

In the last two decades, there have been interesting developments in South and Central America with the creation of organizations there linked to community-based and Indigenous tourism, some of them now with more than 25 years of experience. The following Table 32.3 shows these organizations.

Today, in the Americas, the Organization of American States (OAS), works with participating governments to support community-based and Indigenous tourism development, establishing a platform for community leaders, policy-makers, and business owners to share experiences. Since 2017 OAS has been working in\on this specific issue, beginning with the Vice-Ministry of Tourism in Peru where the program has focused its work on rural community tourism. Since 2019, however, it has changed its direction on Indigenous tourism by working directly with the Ministry of Indian Affairs of the United States, supported by George Washington University.

It is important to mention that the development of Indigenous tourism in the Americas (called in most of the cases community tourism or rural tourism), started at the end of the last century, for example, tourism and Indigenous peoples started in the 1980s linked to ecotourism (Ruiz Ballesteros et al. 2008: 403). In Peru in the mid-1990s many communities in rural areas experienced rapid tourism growth (Mitchel in McCool and Moisey, 2009: 158), and also in Panama, again in the 1990s, where the Guna people created the Statute of Guna Tourism (Pereiro, 2016).

The whole process has been supported by different organizations and marked by important events through the years since, taking small steps, as shown by the following list of the most relevant events performed (Table 32.4).

It is important to emphasize that every country has organized different events about community and Indigenous tourism at national, regional, provincial/departmental, and local levels. The list is extensive but, in particular, the annual events of The ITAC and the AIANTA in the United States should be noted, while other organizations such as RITA in Mexico, FEPTCE in Ecuador,

Table 32.3 Indigenous and community-based tourism organizations in Latin America

Name	Abbreviation	Country	Website
Red Indígena de Turismo de México	RITA	Mexico	www.rita.com.mx
Red Boliviana de Turismo Solidario Comunitario	RED TUSOCO	Bolivia	www.tusoco.com
Federación Plurinacional de Turismo Comunitario del Ecuador	FEPTCE	Equator	Facebook: TurismoComunitarioEc
Red Argentina de Turismo Rural Comunitario	RATuRC	Argentina	Facebook: turismocomunitarioargentina
Asociación Pro-Comunidades Turísticas de Honduras	LARECOTURH	Honduras	www.larecoturh.org
Asociación Nacional de Turismo Indígena de Chile	ANTI	Chile	www.turismoindigena.com
Red de Turismo Comunitario Indígena de Panamá	REDTURI	Panama	www.redturi.org
Red Nacional de Turismo Indígena del Perú	REDNATI	Peru	web.facebook.com/ rednatiperu/?_rdc=1&_rdr
Rede Brasileira de Turismo Solidário e Comunitário	Turisol	Brasil	https://turisol.wixsite.com/ redeturisol
Asociación Nacional de Turismo Indígena de Colombia	ASONTIC	Colombia	www.asontic.org

Table 32.4 Examples of community-based and Indigenous tourism events in the Americas

Event	Year
1st Latin American Community Tourism Encounter, Latin American Sustainable Tourism Network (REDTURS) – International Labour Organization (ILO) (Otavalo, Ecuador)	2001
International Indigenous Tourism Forum (Oaxaca, Mexico)	2002
2nd Latin American Community Tourism Encounter REDTURS – ILO (San Jose, Costa Rica)	2003
1st Encuentro de Turismo Indígena de las Américas (Bariloche, Argentina)	2011
2nd Pacific Asia Indigenous Tourism Forum, Indigenous Tourism British Columbia (ITBC) – WINTA	2015
International Aboriginal Tourism Conference, Tourisme Autochtone Québec (TAQ) – ITAC – WINTA	2015
1st Latin America Community Tourism Encounter (Travolution Foundation)	2017
1st Central America Community Tourism Encounter (Travolution Foundation)	2018
1st Community Rural Tourism Forum, Organization of the Americas States (OAS) – Peru	2018
Indigenous Tourism Forum of the Americas (OAS – George Washington University - Ministry of Indian Affairs US)	2020
1st Indigenous Tourism Forum Latin America (WINTA – virtual)	2020
Community Tourism Encounter (Travolution Foundation – Komu – La Mano del Mono)	2020
3rd Pacific Asia Indigenous Tourism Forum, Chilean Government – WINTA – Chilean Indigenous Tourism Association (ANTI) – Mapuche Tourism Society (STM)	2021

TUSOCO in Bolivia, and RATuRC in Argentina also have yearly meetings but not always as an open event.

This information leads to confirmation that empowerment is the basis of the success of Indigenous tourism, expressed through the creation of grassroots organizations that enable Indigenous peoples working with state agencies on historical demands, requirements, and points of view on tourism development in Indigenous territories. These issues in particular have proven to be key to the success of Indigenous tourism activity, as empowerment can derive from Indigenous tourism governance, including not only the creation of organizations, but also joint work between government agencies that are related to tourism, natural resources, promotion, and Indigenous development generally.

The Chilean Case

According to the 2017 census, 12.8% of the population in Chile replied affirmatively to the question: "Do you consider yourself as belonging to an Indigenous people?". Of this percentage, 79.8% are considered to belong to the Mapuche people, 7.2% Aymara, and 4.1% Diaguita (INE, 2019). The other six Indigenous populations are broken down according to the following percentages: Rapa Nui 0.4%, Lickanantay 1.4%, Quechua 1.5%, Colla 0.9%, Kawéskar 0.1%, and Yagán 0.1%.

Indigenous tourism in Chile is defined as "a tourist activity carried out and directed by communities, families and / or Indigenous peoples, developed in an urban, rural or natural space, historically occupied by Indigenous peoples, combining their customs and traditions, ancestral and contemporary, encouraging a process of cultural exchange with the visitor or the tourist", in Subsecretaría de Turismo de Chile, 2019. This is the most well-known and recognized definition but is being actively debated as different projects require a rethink of the concept in an effort to develop a more accurate definition for the Chilean context. The development of Indigenous tourism in Chile began in the mid-1960s in Rapa Nui (Easter) Island, but it can be considered that for the continental Chilean territory, Indigenous tourism started about 25 years ago, one of the pioneers being the Mapuche – Lafkenche community called Llaguepulli, located in the coast of Araucania Region. This community started working in tourism in 1998, and before that time some communities received visitors sporadically, without intentional tourism planning. Another early development was by the Lickanantay people in the north of Chile in San Pedro de Atacama.

From the Government perspective, since the 1990s, a number of programs have been introduced to assist the development of Indigenous peoples, most notably a program called "*Orígenes*," a multi-phase program for the integrated development of Indigenous Peoples, which ran from 2001 to 2012 and was funded by the Inter-American Development Bank and the Chilean government (De La Maza-Cabrera, 2018). It is also important to mention the "Rural Tourism" program of the Agricultural Development Institute (hereinafter INDAP for Instituto de Desarrollo Agropecuario), which started in 1995, with the objective of generating support and conditions to promote advisory actions, training, and promotion with user companies or potential users dedicated to this field (INDAP). To date, the program has reached 37% of the total beneficiaries served and has created a special sub-program called the "Indigenous Territorial Development Program" (hereinafter PDTI [for Programa de Desarrollo Territorial Indígena]). The PDTI has served nearly 48,000 users (estimated December 31, 2017), belonging to 7 Indigenous villages, in 10 of the 15 regions of Chile (Information retrieved from www.indap.gob.cl/pueblos-originarios). The year 2012 marked an important milestone as the Undersecretariat of Tourism published the "National Tourism Strategy 2012–2020", which defined Indigenous tourism as a priority thematic area. It stated that by 2020, "the Indigenous peoples present in tourist destinations will see tourism as a development path that respects and values the conservation of their identity, their culture and their traditions" (Subsecretaria de turismo de Chile 2019). This definition was very important and began a stronger process of Indigenous tourism development in the country. In February 2022 the new Chilean Tourism Strategy (2022–2030) was published and indigenous tourism was included, the Strategy identifying it as a prioritized tourist experience for the country.

In 2015 the Undersecretariat of Tourism, the National Tourism Service (hereinafter SERNA-TUR [for Servicio Nacional de Turismo]), and the National Corporation for Indigenous Development (hereinafter CONADI [for Corporación Nacional de Desarrollo Indígena]), began to become more deeply involved with the signing of an agreement for the development of projects to support the development of Indigenous tourism at the national level. The most relevant projects include in 2015, the first International Forum for Indigenous Tourism (FITO), an event that was linked strategically to the ATTA Summit aiming to attract people at the international scene, this forum marked an initial point in the new era of Indigenous tourism development in Chile. In 2016, a key action saw the creation of the "First School for Indigenous Tourism Leaders", and this project provided training for Indigenous tourism leaders who later would create the National Indigenous Tourism Organization (ANTI). The other important project that year was the first draft of an action plan for Indigenous Tourism in Chile; while in 2017, there was the design and implementation of integrated Indigenous tourism products. In 2019, three key projects were developed the production of a manual, the design of an action plan for the development of Indigenous tourism strengthening the supply of Indigenous tourism products; and the creation of a guideline for the development of authentic Indigenous tourism. These were intended as decision-making tools for Indigenous peoples and as a guide for Indigenous entrepreneurs who wanted to work in tourism. Most of these projects were led by the Travolution Foundation and financed by government agencies such as the Undersecretariat of Tourism, Sernatur, and Conadi.

Despite the arrival of the COVID pandemic, the development process of Indigenous tourism continued and between 2020 and 2021 two important projects were carried out. The first, led by the Catholic University of Villarrica, aimed at creating an Indigenous tourism certification for the country, a process that included co-design work and joint work with Indigenous tourism leaders who represented the ten Indigenous peoples currently recognized in Chile. The second project was in 2021, when Chile was the host of the third Pacific Asia Indigenous Tourism Conference, an event led by the Chilean Indigenous Tourism Association (ANTI), the Mapuche Tourism Society, and WINTA, supported by Government agencies headed by the Undersecretariat of Tourism. It is important to mention that throughout the process since 2015 the Chilean government and Chilean Indigenous tourism leaders have created an international network, working with ITAC, and WINTA.

As previously explained, Chile has a National Indigenous Tourism Association (formed in 2017), which started conversations with the Undersecretariat of Tourism and was invited to be part of the National Table of Indigenous Tourism, formed in 2016 with the objective of generating coordination and articulation between the public actors that participate in the development of Indigenous tourism and led by the Undersecretariat of Tourism (De La Maza-Cabrera and Calfucura-Tapia, 2021). Participants include the National Tourism Service (SERNATUR), the National Corporation for Indigenous Development (CONADI), Production Development Corporation (CORFO), the Ministry of Health, the Ministry of Transportation, and the National Institute for Agricultural Development (INDAP) through its Rural Tourism program, with other bodies. Until then participation in the National Table of Indigenous Tourism was open and the incorporation of ANTI transformed this entity into a public-private space. This invitation was key, because from that moment on, the Undersecretariat of tourism began to integrate ANTI into most of the various actions aimed at supporting Indigenous tourism in the country. This paved the way for other organizations to be considered, such as the Mapuche Tourism Society.

There are many examples of Indigenous tourism development throughout the Chilean territory, this chapter does not have space to include every single one, so two specific cases are highlighted because they are linked to territorial management through the administration of tourism in emblematic protected areas.

The first is the case of the Lickanantay people in San Pedro de Atacama. The Los Flamencos National Reserve was established in 1990, and in 1997 the Chilean government created the Indigenous Development Area (ADI) Atacama La Grande, recognizing the Likanantay people (Rauch González et al., 2018). Later in 2002, the National Forestry Corporation (CONAF) began to work with the different communities (Ayllus) promoting an associative management model focused on the co-administration of the public use (tourism) in four sectors of the national reserve by eight Lickanantay communities. In addition, some Lickanantay communities manage sectors of great archaeological value for which they have received the support of the Corporation for Indigenous Development (CONADI) and the National Monuments Council (Bustos, 2005).

Lickanantay people have a grassroots organization named Consejo de Pueblos Atacameños, in which there are representatives of each Lickanantay Ayllu (Community). During the years before the pandemic the organization worked very hard making known the importance of respecting the rights and territory of the Lickanantay people, having, in many cases, to go against decisions made not only by non-Indigenous tourism businessmen who work in the territory, but also decisions made by Government agencies responsible for tourism development in the region. Currently, the organization has a person in charge of Lickanantay tourism and continues working with great effect demanding that tourism in their territory deliver benefits to the local residents rather than the negative effects that the activity has shown during the last ten years.

The second case is that of the Rapa Nui People. In 1935 the Chilean State created the Rapa Nui National Park (CONAF, 1997: 57), which covers approximately 43.5% (7,150.88 hectares) of the total area of Rapa Nui Island (usually known as Easter Island). According to official data from the National Forestry Corporation (CONAF), that same year the protected area was declared a National Historical Monument and in 1995 UNESCO declared it a World Heritage Site. Since 1973, the Government of Chile has granted free use to the National Forestry Corporation on Easter Island, with the administration of the park falling to this entity. The tourism activity in Rapa Nui Island started in 1967 and in 1971 the first hotel was built, by 1972 air passenger arrivals were around 4,300 per annum (Porteous, 1980: 137). Since that time tourism has grown exponentially centered on the famous archeological attractions of Rapa Nui culture represented in the statues named Moai found throughout the island.

In 1990, the Rapa Nui people began a territorial claim process. In 2015, they made a request to transfer the lands of the National Park, based on the Pascua Law 16,441 and the Indigenous Law 19,253. In 2016 the Rapa Nui people created an organization named Ma'u Henua Community, which was granted the administration of Rapa Nui National Park in 2017. Since that time the new administration significantly increased profits from tourism. The pandemic severely affected the island which was closed to visitors in March 2020 only reopening in August 2022. During this period of time the organization declared that they wanted to develop a more sustainable form of tourism on the Island, with experiences more directly linked to Rapa Nui people's daily life, aiming to give more benefits to the islanders from Rapa Nui Indigenous Tourism.

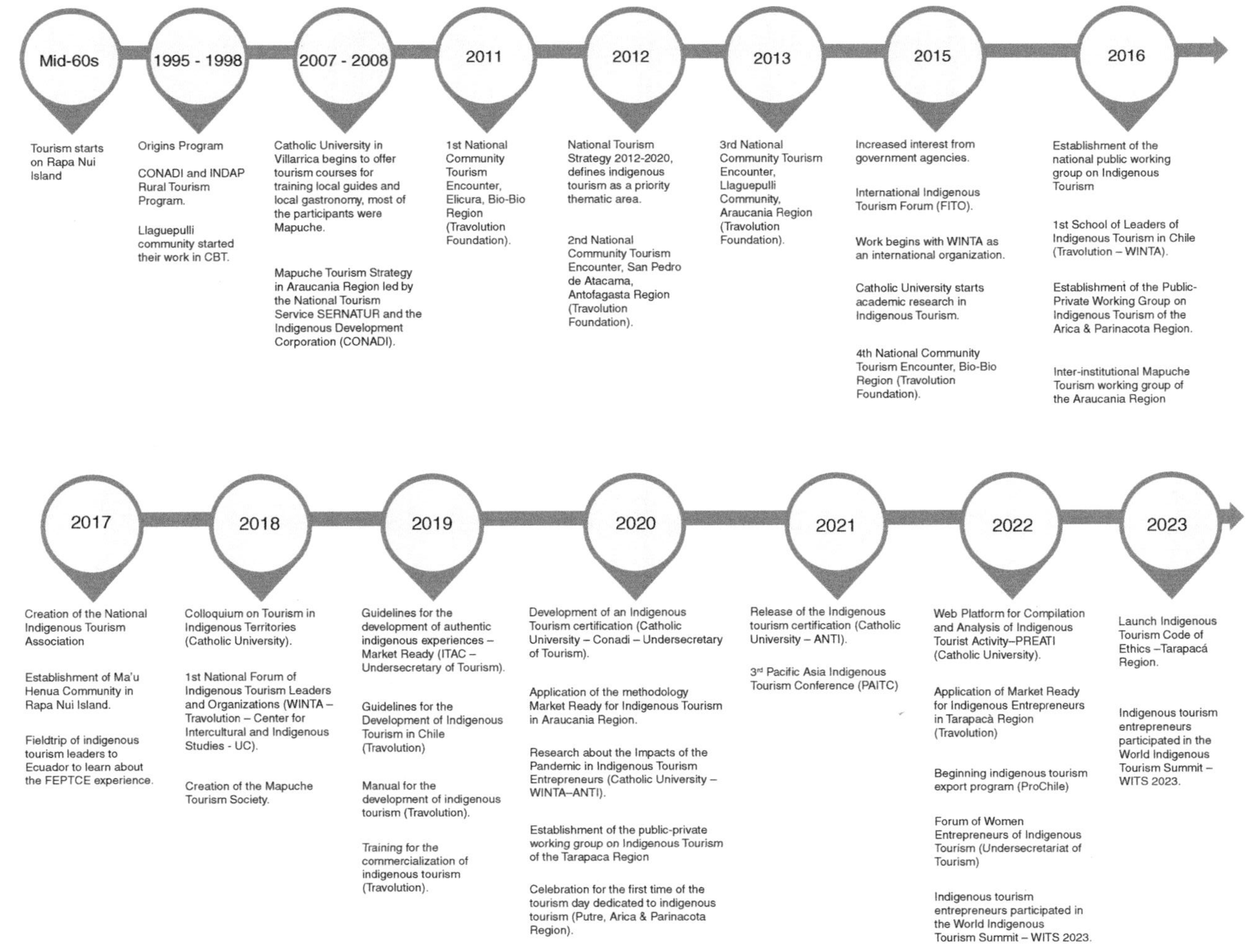

Figure 32.2 Timeline Indigenous tourism development in Chile.

Conclusion

Indigenous tourism in Latin America already has a long history, starting at the end of the 1980s in Peru and in the mid-1990s in Ecuador and Panama. This participation and promotion began with a process from the communities that was later promoted by the International Labor Organization through the initiative called Latin American Sustainable Tourism Network, REDTURS. This initiative worked effectively, managing to leave a mark in several countries and laying the foundations for the relevance of the inclusion of Indigenous tourism as a differentiating cultural element of the offer (through the Otavalo Declaration). It also laid the foundations of the importance of Indigenous self-determination, regarding to determine how, when, and where tourism should be developed in Indigenous territories (through the San Jose Declaration). Subsequently, many key actions have been taken by Government agencies, grassroots organization, NGOs, and Universities.

It is important to recognize that relations between governments and Indigenous organizations that work in tourism go back and forth and often depend on the surrounding political environment in the individual countries. These variations in cooperation can be minimized by creating public policies that allow progress with a clear direction that lasts over time and in order to achieve this goal, grassroots organizations must have the necessary strength and professionalism.

Indigenous tourism development in Chile commenced intentionally with tourism activity on Rapa Nui in the 1960s and by the mid-1990s with the Lickanantay people in San Pedro de Atacama, pioneer Mapuche communities in the Wallmapu (the ancestral Mapuche territory) and the Llaguepulli community (Mapuche – Lafkenche people). The Government also saw the potential of Indigenous tourism in the 1990s, but it was not until 2012 that Indigenous tourism was recognized as a formal tourism experience in the National Tourism Strategy. The year 2015 marked a new era for Indigenous tourism with the development of important projects that allowed the creation of methodologies, strategies, governance, and empowerment for Indigenous people. All this work permitted Chile today to be recognized as one of the countries with considerable development in this area in Latin America and with a very promising future that has to put more emphasis on the activation of Indigenous rights through tourism.

Indigenous tourism in Latin America has great potential and it can be foreseen that it will continue to strengthen and develop. The important thing is that this process does not consider Indigenous tourism is just another type of tourism - it has to recognize that tourism with Indigenous communities includes their rights, and that tourism must be used by those communities as a tool to achieve their objectives. This includes territorial protection and the revitalization of culture and language, reduction of migration and depopulation of communities and villages, enhancement of their culture, and increases in income, in short, the activation of the rights of Indigenous peoples to achieve their long-awaited well-being.

References

Baleva, M. K. A. (2019). *Regaining Paradise Lost: Indigenous Land Rights and Tourism*. Brill: Nijhoff.

Bustos, A. (2005). Hacia un Turismo Intercultural: El Caso Atacameño. *Revista Lider*, 13: 133–150.

Butler, R. & Hinch, T. (1996). *Tourism and Indigenous Peoples*, 1st edition. London: International Thompson Business Press.

Corporación Nacional Forestal (CONAF). (1997). *Plan de Manejo Parque Nacional Rapa Nui*. Chile: Ministerio de Agricultura.

De La Maza-Cabrera, F. (2018). Tourism in Chile's Indigenous Territories: The Impact of Public Policies and Tourism Value of Indigenous Culture. *Latin American and Caribbean Ethnic Studies*, 13(1): 94–111.

De La Maza-Cabrera, F. & Calfucura-Tapia, E. (2021). Turismo y pueblos indígenas: políticas, irrupción y reivindicación en chile. *Chungará (Arica)*, 53(3): 546–552.

Delsing, R. & McCormick, G. (2012). Issues of Land and Sovereignty. The Uneasy Relationship between Chile and Rapa Nui, pp. 54–78 in F. E. Mallon, (Ed.) *Decolonizing Native Histories: Collaboration, Knowledge, and Language in the Americas*. Durham, NC: Duke University Press.

Estadísticas, I. I. N. (2019). *Síntesis de resultados Censo 2017*. Gobierno de Chile: 27.

G Adventures (2017). *Indigenous People and the Travel Industry - Global Good Practice Guidelines*. Available online at https://media.gadventures.com/media-server/dynamic/admin/flatpages/Indigenous_ Tourism_Guidelines_2017.pdf. Accessed 23 September 2023.

Indigenous Tourism Association of Canada (ITAC) (2018). *National Guidelines: Developing Authentic Indigenous Experiences in Canada*. Indigenous Tourism Association of Canada.

International Labour Organisation (ILO) (1989). *Convention 169 Indigenous and Tribal Peoples' Convention*: 9.

Instituto de Desarrollo Agropecuario (INDAP) (2023). *Programa Turismo Rural*. INDAP, Gobierno de Chile.

McCool, S. F. & Moisey, R. N. (2009). *Tourism, Recreation, and Sustainability: Linking Culture and the Environment*. Cambridge, MA: CABI.

OAS (2016). *American Declaration on the Rights of Indigenous Peoples*. Organization of the American States.

Pacific Asia Travel Association (PATA) (2015). *Indigenous Tourism & Human Rights in Asia & the Pacific Region Review, Analysis, & Checklists*. Pacific Asia Travel Association.

Pereiro, X. (2016). A Review of Indigenous Tourism in Latin America: Reflections on an Anthropological Study of Guna Tourism (Panama). *Journal of Sustainable Tourism*, 24(8–9): 1121–1138.

Porteous, J. D. (1980). The Development of Tourism on Easter Island. *Geography*, 65(2): 137–138.

Rauch González, M., Catalán Martina, E., Aguilera Bascur, G., Valenzuela Vergara, I., Maldonado Osorio, S., & Martínez Palma, P. (2018). Gestión intercultural para la conservación en Áreas Silvestres Protegidas del Estado: aprendizajes y desafíos. *Revista Austral de Ciencias Sociales*, (35): 183–204.

REDTURS (2006). *Memoria: IV Encuentro Latinoamericano Códigos de conducta y uso de marcas*. Red de Turismo Sostenible Comunitario para América Latina (REDTURS): 77.

Rethinking Tourism Project (2002). *Declaración del Foro Internacional Indígena de Turismo*. Oaxaca, México.

Ruiz Ballesteros, E., Hernández Ramírez, M., Coca Pérez, A., Cantero, P. A., & Campo Tejedor, A. (2008). Turismo comunitario en Ecuador. *PASOS Revista de turismo y patrimonio cultural*, 6(3): 399–418.

Subsecretaria de turismo de Chile (2019). *Propuesta de lineamientos para el desarrollo del turismo Indígena*. Gobierno de Chile.

United Nations (2002). *Report of the World Summit on Sustainable Development*.

United Nations (2015). *Agenda for Sustainable Development Goals 2030*: 41.

United Nations (2016). 2007 *UN Declaration on the Rights of Indigenous Peoples*. Available online at https://www.un.org/development/desa/Indigenouspeoples/declaration-on-the-rights-of-Indigenous-peoples. html. Accessed 23 September 2023.

UNWTO (2001). *Global Code of Ethics for Tourism*. UNWTO, Madrid.

UNWTO (2002). *Québec Déclaration on Ecotourism*. UNWTO Declarations, Madrid, 12 – N2: 30.

UNWTO (2012). Addendum: *Larrakia Declaration on the Development of Indigenous Tourism*. (Executive Council Ninety-Fourth Session Campeche, Mexico, 23–25 October 2012), UNWTO, Madrid: 3

UNWTO (2019). *Recommendations of the World Committee on Tourism Ethics on Sustainable Development of Indigenous Tourism*. UNWTO, Madrid: 2.

UNWTO (2021). *Inclusive Recovery Guide, Sociocultural Impacts of COVID-19 Issue 4: Indigenous Communities*. UNWTO, Madrid.

UNWTO (2023). *Compendium of Good Practices in Indigenous Tourism – Regional Focus on the Americas*. UNWTO – WINTA. UNWTO, Madrid.

Whitford, M. & Ruhanen, L. (2016). Indigenous Tourism Research, Past and Present: Where to from Here? *Journal of Sustainable Tourism*, 24(8–9): 1080–1099.

WIPO (2017). *Protect and Promote Your Culture: A Practical Guide to Intellectual Property for Indigenous Peoples and Local Communities*: 68.

33

DEVELOPMENT OF THE INAUGURAL QUEENSLAND FIRST NATIONS TOURISM ACTION PLAN 2020–2025 AND THE QUEENSLAND FIRST NATIONS TOURISM COUNCIL

Michelle Whitford, Rhonda Appo, Cameron Costello and Lisa Ruhanen

Introduction

As a tourism destination, Queensland, Australia has a competitive advantage in that it is '…. home to world-class Indigenous tourism experiences and has the unique ability for visitors to experience both Aboriginal and Torres Strait Islander culture' (Year of Indigenous Tourism, 2021:1). To leverage these opportunities, the inaugural Queensland First Nations Tourism Action Plan 2020–2025 and Queensland First Nations Tourism Council (QFNTC) were co-created, driven and developed by First Nations tourism stakeholders, including the authors of this chapter. Two of the authors, Rhonda Appo, Indigenous Program Manager, Queensland Tourism Industry Council (QTIC) and Cameron Costello, Chair, First Nations Tourism Working Group and Deputy Chair: Board of QTIC, continue to be instrumental in the ongoing development of the Queensland First Nations Tourism Action Plan and they are the driving force in the establishment of the QFNTC. The other two authors, in their capacity as working members of the First Nations Tourism Expert Advisor Working Group, have provided ongoing support and expertise to the design, development and implementation of both the Action Plan and the QFNTC.

Importantly, both these initiatives recognise opportunities to leverage the competitive advantage provided by unique cultural heritage and stewardship of country and facilitate the development of a thriving, distinctive and sustainable First Nations tourism sector across Queensland. Therefore, the purpose of this chapter is to share the authors' insights into the crucial role the voices of Aboriginal and Torres Islander peoples of Queensland played in the development and implementation of the inaugural Queensland First Nations Tourism Plan 2020–2025 and the QFNTC.

Indigenous tourism has experienced enormous change throughout the 20th Century and in particular, the first two decades of the 21st Century. Since the mid-20th Century, governments and development organisations, such as the United Nations and World Bank, have worked (to varying

DOI: 10.4324/9781003230335-39

degrees) with Indigenous stakeholders playing a functional role in encouraging the sustainable development of Indigenous tourism. Consequently, around the globe, the sector is increasingly being driven and controlled by Indigenous peoples and communities and more and more often, tourism is being utilised as a way to transform local economies (Ruhanen & Whitford, 2010, 2019) and/or as a platform to revitalise, rediscover and/or reinforce cultural identities (Mackley-Crump, 2016; Travesi, 2018). Moreover, Indigenous tourism has beencontinues to be seen as a key opportunity to promote employment, improve the rights and livelihoods of Indigenous peoples and concomitantly, address a wide range of socio-economic agendas (Fletcher, Pforr, & Brueckner, 2016; Ruhanen & Whitford, 2017; Ruhanen, Whitford, & McLennan, 2015; Whitford & Ruhanen, 2016). Thus globally, there has been a range of policies, strategies and programs established to support Indigenous tourism development for a chronicle of Indigenous people's involvement in tourism, see Whitford and Ruhanen (2016) including The National Aboriginal and Torres Strait Islander Tourism Industry Strategy (NATSITIS) which was published in 1997 in Australia (ATSIC, 1997; Whitford, Bell, & Watkins, 2001) and was arguable, the initial catalyst of sustainable Indigenous tourism development in Australia in the 21st Century.

Indigenous Tourism in Australia

In 2022 in Australia, there are almost 800,000 Aboriginal and Torres Strait Islander peoples who belong to over 500 different Aboriginal and Torres Strait Islander nations (World Vision, 2022). All of these diverse nations are rich in landscape and are connected, to varying degrees, through story, culture, songlines and family. Aboriginal and Torres Strait Islander peoples have been the stewards of Australia's land and waters for over 65,000 years, caring for the environment, nurturing the vast Australian landscapes and employing sustainable cultural practices whilst shaping the history and the national identity of the country. And importantly, over the last 200 years, there has been an ever-increasing demand for the voices of Aboriginal and Torres Strait Islander peoples to be heard (see The Uluru Statement, 2022).

On 30 October 2019, the Minister for Indigenous Australians announced the start of the Indigenous Voice co-design process which provided a mechanism for the voices of Aboriginal and Torres Strait Islander peoples to be heard. In late July 2021, the Indigenous Voice Co-design Process Final Report was submitted to the Australian Government and reiterated that '*Aboriginal and Torres Strait Islander peoples want a greater say on the laws, policies and programs that affect our lives and non-Indigenous Australians support that call*' (Langton & Calma, 2021:ix).

Importantly, the Indigenous Voice co-design process sought to facilitate achieving more sustainable outcomes for Aboriginal and Torres Strait Islander peoples across a wide range of areas including the tourism sector. In Australia, it recognised that First Nations tourism experiences value-add to the Australian tourism narrative alongside core Australian tourism strengths (i.e. coastal, aquatic, food and wine, adventure, nature, wildlife experiences etc) and on the 20th July 2021, the National Indigenous Tourism Advisory Group said that '*…all tourism in Australia is, at its heart, Indigenous tourism*'. The National Indigenous Tourism Advisory Group explained:

As the original, and present, inhabitants of this unceded land, we have shaped the environment, we have nurtured it, carved story in it, sung and danced and cared for every inch of it's vast landscape over the life of our 100,000 year culture. Every beach, desert or forest vista, the very biodiversity that we love and appreciate and share as a tourism experience is the direct product of Aboriginal management. It is still alive and it is still loved as Aboriginal land. This is not a statement of claim, it is a statement of obligation. An obligation that we hold, in custodial responsibility for that land, a responsibility that we want to share with all those who live and travel on our land, so that the system that sustains life can be maintained.

This powerful message comes at a time in Australian history when Aboriginal and Torres Strait Islander peoples are increasingly occupying a place at the table and to leverage wisdom and knowledge that has been gathered over 65,000 plus years, we must listen to the voices that belong to the custodians of Country, the voices of Aboriginal and Torres Strait Islander peoples. This chapter shares insights into the significant and crucial role that the voices of Aboriginal and Torres Islander peoples of Queensland played in the development and implementation of the inaugural Queensland First Nations Tourism Plan 2020–2025 and the QFNTC.

First Nations Tourism in Queensland Australia

The State of Queensland, Australia has an estimated current resident population of more than 5 million people, of whom 221,276 identify as First Nations (approx. 4.6%) (Australian Bureau of Statistics, 2018). This is the second-largest State concentration of First Nations peoples in Australia, after New South Wales (132,708) (Australian Bureau of Statistics, 2018). Furthermore, 'Queensland is home to world-class Indigenous tourism experiences and has the unique ability for visitors to experience both Aboriginal and Torres Strait Islander culture' (Year of Indigenous Tourism, 2021:1) as Queensland is the only State in Australia that is the home of both Aboriginal and Torres Strait Islander cultures. While the First Nations population is dispersed across the State, the majority of First Nations people live in the cities of Cairns, Townsville and Brisbane, the capital city of Queensland (Australian Bureau of Statistics, 2019).

We have known for some time now that Indigenous tourism experiences provide a strong point of differentiation in an incredibly competitive global marketplace (Queensland Tourism Industry Council, 2021a). Queensland's Indigenous tourism sector is measured by the number of visitors that participate in an Indigenous tourism experience during their overnight trip. Prior to the COVID pandemic which commenced in 2019, Indigenous tourism or what is referred to as First Nations tourism in Australia, accounted for 1.15 million international visitors with the UK and the USA being the most popular source markets (Queensland Tourism Industry Council, 2019). The sector grew by 6.4% on average over the five years to December 2019 compared to 6.1% for the wider tourism industry (Tourism and Events Queensland, 2022b) 2018 to 2019 saw a 4% increase from the previous year in those international visitors who enjoyed a First Nations tourism experience in Australia. In 2019, First Nations tourism had an estimated annual value of $5.8 billion (Queensland Tourism Industry Council, 2019). As a result of the COVID pandemic in 2020 and 2021, the tourism sector across the globe weathered the impacts of a complete sector shutdown. Concomitantly however, the enforced hiatus provided tourism stakeholders with an opportunity to listen more acutely to the voices of Aboriginal and Torres Strait Islander peoples and as the tourism sector regroups and moves forward in 2022, we must develop platforms that ensure First Nations voices are a part of all dialogue pertaining to the sustainable design, development, management and implementation of First Nations tourism.

The voice of First Nations people in the tourism sector has been growing louder and stronger over many years but none more so than in the last few years. The First Nations tourism sector in Queensland Australia is supported by three levels of Government; Local Government, State Government and Federal Government. The Federal Government supports First Nations tourism in Queensland through a variety of channels, including a Federal Indigenous Tourism Fund, a Peak Body (i.e. an advocacy group with allied interests) called Indigenous Business Australia and through the Indigenous Land and Sea Corporation. The Queensland State Government entities that support First Nations tourism include the Queensland Government Department of Tourism, Innovation and Sport and a Government-owned corporation called Tourism Events Queensland (TEQ). Additionally, a significant key stakeholder in First Nations tourism in Queensland is called the QTIC, which is a not-for-profit, private sector, membership-based organisation representing the interests of Queensland's tourism and hospitality industry. Other key stakeholders include 13 Regional Tourism Organisations who play a role in marketing and supporting First Nations tourism operators in their regions and 77 Local Government Councils which are increasingly investing in the promotion and development of First Nations tourism as part of their destination experiences. Equally significant stakeholders are the many tourism operators and businesses that have and continue to individually support First Nations employment, training and partnerships.

The Growth of Queensland's First Nations Tourism Sector

Queensland is known as Australia's Sunshine State offering 'incredible diversity of destinations and experiences that truly set Queensland apart as a place to travel for good' (TEQ, 2022b:1). Arguably, Queensland is Australia's most popular tourism destination as it offers tourists experiences at amazing beaches, stunning tropical islands, world heritage-listed rainforests, rivers and reefs including the world-famous Great Barrier Reef. All these destination sites and associated experiences leverage, to varying degrees, the rich, diverse tangible and intangible cultural heritage, languages and traditions of Aboriginal and Torres Strait Islander peoples.

Not surprisingly then, there has beencontinues to be, a range of opportunities for the growth and development of First Nations tourism in Queensland. While the vast number of First Nations tourism products that are currently available in Queensland have been developed over quite a few years, it was in the early 2000s that First Nations tourism really started to gain momentum in the State. Over the ensuing years, a vast amount of significant work has been undertaken by First Nations tourism champions, business operators, Aboriginal and Torres Strait Islander Traditional Owners and key Government stakeholders to build what is now, a much stronger First Nations tourism sector in Queensland.

During the first decade of the 21st Century, the development of Queensland's First Nations tourism sector was sporadic, with piecemeal work being undertaken primarily, by the Queensland Government and/or affiliated bodies (e.g. TEQ, IBA, QTIC). Nevertheless, it was during this period, that the First Nations tourism sector slowly began to morph from a small niche sector of Queensland's tourism industry into an integral and increasingly influential player on both the national and the global tourism stage. Visitor numbers for First Nations tourism experiences slowly started to increase in 2013 and from 2014 to 2019, visitor nights associated with Indigenous tourism in Queensland increased at an annual average rate of 8.4% nights (Ruhanen, Whitford, & Pham, 2020).

Soon after, the number of First Nations tourism visitors steadily grew by an average of 9% per annum and the number of international tourists taking part in Indigenous tourism activity increased by over 40%. The total visitor nights recorded for Queensland in 2018/19 was 4.56 million (Ruhanen et al., 2020) and visitor expenditure on First Nations experiences also rose on average by 8% per year (Dept. Foreign Affairs and Trade, 2019). By 2020, Indigenous tourism represented 1.3% of the total tourism gross state product and achieved a compound annual visitor growth rate of 11.2% per annum c Compared to the national Indigenous market, the Queensland Indigenous tourism market consistently gained increased visitor night share from 16.6% to 18.4% over five years to 2020 (Ruhanen et al., 2020).

The Development of Queensland's Inaugural First Nations Tourism Plan
2020–2025

Arguably, another catalyst for change and positive sector growth in the Queensland First Nations tourism sector occurred in 2008, when QTIC established the Tourism Indigenous Employment Champions Network to leverage (among other things), the considerable scope available to increase the proportion of Indigenous employees engaged in full-time and/or management roles within Indigenous tourism businesses (Ruhanen et al., 2020). The Champions Network focused on increasing the participation of First Nations peoples in the workforce and in the tourism industry and in 2011, QTIC introduced a half day, Indigenous Employment Tourism Forum and hosting about 30 attendees. Under the leadership and management of Rhonda Appo and Cameron Costello, the Champions network grew from strength to strength and in 2017 at the annual Forum, (now known as Destination IQ), participants noted the need for and requested the development of a Queensland First Nations Tourism strategy to guide the development and growth of First Nations tourism across Queensland. It was at this QTIC event, that the seeds were sown for the establishment of Queensland's First Nations Tourism Action Plan 2020–2025 that was to be designed, developed, directed and driven by the First Nations peoples of Queensland.

By the end of 2018, the ripple effect of the first initial conversations pertaining to the development of a First Nations Tourism plan saw growing stakeholder interest emerge across the State. QTIC facilitated the development of the inaugural Queensland First Nations Tourism Action Plan 2020–2025 and during 2018 and much of 2019, the first round of extensive consultations (Nicholls et al., 2016) were undertaken with key First Nations tourism stakeholders and the broader tourism industry across the 13 tourism regions of Queensland.

These consultations produced incredibly rich feedback and the story that emerged from the thematic analysis of consultation data (i.e. interview/discussion notes, recordings) (Walters, 2016) was a 'story of excitement, dreams and aspirations for building a robust, First Nations tourism sector that positions Queensland as the number one tourism destination for First Nations tourism experiences in Australia' (QTIC, 2019:6). The themes that emerged were categorised into benefits, challenges and opportunities (see QTIC, 2019:6–7) and these themes informed the continuous development of the Plan. Additionally, a Strategic Advancement Group constituting eight First Nations members was established to oversee the strategic direction of the Plan and to share insights with Government regarding tourism development opportunities for the Queensland First Nations tourism sector. In December 2019, the Queensland Government's Minister for Tourism, the Honourable Kate Jones MP, launched Queensland's inaugural First Nations Tourism Plan. This was an exciting time for First Nations tourism in Queensland because at the same time, the Queensland First Nations Tourism Action Plan 2020–2025 was launched, the Queensland Government provided $10 million in investment initiatives to support Plan-related activities during this period and also announced 2020 to be the Year of Indigenous Tourism. The premise of the Year of Indigenous Tourism was to recognise and deepen the understanding of local Aboriginal and Torres Strait Islander cultures, knowledges, history and traditions and in turn, showcase and strengthen a growing number of First Nations tourism businesses across the State.

The Queensland First Nations Tourism Action Plan 2020–2025

The Queensland First Nations Tourism Action Plan 2020–2025 is underpinned by the Larrakia Principles which were developed in 2012 at the first Pacific Asia Indigenous Tourism Conference, held in Darwin Australia, on the traditional lands of the Larrakia people

(Queensland Tourism Industry Council, 2021b). It was at this 2012 conference, that 191 delegates representing Indigenous communities from 16 countries, government agencies, the tourism industry and supporting bodies resolved to adopt the principles to guide the development of Indigenous tourism through the Larrakia declaration (see https://un-declaration.narf.org/larrakia-declaration-on-the-development-of-indigenous-tourism/).

Guided by the Larrakia Principles, the Queensland First Nations Tourism Action Plan 2020–2025 is built on the following six Pillars which highlight (among other things), the integral role First Nations voices play in the sustainable development of Indigenous tourism (QTIC, 2019:7):

1 Recognition and Respect:
  Promote recognition and respect for First Nations cultures, stories and connections to and ownership of the country while embracing and reflecting the diversity, aspirations and desires of First Nations peoples and communities.
2 Engagement and Partnerships:
  Encourage the creation of mutually beneficial and strategic partnerships to grow the First Nations tourism sector.
3 Strategic Coordination and Structure:
  Create an entity that gives voice to the First Nations tourism sector and provides advocacy and support.
4 Training, Skill Development and Jobs:
  Develop business capability and capacity development for First Nations tourism businesses to ensure the First Nations tourism sector is driven by a skilled workforce and engaged in quality employment that generates sustainable socio-economic outcomes for First Nations individuals and communities.
5 Authentic Product Development:
  Develop and deliver authentic, quality First Nations products which are export-ready and meet market demand.
6 Marketing and Awareness:
  Position and promote unique, viable, world-class First Nations experiences as a must-do experience while visiting Queensland.

There are goals and objectives under each of these 6 Pillars and to ensure the goals and objectives are met, a 16-member First Nations Tourism Working Group was established to examine and guide the development and implementation of identified actions. The Working Group is representative of the First Nations tourism sector and was brought together by QTIC to drive the direction of the Plan, determine the scope for inclusion and consultation and lead the advocacy for implementation of the First Nations Tourism Action Plan 2020–2025 and establishment of a First Nations tourism Peak Body.

In early 2019, a second round of community and industry consultations (i.e. yarning methodology see Bargallie, 2018; Walker, Fredericks, Mills, & Anderson, 2014) was undertaken across Queensland. The goal of these consultations was to reinforce the co-design and co-development of the Plan by strengthening the voice of First Nations people in Queensland's tourism sector. These consultations also facilitated appropriate and effective implementation of the plan by highlighting the need for tourism industry stakeholders and key Queensland Government Departments to work in partnership with First Nations peoples.

The consultations were also informing the development of celebrations and initiatives for the 2020 Year of Indigenous tourism in Queensland. This significant initiative, however, was short-lived as COVID hit the shores of Australia in March 2020 and the country went into lockdown for the better part of the year. Nevertheless, after successful lobbying by QTIC and Queensland's First Nations tourism business operators, the Queensland State Government agreed to extend the Year of Indigenous Tourism to 2021, a year that also continued to be impacted by extensive domestic lockdowns and closed international borders.

The Development of a First Nations Tourism Peak Body: Queensland First Nations Tourism Council

In 2021 and in response to Strategic Pillar 3 (i.e. strategic coordination and structure) of the Queensland First Nations Tourism Action Plan 2020–2025, another extensive round of inclusive consultations (i.e. yarning methodology (see Bargallie, 2018; Walker et al., 2014) with First Nations stakeholders was undertaken around the State. These consultations were designed to inform the establishment of an independent Peak Body to represent the voice of First Nations tourism in Queensland.

As the inclusive consultations across Queensland gained momentum, there was an ever-increasing groundswell of First Nations peoples representing rainforest peoples, land and sea peoples, desert peoples and coastal peoples, actively supporting the Plan and eagerly adding their voices to the co-creation and co-development of a collective Peak Body that acts as the representative for First Nations tourism in QLD. Through the consultations, the overarching goal of the Peak Body was determined as an entity that not only acts as the collaborative and coordinated voice of Queensland's First Nations tourism sector (PM Glynn Institute, 2021) but also drives the vision of First Nations tourism in Queensland while providing substantial support and advocacy.

The Peak Body, called the QFNTC was formed in late 2021 and officially launched in 2022. It will be a member-driven organization, focused on advocacy for Queensland First Nations tourism businesses. Specifically, QFNTC will:

- Provide a coordinated voice for advocacy, leadership and representation and drive the vision of First Nations tourism in Queensland.
- Facilitate collaboration between First Nations tourism businesses and the tourism industry on destination, business and product development.
- Play a major role in driving the implementation of the Queensland First Nations Tourism Plan in partnership with government, industry and community.

In 2021, Minister Ken Wyatt (MP) (2021:1) noted '*after a difficult year, Indigenous tourism markets are growing, and the Australian Government is committed to investing strategically to meet the current and emerging needs of the sector*'. Therefore, alongside several initiatives, the National Indigenous Tourism Advisory Group [NITAG] was launched in March 2021 to ensure, among other things, that a $40 million Indigenous Tourism Fund is Indigenous led. In November 2021, QTIC hosted Destination IQ and achieved a record attendance number of 300 stakeholders participating in a full-day forum and Destination IQ dinner.

The Drivers for Success

Both the Plan and QFNTC have been designed from the outset to encapsulate the heart and soul of Queensland's First Nations tourism voice in order to achieve a range of goals that will situate Queensland as the number one tourism destination for First Nations tourism experiences in

Australia (QTIC, 2019). The First Nations tourism voice emanates from those First Nations tourism stakeholders who identify as Aboriginal or Torres Strait Islander. The power of Queensland's First Nations tourism development is that it is encased by a movement that has emerged from a bottom-up, inclusive and collaborative approach (Vernon, Essex, Pinder, & Curry, 2005) that has effectively given rise to the development of many organic and authentic partnerships and networks. And as the groundswell gains momentum and maturity, First Nations tourism businesses are identifying gaps in the market and seizing the opportunities that abound. Moreover, they are proactively strengthening their networks, working towards developing tourism clusters, collaborating around tourism product development and discussing the opportunities for leveraging products and experiences across the sector (Whitford & Ruhanen, 2016). They are better understanding the complex nuances of the tourism system (Stainton, 2021; Whitford et al., 2017), the sector at large and the roles that QTIC, TEQ and the QLD Government play.

Arguably, the success of this momentous First Nations tourism movement in Queensland has been partly due to QTICs adoption of a co-design and co-development approach. For instance, all the consultations were led by Rhonda Appo and Cameron Costello and the strategic direction of the Peak Body was guided by the First Nations Working Group. The collaborative and cooperative engagement of key tourism stakeholders has also been critical and must continue in order to facilitate the ongoing sustainable development of Queensland's First Nations tourism sector (Carr, Ruhanen, & Whitford, 2016). Additionally, there can be little argument that government funding and support, alongside the establishment of networks and partnerships have provided necessary levers (i.e. initiatives that can be undertaken in order to drive a desired impact) and continues to present opportunities to leverage participation in the co-creation of initiatives that will arise as a result of Brisbane, Queensland, hosting the 2032 Olympic Games.

It appears reasonable to suggest that these are just some of the fundamental elements that are required to maintain a sustainable First Nations tourism sector. Arguably however, none of these fundamental elements on their own are enough to achieve long-term sustainability without the most important, crucial element of all being included in the mix. This crucial element is preferencing the voice of First Nations tourism stakeholders to ensure that all First Nations tourism activities are driven by the voice of Aboriginal and Torres Strait Islander peoples to ensure the development and delivery of authentic, immersive and highly interactive activities that celebrate Aboriginal and Torres Strait Islander cultural heritage and history, both now and for generations to come.

Indeed, the power of Queensland's First Nations tourism movement is that it is an authentic, collaborative organically grown movement that represents the voice of First Nations peoples. It provides all stakeholders with the opportunity to be heard. It provides all stakeholders with a support mechanism for lobbying and advocacy. And it provides all stakeholders with an opportunity to build First Nations tourism into the heart and soul of all tourism and hospitality sectors across Queensland as we acknowledge that all tourism at its heart, is Indigenous tourism (NIAA, 2021).

Where to From Here?

What does the Queensland First Nations tourism roadmap for 2023–2032 look like as the eyes of the world become increasingly locked onto Queensland as the State commences work on the ten year build-up to the 2032 Olympic Games in a post-pandemic world?

In the first instance, there should be little argument that if the sector is going to facilitate and realise the ongoing sustainable development of First Nations tourism in Queensland, it is critical for Queensland First Nations tourism to continue along the current path and employ a collaborative and cooperative approach that engages key stakeholders and facilitates genuine buy in from those key stakeholders. Concomitantly, the sector must not forget the ongoing need for training, education and funding, alongside the continuing establishment of organically grown networks and partnerships where stakeholders can and will leverage opportunities to participate in the co-creation of initiatives that situate First Nations cultures as an integral part of Queensland life and leverage opportunities that are arising as a result of Queensland hosting the 2032 Olympic Games.

While the 2032 Olympic Games provide a solid goal post with which to guide strategic agendas over the next ten years, the mega event must be viewed, not as the end goal but rather as a stepping-stone towards ongoing, sustainable, authentic and transformative First Nations tourism development in Queensland and beyond. And in the spirit in which this chapter is written, we conclude with the voice of First Nations peoples who remind us:

> To acknowledge that all Tourism is Indigenous Tourism is not to detract from other versions of tourism, with its framing of built attractions and experience as the basis of tourism. Nor is it intended to detract from the tourism value of cities that draw people's gaze away from the landscape, it is instead a gift of strength and ancient linkage that provides the foundation for those businesses and it is what makes Australia unique in the world.
>
> (National Indigenous Tourism Advisory Group, 2021)

Acknowledgment

We acknowledge the traditional owners and custodians of country throughout Australia and acknowledge their continuing connection to land, waters and community.

We pay our respects to the people, the cultures and the elders past, present and emerging.

We acknowledge that many individuals refer to themselves by their clan, mob and/or country.

References

ATSIC (1997). *The National Aboriginal and Torres Strait Islander Tourism Industry Strategy (NATSITIS): A summary / prepared by ATSIC and the Office of National Tourism.* Aboriginal and Torres Strait Islander Commission, Australia. Office of National Tourism, Australia.

Australian Bureau of Statistics (2018). *Estimates of Aboriginal and Torres Strait Islander Australians,* from https://www.abs.gov.au/statistics/people/aboriginal-and-torres-strait-islander-peoples/estimates-aboriginal-and-torres-strait-islander-australians/latest-release

Australian Bureau of Statistics (2019). *Census of Population and Housing: Reflecting Australia - Stories from the Census, 2016* (2071.0), from https://www.abs.gov.au/ausstats/abs@.nsf/Lookup/by%20Subject/2071.0~2016~Main%20Features~Aboriginal%20and%20Torres%20Strait%20Islander%20Population%20-%20Queensland~10003

Bargallie, D. M. (2018). *Maintaining the Racial Contract: Everyday Racism and the Impact of Racial Microaggressions on "Indigenous Employees" in the Australian Public Service.* Queensland University of Technology.

Carr, A., Ruhanen, L., & Whitford, M (2016). Indigenous peoples and tourism: The challenges and opportunities for sustainable tourism. *Journal of Sustainable Tourism, 24*(8–9), 1067–1079.

Dept. Foreign Affairs and Trade (2019). *Indigenous Tourism Surge.* Business Envoy.

Fletcher, C., Pforr, C., & Brueckner, M. (2016). Factors influencing Indigenous engagement in tourism development: An international perspective. *Journal of Sustainable Tourism, 24*(8/9), 1100–1120.

Langton, M., & Calma, T. (2021, July). *Indigenous Voice Co-design Process Final Report to the Australian Government.* Retrieved 2 February 2022, from https://voice.niaa.gov.au/sites/default/files/2021-12/indigenous-voice-co-design-process-final-report_1.pdf

Mackley-Crump, J. (2016). From private performance to the public stage: Reconsidering "staged authenticity" and "traditional" performances at the Pasifika Festival. *Anthropological Forum, 26*(2), 155–176.

Minister Key Wyatt (MP) (2021). *Inaugural Meeting of the New National Indigenous Tourism Advisory Group*. Retrieved 1 April 2021, from https://ministers.pmc.gov.au/wyatt/2021/inaugural-meeting-new-national-indigenous-tourism-advisory-group

NIAA (National Indigenous Australian Agency) (2021). *Indigenous Protected Areas – Commonwealth-funded Indigenous Ranger groups*. Canberra: Commonwealth of Australia

Nicholls, S., Booker, L., Thorpe, K., Jackson, M., Girault, C., Briggs, R., & Caroline Jones, C. (2016). From principle to practice: Community consultation regarding access to Indigenous language material in archival records at the State Library of New South Wales. *Archives and Manuscripts, 44*(3), 110–123.

NITAG (2021). *Indigenous Tourism Fund: Discussion Paper*. Retrieved 20 October, from https://www.niaa.gov.au/indigenous-affairs/economic-development/indigenous-tourism-fund#resources

PM Glyn Institute (2021). *Hearing Indigenous Voices*. Retrieved 10 December 2021, from https://www.pmglynn.acu.edu.au/news/upholding-the-big-ideas-launch/_local/hearing-indigenous-voices

Queensland Tourism Industry Council (2019). *Queensland First Nations Tourism Plan 2020–2025. Voices of Today: Stories for Tomorrow*. Brisbane: QTIC, from http://qticazure.blob.core.windows.net/crmblobcontainer/Version%206%20-%20FNTP%20Final%20Version%20-%20High%20res%20Web.pdf

Queensland Tourism Industry Council (2021a). *First Nations Tourism Peak Body: Discussion Paper*. Brisbane: QTIC, from https://qticazure.blob.core.windows.net/crmblobcontainer/First%20Nations%20Tourism%20Peak%20Body%20Discussion%20Paper_2.pdf

Queensland Tourism Industry Council (2021b). *Larrakia Principles*. Retrieved 7 September 2021, from https://www.qtic.com.au/year-of-indigenous-tourism/first-nations-tourism-potential-plan/Larrakia-Principles/

Ruhanen, L., & Whitford, M. (2017). Indigenous Tourism in Australia: History, trends and future directions. In M. Whitford, L. Ruhanen, & A. Carr (Eds.), *Indigenous tourism: Cases from Australia and New Zealand* (pp. 9–23). Oxford: Goodfellow Publishers.

Ruhanen, L., & Whitford, M. (2019). Cultural heritage and Indigenous tourism. *Journal of Heritage Tourism, 14*(3), 179–191.

Ruhanen, L., Whitford, M., & McLennan, C. (2015). Indigenous tourism in Australia: Time for a reality check. *Tourism Management, 48*, 73–83.

Ruhanen, L., Whitford, M. & Pham, T. (2020). *Queensland Indigenous Tourism Sector Analysis: Research Report prepared for Tourism Events Queensland*. Retrieved 13 March 2022, from https://teq.queensland.com/content/dam/teq/corporate/corporate-searchable-assets/industry/research/special-reports/Queensland-Indigenous-Tourism-Trends.pdf" https://teq.queensland.com/content/dam/teq/corporate/corporate-searchable-assets/industry/research/special-reports/Queensland-Indigenous-Tourism-Trends.pdf

Stainton, D. (2021). *Leiper's Tourism System: A Simple Explanation*. Retrieved 10 December 2021, from https://tourismteacher.com/leipers-tourism-system/

The Uluru Statement (2022). Uluru Statement from the Heart. Retrieved 20 May 2022, from https://ulurustatement.org/the-statement/history/

Tourism and Events Queensland (2022a). Indigenous tourism. Retrieved January 2022, from https://teq.queensland.com/au/en/industry/what-we-do/cruise_indigenous_nature_tourism/indigenous-tourism

Tourism and Events Queensland (2022b). Days like this. Retrieved 23 April 2022, from https://teq.queensland.com/au/en/industry/what-we-do/marketing/current-opportunities/days-like-this

Travesi, C. (2018). Knowing and being known. Approaching Australian indigenous tourism through aboriginal and non-aboriginal politics of knowing. *Anthropological Forum, 28*(3), 275–292.

Vernon, J., Essex, S., Pinder, D., & Curry, K. (2005). Collaborative policymaking: Local Sustainable Projects. *Annals of Tourism Research, 32*(2), 325–345.

Walker, M., Fredericks, B., Mills, K., & Anderson, D. (2014). "Yarning" as a method for community-based health research with Indigenous women: The Indigenous Women's Wellness Research Program. *Health Care for Women International, 35*(10), 1216–1226.

Walters, T. (2016). Using thematic analysis in tourism research. *Tourism Analysis, 21*(1), 107–116.

Whitford, M., Bell, B., & Watkins, M. (2001). Indigenous tourism policy in Australia: 25 Years of rhetoric and economic rationalism. *Current Issues in Tourism, 4*(2–4), 151–181.

Whitford, M., & Ruhanen, L. (2010). Australian indigenous tourism policy: Practical and sustainable policies? *Journal of Sustainable Tourism, 18*(4), 475–496.

Whitford, M., & Ruhanen, L. (2016). Indigenous tourism research, past and present: Where to from here. *Journal of Sustainable Tourism*, 24(8–9), 1–20.

Whitford, M., Ruhanen, L., & Carr, A. (2017). Introduction to Indigenous tourism in Australia and New Zealand. In M. Whitford, L. Ruhanen, & A. Carr (Eds.), *Indigenous Tourism: Cases from Australia and New Zealand* (pp. 1–8). Oxford: Goodfellow Publishers.

World Vision (2022). Know your country. 8 interesting facts about Aboriginal and Torres Strait Islanders, from https://www.worldvision.com.au/global-issues/work-we-do/supporting-indigenous-australia/8-interesting-facts-about-indigenous-australia

Year of Indigenous Tourism (2021). Retrieved 10 December 2021, from https://www.dtis.qld.gov.au/our-work/year-of-indigenous-tourism

34

REFLECTION ON ABORIGINAL TOURISM IN WESTERN AUSTRALIA

A Balancing Act of Opportunity and Challenge

Robert Taylor

WAITOC's journey in fostering and promoting Aboriginal tourism in Western Australia has been a reflective blend of opportunities and challenges. Aboriginal tourism presents a unique chance to showcase our rich cultural heritage and stunning landscapes as the oldest living culture in the world. Still, it also demands careful management to preserve our traditions and environment. WAITOC's core values consider the

> appropriate access and engagement with these traditions, which is a key consideration for developing a sustainable Aboriginal tourism industry in Australia. WAITOC's corporate values encompass both traditional and commercial components but are clearly differentiated as they give priority to the culturalisation of commerce and not the commercialisation of culture.
>
> (WAITOC 2015 np)

Opportunities

The most striking opportunity lies in cultural sharing. Aboriginal tourism allows us to tell our stories, share our history and showcase our traditions from our perspective. It's a powerful platform for cultural education and understanding. This narrative, woven through experiences such as guided walks, art workshops and traditional performances, educates and fosters respect for our culture.

Furthermore, there's a growing international interest in authentic, sustainable travel experiences. Western Australia's vast, unspoiled landscapes and rich Aboriginal heritage position us uniquely in the global tourism market. This aligns with the rising trend of eco-tourism, where visitors seek experiences that are environmentally responsible and culturally respectful.

Economically, Aboriginal tourism is a catalyst for community development. It creates employment opportunities, drives local economic growth and fosters a sense of pride within communities. We also nurture future leaders and entrepreneurs by empowering local Aboriginal people to share

DOI: 10.4324/9781003230335-40

their Culture. In Western Australia, Aboriginal tourism currently contributes $63.8 million to WA's gross product and $41 million to state incomes, employing 516 full-time jobs (Western Australia Tourism 2022).

Challenges

However, these opportunities come with their challenges. One of the primary concerns is over-commercialisation. There's a delicate balance between sharing our culture and maintaining the sanctity and authenticity of our traditions. We must ensure that tourism doesn't dilute or misrepresent our cultural heritage. As the cultural landscape changes in Western Australia with our first names of places starting to appear, we must be aware of the appropriation of our cultures.

Environmental sustainability is another critical challenge. Our landscapes are not just beautiful backdrops; they hold spiritual significance and are integral to our way of life. Tourism development must be sustainable, ensuring minimal impact on these precious ecosystems and considering the Aboriginal communities in these unique places.

Moreover, accessibility and infrastructure in remote areas present practical challenges. Developing tourism experiences that are both authentic and accessible, especially in remote regions, requires thoughtful planning and investment.

Conclusion

As WAITOC continues to lead WA in the Aboriginal tourism space, we will prioritise navigating these opportunities and challenges thoughtfully. We aim to create a sustainable model of Aboriginal tourism that respects and celebrates our culture while contributing positively to Western Australia's economy and global standing. The path is complex but the potential for meaningful impact drives our daily efforts.

Robert Taylor, CEO, WAITOC (2023)

References

Tourism Western Australia (2022). *Aboriginal Tourism Snap Shot 2021–22.* Available online at https://www.tourism.wa.gov.au/Markets-and-research/Specialised-Research-Reports/Pages/WA-Aboriginal-tourism-snapshot.aspx#/ Accessed 12 November 2023

WAITOC (2015). *Ancient Tracks New Journeys.* Available online at https://www.waitoc.com/about-us/corporate/our-documents. Accessed 12 November 2023

35

CONCLUSIONS

Richard Butler and Anna Carr

Introduction

In concluding this volume, it is clear, that despite including contributions from both Indigenous and non-Indigenous authors with very different backgrounds and fields of interest and from all continents, a number of common features and concerns have emerged. Whether any of these are really new, in the sense that they have not been discussed before, is unlikely, for the subject of Indigenous involvement in tourism development has been of academic and non-academic interest for three decades or more, at least since the publication of Valene Smith's *Host and Guests* in 1977. As Whitford and Ruhanen (2016) clearly demonstrated in their seminal paper in the special issue of the *Journal of Sustainable Tourism*, the topic has received a marked increase in attention and research in the last decade, with a much greater involvement of Indigenous authors in publishing their work. This rapid increase in published articles, chapters, and books is reflected in the wider array of examples revealed and issues identified than appeared in the two previous books with similar titles to this one (Butler and Hinch, 1996, 2007). However, several areas have remained relevant and perhaps become even more significant. The second of those publications, like this volume, also had six sections, beginning with Indigenous Knowledge, followed by Indigenous Commerce, Indigenous Environment, Indigenous Culture, Indigenous Community-Based Tourism and Indigenous tourism Policies and Politics.

It is perhaps in that last area, Policies and Politics, that most change has occurred in the intervening period between the books, which bears out the comment made in Hall's chapter (2007) on Politics, Power, and Indigenous Tourism when he wrote "...Few subjects better illustrate the political dimensions of tourism than the issues associated with indigenous tourism" (Hall, 2007: 306). We chose not to have a similarly titled section in this volume, in part because issues relating to, and the ramifications of, policies and politics pervade all sections of this book. Indigenous peoples, supported by those who share their concerns, have put political aspects of the development of tourism at the forefront of their concerns and have argued, with some success, that control over all aspects of development (not just of tourism) that involve Indigenous peoples and their lands, waters and other cultural elements should rest with those people themselves, and that their wishes and needs should be paramount in related decision-making. Success in achieving this aim has not been easy, nor has it been completed, but it is clear, both from the essays here and from the many

DOI: 10.4324/9781003230335-41

changes that have taken place in government policies towards Indigenous peoples throughout the world, that the demands for an appropriate and primary role in the making of policy and decisions for Indigenous peoples are being listened to and acted on in an increasing number of locations.

There are a number of issues and problems that have not been addressed in this volume, partly because aspects of them lie outside the realm that could be covered here, and partly because of a lack of contributions, despite the editors' best efforts. There are clearly gaps in addressing the situation in some parts of the world. There is good coverage through examples from Australasia and the Pacific Islands, from Canada, and to a much lesser degree Asia, the Arctic regions and South America. (In the context of South America, a recent article (in Spanish) in the journal *PASOS*, by Silvestre and Fontana (2023) provides a bibliographic review of literature on Indigenous tourism in Brazil). There is little in the volume specifically on the United States, where Indigenous tourism has a long and problematic history. There it has been linked with the westward movement and settlement of immigrants into mostly unceded Indigenous lands in the 19th century, along with the creation of the ideas or myths of the "conquest of the west" the "wild Indian", and the establishment of national parks and Indian reserves. For a long time, perhaps longer than in many other countries, in the United States, Indigenous tourism was limited to viewing "the natives" with limited involvement from those being "viewed", in much the same way that early tourism to Pacific Islands began (Douglas, 1994). In recent decades much has changed, and Indigenous communities have, in some places, won back rights and ownership, and become engaged in large scale commercial tourism developments, not least in the form of casinos (Carmichael and Jones, 2007). Whether Indigenous owned and operated casinos are truly part of Indigenous tourism is a difficult question. Butler and Hinch (1996: 10) proposed a matrix involving Indigenous control and Indigenous themes as a way of deciding what may be regarded as Indigenous tourism. Where exactly Indigenous casinos fall in that matrix is controversial, some are clearly Indigenous owned and controlled, and some feature a pseudo-indigenous theme as a marketing ploy, but one would not regard them as being Indigenous tourism enterprises in the sense that the term has been used in this volume and others.

Another gap in coverage is the other side of Indigenous tourism, namely tourism *by* Indigenous peoples themselves. It may well be considered inappropriate by some to examine and thus categorise the ethnic, racial, and other characteristics of vacationers, and few have chosen to examine the Indigenous tourist in particular, as Wei et al (2021) note, Indigenous peoples as tourists, i.e., being demanders of tourism rather than fulfilling the traditional role of suppliers of tourism. Such travel nuances have implications for issues such as the demonstration effect of tourist visitation on local residents and the interaction of different but similar cultures. Since the 2007 volume, researchers have attempted to understand the experiences of Indigenous travellers, for example, Peters and Higgins-Desbiolles (2012) study of Indigenous Australians as tourists and a recent study of the motivations of Pacific islanders (Trupp, Pratt, Stephenson, Matatolu and Gibson, 2022). There are other implications, such as potential benefits from sharing ideas and information on ways to deliver Indigenous tourism to both Indigenous and non-Indigenous markets. The World Indigenous Tourism Alliance (WINTA's) hosting of the World Indigenous Tourism Summits(WITS) in Australia (2012), New Zealand (2017), Australia (2023) and Taiwan (2024) have been decisive events in terms of enabling and encouraging the movement of Indigenous tourism professionals to travel the globe. Through WITS, they make connections, share knowledge, give voice to the Indigenous tourism communities worldwide with outcomes including youth mentoring.

This concluding chapter has been divided into three major sections, in addition to this introduction and a final concluding section: knowledge and respect, type of tourism, and control. All are integral to the empowerment and self-determination of Indigenous peoples in tourism. We endeavour, in this last chapter, to present both a summary and a commentary on the issues raised in this volume in relation to these three main themes, but hopefully without repeating the points made in the individual chapters and the linkages between them. The three main themes are closely related and inevitably overlap in terms of content and arguments, because tourism of any type in any situation is a complex arrangement of many different actors and agencies and goals and policies that have evolved in different but linked locations and jurisdictions over time, in some cases, a very long time. Students and researchers in tourism more generally will be familiar with many of the issues raised here, particularly the impacts and implications of tourism development on local communities and environments, for despite its multiple forms, scales and participants, there are many common features of tourism wherever it takes place. In the case of what has become accepted as Indigenous tourism, there are of course additional major characteristics that makes this form of tourism distinct, namely the cultural/spiritual elements, values and characteristics that are imbued in Indigenous peoples worldwide.

Knowledge and Respect

Tribe and Liburd (2016: 55) discussed the role of Indigenous knowledge in their seminal review of The Tourism Knowledge System, and argued that

> The explicit inclusion of Indigenous tourism knowledges in this system offers a site of resistance against possibilities of marginalisation, exploitation and oppression In tourism which may be In the hands of nation states or politically dominant ethnic groups that are actively or passively legitimised by the more established forms of knowledge production.

They note that "Indigenous knowledge is also used to generate competitive advantage" by being able to offer unique insights into Indigenous culture and spirituality, and that it 'speaks back' "to the power, production, categorisation, marginalised, and oppressed position of non-Western cultures, identities and histories". Such a position of weakness means that such knowledge is often viewed as inferior and untrustworthy compared to knowledge obtained using the scientific method. Tribe and Liburd note that this attitude is particularly evident in climate change research, particularly when conditions and events are described with reference to spiritual intervention or phenomena. This is a significant problem given the dimensions of change resulting from climate change, and it illustrates the ways that external impacts upon Indigenous peoples have caused many problems for which Indigenous knowledge has not been accepted as providing meaningful ways of mitigating or adapting to these changes. The recognition and incorporation of Indigenous knowledge when managing tourism activities in all dimensions will have many benefits.

Indigenous knowledge incorporates spatial knowledge in terms of names and descriptions of features such as rivers, mountains, and valleys, and such features will be known by traditional names in the Indigenous language(s). The process of colonisation saw the replacement of Indigenous terms with colonial language words and as a result of such a process, not only the original names were lost, but so too was a wealth of descriptive information that often featured in the original term. As well, the way features were portrayed in maps in Indigenous contexts often does not match the way locations are drawn in scientific maps today. Belyea (1996) noted that, in the context of Amerindian maps,

> Native maps are not crude attempts to render geometric space inferior to contemporary European maps....(but) rely not on fixed positions in space but on a pattern of interconnected

lines…the key to reading the map is …to trace a continuous path from one geographic feature to another.

Thus, while such maps may have appeared worthless to European travellers, they fulfilled their basic geographic purpose of enabling movement from one place to another while incorporating much additional information of value to travellers. The recreation of such maps is rare, but steps are being taken to combine modern geographical techniques (Geographic Information Systems (GIS)) with Indigenous names to produce a GIS platform based on Indigenous knowledge and language (Mamantova and Klyachko, 2022) produced in cooperation with local communities. In New Zealand, Ngāi Tahu have reclaimed their cultural landscape through GIS technology, mapping and recording thousands of traditional place names and associated culturally significant elements in the online atlas, Kā Huru Manu (https://kahurumanu.co.nz/map-stories/; https://kahurumanu.co.nz/map-stories/). While understanding Indigenous languages is probably beyond most tourists, and there would be difficulties in pronunciation in many cases, those are not acceptable reasons for not displaying Indigenous names and terms in Indigenous communities and on maps and guides, where the Indigenous communities wish for this to happen. The descriptions in some chapters (e.g. Chapter 30 by Whitney-Gould) of the way in which material is being presented, in both Indigenous and non-Indigenous languages, demonstrate that local operators realise the opportunity to strengthen the Indigenous language through more frequent use and thus improve the likelihood of its resurgence and survival, as well as providing a unique competitive attraction for visitors. As Matunga et al. (2020: 304) observe "Tourism that provides opportunity for mutual learning, genuine interest and meaningful exchange, is regenerative".

While much of the inappropriate treatment of Indigenous communities has been and still stems from political actions, some intentional, many incidental and ignored or unseen, one of the major threats to Indigenous communities, both in the context of tourism and far more widely, is that of climate change. Global warming and the myriad of other negative impacts on environments relating to this, including rising sea levels, increased turbulence and extreme events, droughts and floods and loss of diversity, represent massive threats to the lifestyles of Indigenous peoples and in some places, to their very survival. Compared to those, the impacts of tourism may be seen as important but less urgent, but while global climate change modification is beyond the ability of Indigenous groups to implement, mitigation of its effects at a local scale is something to which the wealth of Indigenous knowledge can be brought to bear (Scheyvens et al., 2021). As Howitt (2020) argues, Indigenous groups have historically shown great resilience and adaptation to climate induced threats to survival and their ways of life over their long history. As well as being able to survive and prosper in conditions other groups could not, through the use of local knowledge and skills, Indigenous communities and peoples have been able to utilise environmental processes in a sustainable manner, and Howitt argues that this knowledge should be incorporated into current and future approaches to adapting to climate change induced influences. The chapters which have focused on this knowledge show how transferring such wisdom to non-Indigenous visitors can aid in reconciliation and gain respect and be incorporated into environmental management policies and actions to mitigate climate change impacts.

The problems stemming from the ignoring of Indigenous rights and traditions include those impacts from the management of extractive resources, which are often at odds with tourism as well as local communities. Clear-cutting of old growth forests is still a current problem in parts of Canada, often regardless of treaty obligations and international declarations of rights of Indigenous peoples, and mining can have even more severe and long-lasting effects upon both the natural environment and its inhabitants through pollution of air and water, as well as biodiversity loss from the removal of land and water diversions. Such a situation occurred in Oklahoma where improperly managed mine waste has contaminated water for over a century in the homeland of nine tribal nations. As a means of "speaking back" through tourism, what are known as "Toxic Tours" have been offered by the Indigenous communities in that area, taking tourists to witness the damage done to the environment, the disregard of traditional values and rights, and how presenting Indigenous actions to visitors makes them aware of the damage done to the area. Bigby et al. (2023: 407) conclude that. "By leading these toxic tours from indigenous perspectives, across Indigenous lands and waters and in relationships with human and non-human kin alike, these toxic tours can be a potential first step towards deeper relationships of respect, responsibility and stewardship". Their aims mirror those noted in chapters in this volume (Chapters 7, 8 and 9 in particular) where emphasising the contrast between Indigenous views and values and those of external forces impacting Indigenous lands and waters can lead to changing views and reconciliation.

Types of Tourism

What the optimum form of tourism in Indigenous communities is, can probably never be completely determined apart from on a case-by-case basis dependent on location and dominant values. While perhaps the most appropriate form may best be defined as being that desired by the communities involved, there have been more objective attempts at identifying what may be anticipated to be acceptable. Huang and Nguyen (2022) used fuzzy logic in an attempt to model and establish the optimal cultural tourism form for Indigenous tribes (in Taiwan), and after considerable statistical analysis, listed material culture, institutional culture and spiritual culture as being the most important elements identified. They argue that "the vagueness and complexity of the uncertainty and subjectivity in the evaluation process by human thoughts can be entirely quantified" (2022: 10) and point out that as cultural features are symbols, complexity and diversity, their model can" help select the most critical aspects …and enrich decision-making efficiency". Such complex analyses are beyond the capability of many small communities, Indigenous and non-Indigenous; however, emphasising the need for discussion, agreement and coordination before tourism development begins is commendable. As well, almost all forms of development will require working with external partners, both public and private, if tourism is to be more than extremely localised. One example of an alternative approach is that in relation to conservation, as called for by Buschman (2022), arguing for what she termed "co-productive conservation" in Greenland. That approach involves "six iterative and reflexive co-production processes (i.e., co-planning, co-prioritising, co-learning, co-managing, co-delivering, and co-assessing)". Which would embody Indigenous perspectives, knowledge, rights, priorities and livelihoods. She has produced a challenging circular framework for such an approach, which "embrace a process that broadens the evidence base and situates conservation within Indigenous contexts." The emphasis there is clearly on an approach that relies on the integration of different actions within an overall cooperative framework in which the Indigenous viewpoint is at least an equal, if not the dominant partner in such development. Mika and Scheyvens (2022), both contributors to this book, drew on work with Māori tourism operators in Te Awa Tupua (the Whanganui River) to propose an Indigenous tourism model, Kaupapa Tāpoi, that presents a circle form (to represent the natural environment) enclosing elements of peace, justice, sustainability, indigeneity and tribal identity. They opine that whilst "The model privileges Māori language, Māori concepts and Māori symbolism because of the focus on Te Awa Tupua, the Whanganui River, but the principles may be generalisable as foundations for Indigenous tourism elsewhere" (Mika and Scheyvens, 2022: 649).

Throughout the volume, the types of tourism discussed by Indigenous contributors have several features in common, in particular links to their histories, traditions and beliefs, and respect for the environment and the living forms within it. While not all traditional activities may have followed what is now regarded as sustainable use or development (Fennell, 2008), many of the principles of that approach are inherent in current Indigenous tourism offerings. The fact that proponents of many forms of tourism use, or rather misuse, the term sustainable when in fact many elements of those forms of tourism are far from sustainable in reality makes it difficult to argue that Indigenous tourism should follow the triple-bottom line principle of sustainable development, particularly when that terminology is not traditional to Indigenous groups. Priority being given to economic considerations rather than giving social and environmental factors equal importance, as proposed in the principle of sustainable development, has resulted in many problems for Indigenous communities, not only from tourism development. Mika and Scheyven's 2022 paper offers an alternative Indigenous approach that prioritises the environment whilst acknowledging the developmental needs of Indigenous communities.

Ecotourism is a frequently proposed option for many Indigenous communities as the benefits of that form of tourism are argued to be appropriate for such communities, but as Kline and Slocum (2015) pointed out, there is considerable involvement of multinational agencies in the promotion of ecotourism and there needs to be careful assessment of the role of such bodies and their overall compatibility with sustainability in the communities affected. Their assessment of ecotourism developments by such agencies in Africa suggested the need to include four overarching values, the first one being "sensitivity to local needs/culture" (2015: 99). Yahaya et al. (2022: 18) found that the families of Indigenous groups in Ghana generally saw involvement in ecotourism as a secondary activity where the proceeds are used to supply basic needs and concluded that "the sustainability of ecotourism in rural Ghana lies in its positive environmental and economic impacts on the livelihoods of households in eco-tourism communities". While this form of tourism could help wider goals including mitigation of climate change, increased wildlife populations, conservation and beautification of the environment, the population surveyed saw the last two aspects as being the major positive benefits as they had direct positive effects on conserving the environment for future generations (Yahaya, et al., 2022: 29). Ecotourism does bring with it potential problems as noted in Butler and Hinch (1996), including negative reactions of some eco-tourists to specific traditional Indigenous activities such as hunting (Grekin and Milne, 1996).

Elsewhere, Walters and Takamura (2015) discussed some of the options of other approaches and argued for what they called a "decolonised quadruple bottom line", building on the triple bottom line of people, planet and profit representing factors companies should consider in any development. Walters and Takamura drew attention to the absence of Indigenous considerations in such a framework, and in order to decolonise that approach, they proposed a subset consisting of four elements: community, spirituality, sustainability, and entrepreneurship to represent Indigenous values, leading to indigenous innovation. Unlike Mika and Scheyven's 2022 model, Walters and Takamura do not prioritise the environment as the dominant, all-embracing and essential component of their model (it is a feature of two overlapping circles in the Venn diagram). An alternative might be to argue that rather than the traditional triple bottom line, in the case of Indigenous tourism, the fourth element should be entitled Indigenous Control. Alongside people, planet and profit it should be incorporated to ensure that projects are not only truly sustainable (trusting Indigenous peoples will care for the environment), but clearly in line with Indigenous priorities and needs. Figure 35.1 is an illustration of the form such a framework could take, based on a modification of an earlier proposal (Butler, 1974) to recognise the importance of politics and power in all forms of tourism development. Giving equal priority to all four elements would be much more in line with

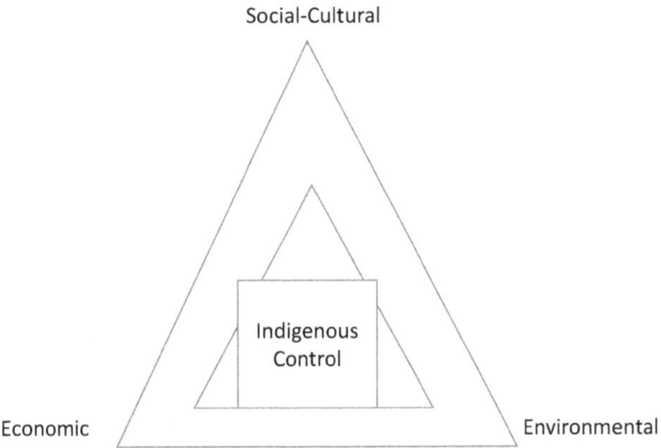

Figure 35.1   Key elements of tourism development.

Indigenous principles and would fit more easily with other, more traditional economic activities of Indigenous communities.

Control

Government interventions, especially when they come from external governments do not always favour or support local and Indigenous views. One example was the recent (summer 2023) actions by the British Parliament in its approval of a ban on the imports of wildlife hunting trophies. This bill (Hunting Trophies (Import Prohibition) Bill) was aimed at reducing the impacts of hunting on endangered species, according to government representatives, and was a good example of conflicts between well-intentioned moves by bodies physically (or mentally) far from the actual setting, running foul of local wishes. The campaign in support of the Bill was vocally aided by over a hundred "celebrities" who saw trophy hunting as undesirable to put it mildly, and hoped to end that activity by not allowing participants to bring trophies home. However, some wildlife organisations, such as the Sustainable Use and Livelihoods Specialist Group of the International Union for the Conservation of Nature (IUCN) and the Species Survival Commission, along with the High Commissioners of six African countries, wrote in opposition based in part on the grounds of the negative impact of such a step on local Indigenous populations. They wrote that providing aid programmes to make up any income lost by the ban was "a 19th century solution to a 21st century problem" (Flanagan, 2023: 8) and noted:

> "Far from saving animals by discouraging hunters, in its present form the bill will undermine a revenue model that gives incentives to local communities in our own countries to maintain wildlife habitats and protect animals from the far bigger threat posed by illegal poaching.
>
> (op cit)

Namibian community leaders signed a letter of protest to the bill, noting that "Africans like me don't have the same attitude towards lions as the average European brought up on the Lion King. We also know the lion as a childkiller, a slaughter of wild stock" (Lawson, 2023: 22). The WWF has argued that the trophy hunting business has been an integral part of Namibia's communal conservation programme aimed at building wildlife populations. An emotional issue such as wildlife trophy hunting does not fit easily with modern western attitudes, and is only one example of how distant voices may attempt to override local and Indigenous preferences. Comments such as that those actions did not "belong to the modern era… and these barbaric policies were not fitting in 'the civilised world'" were "very uncomfortable language for us black Africans who took back control of our nations sixty years ago" (cited by Lawson, 2023: 23) The bill was passed in the House of Commons but failed to pass the House of Lords. Despite calls for it to be proposed again, nothing has happened in the past year.

The generally disgraced policies of subjugation and cultural genocide have mostly disappeared, although, as Viken has noted, along with other Arctic contributors (Chapters 2 and 25) in this volume, cultural appropriation still exists in the case of the Sami for example. The pattern of what has happened to the Sami in Finland mirrored the fate of many Canadian First Nation populations through the residential school policy that existed there until the middle of the 20th century. In 2020 the UN Human Rights Committee rebuked the Finnish government for adding people to the Sami electoral roll without considering the Sami themselves. This was described as "about the right of the Sami to decide who gets to be Sami" according to Kuokkanen, a Sami activist (Moody, 2022) and suggests that inappropriate treatment of Indigenous populations has not disappeared, and as noted elsewhere in this volume, Indigenous peoples are still not recognised as such in some countries, but grouped under the term "ethnic" (Chapter 27). Indigeneity remains a live and controversial issue, and perhaps it is time for a reappraisal and redefinition of the term to enable wider inclusion and recognition.

National parks and many other protected areas in traditional Indigenous areas have traditionally been imposed by governments external to the location involved, both regional and national, and in the process they have often ignored local knowledge, local activities and local needs. Indigenous voices were often ignored in the creation of these parks, and for many years also ignored in the management of them. This has sometimes stemmed from a desire to protect specific environments from the negative effects of industrialisation, urbanisation and resource extraction witnessed in other parts of particular countries, but only rarely has such establishment been for the well-being of the Indigenous inhabitants of many of those areas. As noted earlier, Hall (1998) commented on the similarity between the allocation of otherwise 'worthless land' to either park status or native reservations, and Mason et al. (2022) identified the ignoring of Indigenous rights and traditional uses of the land in the establishment of some Canadian national parks (an issue also highlighted in the chapter on Reflexivity by Butler). The establishment of national parks, like the designation of World Heritage Sites, is often a flag for tourism development built on the unique attributes that mark such places as 'important and worth preserving', at least in the minds of decision-makers who are often far distant. That was certainly the case in the creation of the Rocky Mountain parks in western Canada, when in conjunction with the Canadian Pacific Railroad, the opportunity was seized to develop tourism very explicitly, at the expense of the Indigenous peoples living there, who were relegated to the role of visual attractions (Hart, 1983; Mason, 2014). The establishment of such parks can be positive for Indigenous peoples, but much depends on elements such as co-management and planning in line with Indigenous principles and with careful policies designed to support local communities in appropriate ways (Mika and Scheyvens, 2022). Mabibibi (2021) examined the effectiveness of achieving the Sustainable Development Goals (SDGs) for host communities in Kruger National Park. They recognised the problems the park organisation faced in partnering with local communities, but concluded that 15 of the 17 SDGs had been met in areas including economic, environmental and cultural. Their conclusions suggest that when projects are focused specifically on local community needs and effective partnerships exist, then national parks can serve local Indigenous needs as well as the broader goals of park establishment aimed at a global audience.

Indigeneity and its attributes, especially cultural features, can easily be transformed into stereotyping, a problem from which Indigenous communities have long suffered (Dodson, 2005), primarily in the context of being viewed as exhibits in their own environments. This can lead to tourists expecting Indigenous peoples, whether actively involved in tourism or not, to be wearing specific types of dress and to behave in certain ways, which can result in aversions to Indigenous offerings and images (see Chapters 10 and 18) to false expectations of widespread welcomes to visitors. The issue of ways in which local inhabitants were being portrayed was addressed some time ago by Dann (1995) in his discussion of the common portrayal of destinations and their peoples as being welcoming", an issue also examined by Siever and Matthews (2016). More recently Phillips et al. (2021) have drawn attention to the problems and implications of campaigns such as the promotion of South Pacific islands as places "selling authentic happiness". They point out that the stereotyped image of the "happy native", specifically of the Bula characteristic of Fijians is combined with "romanticised inequality" and serves to "justify and further entrench objective economic inequalities" and has been utilised in attempts to attract "Bulanaires" to invest in Fiji. They cite Scheyvens and Hughes "Poor people in poor locations with poor labour rights (put) on their smiling faces to serve rich guests of former colonial powers and to clean up their mess". While many of the 'rich guests' travelling in the South Pacific islands are not from 'former colonial powers', the point is still valid and a potential problem in many other Indigenous areas. Cohen described the problems arising amongst hunter-gatherer tribes in Thailand from becoming involved in tourism in their traditional areas, an encounter he described as "an encounter between extremes, the parties to the encounter are at the outset complete 'Others' to each other". Cohen goes on to conclude that tourists visiting such tribes are akin to visiting a "human zoo", particularly for those visiting hunter-gatherer communities as they see "man as he was before 'civilisation' or even 'culture'". In an earlier examination of "well-organised and sustainable indigenous tourism" in the Sami communities, Pettersson and Viken (2007: 187) concluded that "The key factor for such a success is responsibility and respect for the culture", which is just as true, if not more so, in the present day.

One area of concern in the context of cultural protection has been the subject of handicrafts and souvenirs, as discussed by Ara and Hoque and Dehrashi and Ghaderi earlier in this volume. Graburn (1976) noted problems with the development of "ethnic and tourist arts" half a century ago and Smith (1996: 295) concisely summed up one issue when she noted

Although some foreign tourists are willing to pay expensive shipping costs …. (for expensive Indigenous items) most visitors want quality items which fit in aircraft luggage, are of durable materials, and are modestly priced in proportion to their size. Above all, the items *must* be Indian made (her emphasis).

Almost inevitably such demands result in changes from the "original' artefacts to meet those conditions. Smith notes miniaturisation as one step and focusing on small items in general is common. Mayuzumi (2021) examined whether the 'ethnic' tourism being practised in Bali which involved selling "traditional culture to the tourist satisfied the requirement for the development of sustainable tourism leading to happiness in the community". Interviews were conducted with Indigenous craftsmen (wood carving and painting) and tourists buying such products. The findings showed that many of the craftsmen had lost their jobs because of craft merchants selling similar goods made elsewhere as local souvenirs, although the craftsmen retained a strong motivation to pass on their skills to the next generation. One result was the establishment of a wood-working programme available to tourists to enable them to learn the techniques, which also allowed the locals to pass on their expertise. In a similar way, on the small Scottish island of Fair Isle, home to the Fair Isle knitting patterns, original hand produced sweaters became too expensive for tourists, and machine produced and finished sweaters have been substituted, and several knitters now offer courses to visitors staying on the island to learn the traditional skill (Butler, 2015). As Ara and Hoque alongside Alonso and Novelli (this volume) discuss, the growth and promotion of Indigenous art and artefacts can be important in terms of empowering Indigenous communities and individuals, in gaining respect, and in promoting globally as well as locally, the significance of Indigenous cultural expression.

Allied to the problems of image and authenticity is the problem that much promotion is still carried out by non-Indigenous agents, often far removed, culturally and physically from the actual locations of the tourism and the people involved. Because of the nature of tourism, where customers, tourists, have to travel to the product (places, peoples, activities) the selling of the tourism product also takes place mostly in the origin areas not the destinations themselves, so the "hosts" may not always know what has been promoted to 'guests', nor what the expectations of visitors may be. This has been a long-standing problem for Indigenous tourism and Indigenous communities but given the relative inability of small communities to promote their tourism opportunities beyond the local region, the most common solution has been to have others do that job. The formation of the (WINTA) and relatively recent development of regional and national agencies by and for Indigenous groups (for example ITAC in Canada discussed in an earlier chapter) can help reduce this problem, and judicious use of social media can also allow small communities to broadcast their message and promotions to a theoretically world-wide market. In the same way, disseminating guidance on appropriate behaviour, identifying suitable and unsuitable places to visit, and clarifying opportunities available to tourists, for example through locally created codes of conduct (as discussed Chapter 11), may become much easier and more widespread. A major problem, however, is that social media, as a relatively uncontrolled form of information dissemination, can cause multiple problems with regards to encouraging unwelcome visitation to specific locations, stereotyping Indigenous characteristics, and misinterpreting messages about appropriate and respectful behaviour sensitive sites.

Concluding Remarks

Many of the problems facing Indigenous peoples stem from the loss of rights and the downplaying or deliberate silencing of Indigenous knowledge. The silencing can come about through ignorance, i.e., lack of knowledge, which may be overcome by Indigenous actions and voices as noted above, and by the ignoring of facts and evidence that might ensure a more appropriate and acceptable management of the human and natural environments in which Indigenous peoples live. Johnston (2000: 89) pointed out that tourism could be of considerable benefit to Indigenous peoples "when Indigenous communities have access to reliable information on the impacts of tourism…and can share strategies for sustainable tourism". In 2023 Scheyvens, Movono and Auckram's study of Fijian communities involved in tourism produced an enticing analysis of how Indigenous peoples' well-being was affected during the pandemic, depicted in the Frangipani Framework: Pacific Peoples' Well-Being Through Tourism. They concluded that "We believe that to build tourism back better, planning and consideration must be done respectfully and be centred around the well-being of destination communities in general, and especially, Indigenous peoples." Tourism can lead to flourishing.

It is important to remember that many of the problems discussed in this volume are not unique to Indigenous peoples, although they may be more critical in the Indigenous context than elsewhere.

Mathieson and Wall (1982), in analysing the impacts of tourism, identified many of the same concerns as noted in these chapters being experienced in many different communities that were experiencing tourism development. Loss of local autonomy, inadequate or absent involvement in decision-making, lack of input into promotion and portrayal of communities, inability to limit the rate and or scale of development, and lack of consideration when external agencies were providing services such as transportation and access to communities. In very few cases have communities of any type been able to control and limit tourism development proposed and supported by external agencies and levels of government. Even where there has been success, it has often been short-lived, as for the example of Calvia (Spain) where attempts to impose local control and move towards a more sustainable form of tourism only lasted a few years (Dodds, 2007). In such situations, Indigenous groups often face greater difficulties in opposing undesired policies and incur more serious cultural impacts from them than do non-Indigenous communities, in part because of the long histories of disadvantage and deprivation.

There is no easy or common solution to such problems. Implementing planning approaches or being inspired by frameworks discussed in this volume requires resourcing, commitment from all parties involved and patience. At the root of many problems is the desire for continual growth and further development, a desire which is not always a priority for those involved in Indigenous tourism, yet they both are vulnerable to and a part of that mindset as participants in the global tourism industry. Traditionally success in tourism destinations has been measured in terms of volume, primarily as demonstrated by increasing numbers of visitors. More has been seen as better, often regardless of the resulting effects, and it is relatively recently that voices (see for example Butler, 2022; Dwyer, 2022) have been raised suggesting and demanding different ways of defining success. At the very least, greater value from tourism in terms of increased economic, environmental and socio-cultural benefits rather than simply more visitors would be an improvement, but that alone is not an acceptable solution. Success in terms of tourism development needs to be focused on satisfaction as experienced by both visitors and local residents, such that both the demand and the supply sides of tourism are happy with the overall experience. It is critical that both parties' benefit, one by a positive experience, and the other from economic benefits with minimal or no social, cultural and environmental costs. It is certain that tourism can never exist without some impacts, to pretend otherwise is wishful thinking. It is possible, however, to use tourism to achieve desired benefits at the expense of only acceptable impacts and costs as perceived by the affected local communities. To achieve such a result requires the communities involved to be clear and in agreement over what is desired from tourism and what impacts and changes, if any, they are prepared to accept, for this agreement to be formulated and expressed before development occurs, and for the local communities to be in a position to direct and control tourism within their communities. Those principles apply to both Indigenous and non-Indigenous communities, although the level of importance of the specific elements involved may be different.

To achieve such a situation requires a different approach to tourism development. In recent years there have been other voices raised calling for such approaches (see for example *Journal of Sustainable Tourism* 2016: 24 and *Tourism Geographies* 2020: 23), including seeking more sustainable and greener forms of tourism, fairer and more equitable development, regenerative tourism, restorative justice focused tourism, and reduced levels (degrowth) of tourism. Most, if not all of these approaches, are of particular relevance to Indigenous tourism.

This volume has not tried to produce answers to the many relevant questions that arise with discussion of Indigenous tourism, but it has, hopefully, provided insights and research informed examples of successful and appropriate approaches to dealing with some of the common problems that beset the subject. There will always, obviously and disappointingly, be failures in communication, in transparency, in equality, equity and justice, and perhaps most of all in policymaking with regards to Indigenous tourism, but that should encourage rather than diminish discussion of such topics in the hope of successful and appropriate resolution of these issues. There clearly remain many other questions to be asked and resolved with respect to the development and operation of Indigenous tourism. Relations with all levels of government, both horizontally and vertically, are critical, as are resolutions of other problems including image perceptions, public attitudes, and Indigenous representation. In dealing with such issues in the future, we might usefully return to Whitney-Gold's powerful analogy of not needing to reinvent the wheel but needing to change the tyre, but looking even more widely, we might usefully conclude that perhaps it is necessary also to learn a new way to share the road or pathway forward.

References

Belyea, B. (1996) Inland journeys, native maps. *Cartographica* 33(2), 1–16.

Bigby, B.C., Jim, R., Hatley, E. (2023) Indigenous-led toxic tours opening pathways for (re)connecting to place, people and all creation. *Australian Journal of Environmental Education* 39(3), 390–409.

Buschman, V.Q. (2022) Framing co-productive conservation in partnership with Arctic Indigenous peoples. *Conservation Biology* 36(6), e13972. https://doi.org/10.1111/cobi.139

Butler, R.W. (1974) The social implications of tourist developments. *Annals of Tourism Research* 2(2), 107–111.

Butler, R.W. (2015) Knitting and more from Fair Isle, Scotland: Small-island tradition and microentrepreneurship. In G. Baldachinno (Ed.) *Entrepreneurship in small island states and territories*, 83–96. Abingdon: Routledge.

Butler, R.W. (2022) Measuring tourism success: Alternative considerations. *World Hospitality and Tourism Themes* 14(1). www.emerald.com/insight/1755-4217.htm

Butler, R., & Hinch, T. (Eds) (1996) *Tourism and Indigenous Peoples*. London: International Thompson Business Press.

Butler, R., & Hinch, T. (Eds) (2007) *Tourism and Indigenous Peoples: Issues and Iimplications*. London: Routledge.

Carmichael, B.A., & Jones, J.I. (2007) Indigenous owned casinos and perceived local community impacts: Mohegan Sun in South East Connecticut, U.S.A. In R. Butler and T. Hinch (Eds) *Tourism and Indigenous Peoples: Issues and Implications*, pp. 95–109. Amsterdam: Elsevier.

Dann, G. (1995) A socio-linguistic approach towards changing tourist imagery. In R. Butler and D. Pearce (Eds) *Change in tourism people, places, processes*. London: Routledge

Dodds, R. (2007) Sustainable tourism and policy implementation: Lessons from the case of Calviá, Spain. *Current Issues in Tourism* 10(4), 296–322.

Dodson, M. (2005) The end in the beginning: Re(de)fining aboriginality. In M. Grossman (Ed.) *Blacklines*, pp. 25–42. Melbourne: Melbourne UP.

Douglas, N. (1994) *They came for savages.* Unpublished PhD thesis. Brisbane, University of Queensland.

Dwyer, L. (2022) Why tourism economists should treat resident well-being more seriously. *Tourism Economics* 29(8). https://doi.org/10.1177/1354816622112808

Fennell, D.A. (2008) Ecotourism and the myth of indigenous stewardship. *Journal of Sustainable Tourism* 16(2), 129–149.

Flanagan, J. (2023) Trophyhunting. *The Times*, June 15, p. 8.

Graburn, N. (1976) *Ethnic and tourist arts; cultural expressions from the fourth world.* Berkeley: University of California Press.

Grekin, J., & Milne, S. (1996) Towards sustainable tourism development: The case of Pond Inlet, NWT. In R. Butler and T. Hinch (Eds) *Tourism and indigenous peoples*, pp. 76–106. London: International Thompson Business Press.

Hall, C.M. (1998) Tourism, national parks and Aboriginal peoples. In R.W. Butler and S.W. Boyd (Eds) *Tourism and national parks*, pp. 56–71. Chichester: John Wiley and Sons.

Hall, C.M. (2007) Politics, power and indigenous tourism. In R. Butler and T. Hinch (Eds) *Tourism and indigenous peoples: Issues and implications*, pp. 305–318. Amsterdam: Elsevier.

Hart, E.J. (1983) *The selling of Canada: The C.P.R. and the beginnings of Canadian tourism.* Banff: Altitude Press.

Howitt, R. (2020) Decolonizing people, place and country: Nurturing resilience across time and place. *Sustainability* 12(15), 5882.

Huang, F.-H., & Nguyen, H. (2022) Selecting optimal cultural tourism for Indigenous tribes by fuzzy MCDM. *Mathematics* 10, 3121. https://doi.org/10.3390/math10173121

Johnston, A. (2000) Indigenous peoples and ecotourism: Bringing Indigenous knowledge and rights into the sustainability equation. *Tourism Recreation Research* 25(2), 89–96.

Kline, C.S., & Slocum, S.L., (2015) Neoliberalism in ecotourism? The new development paradigm of multinational projects in Africa. *Journal of Ecotourism* 14(2–3), 99–112.

Lawson, D. (2023) The celebrity-backed bill will harm the wildlife it claims to protect. *The Sunday Times*, March 12, p. 22.

Mabibibi, M.A. (2021) Successes and challenges in sustainable development goals localisation for host communities around Kruger National Park. *Sustainability* 13(10), 5341.

Mamontova, N., & Klyachko, E. (2022) 'Process toponymy': A GIS-based community-engaged approach to Indigenous dynamic place naming systems and vernacular cartography. *Cartographica* 57(3), 213–225.

Mason, C.W. (2014) *Spirits of the rockies: Reasserting an Indigenous presence in Banff National Park.* Toronto, ON: University of Toronto Press.

Mason, C., Carr, A., Snow, W., Vandermale E., & Philipp, L. (2022) Rethinking the role of Indigenous knowledge in sustainable Mountain Park development in Canada and Aotearoa New Zealand. *Mountain Research and Development* (Special Issue Weaving Together Knowledges—Collaborations in Support of the Wellbeing of Mountain Peoples and Regions) 42(4), A1–A9. https://doi.org/10.1659/mrd.2022.00016

Mathieson, A., & Wall, G. (1982) *Tourism economic, physical and social impacts.* Harlow: Prentice Hall.

Matunga, H., Matunga, H., & Urlich, S. (2020) From exploitative to regenerative tourism. *MAI Journal: A New Zealand Journal of Indigenous Scholarship* 9(3), 295–308. https://doi.org/10.20507/MAIJournal.2020.9.3.10

Mayuzumi, Y. (2021) Is meeting the needs of tourists through ethnic tourism sustainable? Focus on Bali, Indonesia. *Asia-Pacific Journal of Regional Science.* https://doi.org/10.1007/s41685-021-00198-4

Mika J., & Scheyvens, R. (2022) Te Awa Tupua: Peace, justice and sustainability through Indigenous tourism. *Journal of Sustainable Tourism* 30(2–3), 637–657. https://doi.org/10.1080/09669582.2021.1912056

Moody, O. (2022) Finland's leader Sanna Marin promises the truth about 'cultural genocide' of Sami People. *The Sunday Times*, January 1, pp. 6–8.

Peters, A., & Higgins-Desbiolles, F. (2012) De-marginalising tourism research: Indigenous Australians as tourists. *Journal of Hospitality and Tourism Management* 19, 1–9.

Pettersson, R., & Viken, A. (2007) Sami perspectives on Indigenous tourism in northern Europe: commerce or cultural development. In R. Butler and T. Hinch (Eds) *Tourism and indigenous peoples: Issues and implications*, pp. 176–187. Amsterdam: Elsevier.

Phillips, T., Taylor, J., Narain, E., & Chandler, P. (2021) Selling authentic happiness: Indigenous well-being and romanticised inequality in tourism advertising. *Annals of Tourism Research*, 87. https://doi.org/10.1016/j.annals.2020.103115

Scheyvens, R., Carr, A., Movono, A., Hughes, E., Higgins-Desbiolles, F., & Mika, J.P. (2021) Indigenous tourism and the sustainable development goals. *Annals of Tourism Research* 90, 103260.

Scheyvens, R., Movono, A., & Auckram, S. (2023) Pacific peoples and the pandemic: Exploring multiple well-beings of people in tourism-dependent communities. *Journal of Sustainable Tourism* 31(1), 111–130. https://doi.org/10.1080/09669582.2021.1970757

Sievers, B., & Matthews, A. (2016) Beyond whiteness: A comparative analysis of representation of Aboriginality in tourist destination images in New South Wales Australia. *Journal of Sustainable Tourism* 24(8–9), 1298–1314.

Silvestre, R.P., & Fontana, R. de F. (2023) Indigenous tourism in Brazil: A literature review of research published in the period 1999–2021. *PASOS Journal of Tourism and Cultural Heritage* 21(3), 487–501. https://doi.org/10.25145/j.pasos.2023.21.033

Smith, V.L. (1977) *Hosts and guests: The anthropology of tourism.* Philadelphia: University of Pennsylvania Press.

Smith, V. (1996) Indigenous Tourism: The Four Hs. In R. Butler and T. Hinch (Eds) *Tourism and indigenous peoples,* pp. 283–308. London: International Thompson Business Press.

Tribe, J., & Liburd, J. (2016) The tourism knowledge system. *Annals of Tourism Research* 57, 44–61.

Trupp, A., Pratt, S., Stephenson, M.L., Matatolu, I., & Gibson, D. (2022) Representing and evaluating the travel motivations of Pacific Islanders. *International Journal of Tourism Research,* 1–14. https://doi.org/10.1002/jtr.2528

Walters, F., & Takamura, J. (2015) The decolonized quadruple bottom line: A framework for developing indigenous innovation. *Wicazo Sa Review* 30, 77–99. https://api.semanticscholar.org/CorpusID:163094276

Wei, L., Qian, J., & Zhu, H. (2021) Rethinking Indigenous people as tourists: Modernity, cosmopolitanism, and the re-invention of indigeneity. *Annals of Tourism Research* 89, 103200.

Whitford, M., & Ruhanen, L. (2016) Indigenous tourism research, past and present: Where to from here? *Journal of Sustainable Tourism* 24(8–9), 1080–1099.

Yahaya, A-K., Samuel, A., & Abdullah, A. (2022) Sustainable eco-tourism in Ghana: An assessment of environmental and economic impacts in selected sites in the Upper East Region. *Journal of Geography and Regional Planning* 15(2), 18–30.

World Indigenous Tourism Summit
Empowering Opportunity Through Tourism
March 13–16, 2023
Perth, Australia

NOONGAR INDIGENOUS TOURISM RESEARCH PROTOCOL DECLARATIONS AND RECOMMENDATIONS

Preamble

The World Indigenous Tourism Alliance (WINTA) was created from the global, collective aspirations of Indigenous interests in tourism. WINTA is an Indigenous initiative that is based on key human Indigenous rights including the rights to self-determination, self-governance, and the right to maintain and develop our own independent institutions and to develop cross-border relationships critical to maintaining unity and common goals.

WINTA is an Indigenous-led global network of Indigenous and non-Indigenous peoples and organizations who seek to give practical expression to the United Nations Declaration on the Rights of Indigenous Peoples (UNDRIP) through tourism. WINTA guards the Larrakia Declaration, created in 2012 that establishes principles for the development of Indigenous Tourism at a global level based on the internationally recognized UNDRIP. The World Tourism Organization recognized and supported the Larrakia Declaration principles through its Executive Council in 2012.

We embrace the rich Cultural, Ethnic, Aboriginal, and First Nations backgrounds of people involved in tourism. We understand that research in Indigenous Tourism is an important endeavor to guide projects that respect Indigenous protocols and rights, driving the activity toward decolonizing methodologies and offering benefits to Indigenous peoples.

This document is divided in two parts: The first establishes a **Declaration** about Indigenous Tourism research protocols and the second part focuses on **Recommendations** for academia or any agency pursuing a research project or gathering data with and for Indigenous peoples.

Declarations

Declaration 1

Indigenous communities have worldviews and links to traditional lands and ocean spaces that represent unique and diverse cultural expressions. These cultural expressions provide a clear attractor for visitors seeking to experience different geographic locations and their Indigenous communities.

Declaration 2

Institutions with research capability in tourism have an important role to play in advancing tourism research to contribute to the understanding of Indigenous tourism and generate knowledge consistent with best practice requirements of Indigenous Data Sovereignty and Indigenous Data Governance.

Declaration 3

Indigenous natural and cultural heritage, including Indigenous knowledge systems, have been impacted for centuries by colonization and further threatened by global developments including tourism.

Declaration 4

Indigenous peoples have demonstrated that given the opportunity to fully participate in development processes, they can share their worldviews, initiate adaptive responses to cope with external influences and produce tourism opportunities which benefit both visitors and their own communities.

Declaration 5

The evolution of advocacy for protecting the rights of Indigenous peoples, including their knowledge systems, has been accelerated by States and the corporate sector including the tourism industry.

There is increasing recognition of Indigenous peoples' knowledge and the requirements for Indigenous Data Sovereignty and Indigenous Data Governance.

It is hereby resolved to adopt the following:

Recommendations

Researchers and research institutions seeking to produce credible research and data to be used in the course of sustainable Indigenous tourism development should:

1 Ensure that each tourism research project, from start to finish, be carried out in the spirit of equitable partnerships that uphold the tourism rights of Indigenous Peoples as expressed by the Larrakia Declaration and human rights resolutions set out by the United Nations Declaration on the Rights of Indigenous peoples.

    a Commit to partnerships where each project, from start to finish, is built on dialogue, informed consent and respect for Indigenous peoples whose knowledge systems strengthen the vision and value of scientific inquiry.

    b Ensure that tourism research enables Indigenous empowerment and rights in sustainable tourism development as expressed by the *Larrakia Declaration*.

    c Ensure that tourism research fulfills a duty to protect, serve and contribute to Indigenous human rights as set out by the *United Nations Declaration on the Rights of Indigenous Peoples*.

2 Strengthen conventional ethical research standards, research methodologies and practices with engagement principles that enable culturally respectful, diverse, and meaningful Indigenous participation throughout the entire tourism research process.

    a Implement research methodologies and engagement principles grounded by respect, reciprocity, and meaningful Indigenous involvement throughout the entire tourism research process.

    b Recognize Indigenous knowledge holder intellectual property rights and their cultural practices be incorporated into the processes involved in tourism research.

    c Create opportunities for Indigenous capacity building throughout the entire tourism research process to further the development and implementation of Indigenous-driven research now and in the future.

    d Promote the joint creation of research methodologies or incorporate Indigenous methodologies that generate research projects adapted to the reality of the Indigenous people involved in the process and that rescue, revitalize, and incorporate Indigenous knowledge according to the limits imposed by Indigenous people.

    e Encourage the mentoring, remuneration, and capacity building of Indigenous researchers and knowledge holders.

3 Strive to generate research outcomes that contribute genuine change toward tourism sustainability needs identified by Indigenous peoples and their territories, such as ecological, cultural, social, economic, political, and spiritual wellbeing outcomes.

   a Present Indigenous people with deliverables that produce lasting impacts for their empowerment in tourism and their human rights as outlined by the United Nations Declaration on the Rights of Indigenous Peoples.
   b Incorporate Indigenous wisdom and insight to define research data needs, as well as appropriate project objectives and timelines to address the complex tourism sustainability challenges faced by Indigenous people now and in the future.
   c Action the roles tourism research can play to support the efforts of Indigenous peoples to nurture, regenerate, protect, or enrich the lands and waters of our precious Earth.

Ben Sherman, Chairman
World Indigenous Tourism Alliance (WINTA)

Robert Taylor, CEO
Western Australian Indigenous Tourism
Operators Council (WAITOC).

WINTA and WAITOC would like to thank the contributions of:
Johnny Edmonds
Professor Cheryl Kickett-Tucker
Dr Damien Jacobsen
Dr Anna Carr
Dr Freya Higgins-Desbiolles

INDEX